T0183346

Lecture Notes in Artificial Intelligence 12522

Subseries of Lecture Notes in Computer Science

More information about this series at http://www.springer.com/series/1244

Maosong Sun · Sujian Li ·
Yue Zhang · Yang Liu ·
Shizhu He · Gaoqi Rao (Eds.)

Chinese Computational Linguistics

19th China National Conference, CCL 2020
Hainan, China, October 30 – November 1, 2020
Proceedings

Editors
Maosong Sun
Tsinghua University
Beijing, China

Yue Zhang
Westlake University
Hangzhou, China

Shizhu He
Chinese Academy of Sciences
Beijing, China

Sujian Li
Peking University
Beijing, China

Yang Liu
Tsinghua University
Beijing, China

Gaoqi Rao
Beijing Language and Culture University
Beijing, China

ISSN 0302-9743 ISSN 1611-3349 (electronic)
Lecture Notes in Artificial Intelligence
ISBN 978-3-030-63030-0 ISBN 978-3-030-63031-7 (eBook)
https://doi.org/10.1007/978-3-030-63031-7

LNCS Sublibrary: SL7 – Artificial Intelligence

This Springer imprint is published by the registered company Springer Nature Switzerland AG
The registered company address is: Gewerbestrasse 11, 6330 Cham, Switzerland

Preface

Welcome to the proceedings of the 19th China National Conference on Computational Linguistics (CCL 2020). The conference and symposium were hosted online and co-organized by Hainan University, China.

CCL is an annual conference (bi-annual before 2013) that started in 1991. It is the flagship conference of the Chinese Information Processing Society of China (CIPS), which is the largest NLP scholar and expert community in China. CCL is a premier nation-wide forum for disseminating new scholarly and technological work in computational linguistics, with a major emphasis on computer processing of the languages in China such as Mandarin, Tibetan, Mongolian, and Uyghur.

The Program Committee selected 108 papers (74 Chinese papers and 34 English papers) out of 303 submissions for publication. The acceptance rate is 35.64%. The 34 English papers cover the following topics:

- Machine Translation and Multilingual Information Processing (2)
- Fundamental Theory and Methods of Computational Linguistics (3)
- Minority Language Information Processing (4)
- Social Computing and Sentiment Analysis (4)
- Text Generation and Summarization (3)
- Information Retrieval, Dialogue and Question Answering (2)
- Language Resource and Evaluation (4)
- Knowledge Graph and Information Extraction (6)
- NLP Applications (6)

The final program for the 19th CCL was the result of intense work by many dedicated colleagues. We want to thank, first of all, the authors who submitted their papers, contributing to the creation of the high-quality program. We are deeply indebted to all the Program Committee members for providing high-quality and insightful reviews under a tight schedule, and extremely grateful to the sponsors of the conference. Finally, we extend a special word of thanks to all the colleagues of the Organizing Committee and secretariat for their hard work in organizing the conference, and to Springer for their assistance in publishing the proceedings in due time.

We thank the Program and Organizing Committees for helping to make the conference successful, and we hope all the participants enjoyed the first online CCL conference.

September 2020

Maosong Sun
Sujian Li
Yue Zhang
Yang Liu

Organization

Program Committee

Program Chairs

Sujian Li	Peking University, China
Yue Zhang	Westlake University, China
Yang Liu	Tsinghua University, China

Area Co-chairs

Linguistics and Cognitive Science

Gaoqi Rao	Beijing Language and Culture University, China
Timothy O'Donnell	McGill University, Canada

Basic Theories and Methods of Computational Linguistics

Wanxiang Che	Harbin Institute of Technology, China
Fei Xia	University of Washington, USA

Information Retrieval, Dialogue, and Question Answering

Xin Zhao	Renmin University of China, China
Hongzhi Yin	The University of Queensland, Australia

Text Generation and Summarization

Xiaojun Wan	Peking University, China
Jinge Yao	Microsoft Research Asia

Knowledge Graph and Information Extraction

Kang Liu	Institute of Automation, CAS, China
Ruihong Huang	Texas A&M University, USA

Machine Translation and Multilingual Information Processing

Yang Feng	Institute of Computing Technology, CAS, China
Haitao Mi	Ant Group, China

Minority Language Information Processing

Hongxu Hou — Mongolian University, China
Fanglin Wang — Leading Intelligence Corporation, USA

Language Resources and Evaluation

Weiguang Qu — Nanjing Normal University, China
Nianwen Xue — University of Bratis, USA

Social Computing and Sentiment Analysis

Meishan Zhang — Tianjin University, China
Jiang Guo — Massachusetts Institute of Technology, USA

NLP Applications

Qi Zhang — Fudan University, China
Jun Lang — Alibaba, China

Local Arrangement Chairs

Chunjie Cao — Hainan University, China
Ting Jin — Hainan University, China

Evaluation Chairs

Ting Liu — Harbin Institute of Technology, China
Wei Song — Capital Normal University, China

Publications Chairs

Shizhu He — Institute of Automation, CAS, China
Gaoqi Rao — Beijing Language and Culture University, China

Workshop Chairs

Jun Xu — Renmin University of China, China
Xipeng Qiu — Fudan University, China

Sponsorship Chairs

Zhongyu Wei — Fudan University, China
Binyang Li — University of International Relations, China

Publicity Chair

Zhiyuan Liu — Tsinghua University, China

System Demonstration Chairs

Shujian Huang	Nanjing University, China
Zhaopeng Tu	Tencent, China

Student Counseling Chair

Pengyuan Liu	Institute of Computing Technology, CAS, China

Student Seminar Chairs

Jinsong Su	Xiamen University, China
Hongyu Lin	Institute of Computing Technology, CAS, China

Finance Chair

Yuxing Wang	Tsinghua University, China

Organizers

Chinese Information Processing Society of China

Tsinghua University

Hainan University, China

Publishers

Lecture Notes in Artificial Intelligence, Springer

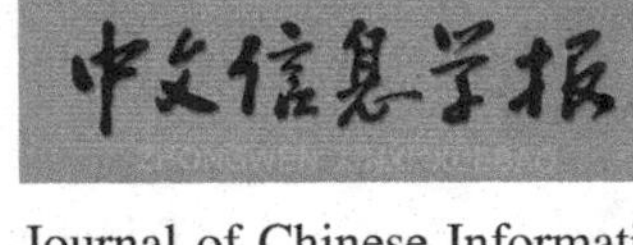

Journal of Chinese Information Processing

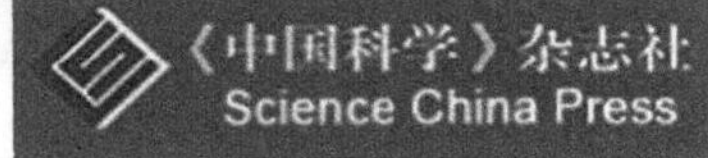

Science China

清华大学学报(自然科学版)
Journal of Tsinghua University (Science and Technology)

Journal of Tsinghua University (Science and Technology)

Sponsoring Institutions

Gold

Silver

CIPOL

Bronze

Contents

Language Resource and Evaluation

Social Computing and Sentiment Analysis

NLP Applications

Fundamental Theory and Methods of Computational Linguistics

A Joint Model for Graph-Based Chinese Dependency Parsing

Xingchen Li, Mingtong Liu, Yujie Zhang(✉), Jinan Xu, and Yufeng Chen

School of Computer and Information Technology, Beijing Jiaotong University, Beijing 100044, China
yjzhang@bjtu.edu.cn

Abstract. In Chinese dependency parsing, the joint model of word segmentation, POS tagging and dependency parsing has become the mainstream framework because it can eliminate error propagation and share knowledge, where the transition-based model with feature templates maintains the best performance. Recently, the graph-based joint model [19] on word segmentation and dependency parsing has achieved better performance, demonstrating the advantages of the graph-based models. However, this work can not provide POS information for downstream tasks, and the POS tagging task was proved to be helpful to the dependency parsing according to the research of the transition-based model. Therefore, we propose a graph-based joint model for Chinese word segmentation, POS tagging and dependency parsing. We designed a character-level POS tagging task, and then train it jointly with the model of [19]. We adopt two methods of joint POS tagging task, one is by sharing parameters, the other is by using tag attention mechanism, which enables the three tasks to better share intermediate information and improve each other's performance. The experimental results on the Penn Chinese treebank (CTB5) show that our proposed joint model improved by 0.38% on dependency parsing than the model of [19]. Compared with the best transition-based joint model, our model improved by 0.18%, 0.35% and 5.99% respectively in terms of word segmentation, POS tagging and dependency parsing.

Keywords: Dependency parsing · Graph-based · Joint model · Multi-task learning

1 Introduction

Chinese word segmentation, part-of-speech (POS) tagging and dependency parsing are three fundamental tasks for Chinese natural language processing, whose accuracy obviously affects downstream tasks such as semantic comprehension, machine translation and question-answering. The traditional method is usually following pipeline way: word segmentation, POS tagging and dependency parsing. However, there are two problems of the pipline way, one is error propagation:

M. Sun et al. (Eds.): CCL 2020, LNAI 12522, pp. 3–16, 2020.
https://doi.org/10.1007/978-3-030-63031-7_1

incorrect word segmentation directly affects POS tagging and dependency parsing, another is information sharing: the tree tasks are strongly related, the label information of one task can help others, but the pipline way cannot exploit the correlations among the three tasks.

Using joint model for Chinese word segmentation, POS tagging and dependency parsing is a solution to these two problems. The previous joint models [7,13,21] mainly adopted a transition-based framework to integrate the three tasks. Based on the standard sequential shift-reduce transitions, they design some extra actions for word segmentation and POS tagging. Although these transition-based models maintained the best performance of word segmentation, POS tagging and dependency parsing, its local decision problem led to the low precision of long-distance dependency parsing, which limited the precision of dependency parsing.

Different from the transition-based framework, the graph-based framework has the ability to make global decisions. Before the advent of neural network, the graph-based framework was rarely applied to the joint model due to its large decoding space to calculate. With the development of neural network technology, the graph-based method for dependency parsing improves rapidly and comes back into researchers' vision. [19] firstly proposed a graph-based unified model for joint Chinese word segmentation and dependency parsing with neural network and attention mechanism, which is superior to the best transition-based joint model in terms of word segmentation and dependency parsing. This work without POS tagging task shows that dependency parsing task is beneficial to Chinese word segmentation.

Chinese word segmentation, POS tagging and dependency parsing are three highly correlated tasks and can improve each other's performance. Dependency parsing is beneficial to word segmentation and POS tagging, while word segmentation and POS tagging are also helpful to dependency parsing, which has been demonstrated by considerable work on the existing transition-based joint model of three tasks. We consider that joint POS tagging task can further improve the performance of dependency parsing. In addition, it makes sense of the model to provide POS information for downstream tasks. For these reasons, this paper proposes a graph-based joint model for word segmentation, POS tagging and dependency parsing. First, we design a character-level POS tagging task, and then combine it with a graph-based joint model for word segmentation and dependency parsing ([19]. As for the joint approach, this paper proposes two ways, one is to combine the two tasks by hard sharing parameters ([3]) and the other is combine the two tasks by introducing tag attention mechanism in the shared parameter layer. Finally, we analyze our proposed models on the Chinese treebank (CTB5) dataset.

2 The Proposed Model

In this section, we introduce our proposed graph-based joint model for Chinese word segmentation, POS tagging and dependency parsing. Through the joint

POS tagging task, we explore the joint learning method among multiple tasks and seek for a better joint model to improve the performance of Chinese dependency parsing further.

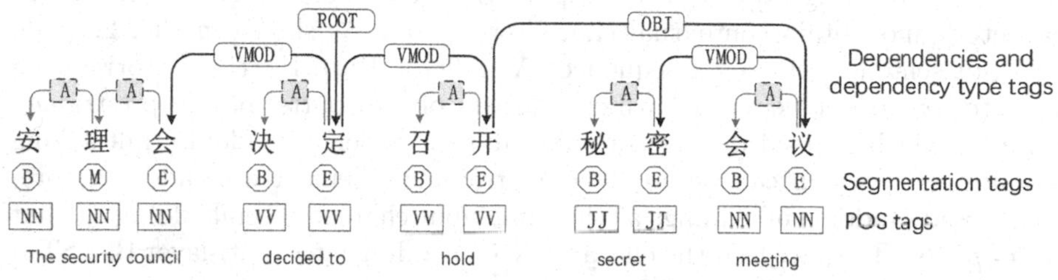

Fig. 1. An example of a character-level dependency tree

2.1 Character-Level Chinese Word Segmentation and Dependency Parsing

This paper refers to [19]'s approach of combining word segmentation and dependency parsing into a character-level dependency parsing task. Firstly, we transform the word segmentation task to a special arc prediction problem between characters. Specifically, we treat each word as a dependency subtree, and the last character of the word is the root node, and for other characters, the next character is its head node. For example, the root node of the dependency subtree of the word "秘密" is "密", and the head node of the character "秘" is "密", which constitutes an intra-word dependency arc of "秘←密". To distinguish it from the dependencies between words, a special dependency label "Append(A)" was added to represent the dependencies between characters within a word. We use the last character in each word (the root node of the dependency subtree) as a representation of this word, and the dependency between words can be replaced by the dependency between last characters of each word. For example, the dependency relationship "安理会←决定" is transformed into "会←定". Figure 1 shows an example of CTB5 dataset being converted to a character-level dependency tree.

2.2 Character-Level POS Tagging

In order to transform the POS tagging into a character-level task, this paper adopts the following rules to convert the POS tag of words into POS tag of each character: the POS tag of each character is the POS tag of the word it is in. In predicting word's POS tag, it is represented by the POS tag of last character of the word. For example, if the predicted POS tag sequence of the word "安理会" is "NN, VV, NN", then the POS tag "NN" of the last character "会" is taken as the POS tag prediction result of the whole word. It is important to note that a word's POS tag is predicted correctly only if the word segmentation is predicted correctly and the last charater's POS tag is also predicted correctly.

2.3 Graph-Based Joint Model for Word Segmentation, POS Tagging and Dependency Parsing

According to Sects. 2 and 2.1, after converting three tasks into two character-level tasks, we designed a shared deep Bi-LSTM network to encode the input characters and obtain contextual character vectors. As shown in Fig. 2, given the input sentence (character sequence) $X = \{x_1, ..., x_n\}$. Firstly, vectorize each character x_i to get vector e_i, which consists of two parts, one is pre-trained vector p_i which is fixed during training, and the other is randomly initializing embeddings s_i which can be adjusted in training. Element-wise adds the pre-trained and random embeddings as the final input characters' embedding e_i, that is $e_i = p_i + s_i$. Then we feed the characters' embedding into multi-layer Bi-LSTM network, and get each character's contextual representation $C = \{c_1, ..., c_n\}$.

$$\overrightarrow{c_i} = \overrightarrow{\mathrm{LSTM}}(e_i, \overrightarrow{c}_{i-1}, \overrightarrow{\theta}); \; \overleftarrow{c_i} = \overleftarrow{\mathrm{LSTM}}(e_i, \overleftarrow{c}_{i+1}, \overleftarrow{\theta}); \; c_i = \overrightarrow{c_i} \oplus \overleftarrow{c_i} \tag{1}$$

After the contextual character vectors are obtained, the character-level POS tagging and dependency parsing are carried out respectively. We adopted the graph-based framework to analyze the character-level dependency parsing task. By taking each character as a node on the graph, and taking the possibility of forming a dependency relationship between characters as a probability directed edge between nodes (from the head node points to the dependency node), we can define dependency parsing as finding a dependency tree with the highest probability that conforms to the dependency grammar on a directed complete graph. The process of dependency parsing contains two subtasks: prediction of dependency relationship and prediction of dependency relationship type.

Prediction of Dependency Relationship: We use $x_i \leftarrow x_j$ to represent the dependency relation between x_i as the dependency node and x_j as the head node. After context encoding, each character obtains a vector representation c_i. Considering that each character has the possibility of being a dependency node and a head node, we use two vectors d_i^{arc} and h_i^{arc} to represent them respectively, and get them from c_i through two different MLP, as shown in formula (2).

$$d_i^{arc} = \mathrm{MLP}_d^{arc}(c_i); \; h_i^{arc} = \mathrm{MLP}_h^{arc}(c_i) \tag{2}$$

To calculate the probability s_{ij}^{arc} of $x_i \leftarrow x_j$, we use biaffine attention mechanism proposed by [5].

$$s_{ij}^{arc} = \mathrm{Biaffine}^{arc}(h_j^{arc}, d_i^{arc}) = h_j^{arc} U^{arc} d_i^{arc} + h_j^{arc} u^{arc} \tag{3}$$

where U^{arc} is a matrix whose dimension is (d_c, d_c), and the d_c is the dimension of vector c_i, u^{arc} is a bias vector. After we get the scores of all head nodes of the i-th character, we select the max score node as its head.

$$s_i^{arc} = [s_{i1}^{arc}, ..., s_{in}^{arc}]; \; y_i^{arc} = \arg\max(s_i^{arc}) \tag{4}$$

Prediction of Dependency Relatoinship Type: After obtaining the best predicted unlabeled dependency tree, we calculate the label scores s_{ij}^{label} for each

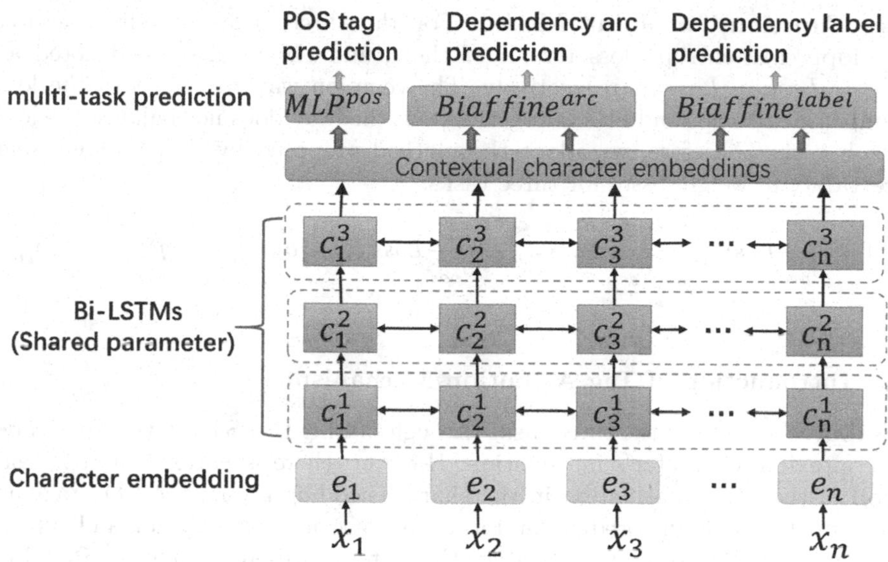

Fig. 2. A joint model of segmentation, POS tagging and dependency parsing with parameter sharing

dependency relationship $x_i \leftarrow x_j$. In our joint model, the arc labels set consists of the standard word-level dependency labels and a special label "A" indicating the intra-dependency within a word. We also use two vectors d_i^{label} and h_i^{label} to represent them respectively, and get them from c_i through two different MLP, and we use another biaffine attention network to calculate the label scores s_{ij}^{label}.

$$d_i^{label} = \mathrm{MLP}_d^{label}(c_i);\ h_i^{label} = \mathrm{MLP}_h^{label}(c_i) \tag{5}$$

$$s_{ij}^{label} = \mathrm{Biaffine}^{label}(h_j^{label}, d_i^{label}) = h_j^{label} U^{label} d_i^{label} + (h_j^{label} \oplus d_i^{label}) V^{label} + b \tag{6}$$

where U^{arc} is a tensor whose dimension is (k, d_c, d_c), k is the number of dependency relationship labels, and V^{label}'s dimension is $(k, 2d_c)$, and b is a bias vector. The best label of the dependency relationship $x_i \leftarrow x_j$ is:

$$y_{ij}^{label} = \arg\max(s_{ij}^{label}) \tag{7}$$

Prediction of POS Tagging: We use multi-layer perceptron (MLP) to calculate the probability distribution of the POS tag for each character.

$$s_i^{POS} = \mathrm{MLP}^{POS}(c_i) \tag{8}$$

The best POS tag of the character x_i is

$$y_i^{POS} = \arg\max(s_i^{POS}) \tag{9}$$

Loss Function for Joint Model: For the three tasks described above, we adopt cross-entropy loss for all of them, and the results are denoted as $Loss_{arc}, Loss_{dep}, Loss_{pos}$ respectively. The common way to deal with the loss of multiple tasks is to add them together, but this way does not balance the loss of each task. Therefore, we adopt the method proposed by [10], that is using uncertainty to weigh losses for three tasks.

$$\mathcal{L}(\theta) = \frac{1}{\delta_{arc}^2} Loss_{arc} + \frac{1}{\delta_{dep}^2} Loss_{dep} + \frac{1}{\delta_{pos}^2} Loss_{pos} + log\delta_{arc}^2 + log\delta_{dep}^2 + log\delta_{pos}^2 \tag{10}$$

2.4 Introduction of Tag Attention Mechanism

The above model joint the three tasks through sharing Bi-LSTM layers to encode the contextual character's information. However, there is no explicit representation of the POS information in the shared encoding layers, the POS tagging task cannot provide the predicted information for word segmentation and dependency parsing. Therefore, we introduce the vector representation of the POS tag and propose the tag attention mechanism (TAM) to integrate the POS information of contextual characters into the vector representation of each character, so that the POS information of the contextual character can also be used in the word segmentation and dependency parsing. This structure is similar to the hierarchically-refined label attention network (LAN) proposed by [4], but we use it to obtain POS information of each layer for subsequent character-level dependency parsing tasks. LAN differs from TAM in that LAN only predicts at the last layer while TAM predicts at each layer. We have tried to predict only at the last layer, but the result of segmentation and dependency parsing is slightly lower than predicting at each layer. The model is shown in Fig. 3.

Firstly, we vectorize the POS tags. Each POS tag is represented by a vector e_i^t, and the represents of the set of POS tags denoted as $E^t = \{e_1^t, ..., e_m^t\}$, which is randomly initialized before model training, and then is adjusted during the model training. Then, we calculate the attention weight between the contextual character vectors and POS tag vectors:

$$\alpha = \text{softmax}(\frac{QK^T}{\sqrt{d_c}}) \tag{11}$$

$$E^+ = \text{Attention}(Q, K, V) = \alpha V \tag{12}$$

$$C^+ = \text{LayerNorm}(C + E^+) \tag{13}$$

where Q, K, V are matrices composed of a set of queries, keys and values. We set $Q = C, K = V = E^t$. The i-th line of α represents the POS tag probability distribution of the i-th character of the sentence. According to this probability distribution α, we calculate the representation of predicted POS tag of each character of the sentence, and it is denoted as E^+. The E^+ is added to the contextual vectors C as the POS tag information. After layer normalization([1], we can obtain the character vectors (C^+) containing the POS information, and

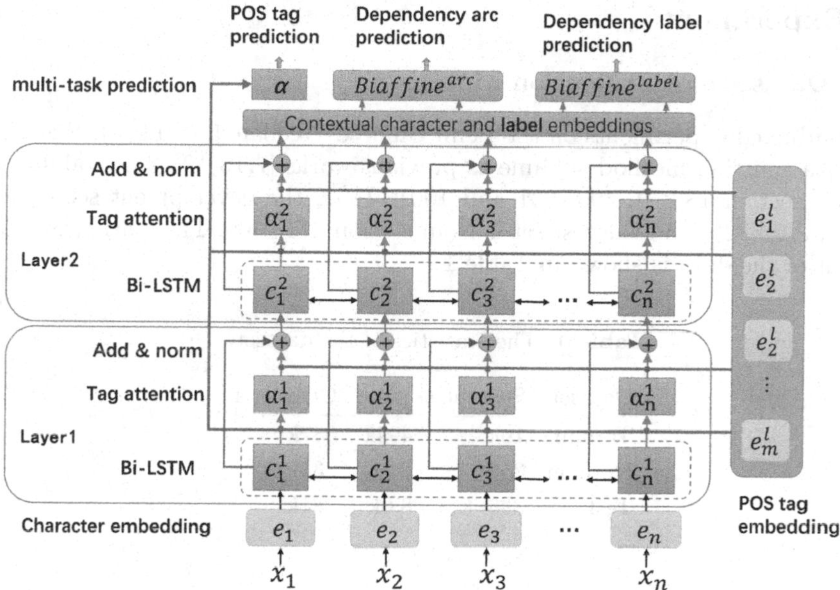

Fig. 3. A joint model of segmentation, POS tagging and dependency parsing with tag attention mechanism

then take it as the input of the next Bi-LSTM layer. After the second layer of Bi-LSTM encoding, each character vector we get will contain every characters' POS information, which can be used by word segmentation and dependency parsing.

When the tag attention mechanism is applied, the i-th line of the calculated attention weight for each layer is the POS tag distribution of the i-th character. Different from the prediction method of POS tagging in previous model, we added the attention weights of all layers as the final POS tag distribution:

$$s_i^{POS} = \sum_j^m \alpha_i^j \tag{14}$$

where, m is the number of layers. The prediction of POS tag is:

$$y_i^{POS} = \arg\max(s_i^{POS}) \tag{15}$$

For word segmentation and dependency parsing, we use the same approach as the previous model. For the losses of three tasks, we also use the same way to calculate it as the previous model.

3 Experiment

3.1 Dataset and Evaluation Metrics

We conducted experiments on the Penn Chinese Treebank5 (CTB-5). We adopt the data splitting method as same as previous works [7,13,19]. The training set is from section 1∼270, 400∼931 and 1001∼1151, the development set is from section 301∼325, and the test set is from section 271∼300. The statistical information of the data is shown in Table 1.

Table 1. The statistics of the dataset.

Dataset	Sentence	Word	Character
Training	16k	494k	687k
Develop	352	6.8K	31k
Test	348	8.0k	81k

Following previous works [9,13,19], we use standard measures of word-level F1 score to evaluate word segmentation, POS tagging and dependency parsing. F1 score is calculated according to the precision P and the recall R as $F = 2PR/(P + R)$ [9]. Dependency parsing task is evaluated with the unlabeled attachment scores excluding punctuations. The output of POS tags and dependency arcs cannot be correct unless the corresponding words are correctly segmented.

3.2 Model Configuration

We use the same Tencent's pre-trained embeddings [17] and configuration as [19], and the dimension of character vectors is 200. The dimension of POS tag vectors is also 200. We use with 400 units for each Bi-LSTM layer and the layer numbers is 3. Dependency arc MLP output size is 500 and the label MLP output size is 100. The dropout rates are all 0.33.

The models are trained with Adam algorithm [11] to minimize the total loss of the cross-entropy of arc predictions, label predictions and POS tag predictions, which using uncertainty weights to combine losses. The initial learning rate is 0.002 annealed by multiplying a fix decay rate 0.75 when parsing performance stops increasing on development sets. To reduce the effects of "gradient exploding", we use gradient clip of 5.0 [16]. All models are trained for 100 epochs.

3.3 Results

We conduct comparison of our models with other joint parsing models. The model shown in Fig. 2 is denoted as Ours and the model shown in Fig. 3 as Ours-TAM (with tag attention mechanism). The comparison models include three types: one is the transition-based joint models with feature templates [7,13,21], the other is the transition-based joint models with neural network [13](4-g, 8-g), and the third is the graph-based model with neural network without POS tagging task [19]. The results are shown in Table 2[1].

Table 2. Performance comparison of Chinese dependency parsing joint models.

Model	Framework	SEG	POS	DEP
Hatori12 [7]	Transition	97.75	94.33	81.56
Zhang14 [21]	Transition	97.67	94.28	**81.63**
Kurita17 [13]	Transition	**98.24**	**94.49**	80.15
Kurita17(4-g) [13]	Transition	97.72	93.12	79.03
Kurita17(8-g) [13]	Transition	97.70	**93.37**	79.38
Yan19 [19][1]	Graph	**98.47**	—	**87.24**
Ours	Graph	98.34	94.60	**87.91**
Ours-TAM	Graph	**98.42**	**94.84**	87.62

From the table, we see that transition-based joint models using feature templates maintain the best performance in word segmentation, POS tagging and dependency parsing for a long time. Although [13](4-g, 8-g) adopted the neural network approach, it still didn't surpass the joint model with feature templates. While, the graph-based joint model [19] obtained the better performance in word segmentation and dependency parsing than all transition-based model.

Our models Ours and Ours-TAM exceeded [19] 0.67 and 0.38% points respectively in dependency parsing, indicating that the POS tag information contributes to dependency parsing. Although they are 0.13 and 0.05% points lower than [19] on word segmentation task respectively, they still exceed the best transition-based joint model with feature templates [13]. [19] does not have POS tagging task, but our models have, and its performance exceeded that of the previous best joint model [13] by 0.11 and 0.35% points respectively, indicating that after the introduction of POS tagging, other tasks such as dependency parsing are also helpful for POS tagging task itself.

3.4 Detailed Analysis

We will further investigate the reasons for the improvement of dependency parsing after the combination of POS tagging task. For a dependency relationship $x_i \leftarrow x_j$, we use $X \leftarrow Y$ to represent its POS dependency pattern, the X is the POS tag of x_i, and the Y is the POS tag of x_j. We calculated the distribution of Y for each X in training set and found that the probability between some X and Y

[1] Yan et al. later submitted an improved version Yan20 [20], and the results of word segmentation and dependency parsing reached 98.48 and 87.86, respectively.

was very high. For example, when X was P(preposition), the distribution of Y was {VV(78.5%), DEG(5.1%), ..., NN(3.1%), ... }. In order to verify whether our models can use these POS informations in training dataset, we calculated the accuracy of each POS dependency patterns in test dataset on our models and the re-implemented model of [19]. The patterns on which the accuracy of our models are better than [19] are shown in left part of Fig. 4.

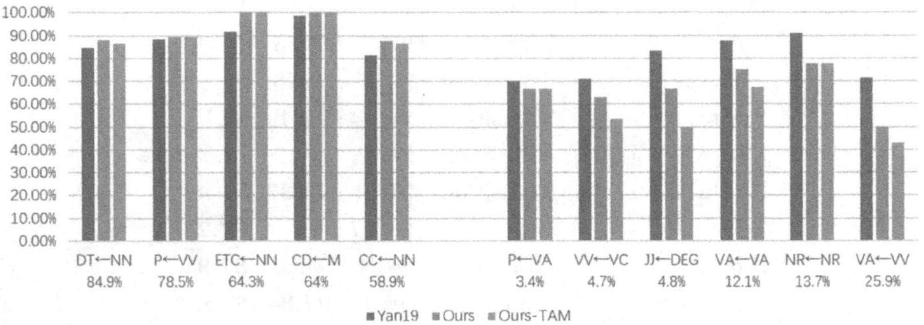

Fig. 4. Comparison of precision on different POS tag patterns before and after joint POS tagging task

Table 3. Head POS distribution

Node POS	Head POS distribution					
DT	NN 84.9%	VV 7.5%	DEG 1.8%	P 1.3%	M 1%	NR 0.8%
P	VV 78.5%	DEG 5.1%	VA 3.4%	VE 3.3%	VC 3.1%	NN 3.1%
ETC	NN 64.3%	NR 22.5%	VV 10.4%	VA 1.6%	VE 0.2%	VC 0.2%
CD	M 64%	NN 20.6%	VV 6.7%	CD 2.7%	DT 1.6%	DEG 1.2%
CC	NN 58.9%	VV 20.5%	NR 7.9%	NT 2.3%	VA 2.1%	M 1.9%

The X of these 5 patterns are {DT, P, ETC, CD, CC}, and the Y's distributions of each X are shown in the Table 3. It is found that all 5 patterns select Y with the highest probability, indicating that our model can fully utilize the POS informations to improve the accuracy of dependencies with these POS dependency patterns. As the example shown in the Fig. 5, when predicting the head node of "从", [19] predicted wrong node "工作", while our models both predicted right node "撤退". The POS tag of "从" is P and the POS tag of correct head node "撤退" is VV whose probability is 78.5%, while the wrong head node 工作 s POS tag is NN whose probability is only 3.1%. Because our models can use these POS informations to exclude the candidate head nodes of low probability POS, thus improving the performance of dependency parsing.

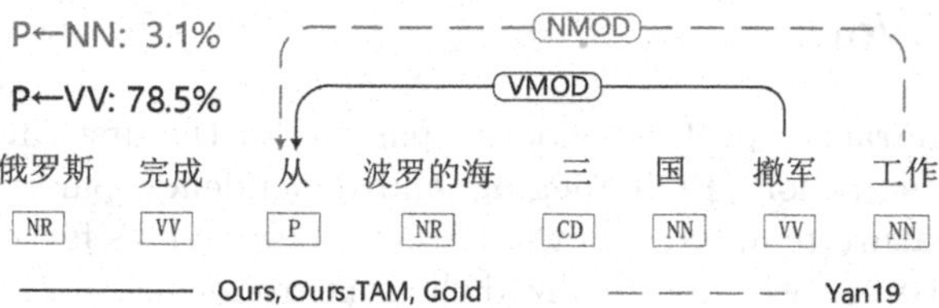

Fig. 5. An example of POS information contributes to dependency parsing

Although Ours-TAM achieved better results in segmentation and POS tagging, the dependency parsing was reduced compared with Ours. The right part of the Fig. 4 shows the patterns on which the accuracy of our models are worse than [19]. It can be found that the dependency probability of these patterns is small, and the addition of POS information actually reduces the accuracy. Therefore, Ours-TAM has better POS information, so the accuracy of these patterns is lower than Ours, thus the overall precision of dependency parsing of Ours-TAM decreases compared with that of Ours.

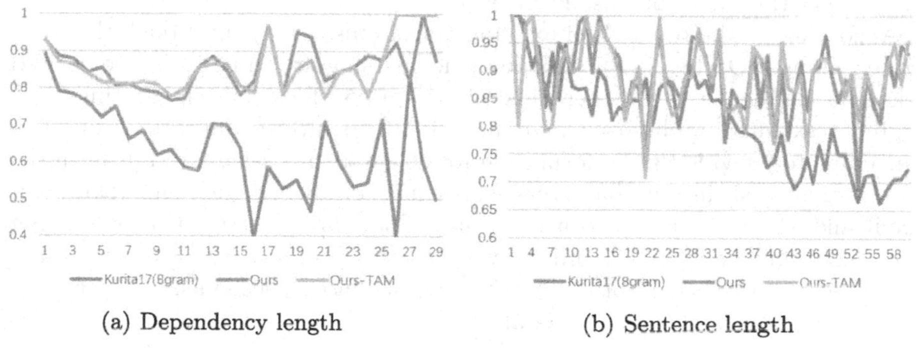

(a) Dependency length (b) Sentence length

Fig. 6. The influence of dependency length and sentence length on dependency parsing

Next, we will investigate the difference between the graph-based joint model and the transition-based joint model in dependency parsing. We compare our graph-based joint models to the transition-based joint model [13] according to dependency length and sentence length respectively. The results are shown in Fig. 6. From the figure, we can see that our proposed joint models on long-distance dependencies have obvious advantages, and the accuracy of the dependency parsing is relatively stable with the increase of sentence length, while the transition-base joint model has an obvious downward trend, which indicates that our graph-based joint model can predict the long-distance dependencies more effectively than transition-based joint model.

4 Related Work

[7] proposed a character-level dependency parsing for the first time, which combines word segmentation, POS tagging and dependency parsing, They combined the key feature templates on the basis of the previous feature engineering research on the three tasks, and realized the synchronous processing of the three tasks. [21] annotated the internal structure of words, and regarded the word segmentation task as dependency parsing within characters to jointly process with three tasks. [13] firstly applied neural network to the charater-level dependency parsing. Although these transition-based joint models achieved best accuracy in dependency parsing, they still suffer from the limitation of local decision.

With the development of neural network, the graph-based dependency parsing models [5,12] using neural networks have developed rapidly. these model fully exploit the ability of the bidirectional long short-term memory network (BiLSTM) [8] and attention mechanism [2,18] to capture the interactions of words in a sentence. Different from transition-based models, the graph-based model can make global decision when predicting dependency arcs, but few joint model adopted this framework. [19] firstly proposed a joint model adopting graph-based framework with neural network for Chinese word segmentation and dependency parsing, but they does not use POS tag.

According to the research of existing transition-based joint model, the word segmentation, POS tagging and dependency parsing are three highly correlated tasks that influence each other. Therefore, we consider that integrating POS tagging task into graph-based joint model [19] to further improve the performance and to provide POS information for downstream tasks. We transform the POS tagging task into a character-level sequence labeling task and then combine it and [19] by using multi-task learning. There are many multi-task learning approaches such as [3,14,15] and [6], we use parameter sharing [3] to realize the joint model, and then improve it with tag attention mechanism. Finally, we analyze the models on the CTB5 dataset.

5 Conclusion

This paper proposed the graph-based joint model for Chinese word segmentation, POS tagging and dependency parsing. The word segmentation and dependency parsing are transformed into a character-level dependency parsing task, and the POS tagging task is transformed into a character-level sequence labeling task, and we use two ways to joint them into a multi-task model. Experiments on CTB5 dataset show that the combination of POS tagging task is beneficial to dependency parsing, and using the POS tag attention mechanism can exploit more POS information of contextual characters, which is beneficial to POS tagging and dependency parsing, and our graph-based joint model outperforms the existing best transition-based joint model in all of these three tasks. In the future, we will explore other joint approaches to make three tasks more mutually reinforcing and further improve the performance of three tasks.

References

1. Ba, J.L., Kiros, J.R., Hinton, G.E.: Layer normalization. arXiv preprint arXiv:1607.06450 (2016)
2. Bahdanau, D., Cho, K., Bengio, Y.: Neural machine translation by jointly learning to align and translate. arXiv preprint arXiv:1409.0473 (2014)
3. Baxter, J.: A Bayesian/information theoretic model of learning to learn via multiple task sampling. Mach. Learn. **28**(1), 7–39 (1997)
4. Cui, L., Zhang, Y.: Hierarchically-refined label attention network for sequence labeling. arXiv preprint arXiv:1908.08676 (2019)
5. Dozat, T., Manning, C.D.: Deep biaffine attention for neural dependency parsing. arXiv preprint arXiv:1611.01734 (2016)
6. Hashimoto, K., Xiong, C., Tsuruoka, Y., Socher, R.: A joint many-task model: growing a neural network for multiple nlp tasks. arXiv preprint arXiv:1611.01587 (2016)
7. Hatori, J., Matsuzaki, T., Miyao, Y., Tsujii, J.: Incremental joint approach to word segmentation, PoS tagging, and dependency parsing in Chinese. In: Proceedings of the 50th Annual Meeting of the Association for Computational Linguistics: Long Papers, vol. 1, pp. 1045–1053. Association for Computational Linguistics (2012)
8. Hochreiter, S., Schmidhuber, J.: Long short-term memory. Neural Comput. **9**(8), 1735–1780 (1997)
9. Jiang, W., Huang, L., Liu, Q., Lü, Y.: A cascaded linear model for joint Chinese word segmentation and part-of-speech tagging. In: Proceedings of ACL-08: HLT, pp. 897–904 (2008)
10. Kendall, A., Gal, Y., Cipolla, R.: Multi-task learning using uncertainty to weigh losses for scene geometry and semantics. In: Proceedings of the IEEE Conference on Computer Vision and Pattern Recognition, pp. 7482–7491 (2018)
11. Kingma, D.P., Ba, J.: Adam: a method for stochastic optimization. arXiv preprint arXiv:1412.6980 (2014)
12. Kiperwasser, E., Goldberg, Y.: Simple and accurate dependency parsing using bidirectional LSTM feature representations. Trans. Assoc. Comput. Linguist. **4**, 313–327 (2016)
13. Kurita, S., Kawahara, D., Kurohashi, S.: Neural joint model for transition-based Chinese syntactic analysis. In: Proceedings of the 55th Annual Meeting of the Association for Computational Linguistics (Volume 1: Long Papers), pp. 1204–1214 (2017)
14. Long, M., Wang, J.: Learning multiple tasks with deep relationship networks. arXiv preprint arXiv:1506.02117 2, 1 (2015)
15. Misra, I., Shrivastava, A., Gupta, A., Hebert, M.: Cross-stitch networks for multi-task learning. In: Proceedings of the IEEE Conference on Computer Vision and Pattern Recognition, pp. 3994–4003 (2016)
16. Pascanu, R., Mikolov, T., Bengio, Y.: On the difficulty of training recurrent neural networks. In: International Conference on Machine Learning, pp. 1310–1318 (2013)
17. Song, Y., Shi, S., Li, J., Zhang, H.: Directional skip-gram: explicitly distinguishing left and right context for word embeddings. In: Proceedings of the 2018 Conference of the North American Chapter of the Association for Computational Linguistics: Human Language Technologies, Volume 2 (Short Papers), pp. 175–180 (2018)
18. Vaswani, A., et al.: Attention is all you need. In: Advances in Neural Information Processing Systems, pp. 5998–6008 (2017)

19. Yan, H., Qiu, X., Huang, X.: A unified model for joint Chinese word segmentation and dependency parsing. arXiv preprint arXiv:1904.04697 (2019)
20. Yan, H., Qiu, X., Huang, X.: A graph-based model for joint Chinese word segmentation and dependency parsing. Trans. Assoc. Comput. Linguist. **8**, 78–92 (2020)
21. Zhang, M., Zhang, Y., Che, W., Liu, T.: Character-level Chinese dependency parsing. In: Proceedings of the 52nd Annual Meeting of the Association for Computational Linguistics (Volume 1: Long Papers), pp. 1326–1336 (2014)

Semantic-Aware Chinese Zero Pronoun Resolution with Pre-trained Semantic Dependency Parser

Lanqiu Zhang, Zizhuo Shen, and Yanqiu Shao(✉)

Beijing Language and Culture University, Beijing, China
zhang_lanqiu@163.com, yqshao163@163.com, blcushzz@gmail.com

Abstract. Deep learning-based Chinese zero pronoun resolution model has achieved better performance than traditional machine learning-based model. However, the existing work related to Chinese zero pronoun resolution has not yet well integrated linguistic information into the deep learning-based Chinese zero pronoun resolution model. This paper adopts the idea based on the pre-trained model, and integrates the semantic representations in the pre-trained Chinese semantic dependency graph parser into the Chinese zero pronoun resolution model. The experimental results on OntoNotes 5.0 dataset show that our proposed Chinese zero pronoun resolution model with pre-trained Chinese semantic dependency parser improves the F-score by 0.4% compared with our baseline model, and obtains better results than other deep learning-based Chinese zero pronoun resolution models. In addition, we integrate the BERT representations into our model so that the performance of our model was improved by 0.7% compared with our baseline model.

1 Introduction

Chinese zero pronoun resolution is a special task of coreference resolution [1]. Its purpose is to find the real referent of the omitted parts with syntactic functions in the text. These omitted parts are usually called zero pronouns, and their real referents are called antecedents. Below is a sentence with zero pronouns:

[][] [*$pro*_1$] [*$pro*_2$] (I have not heard of [her] before, [*pro_1*] heard that [*pro_2*] is a talented beauty.)

In this example, the referent of zero pronoun *pro_1* is "/I", and the referent of zero pronoun *pro_2* is "/her". Since the zero pronoun is not a real word in the text, its resolution is much more difficult than that of the overt pronoun. The existence of zero pronouns poses challenges for machines to automatically understand text.

The existing Chinese zero pronoun resolution models with better performance usually adopt the method of deep learning [2–6]. The deep learning-based methods can make the model automatically extract the task-related distributed representations through end-to-end training, thereby avoiding the problem that traditional machine learning-based methods rely heavily on artificially designed

M. Sun et al. (Eds.): CCL 2020, LNAI 12522, pp. 17–29, 2020.
https://doi.org/10.1007/978-3-030-63031-7_2

feature templates [2]. However, it is difficult for deep learning-based models to encode effective syntactic, semantic and other linguistic information only through end-to-end training. Many deep learning-based Chinese zero pronoun resolution models still use syntactic features extracted from the syntactic parsing tree as a supplement to distributed representations.

Intuitively, semantic information as a higher level linguistic information is also very important to the Chinese zero pronoun resolution task, however few studies have attempted to integrate semantic information into the Chinese zero pronoun resolution model. Therefore, how to effectively integrate semantic information into the Chinese zero pronoun resolution model is a challenging problem. With the development of semantic parsing, the performance of some sentence-level semantic parsers have made remarkable progress, which provides opportunities for the application of sentence-level semantic parsing in other natural language processing tasks.

In this paper, we proposed a semantic-aware Chinese zero pronoun resolution model that integrates the semantic information from pre-trained Chinese semantic dependency graph parser. Chinese semantic dependency graph parsing [7] is a semantic-level dependency parsing task, which is an extension of syntactic dependency parsing. Each node in the semantic dependency graph represents a word in the sentence, and the nodes are connected by directed edges with semantic relationship labels. Figure 1 is an example of a Chinese semantic dependency graph.

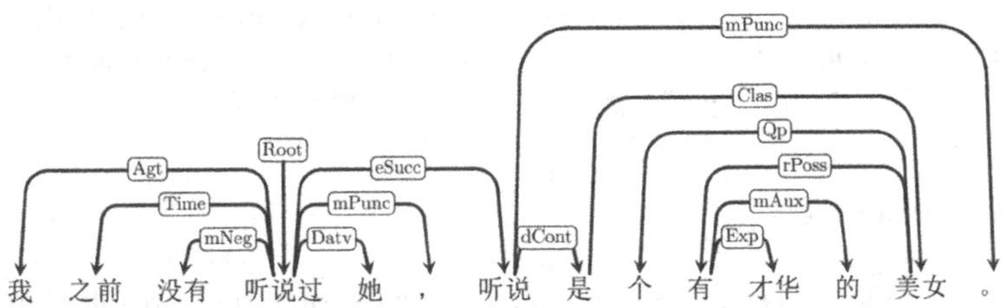

Fig. 1. An example of a Chinese semantic dependency graph

The realization of our model requires two stages. In the first stage, we use the Chinese semantic dependency graph parsing as a pre-training task to obtain a pre-trained semantic dependency graph parser. In the second stage, we feed the sentence which will be processed into the pre-trained semantic dependency graph parser to obtain the semantic-aware representations, and integrate these implicit semantic information into the Chinese zero pronoun resolution model.

We implement a attention-based Chinese zero pronoun resolution model as our baseline model. The experiments on OntoNotes-5.0 dataset show that our proposed Chinese zero pronoun resolution model with pre-trained Chinese semantic dependency parser improves the F-score by 0.4% compared with our baseline model, and obtains better results than other deep learning-based

Chinese zero pronoun resolution models. In addition, we integrate the BERT representations into our model so that the performance of our model was improved by 0.7% compared with our baseline model.

2 Related Work

2.1 Zero Pronoun Resolution

Methods for solving Chinese zero pronoun resolution include rule-based methods, traditional machine learning-based methods, deep learning-based methods, etc. Converse [8] used Hobbs algorithm to traverse the syntactic tree of sentences to find the referent of zero pronoun. Zhao et al. [1] designed more effective manual features for Chinese zero pronoun resolution task, and adopted a decision tree-based method to train supervised model. Kong et al. [9] adopted a tree kernel-based method to model the syntax tree, so that the Chinese zero pronoun resolution model can make full use of the characteristics of the syntax tree. Chen et al. [2] designed a Chinese zero pronoun resolution model based on feed-forward neural network, and represented the zero-pronoun and candidate antecedent by combining manual feature vectors and word vectors, and obtained better performance than traditional machine learning-based methods. Yin et al. [4–6] designed a series of deep learning-based Chinese zero pronoun resolution model, which promoted the application of deep learning to Chinese zero pronoun resolution. Liu et al. [3] transformed the Chinese zero pronoun resolution task into the cloze-style reading comprehension task, and automatically constructed large-scale pseudo-data for the pre-training of their model.

2.2 Pre-training of Syntactic Dependency Parsing

Our method is similar to the method of pre-training of syntactic dependency parser, which has been successfully applied to some natural language processing tasks. Zhang et al. [10] first proposed this method in the task of relation extraction. First, they trained the LSTM-based Biaffine syntactic dependency parser. Then, they extracted implicit syntactic representations from the LSTM layer of the well-trained syntactic dependency parser and integrated these representations into the relation extraction model. Guo et al. [11] and Yu et al. [12] used this method to integrate syntactic representations in the task of target-dependent sentiment analysis and discourse parsing respectively, and verified the effectiveness of this method in these tasks. Zhang et al. [13] systematically studied the application of this method in the task of machine translation. Their experimental results show that this method obtains a more significant improvement than other methods such as Tree-Linearization and Tree-RNN in the task of machine translation. Jiang et al. [14] applied this method to the task of Universal Conceptual Cognitive Annotation(UCCA) [15]. Inspired by the method of integrating pre-trained information in ELMo [16], They made a weighted sum for the output of different LSTM layers of syntactic dependency parser. Their experimental results show that the method of fine-tuning pre-trained syntactic dependency parser improves the performance of UCCA model significantly.

3 Method

Given the success of the method of pre-training of syntactic dependency parser in some natural language processing tasks, we adopt a similar method to take the Chinese semantic graph dependency parsing as a pre-training task, and apply this method to Chinese zero pronoun resolution task.

Our proposed Chinese zero pronoun resolution model with pre-trained Chinese semantic dependency parser is composed of two parts, one is the pre-trained Chinese semantic dependency graph parser and the other is the Chinese zero pronoun resolution model. Specifically, The Chinese semantic dependency graph parser consists of two parts: BiLSTM-based encoder and Biaffine-based decoder. The Chinese zero pronoun resolution model consists of three parts: the zero pronoun module (ZP Module), the candidate antecedents module (CA Module) and the discrimination module. In addition, in order to obtain sentence-level semantic representations, we also used a CNN-based sentence representation extractor (SR Extractor).

For a sentence to be processed, the representations of each word will be feed into the pre-trained Chinese semantic dependency graph parser, so that each word can obtain the semantic-aware representations containing the information of semantic dependency graph. Then, the semantic-aware representations will be integrated into the Chinese zero pronoun resolution model to perform the subsequent processing. The overall architecture of our proposed model is shown in Fig. 2:

3.1 Semantic Dependency Graph Parser

For the semantic dependency graph parser, we adopt 3-layer BiLSTM network and Biaffine network as encoder and decoder. The Biaffine-based parser has achieved the state of the art performance in some tasks related to semantic dependency graph parsing [17,18].

In the process of pre-training, we first use the concatenation of word vector, part of speech vector and character-level vector to represent a word. Then, we feed the word representations into the encoder to obtain the context-aware representations. Finally, we feed the context-aware representations of the word into the decoder to calculate the score of the dependency arc in the semantic dependency graph. The complete calculation process of the semantic dependency graph parser is shown in the following formulas:

$$w_t = [e_t^{(word)}; e_t^{(pos)}; e_t^{(char)}] \tag{1}$$

$$h_t = BiLSTM(w_t, h_{t-1}) \tag{2}$$

$$s_t^{(H,D)} = Biaffine(h_t^H, h_t^D) \tag{3}$$

where w_t means the word representations, h_t means the context-aware representations, $s_t^{(H,D)}$ means the score of the dependency arc, h_t^H and h_t^D mean context-aware representations of the head word and the dependent word respectively.

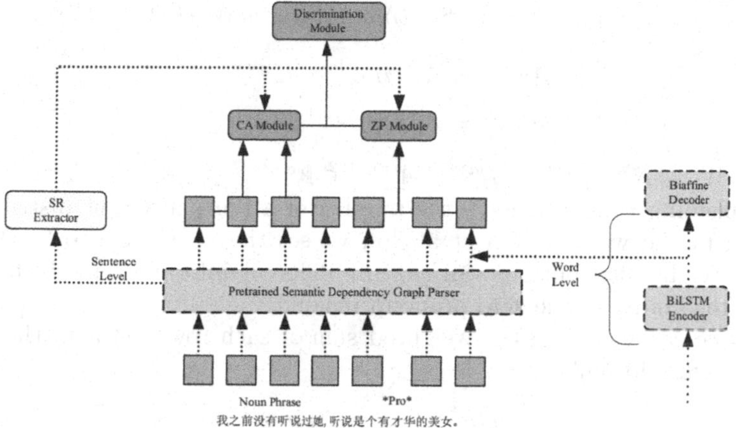

Fig. 2. Chinese zero pronoun resolution model with pre-trained semantic dependency graph parser

3.2 Zero Pronoun Module

According to the work of Yin et al. [6], we use BiLSTM network and self-attention mechanism to encode the preceding and following text of the zero pronoun. The purpose of using the self-attention mechanism is to obtain the attention weight distribution of the preceding and following word sequence. In this way, we can get the more powerful zero pronoun representations.

For a given anaphoric zero pronoun w_{zp}, we use $Context^{(pre)} = (w_1, w_2, ..., w_{zp-1})$ to denote the preceding word sequence of the zero pronoun, and use $Context^{(fol)} = (w_{zp+1}, w_{zp+2}, ... w_n)$ to denote the following word sequence of the zero pronoun. Each word w_t in the sentence is represented by the pre-trained word embedding.

In order to encode the contextual information of the word sequence, we first use two different 1-layer BiLSTM networks to separately process the preceding word sequence and the following word sequence:

$$h_t^{(pre)} = BiLSTM^{(pre)}(w_t, h_{t-1}^{(pre)}) \tag{4}$$

$$h_t^{(fol)} = BiLSTM^{(fol)}(w_t, h_{t-1}^{(fol)}) \tag{5}$$

After that, we can obtain the preceding and following hidden vectors of the zero pronoun $h_t^{(pre)}$ and $h_t^{(fol)}$ from the LSTM networks. We use $H^{(pre)}$ to denote the matrix which is concatenated by all preceding hidden vectors, and use $H^{(fol)}$ to denote the matrix which is concatenated by all following hidden vectors. Where $H^{(pre)} \in \mathbb{R}^{n^{(pre)} \times d}$, $H^{(fol)} \in \mathbb{R}^{n^{(fol)} \times d}$, $n^{(pre)}$ and $n^{(fol)}$ means the number of words in the preceding and following word sequence respectively. d means the dimension of the hidden vectors.

The matrix $H^{(pre)}$ and $H^{(fol)}$ will be feed into the affine-based attention layers $Affine^{(pre)}$ and $Affine^{(fol)}$ to calculate the attention weight distribution of their associated sequences:

$$Affine(H) = Softmax(W_2 tanh(W_1 H^T)) \tag{6}$$

$$A^{(pre)} = Affine^{(pre)}(H^{(pre)}) \tag{7}$$

$$A^{(fol)} = Affine^{(fol)}(H^{(fol)}) \tag{8}$$

where $W_1 \in R^{h \times d}$, $W_2 \in R^{a \times h}$, $A^{(pre)} \in \mathbb{R}^{a \times n^{(pre)}}$, $A^{(fol)} \in \mathbb{R}^{a \times n^{(fol)}}$. It is worth explaining that a denotes the number of attention weight distributions. According to the work of Yin et al. [6], we set the value of a to 2. Different attention weight distributions can capture different information, which further enhances the ability of the zero pronoun module.

Then, we can calculate the weighted sum of each row vector in the matrix by the following formula:

$$h_{zp}^{(pre)} = A^{(pre)} H^{(pre)} \tag{9}$$

$$h_{zp}^{(fol)} = A^{(fol)} H^{(fol)} \tag{10}$$

where $h_{zp}^{(pre)} \in R^{a \times d}$, $h_{zp}^{(fol)} \in R^{a \times d}$, If a is not equal to 1, We need to calculate the average of its row vectors.

At last, We take the concatenation of these two vectors as the final zero pronoun representations:

$$h_{zp} = [h_{zp}^{(pre)}; h_{zp}^{(fol)}] \tag{11}$$

3.3 Candidate Antecedents Module

When building the candidate antecedents module, we need to consider two types of the key information for the candidate antecedents. The first type of information is the context information of the candidate antecedents, and the second type of information is the interactive information between the zero pronoun and the candidate antecedents. Inspired by previous work [19], we use the context-aware boundary representations to capture context information and use attention mechanism to capture interactive information.

The candidate antecedent is usually a noun phrase composed of several words. So, we use $NP = (np_1, np_2, \ldots, np_n)$ to denote the set of all candidate antecedents for a given zero pronoun w_{zp}, and use $np_t = (w_i, w_2, \ldots, w_j)$ to denote a candidate antecedent within the set. First, we feed the pre-trained word vectors into the 1-layer BiLSTM network to obtain the context-aware representations of each word:

$$h_t = BiLSTM(w_t, h_{t-1}) \tag{12}$$

Apparently, we can get the sequence of the context-aware representations $np_t = (h_i, h_2, \ldots, h_j)$ from the outputs of the BiLSTM, where h_i means the start of the candidate antecedent, and h_j means the end of the candidate antecedent. We use h_i and h_j as the context-aware boundary representations of the candidate antecedents.

Then, we use a simple and effective scaled dot-product-based attention layer to calculate the weight distribution of the words in the candidate antecedent. We regard the zero pronoun representations h_{zp} as the query term, and regard the context-aware representations of all words in the candidate antecedent as the key term and value term. For simplicity in formula expression, we use the matrix H_{np} to denote he key term and value term:

$$h_{np}^{(attn)} = Softmax(\frac{h_{zp}H_{np}^T}{\sqrt{d_{np}}})H_{np} \tag{13}$$

where $h_{zp} \in \mathbb{R}^{d_{zp}}$, $H_{np} \in \mathbb{R}^{n \times d_{np}}$, $d_{zp} = d_{np}$, n denotes the number of words in the candidate antecedent. d_{np} denotes the dimension of context-aware representations of all words in the candidate antecedent. d_{zp} denotes the dimension of zero pronoun representations. $h_{np}^{(attn)}$ is the weighted sum the context-aware representations of all words in the candidate antecedent, where $h_{np}^{(attn)} \in \mathbb{R}^{d_{np}}$.

Finally, we take the concatenate of h_i, h_j, and $h_{np}^{(attn)}$ as the final representations of each candidate antecedent.

$$h_{np} = [h_i; h_j; h_{np}^{(attn)}] \tag{14}$$

3.4 Discrimination Module

After obtaining the representations of the zero pronoun and all candidate antecedents of this zero pronoun, we can feed these representations into the discrimination module to predict the real referent of the current zero pronoun.

For the discrimination module, this paper uses a bilinear function to calculate the probability distribution of all candidate antecedents of the current zero pronoun.

$$P(np_t|w_{zp}) = Softmax(h_{zp}UM_{np}^T + b) \tag{15}$$

$$\sum_{t=1}^{m} P(np_t|w_{zp}) = 1 \tag{16}$$

The parameters of the bilinear function are U and b, where $U \in \mathbb{R}^{k \times k}$, $b \in \mathbb{R}^k$, k denotes the dimension of the input vector of the bilinear function. h_{zp} denotes the zero pronoun representations. M_{np} denotes the matrix of all candidate antecedents of the current zero pronoun, where $h_{zp} \in \mathbb{R}^{1 \times k}$, $M_{np} \in \mathbb{R}^{m \times k}$. m denotes the number of all candidate antecedents of the current zero pronoun.

Given the probability distribution of all candidate antecedents of the current zero pronoun, we select the candidate antecedent with the highest probability as the real referent of the current zero pronoun.

3.5 The Integration of Semantic Representations

To make better use of the semantic representations from the pre-trained semantic dependency graph parser, We integrate the semantic representations of word-level and sentence-level into the Chinese zero pronoun resolution model.

Inspired by the work of Jiang et al. [14], we first extract all output vectors from the BiLSTM-based encoder of the pre-trained semantic dependency graph parser and then use a set of trainable parameters to weighted sum these vectors to obtain the final semantic representations. We use h_t^{sem} to denote the semantic representations of a word. This process is formally denoted by the following formula:

$$h_t^l = BiLSTM^{(l)}(w_t, h_{t-1}) \tag{17}$$

$$h_t^{(sem)} = \sum_{l=1}^{L} \alpha_l h_t^l \tag{18}$$

where w_t is the original word representations L is the layer number of the Bi-LSTM-based encoder, and α_l is the normalized weight of each layer.

For the integration of the word-level semantic representations, we simply concatenate the semantic representations of each word with its original word representations:

$$w_t^{(sem)} = [w_t; h_t^{(sem)}] \tag{19}$$

For the integration of the sentence-level semantic representations, We use the CNN-based sentence-level semantic representations extractor to perform 2-dimensional convolution and hierarchical pooling operations on the sentence sequence. Hierarchical pooling [20] is a combination of average pooling and max-pooling, which has better ability to capture word-order information. We use S_1^n to denote a sentence sequence with n words. This process is shown in the following formulas:

$$s^{(sem)} = Pooling(Convolution(S_1^n)) \tag{20}$$

After we obtain the sentence-level semantic representations, we integrate it into the zero pronoun module and the candidate antecedent module. We use two different multi-layer perceptrons to transform sentence-level semantic representations into zero pronoun-related and candidate antecedent-related representations. In this way, even if the zero pronoun and candidate antecedent are in the same sentence, these sentence-level semantic representations are different. This process is shown in the following formulas:

$$h_{zp}^{(sem)} = MLP^{(zp)}(s^{(sem)}) \tag{21}$$

$$h_{np}^{(sem)} = MLP^{(np)}(s^{(sem)}) \tag{22}$$

Finally, the zero pronoun representations and candidate antecedent representations that are integrated into the semantic representations can be formalized as:

$$h_{zp} = [h^{(pre)}; h^{(fol)}; h^{(sem)}] \tag{23}$$

$$h_{np} = [h_i; h_j; h_{np}^{(attn)}; h^{(sem)}] \tag{24}$$

3.6 Training Objective

The training objective is defined as:

$$Loss = -\Sigma_{zp} logP(np_t|w_{zp}) \tag{25}$$

where zp means the number of all anaphoric zero pronouns in the training set.

4 Experiment

4.1 Dataset and Resource

We conduct our experiments on the OntoNotes-5.0 dataset[1] which consists of document-level text selected from 6 domains: Broadcast News (BN), Newswire (NW), Broadcast Conversation (BC), Web Blog (WB), Telephone Conversation (TC) and Magazine (MZ). The training set has 1391 documents, a total of 36487 sentences and 12111 zero pronouns; The development set has 172 documents with a total of 6083 sentences and 1713 zero pronouns. The pre-trained word embedding used in Chinese zero pronoun resolution are trained by Word2Vec algorithm on Chinese Gigawords[2]. For Pre-training the Chinese semantic dependency graph parser, we use the SemEval-2016 Task 9 dataset[3]. For BERT related experiments, We use the Chinese Bert-base model, which has been pre-trained by the Google[4].

4.2 Evaluation Measures

We adopt the Recall, Precision and F-score (denoted as F) as the evaluation metrics of our Chinese zero pronoun resolution model. More specifically, recall, precision and F are defined as:

$$P = \frac{the\ number\ of\ zero\ pronouns\ predicted\ correctly}{the\ number\ of\ all\ predicted\ zero\ pronouns} \tag{26}$$

$$R = \frac{the\ number\ of\ zero\ pronouns\ predicted\ correctly}{the\ number\ of\ zero\ pronouns\ labeled\ in\ all\ datasets} \tag{27}$$

$$F = \frac{2PR}{P+R} \tag{28}$$

[1] http://catalog.ldc.upenn.edu/LDC2013T19.
[2] https://catalog.ldc.upenn.edu/LDC2003T09.
[3] https://github.com/HIT-SCIR/SemEval-2016.
[4] https://github.com/google-research/bert.

4.3 Hyperparameters

For Zero Pronoun Module, the hidden dimension of the LSTM is 128 and the output dimension of the affine-based attention layer is 128. For Candidate Antecedents Module, the hidden dimension of the LSTM is 128 and the output dimension of the scaled dot-product-based attention layer is 128. For all pre-trained representations, we convert the final input dimension to 256. For all LSTM, dropout rates are set to 0.33. For other neural network, dropout rates are set to 0.5. For training, the model is optimized by the Adam algorithm with the initial learning rate 0.003.

4.4 Main Experiments

We chose three deep learning-based Chinese zero pronoun resolution model implemented by Yin et al. as reference: Deep Memory Network-based Chinese zero pronoun resolution model [4] (DMN-ZP Model), Self-attention-based Chinese zero pronoun resolution model [6] (SA-ZP Model) and Deep Reinforcement Learning-based Chinese zero pronoun resolution model [5] (DRL-ZP Model).

We evaluate the performance of our Chinese zero pronoun resolution model on OntoNotes-5.0 development dataset with two different model settings: Chinese zero pronoun resolution model without pre-trained Chinese semantic dependency graph parser (Our Baseline Model), Chinese zero pronoun resolution model with pre-trained Chinese semantic dependency graph parser (Our Semantic-aware Model). The specific experimental results are shown in Table 1:

Table 1. Comparison of different Chinese zero pronoun resolution models

Model	NW(84)	MZ(162)	WB(284)	BN(390)	BC(510)	C(283)	Overall
DMN-ZP model	48.8	46.3	59.8	58.4	53.2	**54.8**	54.9
DRL-ZP model	63.1	50.2	63.1	56.7	57.5	54	57.2
SA-ZP model	**64.3**	52.5	62	**58.5**	57.6	53.2	57.3
Our baseline model	63.3	51.5	61.8	58.2	57.5	53.1	57.2
Our semantic-aware model	64.3	**52.7**	**63.3**	58.3	**58.8**	53.1	**57.6**

Compared with the baseline model, our semantic-aware model has achieved a 0.4 % improvement in F-score. Compared with previous deep learning-based models, the performance of our semantic-aware model is the best. According to the experimental results in various fields, we found that our semantic-aware model obtains the highest F-score in the MZ, BC and WB fields. Among them, the improvement of our semantic-aware model in the BC field is the most obvious. However, in the field of NW, BN and TC, the performance of our semantic-aware model has no advantage. One possible reason for this phenomenon is that the performance of the semantic dependency graph parser in these three fields is relatively poor, and it cannot provide valuable semantic information to the task of Chinese zero pronoun resolution.

4.5 Ablation Experiment

In order to further verify the effectiveness of our model, we tested the performance of models using the word-level and sentence-level integration method through ablation experiments. According to the experimental results in Table 2, we found that both integration methods can improve the performance of our model, and when both integration methods are used simultaneously, the performance of our model is optimal. The word-level integration method can only focus on the semantic information within the same sentence, while the sentence-level integration method has the ability to focus on the difference in sentence-level semantic information between different sentences. Therefore, the word-level integration method may be more suitable for the case where the zero pronoun and the candidate antecedent are in the same sentence, and the sentence-level integration method is more suitable for the case where the zero pronoun and the candidate antecedent are in different sentences. It is the complementarity of these two methods that makes the performance of our model continuously improved.

4.6 Integration with BERT

BERT [21] is a pre-trained language model with strong capabilities and wide application. Many BERT-based natural language processing models have achieved the state of the art performance. In order to verify the effectiveness of our model after integrating the BERT representations, we compared and analyzed the following four sets of experiments: Baseline Model without BERT, Baseline model with BERT, Semantic-aware Model without BERT, Semantic-aware Model with BERT. It is worth noting that the method of integrating BERT information is the same as the method of integrating semantic dependency graph information. The specific experimental results are shown in Table 3:

Table 2. Ablation experiment results

Model	Overall
Baseline model	57.2
Sematic-aware model (Sentence-Level)	57.3
Sematic-aware model (Word-Level)	57.5
Sematic-aware model	**57.6**

According to the experimental results in the Table 3, we can see that the performance of the Semantic-aware Model with BERT is the best. This shows that BERT information and semantic dependency graph information have certain complementarity in the Chinese zero pronoun resolution task. But by comparing the performance of the Semantic-aware Model without BERT and Baseline model with BERT, We can see that the BERT information contributes more to the Chinese zero pronoun resolution task than the semantic dependency graph

Table 3. Integration with BERT

Model	Overall
Baseline model without BERT	57.2
Baseline model with BERT	57.7
Semantic-aware model without BERT	57.6
Semantic-aware model with BERT	**57.9**

information. In addition, we can also see that BERT information improves the Baseline Model more than the Semantic-aware Model. This shows that the BERT model may encode part of the semantic information of the semantic dependency graph. Based on the above analysis, we hope that in the future research, we can further integrate the semantic dependency graph and even the information of semantic role labeling on the basis of the BERT model, so as to further enhance the ability of the BERT model in the Chinese zero pronoun resolution task.

4.7 Conclusion

This paper proposes a semantic-aware Chinese zero pronoun resolution model with pre-trained semantic Dependency Parser. In order to effectively integrate semantic information from the pre-trained semantic dependency graph parser, We integrate semantic representations into the Chinese zero pronoun resolution model at two levels: word level and sentence level. The experimental results show that our proposed model achieves better performance than other deep learning-based models. In addition, we find that BERT information and semantic dependency graph information have certain complementarity in the Chinese zero pronoun resolution task. After our model is enhanced with the BERT representations, its performance has been further improved. In future research, we will explore the integration of BERT information and semantic dependency graph information to provide richer information for Chinese zero-finger resolution tasks.

Acknowledgements. This research project is supported by the National Natural Science Foundation of China (61872402), the Humanities and Social Science Project of the Ministry of Education (17YJAZH068) Science Foundation of Beijing Language and Culture University (supported by the Fundamental Research Funds for the Central Universities) (18ZDJ03) the Open Project Program of the National Laboratory of Pattern Recognition (NLPR).

References

1. Zhao, S., Ng, H.T.: Identification and Resolution of Chinese Zero Pronouns: A Machine Learning Approach, pp. 541–550 (2007)
2. Chen, C., Ng, V.: Chinese Zero Pronoun Resolution With Deep Neural Networks, vol. 1, no. 778–788 (2016)

3. Liu, T., Cui, Y., Yin, Q., Zhang, W., Wang, S., Guoping, H.: Generating and Exploiting Large-scale Pseudo Training Data for Zero Pronoun Resolution, vol. 1, pp. 102–111 (2017)
4. Yin, Q., Zhang, Y., Zhang, W., Liu, T.: Chinese Zero Pronoun Resolution with Deep Memory Network, pp. 1309–1318 (2017)
5. Yin, Q., Zhang, Y., Zhang, W., Liu, T., Wang, W.Y.: Deep Reinforcement Learning for Chinese Zero Pronoun Resolution, vol. 1, pp. 569–578 (2018)
6. Yin, Q., Zhang, Y., Zhang, W., Liu, T., Wang, W.Y.: Zero Pronoun Resolution With Attention-based Neural Network, pp. 13–23 (2018)
7. Che, W., Shao, Y., Liu, T., Ding, Y.: SemEval-2016 task 9: Chinese semantic dependency parsing. In: Proceedings of the 10th International Workshop on Semantic Evaluation (SemEval-2016), pp. 1074–1080. Association for Computational Linguistics (2016)
8. Converse, S.P., Palmer, M.S.: Pronominal Anaphora Resolution in Chinese. University of Pennsylvania (2006)
9. Kong, F., Zhou, G.: A Tree Kernel-based Unified Framework for Chinese Zero Anaphora Resolution, pp. 882–891 (2010)
10. Zhang, M., Zhang, Y., Fu, G.: End-to-end neural relation extraction with global optimization. In: Proceedings of the 2017 Conference on Empirical Methods in Natural Language Processing, pp. 1730–1740, Copenhagen, Denmark. Association for Computational Linguistics (2017)
11. Gao, Y., Zhang, Y., Xiao, T.: Implicit Syntactic Features for Targeted Sentiment Analysis, p. 9 (2017)
12. Yu, N., Zhang, M., Fu, G.: Transition-Based Neural RST Parsing With Implicit Syntax Features, pp. 559–570 (2018)
13. Zhang, M., Li, Z., Fu, G., Zhang, M.: Syntax-Enhanced Neural Machine Translation With Syntax-aware Word Representations, pp. 1151–1161 (2019)
14. Jiang, W., Li, Z., Zhang, M.: Syntax-enhanced ucca semantic parsing. Beijing Da Xue Xue Bao **56**(1), 89–96 (2020)
15. Abend, O., Rappoport, A.: Universal Conceptual Cognitive Annotation (UCCA), p. 11 (2013)
16. Peters, M.E.: Deep Contextualized Word Representations (2018)
17. Dozat, T., Manning, C.D.: Simpler but more accurate semantic dependency parsing. In: Proceedings of the 56th Annual Meeting of the Association for Computational Linguistics (Volume 2: Short Papers), pp. 484–490, Melbourne, Australia. Association for Computational Linguistics, July 2018
18. Shen, Z., Li, H., Liu, D., Shao, Y.: Dependency-gated cascade biaffine network for Chinese semantic dependency graph parsing. In: Tang, J., Kan, M.-Y., Zhao, D., Li, S., Zan, H. (eds.) NLPCC 2019. LNCS (LNAI), vol. 11838, pp. 840–851. Springer, Cham (2019). https://doi.org/10.1007/978-3-030-32233-5_65
19. Lee, K., He, L., Lewis, M., Zettlemoyer, L.: End-to-End Neural Coreference Resolution (2017)
20. Shen, D., et al.: Baseline Needs More Love: On Simple Word-Embedding-Based Models and Associated Pooling Mechanisms (2018)
21. Devlin, J., Chang, M., Lee, K., Toutanova, K.: Bert: pre-training of deep bidirectional transformers for language understanding. arXiv Computation and Language (2018)

Improving Sentence Classification by Multilingual Data Augmentation and Consensus Learning

Yanfei Wang, Yangdong Chen, and Yuejie Zhang(✉)

School of Computer Science, Shanghai Key Laboratory of Intelligent Information Processing, Fudan University, Shanghai 200433, China
{17210240046,19110240010,yjzhang}@fudan.edu.cn

Abstract. Neural network based models have achieved impressive results on the sentence classification task. However, most of previous work focuses on designing more sophisticated network or effective learning paradigms on monolingual data, which often suffers from insufficient discriminative knowledge for classification. In this paper, we investigate to improve sentence classification by multilingual data augmentation and consensus learning. Comparing to previous methods, our model can make use of multilingual data generated by machine translation and mine their language-share and language-specific knowledge for better representation and classification. We evaluate our model using English (i.e., source language) and Chinese (i.e., target language) data on several sentence classification tasks. Very positive classification performance can be achieved by our proposed model.

Keywords: Sentence classification · Multilingual data augmentation · Consensus learning

1 Introduction

Sentence classification is a task of assigning sentences to predefined categories, which has been widely explored in past decades. It requires modeling, representing and mining a degree of semantic comprehension, which are mainly based on the structure or sentiment of sentences. This task is important for many practical applications, such as product recommendation [5], public opinion detection [24], and human-machine interaction [3], etc.

Recently, deep learning has achieved state-of-the-art results across a range of Computer Vision (CV) [15], Speech Recognition [7], and Natural Language Processing tasks (NLP) [11]. Especially, Convolutional Neural Network (CNN) has gained great success in sentence modelling. However, training deep models requires a great diversity of data so that more discriminative patterns can be mined for better prediction. Most existing work on sentence classification focuses

Y. Wang and Y. Chen—Equal contribution.

M. Sun et al. (Eds.): CCL 2020, LNAI 12522, pp. 30–42, 2020.
https://doi.org/10.1007/978-3-030-63031-7_3

on learning better representation for a sentence given limited training data (i.e., *source language*), which resorts to design a sophisticated network architecture or learning paradigm, such as attention model [31], multi-task learning [20], adversarial training [19], etc. Inspired by recent advances in Machine Translation (MT) [30], we can perform an input data augmentation by making use of multilingual data (i.e., *target language*) generated by machine translation for sentence classification tasks. Such generated new language data can be used as the auxiliary information, and provide the additional knowledge for learning a robust sentence representation. In order to effectively exploit such multilingual data, we further propose a novel deep consensus learning framework to mine their language-share and language-specific knowledge for sentence classification. Since the machine translation model can be pre-trained off-the-shelf with great generalization ability, it is worth noting that we do not directly introduce other language data comparing to existing methods in the training and testing phase.

Our main contributions are of two-folds: 1) We first propose utilizing multilingual data augmentation to assist sentence classification, which can provide more beneficial auxiliary knowledge for sentence modeling; 2) A novel deep consensus learning framework is constructed to fuse multilingual data and learn their language-share and language-specific knowledge for sentence classification. In this work, we use English as our source language and Chinese/Dutch as the target language from an English-Chinese/Dutch translator. The related experimental results s how that our model can achieve very promising performance on several sentence classification tasks.

2 Related Work

2.1 Sentence Classification

Sentence classification is a well-studied research area in NLP. Various approaches have been proposed in last a few decades [6,29]. Among them, Deep Neural Network (DNN) based models have shown very good results for several tasks in NLP, and such methods become increasing popular for sentence classification. Various neural networks are proposed to learn better sentence representation for classification. An influential one is the work of [13], where a simple Convolutional Neural Network (CNN) with a single layer of convolution was used for feature extraction. Following this work, Zhang et al. [36] used CNNs for text classification with character-level features provided by a fully connected DNN. Liu et al. [20] used a multi-tasking learning framework to learn multiple related tasks together for sentence classification task. Based on Recurrent Neural Network (RNN), they utilized three different mechanisms of sharing information to model text. In practice, they used Long Short-Term Memory Network (LSTM) to address the issue of learning long-term dependencies. Lai et al. [16] proposed a Recurrent Convolutional Neural Network (RCNN) model for text classification, which applied a recurrent structure to capture contextual information and employed a max-pooling layer to capture the key components in texts. Jiang et al. [10] proposed a text classification model based on deep belief network and softmax regression. In their model, a deep belief network was introduced to solve the sparse high-dimensional matrix computation problem of text data. They then used softmax regression to classify the text. Yang et al. [31] used Hierarchical Attention Network (HAN) for document classification in their model, where a hierarchical structure was introduced to mirror the hierarchical structure of documents, and two levels of attention mechanisms were applied both at the word and sentence level.

Another direction of solutions for sentence classification is to use more effective learning paradigms. Yogatama et al. [33] combined Generative Adversarial Networks (GAN) with RNN for text classification. Billal et al. [1] solved the problem of multi-label text classification in semi-supervised learning manner. Liu et al. [19] proposed a multi-task adversarial representation learning method for text classification. Zhang et al. [35] attempted to learn structured representation of text via deep reinforcement learning. They tried to learn sentence representation by discovering optimized structures automatically and demonstrated two attempts of Information Distilled LSTM (ID-LSTM) and Hierarchically Structured LSTM (HS-LSTM) to build structured representation.

However, these tasks do not take into account the auxiliary language information corresponding to the source language. This auxiliary language can provide the additional knowledge to learn more accurate sentence representation.

2.2 Deep Consensus Learning

Existing sentence classification works [1,10,13,16,33,35,36] mainly focus on feature representation or learning a structured representation [35]. Deep learning based sentence classification models have obtained impressive performance. Those approaches are largely due to the powerful automatic learning and representation capacities of deep models, which benefit from big labelled training data and the establishment of large-scale sentence/document datasets [1,33,35]. However, all of the existing methods usually consider only one type of language information by a standard single language process. Such methods not only ignore the potentially useful information of other different languages, but also lose the opportunity of mining the correlated complementary advantages across different languages. A similar model is [20], which used synthetic source sentences to improve the performance of Neural Machine Translation (NMT). While sharing the high-level multilingual feature learning spirit, the proposed consensus learning model significantly has the following three outstanding characteristics. (1) Beyond the language concatenation based on fusion, our model uniquely considers a synergistic cross-language interaction learning and regularization by consensus propagation. This aims to overcome the challenge of learning discrepancy in multilingual feature optimization. (2) Instead of the traditional single loss design, a multi-loss concurrent supervision mechanism is deployed by our model. This enforces and improves the model's individuality learning power of language-specific feature. (3) Through NMT, we can eliminate some of the ambiguous words and highlight some key words.

3 Methodology

We aim to learn a deep feature representation model for sentence classification based on language-specific input, without any specific feature transformation. Figure 1 depicts our proposed framework, which consists of two stages. The first stage performs multilingual data augmentation from an off-the-shelf machine translator; and the second one feeds the source language data and generated target language data to our deep consensus learning model for sentence classification.

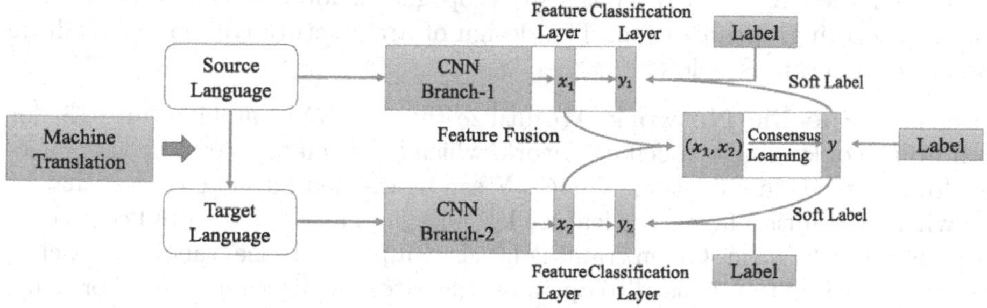

Fig. 1. The framework of our proposed model for sentence classification.

3.1 Multilingual Data Augmentation

Data augmentation is a very important technique in machine learning that allows building better models. It has been successfully used for many tasks in areas of CV and NLP, such as image recognition [15] and MT [35]. In MT, Back-translation is a common data argumentation method [25,39], which allows us to combine monolingual training data. Especially when the existing data is insufficient to learn a discriminative representation for a specific task, the data augmentation methods can be used.

In sentence classification, given an input sentence in one language, we perform data augmentation by translating the sentence to another language using existing machine translation methods. We name the input language as *source language* and the translated language as *target language*. This motivation comes from the recent great advance in NMT [30]. Given an input sentence in source language, we simply call the *Google* Translation API[1] to get the translated data in target language. Comparing to other state-of-art NMT models, the *Google* translator has the advantage of both effectiveness and efficiency in real application scenarios. Since target language is used for multilingual data augmentation and the type of it is not important to the proposed model, we random choose Chinese and Dutch respectively as the target language for multilingual data augmentation, and the source language depends on the language of input sentence.

[1] https://cloud.google.com/translate/.

3.2 Deep Consensus Learning Model

Learning a consensus classification model with the combination of several beneficial information into one final prediction can lead to a more accurate result [2]. Thus we use two languages of data, $\{S_1, S_2, S_3, \cdots, S_{N-1}, S_N\}$ and $\{T_1, T_2, T_3, \cdots, T_{N-1}, T_N\}$, to perform consensus learning for sentence classification. As shown in Fig. 1, our model has three parts: (1) Two branches of language-specific subnetworks for learning the most discriminative features for each language data; (2) One fusion branch responsible for learning the language-share representation with the optimal integration of two kinds of language-specific knowledge; and (3) Consensus propagation for the feature regularization and learning optimization. The design of architecture components will be described in detail as below.

Language-Specific Network. We utilize the *TextCNN* architecture [13] for each branch of language-specific network, which has been proved to be very effective for sentence classification. *TextCNN* can be divided into two stages, that is, one with convolution layers for feature learning, and another with full connected layers for classification. Given training labels of input sentence, the Softmax classification loss function is used to optimize the category discrimination. Formally, given a corpus of sentences of source language $\{S_1, S_2, S_3, \cdots, S_{N-1}, S_N\}$, the training loss on a batch of n sentences can be computed as:

$$L_{S\_brch} = -\frac{1}{n}\sum_{i=1}^{n} \log\left(\frac{\exp\left(w_{y_i}^T S_i\right)}{\sum_{k=1}^{c} \exp\left(w_k^T S_i\right)}\right) \tag{1}$$

where c is the number of categories of sentences; y_i denotes the category label of the sentence S_i; and w is the prediction function parameter of the training category class k. The training loss for target language branch $L_(T\_brch)$ can be computed in the same manner. Meanwhile, since the source language and target language belong to different language spaces, such two branches of language-specific networks are trained with the uniform architecture but different parameters.

Language-share Network. We perform the language-share feature learning from two language-specific branches. For this purpose, we firstly perform the language-share learning by fusing across from these two branches. For design simplicity and cost efficiency, we achieve the feature fusion on the feature vectors from the concatenation layer before dropout in *TextCNN* by an operation of *Concat→FC→Dropout→FC→Softmax*. This produces a category prediction score for input pair (a sentence in source language and its translated one in target language). We similarly utilize the Softmax classification loss L_ST for the language-share classification learning as that in the language-specific branches.

Consensus Propagation. Inspired by the teacher-student learning approach, we propose to regularize the language-specific learning by consensus feedback from the language-share network. More specifically, we utilize the consensus probability $P_{ST} = \left[p_{ST}^1, p_{ST}^2, \cdots, p_{ST}^{c-1}, p_{ST}^c\right]$ from the language-share network as the *teacher* signal (called "*soft label*" versus the ground-truth one-hot "*hard label*") to guide the learning process of all language-specific branches (*student*) concurrently by an additional regularization, which can be formulated in a cross-entropy manner as:

$$\mathcal{H}_S = -\frac{1}{c}\sum_{i=1}^{c}\left(p_{ST}^{i}\ln\left(p_{s}^{i}\right) + \left(1 - p_{ST}^{i}\right)\ln\left(1 - p_{s}^{i}\right)\right) \tag{2}$$

where $P_S = [p_S^1, p_S^2, p_S^3, \cdots, p_S^{c-1}, p_S^c]$ defines the probability prediction over all c sentence classes by the source language branch. Thus the final loss function for the language-specific network can be re-defined via enforcing an additional regularization in Eq. (1).

$$L_S = L_{S\_brch} + \lambda\mathcal{H}_S \tag{3}$$

where λ controls the importance tradeoff between two terms. The regularization terms $\mathcal{H}_T$ and L_T for target language branch can be computed in the same way.

The training of our proposed model proceeds in two stages. First, we rely on training the language-specific network separately, which is terminated by the early stopping strategy. Afterwards, the language-share network and consensus propagation loss are introduced. We use the whole loss defined in Eq. (3) and L_{ST} to train the language-specific network and language-share network at the same time. In the testing time, given an input sentence and its translated sentence, the final prediction is obtained by averaging the three prediction scores from the language-specific networks and the language-share network.

4 Experiment and Analysis

In this section, we investigate the empirical performance of our proposed architecture on five benchmark datasets for sentence classification.

4.1 Datasets and Experimental Setup

The sentence classification datasets include:

(1) ***MR***: This dataset includes movie reviews with one sentence per review, in which the classification involves detecting positive/negative reviews [23].
(2) ***CR***: This dataset contains annotated customer reviews of 5 products, and the target is to predict positive/negative reviews [8].
(3) ***Subj***: This dataset is a subjectivity dataset, which includes subjective or objective sentiments [22].
(4) ***TREC***: This dataset focuses on the question classification task that involves 6 question types [18].
(5) ***SST-1***: This dataset is Stanford Sentiment Treebank, an extension of *MR*, which contains training/development/testing splits and fine-grained labels (very positive, positive, neutral, negative, very negative) [27].

Similar with [13], the initialized word vectors for source language are obtained from the publicly available *word2vec* vectors that were trained on 100 billion words from *Google News*. For target language of Chinese, we retrain the *word2vec* models on *Chinese Wikipedia Corpus*; and for target language of Dutch, we retrain the *word2vec* models on *Dutch Wikipedia Corpus*. In our experiments, we choose the *CNN-multichannel* model variant of *TextCNN* because of its better performance.

4.2 Ablation Study

We first compare our proposed model with several baseline models for sentence classification. Here, we use ***S+T*** to indicate that the model's input contains the source language and the target language. ***T***(*) indicates the type of target language, i.e., ***T(CH)*** indicates that the target language is Chinese, and ***T(DU)*** indicates that the target language is Dutch. Figure 2 and 3 show the comparison results of classification accuracy rate on five benchmark datasets. *CNN(S)* denotes the *CNN-multichannel* model variant of *TextCNN*, which only uses the source language data of English for training and testing. *CNN(T)* is a retrained *TextCNN* model on the translated target language data of Chinese(CH)/Dutch(DU), and the other settings keep the same as *CNN(S)*. *Ours(S+T(*))* denotes our model by combining multilingual data augmentation with deep consensus learning. We can find that *Ours(S+T(*))* performs much better than those baselines, which proves the effectiveness of our framework. It is obvious that multilingual data augmentation can provide the beneficial additional discrimination for learning a robust sentence representation for classification. It is worth noting that *CNN(T)* is even better than *CNN(S)* on *MR*. This indicates that existing machine translation methods can not only keep the discriminative semantics of source language, but also create useful discrimination in target language space.

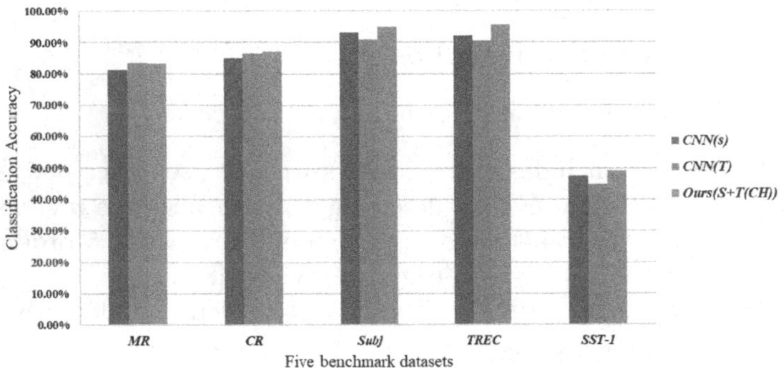

Fig. 2. The comparison results with existing baseline models based on English→ Chinese MT.

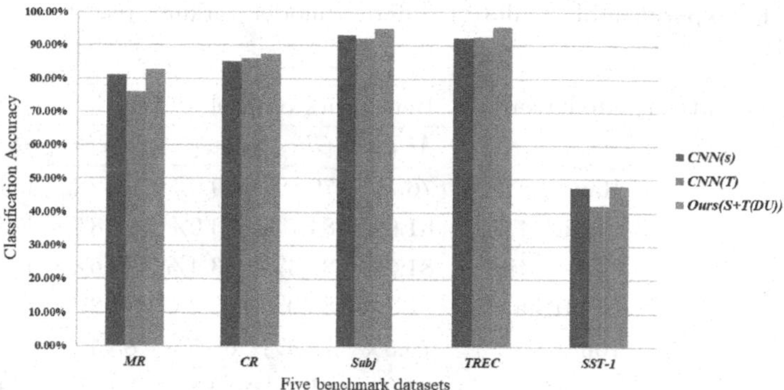

Fig. 3. The comparison results with existing baseline models based on English→ Dutch MT.

Similar to *TextCNN*, we also use several variants of the model to demonstrate the effectiveness of our model. As we know, when lacking a large supervised training set, we usually use word vectors obtained from unsupervised neural language models to initialize word vectors for performance improvement. Thus we use various word vector initialization methods to validate the model.

The different word vector initialization methods include:

(1) **Rand**: All words are randomly initialized and can be trained during training.
(2) **Static**: All words of input language are initialized by pre-trained vectors from the corresponding language *word2vec*. Simultaneously, all these words are kept static during training.
(3) **Non-static**: This is an initialization method same to Static, but the pre-trained vectors can be finetuned during training.
(4) **Multichannel**: This model contains two types of word vector, which are treated as different channels. One type of word vector can be finetuned during training, while the other keeps static. Two types of word vector are initialized with the same word embedding form *word2vec*.

In Table 1, we show the experimental results of different model variants based on English→ Chinese MT. Compared to the source language *S*, the accuracy rates of the target language *T(CH)* classification are partly improved or decreased, which shows the strong dataset dependency. Considering that the proposed *S+T(CH)* model with Multichannel obtains the current optimal results, we choose the model with Multichannel as our final results. Similar to Table 1, we show the experimental results of different model variants based on English→ Dutch MT in Table 2. Combining the experimental results in Tables 1 and 2, we have enough reasons to prove the validity of our consensus learning method.

Table 1. The experimental results of different model variants based on English→ Chinese MT.

Evaluation pattern	Model variant	Benchmark dataset				
		MR	*CR*	*Subj*	*TREC*	*SST*-1
S	Rand	76.1%	79.8%	89.6%	91.2%	45.0%
	Static	81.0%	84.7%	93.0%	92.8%	45.5%
	Non-static	81.5%	84.3%	93.4%	93.6%	48.0%
	Multichannel	81.1%	85.0%	93.2%	92.2%	47.4%
T(CH)	Rand	79.5%	79.8%	88.5%	85.4%	42.5%
	Static	83.0%	81.4%	89.8%	89.4%	43.6%
	Non-static	82.5%	86.4%	90.1%	90.4%	42.9%
	Multichannel	83.4%	86.4%	90.9%	90.4%	44.8%
S+T(CH)	Rand	79.7%	77.2%	92.0%	92.4%	47.1%
	Static	81.8%	86.4%	93.6%	95.0%	47.6%
	Non-static	81.7%	**87.9%**	94.5%	**95.2%**	48.0%
	Multichannel	**83.2%**	87.1%	**95.0%**	**95.6%**	**49.1%**

Table 2. The experimental results of different model variants based on English→ Dutch MT.

Evaluation pattern	Model variant	Benchmark dataset				
		MR	*CR*	*Subj*	*TREC*	*SST*-1
T(DU)	Rand	66.5%	78.5%	85.3%	84.8%	35.3%
	Static	75.0%	82.1%	91.6%	89.0%	40.8%
	Non-static	76.6%	86.6%	92.8%	93.0%	42.9%
	Multichannel	76.0%	86.1%	92.1%	92.6%	42.0%
S+T(DU)	Rand	76.1%	87.1%	89.5%	90.8%	42.6%
	Static	81.6%	85.6%	93.4%	94.8%	46.2%
	Non-static	81.8%	84.0%	93.9%	**95.6%**	46.8%
	Multichannel	**82.8%**	**87.3%**	**95.3%**	**95.6%**	**47.9%**

4.3 Comparison with Existing Approaches

To further exhibit the effectiveness of our model, we compare our approach with several state-of-the-art approaches, including recent LSTM-based models and CNN-based models. As shown in Table 3, it can be concluded that our approach can gain very promising results comparing to these methods. The whole performance is measured by the accuracy rate for sentence classification. We roughly divide the existing approaches into four categories. The first category is the RNN-based model, in which Standard-RNN refers to Standard Recursive Neural Network [27], MV-RNN is Matrix-Vector Recursive Neural Network [26], RNTN denotes Recursive Neural Tensor Network [27], and DRNN represents

Deep Recursive Neural Network [9]. The second category is the LSTM-based model, in which bi-LSTM stands for Bidirectional LSTM [28], SA-LSTM means Sequence Autoencoder LSTM [4], Tree-LSTM is Tree-Structured LSTM [28], and Standard-LSTM represents Standard LSTM Network [28]. The CNN-based model is the third category, in which DCNN denotes Dynamic Convolutional Neural Network [12], CNN-Multichannel is Convolutional Neural Network with Multichannel [13], MVCNN refers to Multichannel Variable-Size Convolution Neural Network [32], Dep-CNN denotes Dependency-based Convolutional Neural Network [21], MGNC-CNN stands for Multi-Group Norm Constraint CNN [38], and DSCNN represents Dependency Sensitive Convolutional Neural Network [34]. The fourth one is based on other methods, in which Combine-skip refers to skip-thought model with the concatenation of the vectors from uni-skip and bi-skip [14], CFSF indicates initializing Convolutional Filters with Semantic Features [17], and GWS denotes exploiting domain knowledge via Grouped Weight Sharing [37]. Especially on *MR*, our model of ***S+T(CH)*** can achieve the best performance by a margin of nearly 5%. This improvement demonstrates that our multilingual data augmentation and consensus learning can make great contributions to such sentence classification task. Through multilingual data augmentation, important words will be retained. The NMT systems can map those ambiguous words in source language to different word units in target language, which can achieve the result of word disambiguation. Essentially, our method can enable CNNs to obtain better discrimination and generalization abilities.

Table 3. The comparison results between the state-of-the-art approaches and ours.

Model	Approach	Benchmark dataset				
		MR	*CR*	*Subj*	*TREC*	*SST*-1
RNN-based model	Standard-RNN [27]	-	-	-	-	43.2%
	MV-RNN [26]	-	-	-	-	44.4%
	RNTN [27]	-	-	-	-	45.7%
	DRNN [9]	-	-	-	-	49.8%
LSTM-based model	bi-LSTM [28]	-	-	-	-	49.1%
	SA-LSTM [4]	80.7%	-	-	-	-
	Tree-LSTM [28]	-	-	-	-	**51.0%**
	Standard-LSTM [28]	-	-	-	-	45.8%
CNN-based model	DCNN [12]	-	-	-	93.0%	48.5%
	CNN-Multichannel [13]	81.1%	85.0%	93.2%	85.0%	47.4%
	MVCNN [32]	-	-	93.9%	-	49.6%
	Dep-CNN [21]	-	-	-	95.4%	49.5%
	MGNC-CNN [38]	-	-	94.1%	95.5%	-
	DSCNN [34]	82.2%	-	93.9%	**95.6%**	50.6%
Model based on other methods	Combine-skip [14]	76.5%	80.1%	93.6%	92.2%	-
	CFSF [17]	82.1%	86.0%	93.7%	93.7%	-
	GWS [37]	81.9%	84.8%	-	-	-
Our model	*Ours (S+T(CH))*	**87.6%**	**87.1%**	**95.0%**	**95.6%**	49.1%
	Ours (S+T(DU))	**82.8%**	**87.3%**	**95.3%**	**95.6%**	47.9%

To further demonstrate the superiority of our proposed model, we also use English as the source language and Dutch as the target language to evaluate the model of ***S+T(DU)***. On the four benchmark datasets of *MR*, *CR*, *Subj*, and *TREC*, our models of ***S+T(CH)*** and ***S+T(DU)*** have both achieved the best results at present.

5 Conclusion and Future Work

In this paper, multilingual data augmentation is introduced to further improve sentence classification. A novel deep consensus learning model is established to fuse multilingual data and learn the language-share and language-specific knowledge. The related experimental results demonstrate the effectiveness of our proposed framework. In addition, our method requires no external data comparing to existing methods, which makes it very practical with good generalization abilities in real application scenarios. In the future, we will try to explore the performance of the model on larger sentence/document datasets. The linguistic features of different languages will be also considered when selecting the target language.

Acknowledgements. This work was supported by National Natural Science Foundation of China (No. 61976057, No. 61572140), and Science and Technology Development Plan of Shanghai Science and Technology Commission (No. 20511101203, No. 20511102702, No. 20511101403, No. 18511105300). Yanfei Wang and Yangdong Chen contributed equally to this work, and were co-first authors. Yuejie Zhang was the corresponding author.

References

1. Billal, B., Fonseca, A., Sadat, F., Lounis, H.: Semi-supervised learning and social media text analysis towards multi-labeling categorization. In: 2017 IEEE International Conference on Big Data (Big Data), pp. 1907–1916. IEEE (2017)
2. Chen, Y., Zhu, X., Gong, S.: Person re-identification by deep learning multi-scale representations. In: Proceedings of the IEEE International Conference on Computer Vision Workshops, pp. 2590–2600 (2017)
3. Clavel, C., Callejas, Z.: Sentiment analysis: from opinion mining to human-agent interaction. IEEE Trans. Affect. Comput. **7**(1), 74–93 (2015)
4. Dai, A.M., Le, Q.V.: Semi-supervised sequence learning. In: Advances in Neural Information Processing Systems, pp. 3079–3087 (2015)
5. Dong, R., O'Mahony, M.P., Schaal, M., McCarthy, K., Smyth, B.: Sentimental product recommendation. In: Proceedings of the 7th ACM Conference on Recommender Systems, pp. 411–414 (2013)
6. Fernández-Delgado, M., Cernadas, E., Barro, S., Amorim, D.: Do we need hundreds of classifiers to solve real world classification problems? J. Mach. Learn. Res. **15**(1), 3133–3181 (2014)
7. Graves, A., Mohamed, A.r., Hinton, G.: Speech recognition with deep recurrent neural networks. In: 2013 IEEE International Conference on Acoustics, Speech and Signal Processing, pp. 6645–6649. IEEE (2013)

8. Hu, M., Liu, B.: Mining and summarizing customer reviews. In: Proceedings of the Tenth ACM SIGKDD International Conference on Knowledge Discovery and Data Mining, pp. 168–177 (2004)
9. Irsoy, O., Cardie, C.: Deep recursive neural networks for compositionality in language. In: Advances in Neural Information Processing Systems, pp. 2096–2104 (2014)
10. Jiang, M., et al.: Text classification based on deep belief network and softmax regression. Neural Comput. Appl. **29**(1), 61–70 (2016). https://doi.org/10.1007/s00521-016-2401-x
11. Kalchbrenner, N., Grefenstette, E., Blunsom, P.: A convolutional neural network for modelling sentences. In: Proceedings of the 52nd Annual Meeting of the Association for Computational Linguistics, pp. 655–665 (2014)
12. Kalchbrenner, N., Grefenstette, E., Blunsom, P.: A convolutional neural network for modelling sentences. In: Proceedings of the 52nd Annual Meeting of the Association for Computational Linguistics (Volume 1: Long Papers), pp. 655–665 (2014)
13. Kim, Y.: Convolutional neural networks for sentence classification. In: Proceedings of the 2014 Conference on Empirical Methods in Natural Language Processing (EMNLP), pp. 1746–1751 (2014)
14. Kiros, R., et al.: Skip-thought vectors. In: Advances in Neural Information Processing Systems, pp. 3294–3302 (2015)
15. Krizhevsky, A., Sutskever, I., Hinton, G.E.: Imagenet classification with deep convolutional neural networks. In: Advances in Neural Information Processing Systems, pp. 1097–1105 (2012)
16. Lai, S., Xu, L., Liu, K., Zhao, J.: Recurrent convolutional neural networks for text classification. In: Twenty-Ninth AAAI Conference on Artificial Intelligence (2015)
17. Li, S., Zhao, Z., Liu, T., Hu, R., Du, X.: Initializing convolutional filters with semantic features for text classification. In: Proceedings of the 2017 Conference on Empirical Methods in Natural Language Processing, pp. 1884–1889 (2017)
18. Li, X., Roth, D.: Learning question classifiers. In: Proceedings of the 19th International Conference on Computational Linguistics, vol. 1, pp. 1–7. Association for Computational Linguistics (2002)
19. Liu, P., Qiu, X., Huang, X.J.: Adversarial multi-task learning for text classification. In: Proceedings of the 55th Annual Meeting of the Association for Computational Linguistics (Volume 1: Long Papers), pp. 1–10 (2017)
20. Liu, P., Qiu, X., Huang, X.: Recurrent Neural Network for Text Classification With Multi-task Learning, pp. 2873–2879 (2016)
21. Ma, M., Huang, L., Zhou, B., Xiang, B.: Dependency-based convolutional neural networks for sentence embedding. In: Proceedings of the 53rd Annual Meeting of the Association for Computational Linguistics and the 7th International Joint Conference on Natural Language Processing (Volume 2: Short Papers), pp. 174–179 (2015)
22. Pang, B., Lee, L.: A sentimental education: sentiment analysis using subjectivity summarization based on minimum cuts. In: Proceedings of the 42nd Annual Meeting on Association for Computational Linguistics, p. 271. Association for Computational Linguistics (2004)
23. Pang, B., Lee, L.: Seeing stars: exploiting class relationships for sentiment categorization with respect to rating scales. In: Proceedings of the 43rd Annual Meeting on Association for Computational Linguistics, pp. 115–124. Association for Computational Linguistics (2005)
24. Pang, B., Lee, L., et al.: Opinion mining and sentiment analysis. Found. Trends® Inf. Retrieval **2**(1–2), 1–135 (2008)

25. Sennrich, R., Haddow, B., Birch, A.: Improving neural machine translation models with monolingual data. In: Proceedings of the 54th Annual Meeting of the Association for Computational Linguistics (Volume 1: Long Papers), pp. 86–96 (2016)
26. Socher, R., Huval, B., Manning, C.D., Ng, A.Y.: Semantic compositionality through recursive matrix-vector spaces. In: Proceedings of the 2012 Joint Conference on Empirical Methods in Natural Language Processing and Computational Natural Language Learning, pp. 1201–1211. Association for Computational Linguistics (2012)
27. Socher, R., et al.: Recursive deep models for semantic compositionality over a sentiment treebank. In: Proceedings of the 2013 Conference on Empirical Methods in Natural Language Processing, pp. 1631–1642 (2013)
28. Tai, K.S., Socher, R., Manning, C.D.: Improved semantic representations from tree-structured long short-term memory networks. In: Proceedings of the 53rd Annual Meeting of the Association for Computational Linguistics and the 7th International Joint Conference on Natural Language Processing (Volume 1: Long Papers), pp. 1556–1566 (2015)
29. Tong, S., Koller, D.: Support vector machine active learning with applications to text classification. J. Mach. Learn. Res. **2**(Nov), 45–66 (2001)
30. Wu, Y., et al.: Google's neural machine translation system: Bridging the gap between human and machine translation. arXiv preprint arXiv:1609.08144 (2016)
31. Yang, Z., Yang, D., Dyer, C., He, X., Smola, A., Hovy, E.: Hierarchical attention networks for document classification. In: Proceedings of the 2016 Conference of the North American Chapter of the Association for Computational Linguistics: Human Language Technologies, pp. 1480–1489 (2016)
32. Yin, W., Schütze, H.: Multichannel variable-size convolution for sentence classification. In: Proceedings of the Nineteenth Conference on Computational Natural Language Learning, pp. 204–214 (2015)
33. Yogatama, D., Dyer, C., Ling, W., Blunsom, P.: Generative and discriminative text classification with recurrent neural networks. arXiv preprint arXiv:1703.01898 (2017)
34. Zhang, R., Lee, H., Radev, D.: Dependency sensitive convolutional neural networks for modeling sentences and documents. In: Proceedings of NAACL-HLT, pp. 1512–1521 (2016)
35. Zhang, T., Huang, M., Zhao, L.: Learning structured representation for text classification via reinforcement learning. In: Thirty-Second AAAI Conference on Artificial Intelligence (2018)
36. Zhang, X., LeCun, Y.: Text understanding from scratch. arXiv preprint arXiv:1502.01710 (2015)
37. Zhang, Y., Lease, M., Wallace, B.C.: Exploiting domain knowledge via grouped weight sharing with application to text categorization. In: Proceedings of the 55th Annual Meeting of the Association for Computational Linguistics (Volume 2: Short Papers), pp. 155–160 (2017)
38. Zhang, Y., Roller, S., Wallace, B.C.: MGNC-CNN: a simple approach to exploiting multiple word embeddings for sentence classification. In: Proceedings of the 2016 Conference of the North American Chapter of the Association for Computational Linguistics: Human Language Technologies, pp. 1522–1527 (2016)
39. Zhang, Z., Liu, S., Li, M., Zhou, M., Chen, E.: Joint training for neural machine translation models with monolingual data. In: Thirty-Second AAAI Conference on Artificial Intelligence (2018)

Information Retrieval, Dialogue and Question Answering

Attention-Based Graph Neural Network with Global Context Awareness for Document Understanding

Yuan Hua[1], Zheng Huang[1,2](✉), Jie Guo[1], and Weidong Qiu[1]

[1] Shanghai Jiao Tong University, Shanghai, China
{isyuan.hua,huang-zheng,guojie,qiuwd}@sjtu.edu.cn
[2] Westone Cryptologic Research Center, Beijing, China

Abstract. Information extraction from documents such as receipts or invoices is a fundamental and crucial step for office automation. Many approaches focus on extracting entities and relationships from plain texts, however, when it comes to document images, such demand becomes quite challenging since visual and layout information are also of great significance to help tackle this problem. In this work, we propose the attention-based graph neural network to combine textual and visual information from document images. Moreover, the global node is introduced in our graph construction algorithm which is used as a virtual hub to collect the information from all the nodes and edges to help improve the performance. Extensive experiments on real-world datasets show that our method outperforms baseline methods by significant margins.

Keywords: Document understanding · Attention · Graph neural network

1 Introduction

Information Extraction [1,10,21] is a widely studied task of retrieving structured information from texts and many inspiring achievements have been made in this field. However, most of these works are generally focusing on extracting entities and relationships from plain texts which are not appropriate to apply directly on document understanding.

Document understanding is the process of automatically recognizing and extracting key texts from scanned unstructured documents and saving them as structured data. Document understanding was already introduced in a competition of ICDAR 2019, where the goal was to detect texts in documents and extract key texts from receipts and invoices. In this work, we focus on document understanding which is mainly about key information extraction from scanned unstructured documents. The following paragraphs summarize the challenges of the task and the contributions of our work.

M. Sun et al. (Eds.): CCL 2020, LNAI 12522, pp. 45–56, 2020.
https://doi.org/10.1007/978-3-030-63031-7_4

1.1 Challenges

Document understanding is a challenging task and there are little research works published in this topic so far. Although it seems that traditional named entity recognition networks or layout analysis networks are related to this topic, none of the existing research can fully address the problems faced by document understanding.

Firstly, context requires balance. The key cue of the entities usually appears in their neighbors and too much context will add noise and increase problem dimensionality making learning slower and more difficult. As shown in Fig. 1, in order to identify the label of *$11900*, the text *Total* on its left side is good enough for the model to recognize its tag correctly. Instead of increasing the recognition accuracy, too much context like *Tax*, *Subtotal* will lead the performance even worse. Appropriate context is very problem specific and we need to get this relationship by training.

Secondly, it is not adequate to represent the semantic meaning in documents by using text alone. For example, there can be multiple date related entities in one document such as *due date* and *purchase date.* It is difficult for the model to distinguish them only by textual information. Thus, more information like visual information or layout information also needs to be considered at the same time.

Thirdly, the positional cue is critical sometimes. An example is shown in the right side of Fig. 1. As for the entity *Vender Name*, it appears at the top of the document in most cases. The model will benefit from it if it can leverage this information.

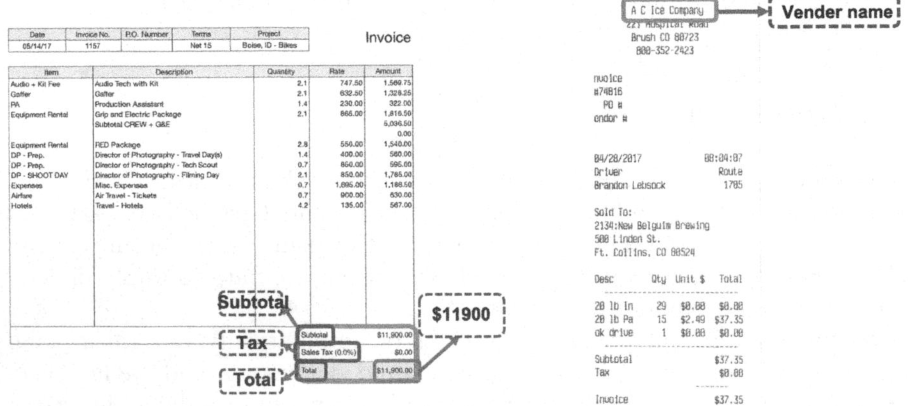

Fig. 1. Examples of Documents and example entities to extract.

1.2 Contributions

In this work, we present a novel method that achieves the document understanding problem as a node classification task. The method first computes a text embedding and an image embedding for each text segment in the document. Then graph construction algorithm will use the coordinates of bounding boxes to generate a unique graph for each document. In order to leverage positional cue effectively, the global node is first proposed in document understanding field which represents the universal context of the current document. Finally, the graph attention network will combine textual information with visual information and the positional cue for information extraction.

The main contributions of this paper can be summarized as follows: 1) we propose a graph construction algorithm to generate a unique graph for each document and achieve the document understanding task as a graph node classification task; 2) the proposed model can capture global context information and local compositions effectively; 3) extensive experiments have been conducted on real-world datasets to show that our method has significant advantages over the baseline methods.

2 Related Works

Several rule-based document understanding systems were proposed in [2,3,14]. Laura et al. [2] presented a case for the importance of rule-based approaches to industry practitioners. SmartFix by Andreas et al. [3] employs specific configuration rules designed for each template. The study by Schuster et al. [14] offers a template matching based algorithm to solve the document understanding problem and plenty of templates have to be constructed and maintained to deal with different situations. However, rule-based methods rely heavily on the predefined templates or rules and are not scalable and flexible for most document understanding problems since documents in real life have no fixed layout. Furthermore, updating the templates or rules requires a lot of effort.

A recent study by Zhao et al. [20] proposed Convolutional Universal Text Information Extractor (CUTIE). CUTIE treats the document understanding task as an image semantic segmentation task. It applies convolutional neural networks on gridded texts where texts are semantical embeddings. However, this work only uses text-level features and doesn't involve image-level features.

Inspired by BERT [4], Xu et al. [18] proposed LayoutLM method. It applies BERT architecture for the pre-training of text and layout. Although LayoutLM uses image features in the pre-training stage and it performs well on several downstream tasks, the potential relationship between two text segments hasn't been taken into consideration. In addition, sufficient data and time are required to pre-train the model inefficiently.

Since graph neural networks [9,13,17] have shown great success in unstructured data tasks, more and more research works are focusing on using GNN to tackle the document understanding problem. Liu et al. [11] presented a GCN-based method for information extraction from document images. It is a work attempting to extract key information with customized graph convolution model. However, prior knowledge and extensive human efforts are needed to predefine task-specific node and edge representations. One study by Yu et al. [19] explores the feature fusion of textual and visual embeddings by GNN. This work differs from ours because it still treats the document understanding task as the sequence tagging problem and uses a bi-directional LSTM model to extract entities which has already been proved to have limited ability to learn the relationship among distant words.

3 Proposed Method

This section demonstrates the architecture of our proposed model. To extract textual context, our model first encodes each text segment in the document by pre-trained BERT model as its corresponding text embeddding. Then using multiple layers of CNN to get its image embedding. The combination of these two types of embeddings will generate unique global node representation and various local node representations. These node representations contain both visual context and textual context and will be used as node input to the graph attention network. Our model transforms the document understanding task into a node classification problem by taking both local context and global context into account.

3.1 Feature Extraction

Figure 2 is the overall workflow of feature extraction. As shown in Fig. 2, we calculate node representations for both global nodes and local nodes where global nodes capture universal information and local nodes extract internal information. Different from the existing information extraction models that only use plain text features, we also use image features to obtain morphology information to our model.

Text Feature Extraction. We use pre-trained BERT model to generate text embeddings for capturing both global and local textual context. For a set of text segments in the document, we concatenate them by their coordinates from left to right and from top to bottom to generate a sequence. Given a sequence $seq_i = (w_1^{(i)}, w_2^{(i)}, ..., w_n^{(i)})$, text embeddings of a sequence seq_i are defined as follows

$$TE_{0:n}^{(i)} = BERT(w_{0:n}^{(i)}; \Theta_{BERT}) \quad (1)$$

where $w_{0:n}^{(i)} = [w_0^{(i)}, w_1^{(i)}, ..., w_n^{(i)}]$ denotes the input sequence padding with $w_0^{(i)} = [CLS]$. $[CLS]$ is a specific token to capture full sequence context which is introduced in [4]. $TE_{0:n}^{(i)} = [TE_0^{(i)}, TE_1^{(i)}, ..., TE_n^{(i)}] \in \mathbf{R}^{n*d_{model}}$ denotes the output sequence embeddings and d_{model} is the dimension of the model. $TE_k^{(i)}$ represents the k-th output of pre-trained BERT model for the i-th document.

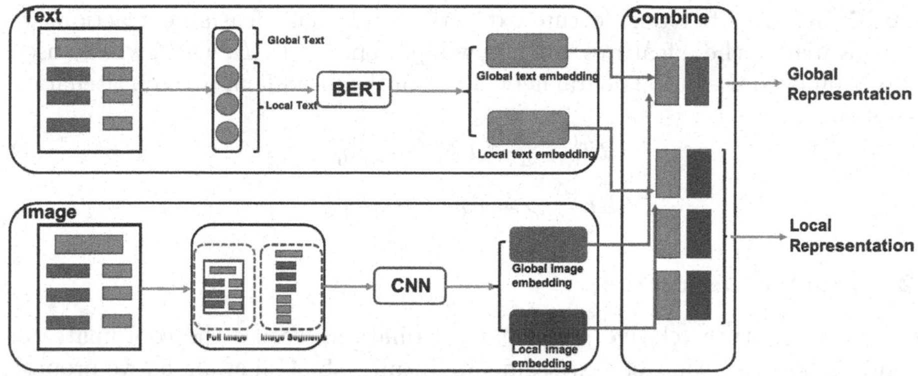

Fig. 2. Workflow of feature extraction.

Θ_{BERT} represents the parameters of pre-trained BERT model. Each text segment of a text sequence is encoded independently and we can get global text embedding and local text embedding simultaneously, defining them as

$$TE_{Global}^{(i)} = TE_0^{(i)} \tag{2}$$

$$TE_{Local}^{(i)} = [TE_1^{(i)}, TE_2^{(i)}, ..., TE_n^{(i)}] \tag{3}$$

Image Feature Extraction. For image embedding generation, we using CNN for catching both global and local visual information. Given a set of image segments cropped by bounding boxes $seg_i = (p_1^{(i)}, p_2^{(i)}, ..., p_n^{(i)})$, image embeddings of segments seg_i are defined as follows

$$IE_{0:n}^{(i)} = CNN(p_{0:n}^{(i)}; \Theta_{CNN}) \tag{4}$$

where $p_{0:n}^{(i)} = [p_0^{(i)}, p_1^{(i)}, ..., p_n^{(i)}]$ denotes the input image segments appending with $p_0^{(i)} = full\_image$. We use $p_0^{(i)}$ to capture global morphology information of the document image. $p_k^{(i)} \in \mathbf{R}^{H*W*3}$ represents k-th image segment of i-th document and H means height of the image, W means width of the image. $IE_{0:n}^{(i)} = [IE_0^{(i)}, IE_1^{(i)}, ..., IE_n^{(i)}] \in \mathbf{R}^{n*d_{model}}$ denotes the output image embeddings and d_{model} is the dimension of the model. In our work, we use classic ResNet model [6] as backbone to extract image features and a full connected layer is used to resize output to d_{model} dimension. $IE_k^{(i)}$ represents the k-th output of CNN model for the i-th document. Θ_{CNN} represents the parameters of CNN model. Each image segment is encoded independently and we can get global image embedding and local image embedding synchronously, defining them as

$$IE_{Global}^{(i)} = IE_0^{(i)} \tag{5}$$

$$IE_{Local}^{(i)} = [IE_1^{(i)}, IE_2^{(i)}, ..., IE_n^{(i)}] \tag{6}$$

Combination. After text feature extraction and image feature extraction, we can concatenate these features into a new representation RE, which will be used as node input to the graph neural network. $\oplus$ in the formula means concatenation operation.

$$RE_{Global}^{(i)} = TE_0^{(i)} \oplus IE_0^{(i)} \tag{7}$$

$$RE_{Local}^{(i)} = TE_{1:n}^{(i)} \oplus IE_{1:n}^{(i)} \tag{8}$$

3.2 Graph Construction

In order to capture relative positional information, we use the coordinates of bounding boxes to connect text segments. Inspired by Gui et al. [5], we propose the global node mechanism which is used as a virtual hub to capture long-range dependency and high-level features.

The whole document is converted into a directed graph, as shown in Fig. 3, where each node represents a text segment and the connection between two nodes can be treated as an edge. Given a set of text segments inside a document, first of all, we need to merge these text segments into different lines based on their bounding boxes' coordinates. To be more specific, if the overlap of the two text segments on the vertical axis exceeds 60%, the two text segments are considered to belong to the same line. In order to capture layout information, we build connection for each text segment in the same line. In addition, an extra connection is built between current text segment and every text segments in its previous line.

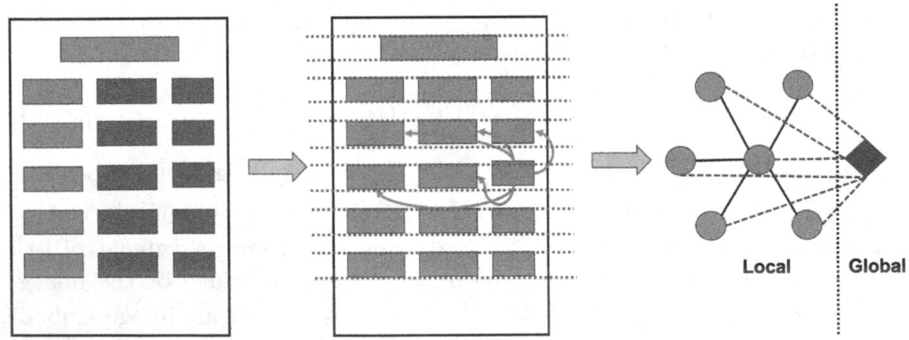

Fig. 3. Illustration of graph construction.

To capture global information, we add a global node to connect each local node. The global node is used as a virtual hub to collect universal information from all the nodes inside the graph. Since all internal nodes are connected with global node which means every two non adjacent nodes are two-hop neighbors, universal information can be distributed to these local nodes through such connections.

3.3 Recurrent-Based Aggregate and Update

Attention-based graph neural network [17] is applied to fuse multiple information in the graph, as shown in Fig. 4. In our model, graph convolution is defined based on the self-attention mechanism and aggregation and update of global node and local node are treated equally.

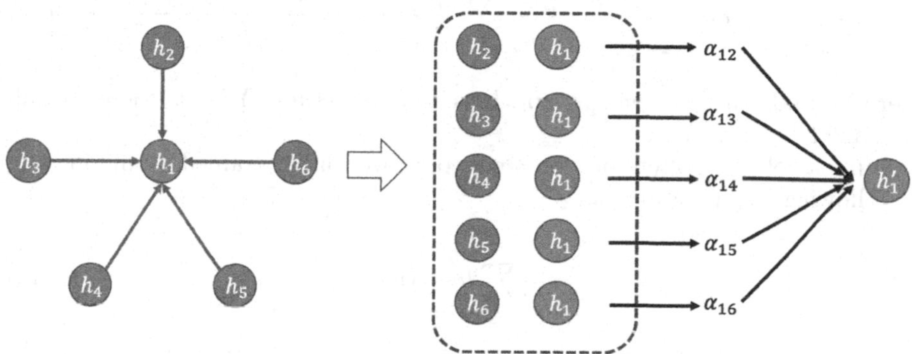

Fig. 4. Aggregation in Graph Neural Network.

Given a node v_i and its hidden state h_i which is initialized by RE, the output embedding of node v_i can be calculated by self-attention mechanism as the follows

$$\boldsymbol{h}_i' = \sigma(\sum_{j \in N_i} \alpha_{ij} W \boldsymbol{h}_j) \tag{9}$$

where $\boldsymbol{h}_i'$ is the aggregation and update of $\boldsymbol{h}_i$ and $\boldsymbol{h}_j$ is the hidden state of node v_i's neighbour v_j. σ is an activation function and α_{ij} is the attention coefficient which indicates the importance of node j's features to node i. The coefficients computed by the attention mechanism can be expressed as:

$$\alpha_{ij} = \frac{exp(LeakyReLU(V^T[Wh_i \oplus Wh_j]))}{\sum_{k \in N_i} exp(LeakyReLU(V^T[Wh_i \oplus Wh_k]))} \tag{10}$$

where W and V are trainable parameters. We apply the LeakyReLU nonlinearity (with negative input slope $\alpha = 0.2$) to avoid the "dying ReLU" problem.

Similarly to Vaswani et al. [16], we also employ multi-head attention to improve the performance of our model. K attention mechanisms execute independently and their features are concatenated in the end. The final representation is as the follows and $\oplus$ in the formula means concatenation operation:

$$\boldsymbol{h}_i' = \overset{K}{\underset{k=1}{\oplus}} \sigma(\sum_{j \in N_i} \alpha_{ij}^k W^k \boldsymbol{h}_j) \tag{11}$$

3.4 Decoding and Information Extraction

A conditional random field (CRF) is used to generate a family of conditional probability for the sequence. Given the sequence of final node states $h_{1:n}^{final} = [h_1^{final}, h_2^{final}, ..., h_n^{final}]$, and the probability of a label sequence $\hat{y} = [\hat{l}_1, \hat{l}_2, ..., \hat{l}_n]$ can be defined as the follows

$$p(\hat{y}|s) = \frac{exp(\sum_{i=1}^{n} W_{(l_{i-1},l_i)} h_i^{final} + b_{(l_{i-1},l_i)})}{\sum_{y' \in Y(s)} exp(\sum_{i=1}^{n} W_{(l'_{i-1},l'_i)} h_i^{final} + b_{(l'_{i-1},l'_i)})} \tag{12}$$

where W and b are the weight and bias parameters and $Y(s)$ is the set of all arbitrary label sequences.

Our model parameters of whole networks are jointly trained by minimizing the following loss function as:

$$L = -\sum_{i=1}^{N} log(p(y_i|s_i)) \tag{13}$$

Decoding of CRF layer is to search the output sequence y^* having the highest conditional probability for testing.

$$y^* = \underset{y \in Y(s)}{argmax}\, p(y|s) \tag{14}$$

Viterbi algorithm is used to calculate the above equations, which can improve algorithm operation efficiency.

4 Experiments

We use Pytorch framework to implement our experiments on a GTX 1080Ti GPU and apply our model for information extraction from two real-world datasets.

4.1 Datasets

We conduct experiments on two document understanding datasets. **(1) Contract Dataset:** Contract Dataset is a dataset from Alibaba Tianchi Competition. The dataset contains six types of named entities: Party A, Party B, Project Name, Contract Name, Contract Amount and Consortium Members. This dataset has both the original PDF format documents and annotation files of target named entities. The train set consists of 893 contracts and test set consists of 223 contracts. **(2) SROIE:** SROIE is composed of scanned receipt images and is annotated with 4 types of named entities: Company, Address, Date and Total. The train set consists of 627 receipt images and test set consists of 347 receipt images.

4.2 Implementation Details

We use the Adam [8] as the optimizer, with a learning rate of 3e-6 for all datasets. We employ the Dropout [15] with a rate of 0.5 for node aggregation and update. In the feature extraction part, the text feature extractor is pre-trained BERT model and the hyper-parameter of BERT used in our paper is same as [4]. The dimension of text embedding is 512. The image feature extractor is ResNet-50 model and the hyper-parameter of ResNet-50 used in our paper is same as [6]. We add a full connected layer after ResNet-50 to resize the output dimension to 512. Then the combination of text embeddings and image embeddings is applied as the input of the graph neural network. We apply 3 graph attention layers with 24 multi-heads and the dimension of hidden state is 1024. The standard F1 score is used as evaluation metrics.

4.3 Evaluation

We compare the performance of our model with Bi-LSTM-CRF [7] and BERT-CRF [4]. Bi-LSTM-CRF uses Bi-LSTM architecture to extract text information and a CRF layer to get tags. BERT-CRF applies BERT model as backbone to replace Bi-LSTM model and also a CRF layer after to extract entities. The input text sequence is generated by text segments concatenated from left to right and from top to bottom according to [12].

Table 1. F1-score performance comparisons from contract dataset.

Entities	Bi-LSTM-CRF	BERT-CRF	Our model
Party A	72.2	75.3	79.1
Party B	83.5	84.2	88.4
Project Name	65.6	68.3	74.8
Contract Name	69.2	71.5	80.2
Contract Amount	86.3	89.8	92.3
Consortium Members	45.2	46.1	54.6
Macro Average	70.3	72.5	**78.2**

4.4 Result

We report our experimental results in this section. Table 1 lists the F1 score of each entity of contract dataset. Macro-averages in the last row of the table are the averages of the corresponding columns, indicating the overall performance of each method on all entity types. In the contract scenario, as can be seen from Table 1, our model outperforms Bi-LSTM-CRF by 12% in F1 score and leads to a 8.00% increment of F1 score over BERT-CRF model. Moreover, our model

Table 2. F1-score performance comparisons from SROIE dataset.

Entities	Bi-LSTM-CRF	BERT-CRF	Our model
Company	85.1	86.8	93.5
Address	88.3	89.1	94.6
Date	94.2	96.2	97.3
Total	83.5	84.7	92.1
Macro Average	87.8	89.2	**94.4**

outperforms the two baseline models in all entities. Further analysis shows that our model makes great improvements in those entities like *Contract Name* and *Project Name*. These entities have conspicuous layout features and morphological features which can't be captured by text alone models.

Furthermore, as shown in Table 2, our model shows significant improvement over the baseline methods on SROIE dataset. Compared with the existing Bi-LSTM-CRF model and BERT-CRF model, our model gives the best results by a large margin. These results suggest that, compared to previous text alone methods, our model is able to extract more information from the document to learn a more expressive representation through graph convolutions.

4.5 Ablation Studies

To study the contribution of each component in our model, we conduct ablation experiments on both two datasets and display the results in Table 3. In each study, we exclude visual features and the use of global node respectively, to see their impacts on F1 scores on both two datasets.

Table 3. Ablation studies of individual component.

Configurations	Contract dataset	SROIE dataset
Full model	78.2	94.4
W/o visual feature	75.3	90.1
W/o global node	76.7	92.3

As described in Table 3, when we remove visual features, the result drops to the F1 score of 75.3 on contract dataset and 90.1 on SROIE dataset. This indicates that visual features can play an important role in addressing the issue of ambiguously extracting key information. Furthermore, the results show that the model's performance is degraded if the global node is removed, indicating that global connections are useful in the graph structure.

5 Conclusions and Future Works

This paper studies the problem of document understanding. In this work, we present a novel method that takes global context into account to refine the graph architecture on the complex documents. The explanatory experiments suggest that our proposed model is capable of extracting more information from documents to learn a more expressive representation through attention-based graph convolutions. We hope that our research will serve as a base for future studies on document understanding. Furthermore, we intend to extend our model to other document related tasks, such as document classification or document clustering.

Acknowledgements. This work was supported by The National Key Research and Development Program of China under grant 2017YFB0802704 and 2017YFB0802202.

References

1. Akbik, A., Bergmann, T., Vollgraf, R.: Pooled contextualized embeddings for named entity recognition. In: Proceedings of the 2019 Conference of the North American Chapter of the Association for Computational Linguistics: Human Language Technologies, Volume 1 (Long and Short Papers), pp. 724–728 (2019)
2. Chiticariu, L., Li, Y., Reiss, F.: Rule-based information extraction is dead! long live rule-based information extraction systems! In: Proceedings of the 2013 Conference on Empirical Methods in Natural Language Processing, pp. 827–832 (2013)
3. Dengel, A.R., Klein, B.: *smartFIX*: a requirements-driven system for document analysis and understanding. In: Lopresti, D., Hu, J., Kashi, R. (eds.) DAS 2002. LNCS, vol. 2423, pp. 433–444. Springer, Heidelberg (2002). https://doi.org/10.1007/3-540-45869-7_47
4. Devlin, J., Chang, M.W., Lee, K., Toutanova, K.: Bert: pre-training of deep bidirectional transformers for language understanding. arXiv preprint arXiv:1810.04805 (2018)
5. Gui, T., et al.: A lexicon-based graph neural network for Chinese NER. In: Proceedings of the 2019 Conference on Empirical Methods in Natural Language Processing and the 9th International Joint Conference on Natural Language Processing (EMNLP-IJCNLP), pp. 1039–1049 (2019)
6. He, K., Zhang, X., Ren, S., Sun, J.: Deep residual learning for image recognition. In: Proceedings of the IEEE Conference on Computer Vision and Pattern Recognition, pp. 770–778 (2016)
7. Huang, Z., Xu, W., Yu, K.: Bidirectional LSTM-CRF models for sequence tagging. arXiv preprint arXiv:1508.01991 (2015)
8. Kingma, D.P., Ba, J.: Adam: a method for stochastic optimization. arXiv preprint arXiv:1412.6980 (2014)
9. Kipf, T.N., Welling, M.: Semi-supervised classification with graph convolutional networks. arXiv preprint arXiv:1609.02907 (2016)
10. Lample, G., Ballesteros, M., Subramanian, S., Kawakami, K., Dyer, C.: Neural architectures for named entity recognition. arXiv preprint arXiv:1603.01360 (2016)
11. Liu, X., Gao, F., Zhang, Q., Zhao, H.: Graph convolution for multimodal information extraction from visually rich documents. arXiv preprint arXiv:1903.11279 (2019)

12. Palm, R.B., Winther, O., Laws, F.: Cloudscan-a configuration-free invoice analysis system using recurrent neural networks. In: 2017 14th IAPR International Conference on Document Analysis and Recognition (ICDAR), vol. 1, pp. 406–413. IEEE (2017)
13. Scarselli, F., Gori, M., Tsoi, A.C., Hagenbuchner, M., Monfardini, G.: The graph neural network model. IEEE Trans. Neural Netw. **20**(1), 61–80 (2008)
14. Schuster, D., et al.: Intellix-end-user trained information extraction for document archiving. In: 2013 12th International Conference on Document Analysis and Recognition, pp. 101–105. IEEE (2013)
15. Srivastava, N., Hinton, G., Krizhevsky, A., Sutskever, I., Salakhutdinov, R.: Dropout: a simple way to prevent neural networks from overfitting. J. Mach. Learn. Res. **15**(1), 1929–1958 (2014)
16. Vaswani, A., et al.: Attention is all you need. In: Advances in Neural Information Processing Systems, pp. 5998–6008 (2017)
17. Veličković, P., Cucurull, G., Casanova, A., Romero, A., Lio, P., Bengio, Y.: Graph attention networks. arXiv preprint arXiv:1710.10903 (2017)
18. Xu, Y., Li, M., Cui, L., Huang, S., Wei, F., Zhou, M.: Layoutlm: pre-training of text and layout for document image understanding. arXiv preprint arXiv:1912.13318 (2019)
19. Yu, W., Lu, N., Qi, X., Gong, P., Xiao, R.: Pick: processing key information extraction from documents using improved graph learning-convolutional networks. arXiv preprint arXiv:2004.07464 (2020)
20. Zhao, X., Niu, E., Wu, Z., Wang, X.: Cutie: learning to understand documents with convolutional universal text information extractor. arXiv preprint arXiv:1903.12363 (2019)
21. Zheng, S., Wang, F., Bao, H., Hao, Y., Zhou, P., Xu, B.: Joint extraction of entities and relations based on a novel tagging scheme. arXiv preprint arXiv:1706.05075 (2017)

Combining Impression Feature Representation for Multi-turn Conversational Question Answering

Shaoling Jing[1,2,3](✉), Shibo Hong[2], Dongyan Zhao[1], Haihua Xie[2], and Zhi Tang[1]

[1] Wangxuan Institute of Computer Technology, Peking University, Beijing 100871, China
{jingshaoling,zhaody,tangzhi}@pku.edu.cn
[2] State Key Laboratory of Digital Publishing Technology, Peking University Founder Group Co. LTD., Beijing, China
{hongshibo,xiehh}@founder.com
[3] Postdoctoral Workstation of the Zhongguancun Haidian Science Park, Beijing, China

Abstract. Multi-turn conversational Question Answering (ConvQA) is a practical task that requires the understanding of conversation history, such as previous QA pairs, the passage context, and current question. It can be applied to a variety of scenarios with human-machine dialogue. The major challenge of this task is to require the model to consider the relevant conversation history while understanding the passage. Existing methods usually simply prepend the history to the current question, or use the complicated mechanism to model the history. This article proposes an impression feature, which use the word-level inter attention mechanism to learn multi-oriented information from conversation history to the input sequence, including attention from history tokens to each token of the input sequence, and history turn inter attention from different history turns to each token of the input sequence, and self-attention within input sequence, where the input sequence contains a current question and a passage. Then a feature selection method is designed to enhance the useful history turns of conversation and weaken the unnecessary information. Finally, we demonstrate the effectiveness of the proposed method on the QuAC dataset, analyze the impact of different feature selection methods, and verify the validity of the proposed features through visualization.

Keywords: Conversational Question Answering · Feature representation · Machine reading comprehension

1 Introduction

Conversational Question Answering (ConvQA) is a new question answering task that requires a comprehension of the context, which has recently received more

M. Sun et al. (Eds.): CCL 2020, LNAI 12522, pp. 57–69, 2020.
https://doi.org/10.1007/978-3-030-63031-7_5

and more attention [1–5]. Since conversation is one of the most natural ways for humans to seek information, it carries over context through the dialogue flow. Specifically, we ask other people a question, dependending on their answer, we follow up with a new question, and second answer with additional information will be given based on what has been discussed [6]. Therefore, multi-turn conversational question answering is formed in this way. It can be used in many fields as a personal assistant systems, such as, customer service, medical, finance, education, etc. Moreover, with the rapid development of artificial intelligence technology in theory and practical applications, many personal assistant products have been launched in the market, such as Alibaba AliMe, Apple Siri, Amazon Alexa, etc. Although these assistants are capable to cover some simple tasks, they cannot handle complicated information-seeking conversations that require multiple turns of interaction [3].

In the tasks of two recent multi-turn ConvQA datasets, CoQA [6] and QuAC [7], given a passage, a question, and the conversation context preceding the question, the task is to predict a span of passage as the answer or give an abstractive answer based on the passage. So the machine has to understand a text passage and conversation history to answer a series of questions. Each conversation in the QuAC dataset is obtained by two annotators playing the roles of teacher (information-provider) and student (information-seeker) respectively. During the conversation, the student only has access to the heading of passage and tries to learn about a hidden Wikipedia passage by asking a sequence of freeform questions. The teacher answers the question by providing a span of text in the passage, as in existing reading comprehension tasks SQuAD [8], and gives the dialog acts which indicate the student whether the conversation should follow up. The CoQA has abstractive answers involving adding a pronoun (Coref) or inserting prepositions and changing word forms (Fluency) to existing extractive answers [9]. Both datasets contain yes/no questions and extractive answers. Compared with the CoQA[1], the QuAC[2] setting is similar to a user query on search engines. Therefore, this article intends to use the QuAC dataset for ConvQA experiments.

Most existing multiple turns of question answering methods [1,3,9,10] emphasize the influence of historical context on current questions. However, there is a great lack of public studies on selecting or re-weighting of the conversation history turns, and re-representing the current questions and passages. Therefore, this paper proposes an impression feature combined with conversational history. Specifically, we propose a multi-turn conversational question answering model combining with impression features. In order to learn the useful information from the conversation history, we separately calculate the word-level inter attention and turn inter attention from the conversation history to the current question and the passage. Then the learned representation is used as impression feature and fed to BERT [11] with other inputs. The final representation is used to predict the answers.

[1] https://stanfordnlp.github.io/coqa/.
[2] http://quac.ai/.

Therefore, the contributions are as follows:

(1) Design an impression feature representation. This feature helps the model to learn more accurate information from the context of the historical conversation turns and assists the model in understanding passage and conversation, which provides new insights to the ConvQA task.
(2) Adapt different feature selection methods to verify the impact of the proposed impression feature representation on the model.
(3) A multiple turn conversational question answering model combining impression features is proposed.

2 Related Work

ConvQA is closely related to Machine Reading Comprehension (MRC) and conversational system.

The ConvQA task is similar to the machine reading comprehension task [8], but the major difference from MRC is that the questions in ConvQA are organized in conversations [3], such as CoQA [6], QuAC [7]. Some questions rely on the historical questions or answers through pronouns. However, the questions of traditional MRC datasets (such as SQuAD [8] and SQuAD2.0 [12]) are independent of each other and have no relevance. Compared with the traditional MRC task, multi-turn ConvQA based on MRC adds multiple turns of conversation history to the original MRC task, making the ConvQA task more suitable for human daily conversation habits.

The existing methods for ConvQA in [2] and [3] determine whether the token in the question and the passage appear in each round of the historical conversation, and take the distance from the history turn of answers to the current question as the relative position, finally use the embedding of the relative position as an input of BERT encoder [11]. These methods are simple and effective, but they are not applicable to some no span-based answers. Because the token in the abstractive answer may be synonymous with a word in the historical answer, not the same word. In this case, the relative position is invalid. Moreover, a large amount of redundant information may also be introduced, and there may be a possibility of over-learning. Therefore, this paper focuses on how to select historical context and integrate its information into current question and passage.

ConvQA is very similar to the Background Based Conversations (BBCs) which recently proposed in the field of conversational systems. The latter is proposed to generate a more informative response based on unstructured background knowledge. But most of the research is aimed at topic-specific field [4], such as the conversation for movies [13,14] and diverse set of topics of Wikipedia [15]. Therefore, question answering based on reading comprehension and BBCs, these two tasks have in common that when responsing to each current sentence, not only the passage or background, but also the historical conversational context must be considered. The difference is that the former pays more attention to the ability of the model to understand the passage. The latter pays more attention to the ability of the model to understand the conversational context.

In terms of model structure, RNN-based structure and BERT-based model [11] have certain effectiveness on ConvQA, MRC and BBCs tasks. The RNN-based model [1] can learn the impact of historical questions and answers on the current question and passage, but it cannot learn the deep bidirectional context representation. The BERT-based model is proved to greatly improve the performance of ConvQA [2,3], but it lacks reasonable integration into the history turns of conversation. Therefore, this paper proposes a method to model the history turns of questions and answers, generate impression features, and integrate them into the current question and passage to improve model performance.

3 Our Approach

3.1 Task and Notations Definition

The ConvQA task is defined as [6] and [7], given a passage x, the k-th question q_k in the conversation and the history conversation H_k preceding q_k, the task is to predict the answer a_k to the question q_k. There are only extractive answers in dataset QuAC [7]. So the task is to predict the text span a_k within passage x. For the question q_k, there is $k-1$ turns of history conversation, and i-th turn of history conversation H_k^i includes a question q_i and its groundtruth answer a_i, which is $H_k^i = \{q_k^i, a_k^i\}_{i=1}^{k-1}$.

In order to ensure that the latter part of the long passage can be learned by the model, we divide the given passage x into N parts with sliding window following the previous work [11], it is denoted as $x = \{x_n\}_{n=1}^{N}$ and $x_n = \{x_n(t)\}_{t=1}^{T}$, where $x_n(t) \in \mathbb{R}^h$ refers to the representation of the t-th token in x_n, T is the sequence length and h is the hidden size of the token representation. The k-th question is denoted as $q_k = \{q_k(j)\}_{j=1}^{J}$, $q_k \in \mathbb{R}^{J\times h}$, where $q_k(j) \in \mathbb{R}^h$ refers to j-th token in q_k and J is the maximum question length. All $k-1$ turns of history question and answer sequences are represented as $H_k = \{H_k^i\}_{i=1}^{I}$, $H_k \in \mathbb{R}^{I\times M\times h}$, where I is the maximum number of history turns for all conversations. The i-th turn history conversation of the k-th question is denoted as $H_k^i = \{H_k^i(m)\}_{m=1}^{M}$, $H_k^i \in \mathbb{R}^{M\times h}$, where $h_k^i(m) \in \mathbb{R}^h$ is m-th token in H_k^i and M is the maximum length of history questions and answers.

3.2 Impression Feature Representation

Multiple NLP tasks obtained state-of-the-art results by using pre-trained language model BERT, which learned the deep bidirectional representations through transformer [16]. Adaptive to this paper, the encoder of BERT model encodes the question q_k, the passage x and the proposed Impression Feature (ImpFeat) that attend the conversational histories H_k into contextualized representation, which is shown in Fig. 1. The input sequences composed of token-level questions q_k and passages x_n are fed into the BERT model. Then the BERT encoder generates the token-level contextualized representation based on the token embedding, segment embedding, position embedding and the proposed impression feature (the different color row in the orange dotted lines of Fig. 1). Finally, based on the output representation, the answer span predictor calculate the probability of each token as the beginning and end of the answer. Among them, the proposed impression feature (red-cyan row in the orange dotted frame) generation is detailed in Fig. 2.

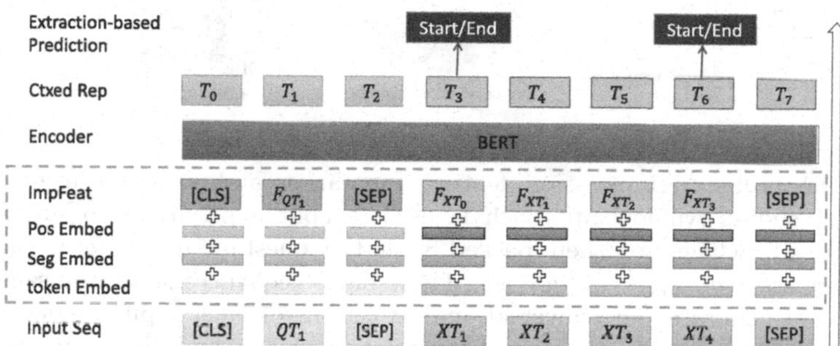

Fig. 1. Our model with ImpFeat. It mainly reveals the process from the input of questions and passages (the light yellow-green row) to the contextualized representation (the pink-purple row), and then to the generation of answers (navy blue). This process includes the steps of inputting sequences, making features (marked by orange-dotted lines), BERT encoding, and predicting answers. The method of generating ImpFeat (red-cyan row in the right of Fig. 2) from input sequence (the light yellow-green row in the left of Fig. 2) is detailed in Fig. 2. (Color figure online)

As shown in Fig. 2, the generation of impression features mainly includes two stages, word-level inter attention and turn inter attention. An input sequence contains a question q_k and a sub-passage x_n. For convenience, q_k is used as the representative of the input sequence in the following formula. The calculation method of the sub-passage x_n is the same as it. So the generation process is as follows.

Step 1: we follow word-level inter attention in the previous work [1] to compute the attended vector from history turns of questions and answers to the input sequence. The relevance score matrix between j-th token of the current question and m-th history questions or answers is defined as Eq. 1:

$$r_j^i(m) = \tanh(Uq_k(j))D\tanh(UH_k^i(m)) \tag{1}$$

where, $r \in \mathbb{R}^{J\times I\times M}$, $D \in \mathbb{R}^{d\times d}$ is a diagonal matrix, and $U \in \mathbb{R}^{d\times h}$, d is the attention hidden size. The word-level attentive weight of m-th token in i-th history conversation to the j-th token of the current question q_k is represented as $\hat{\alpha}_j^i(m)$:

$$\hat{\alpha}_j^i(m) = \frac{e^{r_j^i(m)}}{\sum_{i'=1}^{I}\sum_{m'=1}^{M} e^{r_j^i(m)}} \tag{2}$$

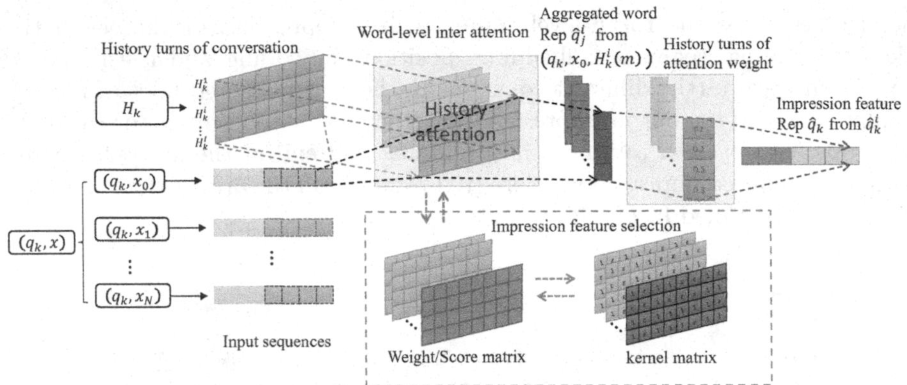

Fig. 2. The proposed impression feature generation and selection using history attention. A sliding window approach is used to split a passage into sub-passages $(x_0, x_1, \cdots, x_N)$, which are then packed with the question q_k to form the input sequences $(q_k, x_0), (q_k, x_1), \cdots, (q_k, x_N)$. These input sequences share the same question. Then we generate the conversation history H_k of each input sequence. Take (q_k, x_0) for illustration, we did word-level inter attention and turn inter attention respectively. Word-level inter attention is applied to calculate attention $\hat{q}_k^i$ from each token of the conversational history to each token of the input sequence. Then turn inter attention is calculated from different history turns of conversation to the input sequence. In addition, we also make feature selection (in the blue dotted lines) for the obtained historical memory in word-level inter attention stage to make the memory is selective. (Color figure online)

Therefore, the aggregated word-level representation of all tokens in i-th history turn of conversation to the j-th token of the current question is represented as $\hat{q_j^i}$:

$$\hat{q}_j^i = \sum_{m=1}^{M} \hat{\alpha}_j^i(m) H_k^i(m) \tag{3}$$

Step 2: To learn the attention from different history turns of conversation to the input sequence, i.e. history turn inter attention, we learn an attention vector $D \in \mathbb{R}_I$ to compute attention weight from aggregated representation of i-th history turn of conversation to the current question. Initialize the weight matrix D with random values, then we get:

$$\hat{w}_i = \frac{e^{\hat{q}_j^i \cdot D}}{\sum_{i'=1}^{I} e^{\hat{q}_j^{i'} \cdot D}} \tag{4}$$

Further, the ImpFeat representation of all tokens of all history turns of conversation to the input question is denoted as $\hat{q}_k(j)$:

$$\hat{q}_k(j) = \sum_{i=1}^{I} \hat{w}_i \hat{q}_j^i \tag{5}$$

Step 3: To learn the attention within the tokens of the input question and passage, self-attention in Transformer structure [16] is applied here. So $\hat{q}_k(j)$ is refered as impression feature representation, and is merged with the token embedding, segment embedding and position embedding as the input of BERT.

The proposed two attention methods, and the self-attention in Transformer [16] respectively learn the attention from the tokens of history conversation to the input sequence, the attention from history turns to the input sequences, and the attention within the input sequence. So the model learns the historical information from different dimensions. Just like human reading, the model has a deep impression on historical information, which is why we express the learned representation as the impression feature. In addition, we also make feature selection for the obtained historical memory in word-level inter attention stage to make the memory is selective.

3.3 Impression Feature Selection

In order to verify whether the attention learned above is effective, and remove some redundant information. In step1, we use a kernel matrix to disturb the weights learned by the input sequence and history turns of conversation. Make

$$r_j^i = \sum_{m=1}^{M} r_j^i(m) \tag{6}$$

Then we sort r_j^i for each token of input sequence, select the historical turn number corresponding to the top s of r_j^i as the selected useful turn, which is represented as $r_j^{s'}$, $0 \leq s' \leq I$, and generate the corresponding kernel matrix :

$$a = \{a_j^i(m\}_{1 \leq i \leq I, 1 \leq m \leq M}, a_j^i(m) = \begin{cases} 1, & \text{if } i = s' \\ \epsilon, & \text{otherwise} \end{cases} \tag{7}$$

where, ϵ is equals to a very small value, it is 0.001 in this paper. s is from 3 to 5 in this paper. $a_j^{s'}(m) = 1$ for all m in the s'-th turn. The new weight matrix after selection is represented as:

$$\alpha_j^i(m) = \hat{\alpha}_j^i(m) \cdot a_j^i(m) \tag{8}$$

where, $\alpha_j^i(m)$ represents that which history turns of conversation are more useful to the input sequence. Then we use the new weight matrix $\alpha_j^i(m)$ to replace $\hat{\alpha}_j^i(m)$ in Eq. (3), the q_k after adding impression feature selection is represented as:

$$q_j^i = \sum_{i=1}^{I} \alpha_j^i(m) H_k^i(m) \tag{9}$$

At last, use Eq. (9) and Eq. (5) to recalculate the ImpFeat representation.

4 Experiments

4.1 Data Description

The QuAC [7] dataset mentioned in the introduction is used for our experiment. It is a large-scale dataset contained more than 8,850 conversations and 98,400 questions. Statistics for this dataset is summarized in Table 1, we can only access the training and validation data.

Table 1. Statistics of QuAC dataset.

Items	Training data	Validation data
Number of passages	6,843	1,000
Number of dialogs	11,567	1,000
Number of questions	83,568	7,354
Average questions per dialogs	7.2	7.4
Average tokens per passage	396.8	440.0
Average tokens per question	6.5	6.5
Average tokens per answer	15.1	12.3
Min/Avg/Med/Max history turns per question	0/3.4/3/11	0/3/5/3/11
% unanswerable	20.2	20.2

4.2 Experimental Setup

Competing Methods. The methods with published papers on QuAC leaderboard[3] are considered as baselines. To be specific, the competing methods are:

BiDAF++ [7,17]: BiDAF++ is a re-implementation of a top-performing SQuAD model [17], which augments bidirectional attention flow (BiDAF) [18] with self-attention and contextualized embeddings.

BiDAF++ w/2-ctx [7]: Based on BiDAF++, BiDAF++ w/r-ctx consider the context(ctx) from the previous r QA pairs. When $r = 2$, the model reached the best performance.

FlowQA [10]: This model incorporate intermediate representations generated during the process of answering previous questions, thus it integrates the latent semantics of the conversation history more deeply than approaches that just concatenate previous questions/answers as input.

BERT [2]: A ConvQA model with BERT is implemented and without any history modeling. We re-implement the model with batch size as 12 and marked with BERT_BZ12.

[3] http://quac.ai/.

BERT + PHQA [2]**:** Based on BERT, this model adds conversation history by prepending history turn(s) to the current question. Here, PHQA prepends both history questions and answers. **BERT + PHA** prepends answers only.

BERT + HAE [2]**:** This approach model the conversation history by adding history answer embedding that denote whether a token is part of history answers or not.

BERT + PosHAE [3]**:** Based on BERT + HAE, This model learn position information of history turns by setting the distance from the historical turn to the current turn.

BERT + Att_PHQA: We implement a BERT-based ConvQA model that encode attention of history questions and answers (Att_PHQA), where, attention is computed from the prepended previous r QA pairs $(q_k, q_{k-1}, a_{k-1}, \cdots, q_1, a_1)$ to the input sequence (q_k, x_n). Here $r = 2$, i.e. $(q_k, q_{k-1}, a_{k-1}, q_{k-2}, a_{k-2})$.

BERT + Att_PHA: A BERT-based ConvQA model that encode attention of history answers only, where the prepended previous history is formed by $(q_k, a_{k-1}, a_{k-2}, \cdots, a_1)$. we set max answer length as 35 since it gives the best performance under this setting.

BERT + ImpFeat w/r-ctx: This is the solution we proposed in Sect. 3. The history turns of conversation H_k from the previous r QA pairs.

Hyper-parameter Settings and Implementation Details. In order to compare with methods similar to this article, such as BERT + HAE [2], BERT + posHAE [3], most of our experimental setting are the same as paper [3], such as Tensorflow[4], v0.2 QuAC data, and BERT-Base Uncased model with the max sequence length of 384. The difference is that the batch size is set to 12, and the max answer length is set to 35 in BERT+ Att_PHA. The total training steps is set to 58000. Experiments are conducted on a single NVIDIA TESLA V100 GPU.

Evaluation Metrics. The QuAC challenge provides two evaluation metrics, word-level F1 and human equivalence score (HEQ) [7]. Word-level F1 evaluates the overlap between prediction and references. HEQ is used to check if the system's F1 matches or exceeds human F1. It has two variants: (1) the percentage of questions for which this is true (HEQ-Q), and (2) the percentage of dialogs for which this is true for every question in the dialog (HEQ-D).

4.3 Experimental Results and Analysis

Main Evaluation Results. The results on the validation sets are reported in Table 2. To implement the method of this article, we re-implement the BERT-based question answering model on the QuAC dataset, and set the batch size as 12. The result is slightly smaller 1% than the result in paper [2], which is caused by the different hyperparameters setting. Moreover, we summarize our

[4] https://www.tensorflow.org/.

observations of the results as follows: (1) BERT + Att_PHA brings a significant improvement compared with BERT + PHA. This shows the advantage of using attention and suggests that making attention from history answer to the current question and passage plays an important role in conversation history modeling. (2) Computing attention with PHQA and PHA are both effective. BERT + Att_PHA achieves a higher performance compared to BERT + Att_PHQA, which indicates that all history answers contribute more information to the model than just the previous two turns of conversation history. (3) Our model (BERT + ImpFeat) obtains a substantially significant improvements over the BERT + Att_PHA model, but suffer the poor performance than FlowQA and BERT + PosHAE. One possible reason is that the impression feature has learned the token relevance from the context history to the current and passage, but it seems that there is still lack of topic flow and positional information of the conversation history, so that there is not enough improvement. (4) BERT + ImpFeat w/4-ctx outperform BERT + ImpFeat w/11-ctx, which indicates that the number of history pairs still affect the performance of the model, but four turns of context history may not be optimal result since we have not yet do experiments for all different history turns.

Table 2. Evaluation results on QuAC. Validation result of BiDAF++, FlowQA are from [7] and [10]. "-" means a result is not available.

Models	F1	HEQ-Q	HEQ-D
BiDAF++	51.8	45.3	2.0
BiDAF++ w/2-ctx	60.6	55.7	5.3
FlowQA	**64.6**	-	-
BERT	54.4	48.9	2.9
BERT + PHQA	62.0	57.5	5.4
BERT + PHA	61.8	57.5	4.7
BERT + HAE	63.1	58.6	6.0
BERT + PosHAE	**64.7**	**60.7**	**6.0**
BERT_Batchsize12	53.26	46.15	2.6
BERT + Att_PHQA	54.3	47.45	2.2
BERT + Att_PHA	62.48	57.74	5.3
BERT + ImpFeat w/11-ctx	63.02	58.54	**6.2**
BERT + ImpFeat w/4-ctx	**63.67**	**59.17**	5.9

Ablation Analysis. In order to verify whether the proposed impression feature selection method is effective, we set different selection methods for comparison. Specifically, we randomly set the element of a in Eq. (7) to 1 or ϵ, then predict the answer. The results in Table 3 shows that after removing or replacing our feature selection method, the model performance drops significantly, indicating the importance of our proposed selection method.

Table 3. Results for ablation analysis. "w/o" means to remove or replace the corresponding component.

Models	F1	HEQ-Q	HEQ-D
BERT + ImpFeat w/4-ctx	63.67	59.17	5.9
w/o ImpFeat Selection	62.06	57.49	5.5
w/o Random Selection	23.75	23.02	0.6

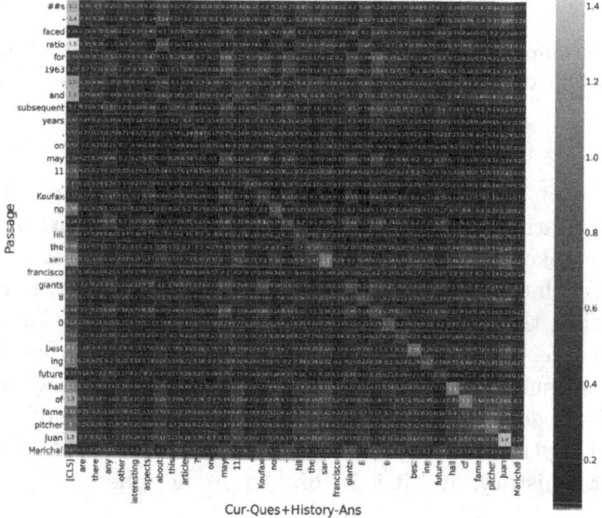

Fig. 3. The heatmap of attention score from the current question and conversation history (Cur-Ques + History-Ans) to the passage. The first cloumn is the aggregated scores, the second to ninth tokens on the horizontal axis indicate the ninth current question, and the remaining tokens represent a part of the answer of the sixth turn conversation history. The vertical axis represents parts of passage tokens.

Impression Feature Analysis. To further analyze the impression feature, we randomly select an example and visualize the relationship between current question, passage, and conversation history, as shown in Fig. 3 and 4, respectively. In Fig. 3, the passage is from "..., faced ratio for 1963, and subsequent years. On May 11, Koufax no-hit the San Francisco Giants 8-0, besting future Hall of Fame pitcher Juan Marichal–himself a no-hit pitcher a month later, ...". The current question is from "Are there any other interesting aspects about this article?", and the sixth turn of history answer is parts of the passage. We can see

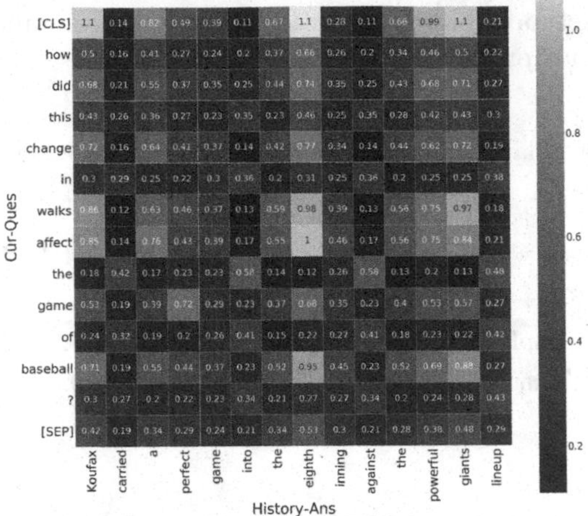

Fig. 4. The heatmap of attention score from the conversational history answer (History-Ans) to the current question (Cur-Ques). The first row is the aggregated scores.

that the tokens that are more relevant to the passage have a higher score and the stronger correlation, their corresponding color are redder, even white. On the contrary, the tokens that are less relevant to the passage have a lower score and the worse correlation, their corresponding color are darker. Furthermore, we can clearly see that there is a diagonal score that is generally large, because its answer exactly corresponds to the original answer. Besides, from Fig. 4, we can see that the tokens such as "powerful", "graints" in history answers are more relevant to the tokens "change", "walks", "affect" and "basketball" in the current question, indicating that the impression feature has learned relevant information from conversation history, and it is helpful to predict answers.

5 Conclusion and Future Work

Based on the general framework for ConvQA, we propose a new feature named impression feature, and combine the proposed feature with token embedding, position embedding and segment embedding as the input of BERT encoder. Then we introduce an impression feature selection method to select the important history information. Extensive experiments show the effectiveness of our method. Finally, we perform an in-depth analysis to show the different attention methods under different setting. Future work will consider to integrate multi-oriented information and a free-form answer type for ConvQA.

Acknowledgments. We thank all people who did human evaluation. This work are funded by China Postdoctoral Science Foundation (No.2019M660578), National Key Research and Development Program (No.2019YFB1406302), and Beijing Postdoctoral Research Foundation (No.ZZ2019-93).

References

1. Zhu, C., Zeng, M., Huang, X.: SDNet: contextualized attention-based deep network for conversational question answering. CoRR, abs/1812.03593 (2018)
2. Qu, C., Yang, L., Qiu, M., Bruce Croft, W., Zhang, Y., Iyyer, M.: Bert with history answer embedding for conversational question answering. In: SIGIR 2019: Proceedings of the 42nd International ACM SIGIR Conference on Research and Development in Information, pp. 1133–1136 (2019)
3. Qu, C., et al.: Attentive history selection for conversational question answering. In: Proceedings of the 28th ACM International Conference on Information and Knowledge Management, pp. 1391–1400, November 2019
4. Meng, C., Ren, P., Chen, Z., Monz, C., Ma, J., de Rijke, M.: Refnet: a reference-aware network for background based conversation. arXiv preprint arXiv:1908.06449 (2019)
5. Pruthi, D., Gupta, M., Dhingra, B., Neubig, G., Lipton, Z.C.: Learning to deceive with attention-based explanations. In: The 58th Annual Meeting of the Association for Computational Linguistics (ACL), July 2020
6. Reddy, S., Chen, D., Manning, C.D.: CoQA: a conversational question answering challenge. Trans. Assoc. Comput. Linguist. **7**, 249–266 (2019)
7. Choi, E., et al.: QuAC: question answering in context. In: Proceedings of the 2018 Conference on Empirical Methods in Natural Language Processing (2018)
8. Rajpurkar, P., Zhang, J., Lopyrev, K., Liang, P.: SQuAd: 100,000+ questions for machine comprehension of text. In: Proceedings of the 2016 Conference on Empirical Methods in Natural Language Processing (2016)
9. Yatskar, M.: A qualitative comparison of CoQA, SQuAD 2.0 and QuAC. arXiv preprint arXiv:1809.10735 (2018)
10. Huang, H.-Y., Choi, E., Yih, W.: FlowQA: grasping flow in history for conversational machine comprehension. CoRR, abs/1810.06683 (2018)
11. Devlin, J., Chang, M.-W., Lee, K., Toutanova, K.: Bert: pre-training of deep bidirectional transformers for language understanding. arXiv preprint arXiv:1810.04805 (2018)
12. Rajpurkar, P., Jia, R., Liang, P.: Know what you don't know: unanswerable questions for squad. In: Proceedings of the 56th Annual Meeting of the Association for Computational Linguistics (Volume 2: Short Papers) (2018)
13. Moghe, N., Arora, S., Banerjee, S., Khapra, M.M.: Towards exploiting background knowledge for building conversation systems. In: Proceedings of the 2018 Conference on Empirical Methods in Natural Language Processing (2018)
14. Zhou, K., Prabhumoye, S., Black, A.W.: A dataset for document grounded conversations. In: Proceedings of the 2018 Conference on Empirical Methods in Natural Language Processing (2018)
15. Dinan, E., Roller, S., Shuster, K., Fan, A., Auli, M., Weston, J.: Wizard of Wikipedia: Knowledge-powered conversational agents. arXiv preprint arXiv:1811.01241 (2018)
16. Vaswani, A., et al.: Attention is all you need. In: Advances in Neural Information Processing Systems, pp. 5998–6008 (2017)
17. Peters, M.E.: Deep contextualized word representations. arXiv preprint arXiv:1802.05365 (2018)
18. Seo, M., Kembhavi, A., Farhadi, A., Hajishirzi, H.: Bidirectional attention flow for machine comprehension. arXiv preprint arXiv:1611.01603 (2016)

Text Generation and Summarization

Chinese Long and Short Form Choice Exploiting Neural Network Language Modeling Approaches

Lin Li[1,2](✉), Kees van Deemter[1], and Denis Paperno[1]

[1] Utrecht University, Utrecht, The Netherlands
{l.li1,c.j.vandeemter,d.paperno}@uu.nl
[2] Qinghai Normal University, Xining, China

Abstract. Lexicalisation is one of the most challenging tasks of Natural Language Generation (NLG). This paper presents our work in choosing between long and short forms of elastic words in Chinese, which is a key aspect of lexicalisation. Long and short forms is a highly frequent linguistic phenomenon in Chinese such as 老虎-虎 *(laohu-hu, tiger)*. The choice of long and short form task aims to properly choose between long and short form for a given context to producing high-quality Chinese.

We tackle long and short form choice as a word prediction question with neural network language modeling approaches because of their powerful language representation capability. In this work, long and short form choice models based on the-state-of-art Neural Network Language Models (NNLMs) have been built, and a classical n-gram Language Model (LM) is constructed as a baseline system. A well-designed test set is constructed to evaluate our models, and results show that NNLMs-based models achieve significantly improved performance than the baseline system.

Keywords: Lexical choice · Long and short form · Language modeling · N-gram model · BERT · ERNIE

1 Introduction

The long and short form of an elastic word refers to words have different word length (i.e. number of syllables) but share at least one identical word meaning such as 丢失-丢 *(diushi-diu, lose)*. Duanmu [1] points out that as high as 80% percent of Chinese words has both long and short forms, therefore Chinese speakers need to make the choice between long and short forms during daily communication. Like human speakers and writers, the long and short form choice task also needs to be carefully resolved for various domain including Natural Language Generation [2], Machine Translation [3], and Style Transfer [4].

The first author of this paper received support from Grant 2016-ZJ-931Q, 2019-GX-162, and 61862055, which is gratefully acknowledged.

M. Sun et al. (Eds.): CCL 2020, LNAI 12522, pp. 73–80, 2020.
https://doi.org/10.1007/978-3-030-63031-7_6

In this work, we focus on long and short forms that share at least one same word meaning and one same morpheme, but compose of different number of syllables. The long and short form choice task is formulated as Fill-in-the-blank (FITB) task [2,5], whose goal is to select a missing word for a sentence from a set of candidates. A FITB example used in this work is shown in Table 1.

Table 1. A long and short form choice FITB question example.

Sentence	Long Form	Short Form
她去日本旅游时，必逛各种免税 ________[3]。 *When travels to Japan, she must go to duty free________.*	*(1)* 商店 *(1) shop*	*(2)* 店 *(2) shop*

The lexical choice is difficult in the context of long and short forms for most language processing systems due to the identical word sense leading to their preceding and subsequent contexts are too similar to providing distinguishing information. To address this problem, we investigate in learning language representation by LMs to making elegant choice of long and short forms. This paper makes the following contributions: (1) propose long and short form choice models by making use of language modeling approaches LSTM-RNN LM and pre-trained LM (BERT [6] and ERNIE [7]) (2) to compare the performance of different LMs, constructing a well-designed test set for long and short form choice task.

The remainder of this paper is organized as follows. In Sect. 2, we discuss related work. Section 3 describes the language modeling methods we have used for our research and introduce our models. Section 4 presents our experimental results. We conclude with a discussion in Sect. 5.

2 Related Work

A lot of words can be expressed by either a long form or a short form [8], for instance, elastic word, abbreviation, reduplication. In this work, we focus on the choice of long and short form of elastic words, that is, to choose between the long form (disyllabic) and short form (monosyllabic) of an elastic word that shares one morpheme and at least one same word meaning, and are interchangeable in some contexts [9]. Previous work [1,9–11] show that as high as 90% Chinese word has long and short forms, which is a key issue in Chinese lexical choice. Li et al. [12] investigated the problem of long and short form choice through human and corpus-based approaches, whose results support the statistical significant correlation between word length and the predictability of its context. Most previous work investigate the distribution and preference of long and short form based on corpus. It is still an open question to automated choose between long and short forms for a given context.

We framed the choosing between long and short forms as a FITB task proposed by Edmonds [13] in English near-synonyms choice. Unsupervised statistical approaches were applied to accomplish FITB task in near-synonym choice, for instance, Co-occurrence Networks [13] and Pointwise Mutual Information (PMI) [14] were used to build up near-synonym choice model separately. Wang and Hirst [15] explore lexical choice problem by capturing high dimensional information of target words ant their contexts thorough Latent Semantic Space.

Language models have obtained excellent performance in many language processing tasks, thus they have been also used to tackle the lexical choice task. A 5-gram language model [16] was trained from a large-scale Web corpus to choosing among English near-synonyms, following which Yu et al. [17] implemented n-gram language model to Chinese near-synonym choice. N-gram model shows a better accuracy than PMI in near-synonym choice which is similar to our task. Neural Language Models overcome the limitation of n-gram language model by its powerful capability of long-range dependency. Recurrent Neural Networks (RNN) [18] and it variation Long-short Term Memory (LSTM) [19]. Zweig et al. [5] tackled the sentence completion problem with various approaches like language models. NNLMs achieved a better performance in these work, whose improvement can be attributed to its capability of capturing global information.

3 Long and Short Form Choice via Language Models

Language modeling is an effective approach to solve the task by computing occurrence probability of each candidate words. Given a context, the best long and short form can be chosen according to the probability acquired from language models. The state-of-the-art language modeling techniques and apply them to our task is described in this section.

3.1 N-Gram Language Model

An input sentence S contains n words, i.e.,

$$S = \{w_1 w_2 w_3 ... w_i ... w_{n-2} w_{n-1} w_n\} \tag{1}$$

where w_i (i^{th} word of the sentence), denotes the lexical gap. The candidate words for the gap is $w_i = \{w_{long}, w_{short}\}$. Our task is to choose the w_i that best matches with the context.

N-gram language model, a classical probability language model, has succeeded in many previous work [5,16,17] by capturing contiguous word associations in given contexts. A n-gram smoothed model [16] for long/short word choice is used as our baseline model, whose key idea of acquiring the probability of a string is defined as follow:

$$P(S) = \prod_{i=1}^{p+1} P(w^i | w_{i-n+1}^{i-1}) = \prod_{i=1}^{p+1} \frac{C(w_{i-n+1}^{i}) + M(w_{i-n+1}^{i-1}) P(w_i | w_{i-n+2}^{i-1})}{C(w_{i-n+1}^{i-1}) + M(w_{i-n+1}^{i-1})} \tag{2}$$

$$M(w_{i-n+1}^{i-1}) = C(w_{i-n+1}^{i-1}) - \sum_{w_i} C(w_{i-n+1}^{i}) \tag{3}$$

where p is the number of words in the input sentence, i is the word position, $C(w_{i-n+1}^{i})$ and $C(w_{i-n+1}^{i-1})$ denotes the occurrence of the n-gram in the corpus, $P(w_i|w_{i-n+2}^{i-1})$ is the probability of w_i occurs given the words w_{i-n+1}^{i-1}, missing count $M(w_{i-n+1}^{i-1})$ is defined as 2.

The lexical gap of the input sentence S is replaced by long and short form separately, as follow:

$$S_1 = \{w_1 w_2 w_3 ... w_{long} ... w_{n-2} w_{n-1} w_n\}$$
$$S_2 = \{w_1 w_2 w_3 ... w_{short} ... w_{n-2} w_{n-1} w_n\}$$

Equation 1 is used to calculate $P(S_1)$ and $P(S_2)$, and take the target word in the sentence with higher probability as result. A disadvantage of n-gram model is not capable of maintaining long distance dependencies that play important role on long/short word choice. Hence, we proposed a neural language model to accomplish our task.

3.2 Recurrent Neural Networks (RNNs) Language Model

N-gram LM assigns probabilities to sentences by factorizing their likelihood into n-grams, whose modeling ability is limited because of data sparsity and long-distance dependency problem. NNLM have been proposed to model NL by [20], and outperform N-gram LM in many tasks [18,19] due to its ability of (1) each word w is represented as a low-dimensional density vector (2) retain long-span context information, which is failed captured by n-gram language model.

Recurrent Neural Networks (RNNs) have shown impressive performances on many sequential modeling tasks, thus we hypothesize that the performance of long/short form choice can be improved by adopting RNNs LM. Training a RNNs LM is difficult because of the vanishing and exploding gradient problems. Several variants of RNNs have been proposed to tackle with these two problems, among which Long Short-Term Memory is one of the most successful variants. In this work, we employ LSTM-RNNLMs to solve long/short form choice question. The LSTM adopted in this work is described as follows:

$$\begin{aligned}
i_t &= \sigma(U_i x_t + W_i s_{t-1} + V_i c_{t-1} + b_i) \\
f_t &= \sigma(U_f x_t + W_f s_{t-1} + V_f c_{t-1} + b_f) \\
g_t &= f\,(U x_t + W s_{t-1} + V c_{t-1} + b) \\
c_t &= f_t \odot c_{t-1} + i_t \odot g_t \\
o_t &= \sigma\,(U_o x_t + W_o s_{t-1} + V_o c_t + b_o) \\
s_t &= o_t \cdot f\,(c_t) \\
y_t &= g\,(V s_t + M x_t + d)
\end{aligned}$$

where x_t is input vector and y_t is output vector at time step t, i_t, f_t, o_t are input gate, forget gate and output gate respectively. c_{t-1} is the internal memory of unit, s_{t-1} is the LSTM hidden state at the previous time step. The uppercase (e.g., U_i and W) are weight matrices, the lowercase (e.g., b_i and b) is bias. f is the activation function and σ is the activation function for gates. The symbol $\odot$ is the Hadamard product or element-wise multiplication. Because of the architecture of LSTM-RNNLMs, the model has the potential to model long-span dependency.

3.3 Pre-trained Language Models

Language modeling aims to predict a distribution over a large scale of vocabulary items, by which solving the long/short form choice is a hard objective for our LSTM-RNNs acquired by limited size of training set and computation resource. We have a implicit assumption that the use of a powerful pre-trained language model is helpful to our task. Large-scale language models have achieved great success in many different Natural Language Understanding tasks. In this work, we focus on tackle our research question two very largely publicly LMs BERT and ERNIE.

LSTM-RNN LMs usually use the n preceding words as input to predict the next word $n + 1$, which cannot capture subsequent words of the word $n + 1$. BERT tackle this problem by retaining information of all the words in some fixed length sequence. Thus, we re-implemented BERT as a long and short form predictor to assign probability for a target word in a given context. BERT's model architecture is a multi-layer bidirectional Transformer encoder, whose success can be largely attributed to its Multi-Head Attention mechanism. By the attention mechanism, BERT is able to solving problems by learning the best representation through computing a weighted sum of the values of all words. The BERT-Base Chinese model adopted in this work is trained on a large scale of Chinese Simplified and Traditional corpus (based on an architecture of 12 layers, 768 hidden units, 12 heads, and 110M parameters). We tested the Bert with the methodology we used to test LSTM-RNNs.

ERNIE is a knowledge integration language representation model for Chinese, whose language representation is enhanced by using entity-level and phrase-level masking strategies in addition to a basic-level masking strategy. ERNIE has the same model structure as BERT-base, which uses 12 Transformer encoder layers, 768 hidden units and 12 attention heads.

4 Experiments and Results

Our baseline is a smoothed 4-gram language model, described in Sect. 3.1. In our training data set, we keep the words occurring at least 50 times, and filter out 2-gram, 3-gram, and 4-gram that occur less than three times. For the model based on LSTM-RNN LM, we set the word embeddings as 300, the LSTM hidden states as 128, sentence max length as 50, and learning rate as 0.1.

4.1 Data Resources

A large scale corpus is used in this work, which is Chinese online news in June 2012 (approximately contains 64M Chinese words)[1]. We split the corpus into two parts: 90% of the corpus is used for training and 10% for testing. The same training set is employed to train the 5-gram LM and LSTM-RNN LM, which ensure the comparability of these two models.

To test our models, we carefully construct a test set based on the corpus. Firstly, we randomly choose 175 different long/short forms from. Then, 6 sentences for each of these long/short forms are extracted from the corpus, in which the sentences contain the same number of long and short forms. Finally, we get a test set by slightly editing these sentences manually, which consists of 1050 sentences.

4.2 Results

Table 2 summarizes our results tested by the identical test set, which shows that all our models based on NNLMs approaches perform better than the baseline model. The improvement in accuracy of LSTM-RNN is 3.43%; the accuracy has been improved 10.96% by adopting BERT; and ERNIE performs the best in our task whose accuracy reaches 82.67%. Our results show that NNLMs is more capable than Ngram LM in long and short form choice task. We think our model based on LSTM-RNNs LM is not as well-performed as the two pre-trained NNLMs is because of its simpler neural network architecture and a smaller training set.

4.3 Post-hoc Analysis

According to semantic relation of the two morphemes of long forms, the long and short forms can be categorized into 7 groups [12]. The X-XX category refers to reduplicated long and short forms such as 妈妈-妈 *(mama-ma, mother)* or 仅仅-仅 *(jinjin-jin, only)*. All our models perform very well in predicting X-XX especially 5-gram LM performing the best, which suggests that the local context makes more contribution to the reduplication form choice than to other categories. Comparing with other categories of long and short forms, our models based on LSTM-RNN and ERNIE obviously perform bad in X-0X category, whose accuracy of this X-0X[2] is significant lower than the average accuracy (20.00% and 14.33% respectively). We think this is due to the comparatively low frequency of X-0X according to observation of our train set for LSTM-RNN LM.

---

[1] https://www.sogou.com/labs/resource/cs.php.

[2] X-0X refers the long and short form like 小麦-麦 *(xiaomai-mai, wheat)*.

Table 2. Accuracy of language modeling methods tested by identical data set.

Method	5-gram	LSTM-RNN	BERT	ERNIE
X-X'X	60.67%	77.33%	82.67%	88.00%
X-X0'	59.33%	78.00%	82.67%	78.67%
X-XY	62.67%	73.33%	82.67%	90.67%
X-0'X	66.67%	75.33%	75.33%	86.00%
X-XX	96.67%	88.00%	84.67%	87.33%
X-0X	71.33%	53.33%	76.00%	68.00%
X-X0	72.00%	68.00%	82.00%	80.00%
Accuracy	69.90%	73.33%	80.86%	82.67%

5 Conclusion

In this paper, we have investigated methods for answering long short form choice question. This question is significant because it is a key aspect of lexical choice which is still not well solved by many language processing systems. Through this work, we find that both all NNLM-based models do obviously outperform than Ngram LM. And our results show that all models perform very well in X-XX category but not very well in X-0X category. Our future work will be in the direction of eliminating the bias from NNLMs. Human evaluation for long and short form choice models also will be our further research content.

References

1. San, D.: How many Chinese words have elastic length. Eastward flows the Great river: Festschrift in honor of Prof. William S.-Y. Wang on his 80th birthday, pp. 1–14 (2013)
2. Inkpen, D.Z., Hirst, G.: Near-synonym choice in natural language generation. In: Recent Advances in Natural Language Processing, vol. 3, pp. 141–152 (2004)
3. Nguyen, T.Q., Chiang, D.: Improving lexical choice in neural machine translation. arXiv preprint arXiv:1710.01329 (2017)
4. Fu, Z., Tan, X., Peng, N., Zhao, D., Yan, R.: Style transfer in text: exploration and evaluation. In: Thirty-Second AAAI Conference on Artificial Intelligence (2018)
5. Zweig, G., Platt, J.C., Meek, C., Burges, C.J.C., Yessenalina, A., Liu, Q.: Computational approaches to sentence completion. In: Proceedings of the 50th Annual Meeting of the Association for Computational Linguistics: Long Papers-Volume 1, pp. 601–610. Association for Computational Linguistics (2012)
6. Devlin, J., Chang, M.W., Lee, K., Toutanova, K.: Pre-training of deep bidirectional transformers for language understanding. CoRR, abs/1810.04805 (2018)
7. Sun, Y., et al.: ERNIE 2.0: a continual pre-training framework for language understanding. arXiv preprint arXiv:1907.12412 (2019)
8. Packard, J.L.: The Morphology of Chinese: A Linguistic and Cognitive Approach. Cambridge University Press, Cambridge (2000)

9. Duanmu, S., Dong, Y.: Elastic Words in Chinese. The Routledge Encyclopedia of the Chinese Language, pp. 452–468 (2016)
10. Guo, S.: the function of elastic word length in Chinese. Yen Ching Hsueh Pao **24**, 1–34 (1938)
11. Huang, L., Duanmu, S.: a quantitative study of elastic word length in modern Chinese. Linguist. Sci. **12**(1), 8–16 (2013)
12. Li, L., van Deemter, K., Paperno, D., Fan, J.: Choosing between long and short word forms in mandarin. In: Proceedings of the 12th International Conference on Natural Language Generation, pp. 34–39 (2019)
13. Edmonds, P.: Choosing the word most typical in context using a lexical co-occurrence network. In: Proceedings of the eighth conference on European chapter of the Association for Computational Linguistics, pp. 507–509. Association for Computational Linguistics (1997)
14. Inkpen, D.: A statistical model for near-synonym choice. ACM Trans. Speech Lang. Process. (TSLP) **4**(1), 1–17 (2007)
15. Wang, T., Hirst, G.: Near-synonym lexical choice in latent semantic space. In: Proceedings of the 23rd International Conference on Computational Linguistics, pp. 1182–1190. Association for Computational Linguistics (2010)
16. Islam, A., Inkpen, D.: Near-synonym choice using a 5-gram language model. Res. Comput. Sci. **46**, 41–52 (2010)
17. Yu, L.-C., Chien, W.-N., Chen, S.-T.: A baseline system for Chinese near-synonym choice. In: Proceedings of 5th International Joint Conference on Natural Language Processing, pp. 1366–1370 (2011)
18. Mirowski, P., Vlachos, A.: Dependency recurrent neural language models for sentence completion. arXiv preprint arXiv:1507.01193 (2015)
19. Tran, K., Bisazza, A., Monz, C.: Recurrent memory networks for language modeling. arXiv preprint arXiv:1601.01272 (2016)
20. Mikolov, T., Karafiát, M., Burget, L., Cernocký, J., Khudanpur, S.: Recurrent neural network based language model. In: INTERSPEECH, pp. 1045–1048 (2010)

Refining Data for Text Generation

Qianying Liu[1,2], Tianyi Li[1], Wenyu Guan[1], and Sujian Li[1(✉)]

[1] Key Laboratory of Computational Linguistics, MOE, Peking University, Beijing, China
{litianyi01,guanwy,lisujian}@pku.edu.cn

[2] Graduate School of Informatics, Kyoto University, Kyoto, Japan
ying@nlp.ist.i.kyoto-u.ac.jp

Abstract. Recent work on data-to-text generation has made progress under the neural encoder-decoder architectures. However, the data input size is often enormous, while not all data records are important for text generation and inappropriate input may bring noise into the final output. To solve this problem, we propose a two-step approach which first selects and orders the important data records and then generates text from the noise-reduced data. Here we propose a learning to rank model to rank the importance of each record which is supervised by a relation extractor. With the noise-reduced data as input, we implement a text generator which sequentially models the input data records and emits a summary. Experiments on the ROTOWIRE dataset verifies the effectiveness of our proposed method in both performance and efficiency.

Keywords: Data-to-text generation · Sequence-to-sequence · Model Efficiency

1 Introduction

Recently the task of generating text based on structured data has attracted a lot of interest from the natural language processing community. In its early stage, text generation (TG) is mainly accomplished with manually compiled rules or templates, which are inflexible and mainly based on expert knowledge [4,5,11]. With the development of neural network techniques, especially sequence-to-sequence (seq2seq) models, generating short descriptive texts from structured data has achieved great successes, including generating wikipedia-style biographies [6,14] and restaurant introductions [8].

However, the task of generating long text, such as generating sports news from data, still fails to achieve satisfactory results. The existing models often forge fake context, lose sight of key facts and display inter-sentence incoherence [16]. For the sports news generation task, one challenging problem is that the input records are both large and noisy. Specifically, the inputted box scores, which contains hundreds of data records, belong to 40 different categories, such as fouls, three-pointer, starting position and so on. Meanwhile, not all of the

M. Sun et al. (Eds.): CCL 2020, LNAI 12522, pp. 81–94, 2020.
https://doi.org/10.1007/978-3-030-63031-7_7

inputted records are reflected in the sports news, and there exists a serious non-parallelism between data records and texts. According to our statistics for 3000 parallel sports news and its data records which is shown in Table 1 and Fig. 1, an average of only 19.3 data records out of 670.6 are mentioned in the summaries on average, namely only less than 5% of the data records are reflected in the human written news and rest 95% of them may bring noise into the model. Such large and noisy input has also caused the parameter amount of the embedding and encoder layer to be enormous, which leads to massive memory usage and limits the computation speed. In such situation, it is essential to refine data records and choose those important information before generating the final text.

Table 1. Statistics of data records in 3000 sports news.

Object	Number
Average data records mentioned	19.30
Average data records in box data	670.65
Average summary length	348.93
Types of data records	40

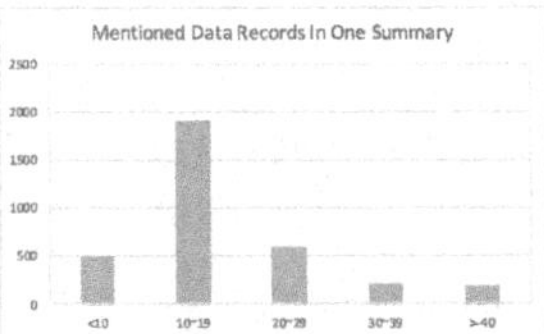

Fig. 1. Statistics of data records mentioned in 3000 sports news. The horizontal axis stands for summary numbers and the vertical axis stands for data record numbers.

In addition, sport news is far more complex than short descriptive text in that they need to consider overall coherence [1]. For example, it would be weird if there is an abrupt topic change between neighboring sentences. If we just pour all the data records with no order into a model, it would be difficult for the summarization model to learn content planning by itself. Thus, it is a good practice to order the data records before text generation.

As stated above, in this paper, we propose to refine data records for the data-to-text generation task by training a model to select an appropriate subset of data records, which carries the key facts of the game, and further to plan an appropriate order for the selected records. This is also similar to the action of human writers who usually plan the important information to include before they write their articles.

Next, one key problem is to label the important records which would be time consuming and expensive. To solve this problem, inspired by Wiseman et al. [16] which used an information extraction (IE) system for evaluation and Mintz et al. [7] which used distance learning for relation extraction, we build an IE system based on distant supervision. The IE system extracts relations from gold text, matches them to the corresponding data records and its results can then be used to supervise the process of content selection and planning. Then, we design a ranking unit to learn which data records are selected and in what order they appear. Here we choose to use the learning-to-rank (L2R) method instead of a classifier, because there exists heavy imbalance between positive and negative instances. We also design a rule-based model to further help select the data records. We rank each data record by an overall score based on the two rankers and rule-based system. Finally, we feed the selected and ordered records, which not only the noise and the input size is reduced but also the content is planned, to the generator to obtain the summaries. In this way memory usage could be largely reduced, thus the training process could be accelerated.

We evaluate our method on the ROTOWIRE dataset [16]. The results show how our system improves the model's ability of selecting appropriate context and ordering them. While we achieve comparable BLEU score, the efficiency of the model is greatly improved.

2 Related Work

Data-to-text generation has been an important topic of natural language generation for decades. Early approaches mainly use templates and rules to perform content selection and surface realization [4,5,11]. These models have good interpretability and controllability, but the generated content often have problems in terms of diversity and consistency.

Recently, neural network techniques have greatly improved the results of generating short descriptive text from data. The E2E dataset [6] stated the task of generating natural language descriptive text of the restaurants from structured information of the restaurants. The Wikibio dataset [8] gives the infobox of wikipedia as the input data and the first sentence of the corresponding biography as output text. Various approaches have achieved good results on these two datasets which considered content selection and planning. Sha et al. [14] proposed a method that models the order of information via link-based attention between different types of data records. Perez-Beltrachini and Lapata [9] introduce a content selection method based on multi-instance learning.

Generating sport news summaries on the other hand,is more challenging because not only the output text is longer and more complex, but also the input data records are numerous and diversed. Wiseman et al. [16] proposed the ROTOWIRE data set and gave baselines model based on end-to-end neural networks with attention and copy mechanism, these models often overlook key facts, repeatedly output the same information and make up irrelevant content. Puduppully et al. [10] designed a system that uses gate mechanism and pointer network to select and plan the content. They only used the IE system to guide content planning, while we let the IE system guide both content selecting and planning. Meanwhile our system is lighter and has higher efficiency since we only feed the neural network with a small subset of the large set of data records.

3 Model

Our model consists of three modules: information extraction, data refining (record selection and planning) and text generation. Figure 2 is a brief flow chart showing the pipeline of our model, which illustrates the data flow and how the models are trained.

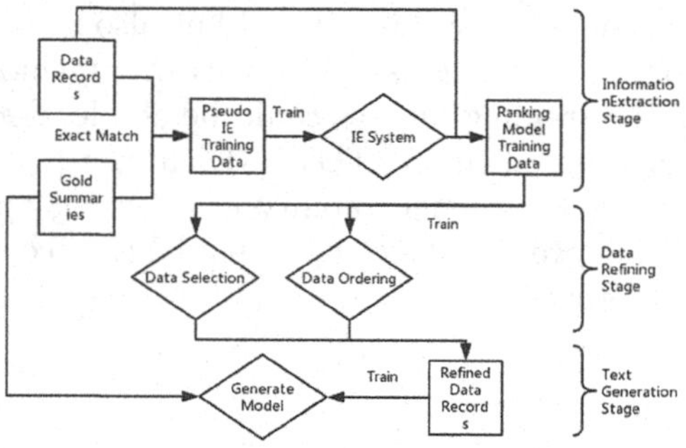

Fig. 2. A brief flow graph of our model.

3.1 Information Extraction

This module aims to provide supervision for data refining and text generation, and is only used during training. We build a relation extractor similar to Wiseman et al. [16], who used a relation extractor for automatic evaluation. We do not have human-annotated data for this specific domain, but this relation extractor can be trained by distance learning [7], which uses exact match between candidate entity-value pairs and data records to build pseudo training data. For example, from a sentence *A scored 4 points and B scored 8 points*, which has two entities {*A*, *B*} and two values {*4*, *8*}, we can extract 4 candidate entity-value pairs {*(A, 4)*, *(A, 8)*, *(B, 4)*, *(B, 8)*}. Then we compare them with the original data records and check whether these candidate pairs match with data records. In this example we can find *(A, 4, PTS)* and *(B, 8, PTS)* in the original data records, so we label the candidate pairs as {*(A, 4, PTS)*, *(A, 8, norel)*, *(B, 4, norel)*, *(B, 8, PTS)*}, where *norel* is the label that stands for no relationship and form the pseudo data. To be noticed, there might be multiple data records that match with the candidate pair, so the training data here is multi-labeled. The reason why we use an IE system instead of using the pseudo data straight away is because with the help of context information, the IE system can make better decisions and generalize better than the exact-match method.

To train the IE system, we cast the relation extraction task into a classification problem by modeling whether an entity-value pair in the same sentence has relation or not [12,17]. We use neural network to train the relation extractor and ensemble various models to further improve the performance. Formally, given an input sentence $\mathbf{x} = \{x_t\}_{t=1}^{n}$ which contains an entity-value candidate pair *(r.E, r.M)*, we first embed each word into a vector e_t^W. The embedding is then concatenated with two position embedding vectors e_t^E and e_t^V, which stands for the distance between the word and the entity and the value. Then the final word embeddings $e_t = concat\left\{e_t^W, e_t^E, e_t^V\right\}$ are fed into a bi-directional long short-term memory network (BiLSTM) or a convolutional neural network (CNN) to model the sequential information.

$$\begin{aligned} h_t &= BiLSTM(e_t, h_{t-1}, h_{t+1}) \\ h_{LSTM} &= h_n \end{aligned} \tag{1}$$

$$h_{CNN} = CNN(concat\left\{e_t\right\}_{t=1}^{n}) \tag{2}$$

After encoding the sentence, we use multilayer perceptron network (MLP) with a rectified linear unit (ReLU) as active function to make classification decisions and maintain the model's prediction of the candidate pair $r.T$. To be minded, the output $r.T$ is a vector where each position indicates whether the candidate pair is aligned with the data record at this position. Since there could be multiple labels, the output vectors are not distributions.

$$r.T = ReLU(Wh + b) \tag{3}$$

Because the training data is multi-labeled, we use negative marginal log likelihood as the loss function, namely each position is optimized toward 1 if positive and 0 if negative. We then map the positive candidate pairs back to the data records as silver training labels for the next stage. If a positive candidate pair *(entity, value, r.T)*, which is extracted from the xth sentence, is also in the data records, we label this data record as *Appeared in the xth sentence of the summary*.

3.2 Data Refining

In this module, we use two ranking models to refine the data records. These two rankers have different targets to optimize and separately perform content selection and ordering.

For content selection, we use both ListNet [2] and rule-based methods to select data records. The training data of this stage is seriously imbalanced: more than 95% of the input data records do not appear in the summaries and are labeled as negative. This makes it difficult for classification models to achieve good results. So here we use the L2R method to perform content selection. Instead of a point-wise loss function, which looks at a single example at a time,

Table 2. The details of features used for the ranking unit.

Feature	Type	Explanation
Record type	One hot	The one-hot representation of record type (i.e. PTS)
Is team	Value	Boolean of team or player
Home visit	Value	Boolean of home or visit team
Win lose	Value	Boolean of win or loss
Win ratio	Value	The win ratio of previous matches
Lose ratio	Value	The lose ratio of previous matches
Team performance	Values	All values of the team (i.e. PTS, PTS_QTR1, FG_PCT)
Player performance	Values	All values of the player. Zeros if it is team record
Start position	One hot	The start position of player. Zeros if it is team record
Pair value	Value	The value of **f**, if not a number then 0
N/A	Value	Whether the value is N/A
Team rank	Values	Whether the team value is larger that the other
Player rank	Values	The rank of each record type of this player

pair-wise and list-wise loss functions try to come up with the optimal ordering of a pair or a list of examples. In this stage we use ListNet, which optimizes a list-wise loss function, so the data imbalance problem can be relieved. Given a list of data records $\mathbf{r} = \{r_k\}_{k=1}^{n} = \{r.E_k, r.M_k, r.T_k\}_{k=1}^{n}$, we design hand-craft features and form a feature vector f_k for each data record as the input of the ranking model. We give the details of the features in the Table 2. Then the ranking model assigns a score s_k^S to each data record.

$$s_k^S = ListNet(f_k) \tag{4}$$

During inference stafe, we use a hyper-parameter threshold α tuned on the validation set to choose data records.

The rules are designed based on common sense and statistics of basketball news. We observe that several types of data records are chosen mainly according to whether the data record's value is larger than a specific threshold. Some other type of data records always appear in pairs, such as FTA and FTM. We give a table of details of the rules in the Table 3.

For content ordering, we use a pair-wise L2R method RankBoost [3] to reorder the selected data records. While training, we use the subset of data records $\mathbf{r} = \{r_k | r_k.t \neq negative\}$ to train this model. When we perform infer-

Table 3. The details of rules for the ranking unit. 'all' stands for choosing all records of this type of data. 'bar' stands for choosing the data records which value is larger than the threshold.

Type	Rule	Threshold
TEAM-PTS	All	\
TEAM-WINS	All	\
TEAM-LOSSES	All	\
AST	Bar	9
PTS	Bar	11
REB	Bar	9
TEAM-FG3_PCT	Bar	45
TEAM-FG_PCT	Bar	10

ence, the output of the content selecting unit is used as the input. We similarly embed r_k into a feature vector f_k and then use RankBoost to assign a score s_k^O to each r_k.

$$s_k^O = RankBoost(f_k) \tag{5}$$

We use s_k^O to reorder $\{r\}$ into $\{r^O\}$ and feed this ordered list of data records to the text generation module.

3.3 Text Generation

In the text generation module, we use a sequence-to-sequence encoder-decoder system to generate the summaries [15]. Given a list of data records $r^O = [r_k^O]_{k=1}^n$. We map these data records to a feature vector e_k by embedding $r.E$, $r.T$ and $r.M$ and concatenate the three embedding vectors and then use one layer of MLP to merge them into the final embedding vector.

The embeddings are then fed into the encoder, which is a BiLSTM to sequentially model the input and maintain the encoder output vectors hidden states h_t.

$$h_t = [h_t^f; h_t^b] = BiLSTM(e_t, h_{t-1}^f, h_{t+1}^b) \tag{6}$$

The decoder is built based on the Gated Recurrent Network (GRU). At each time step the decoder receives an input e_t^d and calculates the output vector s_t^d. Meanwhile it updates its own hidden state h_t^d.

$$s_t^d, h_t^d = GRU(e_t^d, h_{t-1}^d) \tag{7}$$

Here we implement the attention mechanism, conditional copy mechanism and coverage mechanism to further improve the model's performance.

Attention and Coverage. The attention at each step is calculated similar to See et al. [13], which is called perception attention. To calculate the attention weight between the hidden state of the decoder h_t^d and one output of the encoder h_i, we map the two vectors to fix size vectors seperately by two MLPs W_a and U_a with trainable bias b_a as h_{a_i}. Then we use a trainable vector v_a and dot multiply it with $tanh(h_{a_i})$ as the attention score s_{t_i}. At last we calculate the softmax over attention scores $\{s_{t_i}\}_{i=0}^n$ as the attention weights $\{a_{t_i}\}_{i=0}^n$. We finally dot-multiply the attention weights $\{a_{t_i}\}_{i=0}^n$ with the encoder outputs $\{h_i\}_{i=0}^n$ and sum them as the final attention vector $h_{t_{attn}}$.

$$h_{a_i} = W_a h_t^d + U_a h_i + b_a \tag{8}$$

$$s_{t_i} = v_a^T tanh(h_{a_i}) \tag{9}$$

$$a_{t_i} = softmax(s_{t_i}) = \frac{exp(s_{t_i})}{\sum_j exp(s_{t_j})} \tag{10}$$

$$h_{t_{attn}} = \sum_{i=0}^{n} a_{t_i} h_i \tag{11}$$

We also found that model often tends to repeatedly write about the same information, so we introduce coverage mechanism here to relief this problem. The key idea of coverage is to reduce the probability of paying attention to the information that is already generated.

If the sum of the previous attention weights is very high, there is a high probability that the information of this position is already generated. So in coverage model, we maintain a coverage score c_{t_i} for each encoder position at each decoder timestep, which is the sum of the attention weight of the previous timesteps $\{a_{t'_i}\}_{t'=0}^{t-1}$.

$$c_{t_i} = \sum_{t'=0}^{t-1} a_{t'_i} \tag{12}$$

We then modify the previous attention score with this coverage score. We assign a trainable weight vector w_c to c_{t_i} and sum it with h_{a_i} to maintain the adapted attention score.

$$s_{t_i} = v_a^T tanh(h_{a_i} + w_c c_{t_i}) \tag{13}$$

Conditional Copy. The copy mechanism has shown great effectiveness as an augmentation of encoder-decoder models recently. At each step the model uses an additional variable z_t to choose to copy or generate a word. The model either copies a word from the input sequence or generates a word from the vocabulary at step t.

Although both $r_k.E$ and $r_k.M$ may appear in the summaries, we only consider the probability of copying $r_k.M$. Instead of directly marginalizing out the latent-variable z_t, when we train the model we assume that any word y_t that appears both in the source data records and the summary is copied, so that we can jointly optimize the negative log-likelihood of y_t and z_t. To be noticed, there might be not only one $r_k.M$ that matches with y_t. Because our input data shares the same sequential order with the information mentioned in the summaries, we map the values from the start of the data records and skip the ones that are already mapped to align the records and copied values.

$$y = \begin{cases} p_{copy}(y_t|z_t; y_{1:t-1}; h_{1:n})p(z_t|y_{1:t-1}; h_{1:n}); \\ \qquad z_t = 1 \\ p_{generate}(y_t|z_t; y_{1:t-1}; h_{1:n})p(z_t|y_{1:t-1}; h_{1:n}); \\ \qquad z_t = 0 \end{cases}$$

We use the attention weights explained previously as the distribution $p_{copy}(y_t|z_t; y_{1:t-1}; h_{1:n})$. We concatenate the decoder input e_t^d, the decoder output s_t^d and the attention vector $h_{t_{attn}}$ and feed them into one MLP layer with sigmoid to model $p(z_t|y_{1:t-1}; h_{1:n})$.

4 Experiments and Results

4.1 Dataset

Here we use the ROTOWIRE dataset [16], which contains 3378 data-text pair in the training data. In addition to BLEU, this data set provides three automatic evaluation metrics, which are content selection (CS), relation generation (RG), and content ordering (CO). The first primarily targets "what to say" while the latter two metrics target "how to say". These three metrics are calculated based on an information extraction system that serves to align entity-mention pairs in the text with data records. We use the code released by Wiseman et al. [16] to maintain the evaluation scores of our model.

4.2 Performance

The results of our model and other baseline systems are shown in Table 4.

From the results we can see the effectiveness of our model, since it has significantly improved all the content evaluation metrics. Thus we can say refining the input data can help the model to be faithful to the input (RG), select good content (CS) and order them considering overall coherence (CO). We can see the BLEU score of our model is slightly lower than the baseline models. We think this is acceptable in trade of the great improvement of other evaluation scores.

Table 4. The results of text generation on validation set and test set. CC stands for conditional copy, JC stands for joint copy, TVD stands for the total variation distance loss, Rec stands for reconstruction losses, R stands for ranking.

Model	RG		CS		CO	BLEU
	P	#	P	R	DLD	
Validation set						
Template [16]	**99.35**	**49.7**	18.28	**65.52**	12.2	6.87
CC [16]	71.07	12.61	21.90	27.27	8.70	**14.46**
JC + TVD + Rec [16]	57.51	11.41	18.28	25.27	8.05	12.04
CC + R	76.86	16.43	**31.20**	38.94	**14.98**	13.27
Test set						
Template [16]	**99.30**	**49.61**	18.50	**64.70**	8.04	6.78
CC [16]	71.82	12.61	21.90	27.16	8.68	**14.49**
JC + TVD +Rec [16]	60.27	9.18	23.11	23.69	8.48	12.96
CC + R	75.12	16.90	**32.79**	39.93	**15.62**	13.46

5 Analysis

5.1 Content Selection

The results of our content selection and ordering models on the valid set are shown in Table 5. The results can prove our models' ability of refining data. We can see, because of imbalanced training data, ranking models with a threshold can significantly out perform classification models.

Table 5. The results of content selection and data ordering on the valid set.

Model	P	R	F1
ListNet	18.08	26.93	21.63
SVM	10.76	21.27	14.29
Random Forest	9.63	55.36	16.41
ListNet + Rule	59.02	59.98	59.50

5.2 Model Efficiency

Our model also significantly improves the efficiency of the model. We show the comparison of our model and CC [16] model in Table 6 and Fig. 3. Our model significantly reaches convergence faster and uses less memory and time to train. The parameter in the embedding and encoder layer is greatly reduced due to the refining of the input. For case #1 and #2, we can see the GPU memory usage

Table 6. The results of original input and refined order input. 'emb' and 'hid' stands for embedding and hidden dimensions. 'bs' stands for batch size. 'GPU' stands for the maximum memory used on GPU. 'time' stands for the time used for every epoch, the unit is minute.

#	Select	emb	hid	bs	GPU	Time
1	False	600	600	2	9275	214
2	True	600	600	2	2163	45
3	True	600	600	16	10375	8
4	True	1200	1200	12	10525	16

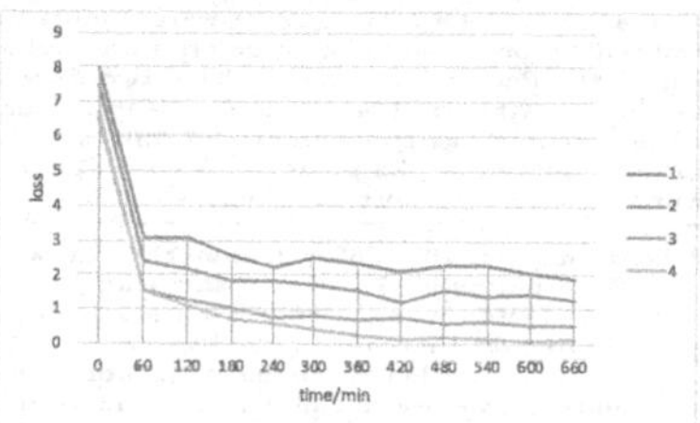

Fig. 3. Statistics of how the loss changes over time. The number labels of the poly-lines match with the order in Table 4.

and the time for each epoch is greatly reduced, which leads to faster convergence of the model. In case #3 and #4, we show that by refining the input, we can allow larger batch size, embedding size and hidden state size for the model to further boost the performance. While the architecture of the generator of our model and CC is similar, we show refining the input can greatly improve the model's efficiency.

5.3 Case Study

Here we show one example of the pipeline on the validation set in Fig. 4. We show the triples extracted by the IE system, triples extracted by the refining unit the gold text and the final generated text.

From this example we can see, the IE system has a strong ability of extracting relation pairs from the gold text. The IE system missed two information pairs which are (Pacers,35,TEAM-TEAM-PTS_QTR3) and (Knicks,12,TEAM-TEAM-PTS_QTR3), but succeeded in all other pairs, ending with an accuracy of 87.5% in this example.

The refining system shows a high precision comparing to the gold reference, covering 12 out of 16 triples.

The generated text is very faithful to the refined input at the first 5 sentences, but began making up false information when it tries to generate facts not given by the refined input. 2 - 3 3Pt, 3 - 3 FT are fake information about Jose Calderon where the corresponding information is not selected by the refining system.

IE (New York Knicks, 82, TEAM-PTS), (New York Knicks, 9, TEAM-WINS), (Indiana Pacers, 31, TEAM-LOSSES), (Indiana Pacers, 17, TEAM-WINS), (Indiana Pacers, 103, TEAM-PTS), (New York Knicks, 38, TEAM-LOSSES), (Roy Hibbert, 18, PLAYER-PTS), (Roy Hibbert, 10, PLAYER-REB), (Carmelo Anthony, 7, PLAYER-FGM), (Carmelo Anthony, 16, PLAYER-FGA), (Carmelo Anthony, 18, PLAYER-PTS), (Rodney Stuckey, 22, PLAYER-PTS), (Rodney Stuckey, 13, PLAYER-FGA), (Rodney Stuckey, 8, PLAYER-FG)

Refine (Knicks, 38, TEAM-LOSSES), (Pacers, 31, TEAM-LOSSES), (Knicks, 9, TEAM-WINS), (Pacers, 17, TEAM-WINS), (Knicks, 42, TEAM-FG_PCT), (Pacers, 53, TEAM-FG_PCT), (Pacers, 33, TEAM-FG3_PCT), (Knicks, 31, TEAM-FG3_PCT), (Knicks, 82, TEAM-PTS), (Pacers, 103, TEAM-PTS), (Carmelo Anthony, 18, PLAYER-PTS), (Carmelo Anthony, 16, PLAYER-FGA), (Carmelo Anthony, 7, PLAYER-FGM), (Carmelo Anthony, 2, PLAYER-FG3M), (Carmelo Anthony, 4, PLAYER-FG3A), (Ian Mahinmi, 10, PLAYER-REB), (Carmelo Anthony, 25, PLAYER-MIN), (Lou Amundson, 17, PLAYER-PTS), (Rodney Stuckey, 22, PLAYER-PTS), (Jose Calderon, 9, PLAYER-PTS), (Jose Calderon, 28, PLAYER-MIN), (Jose Calderon, 4, PLAYER-FGM), (Jose Calderon, 7, PLAYER-FGA), (Carmelo Anthony, 2, PLAYER-FTM), (Carmelo Anthony, 2, PLAYER-FTA), (Carmelo Anthony, 1, PLAYER-REB), (Carmelo Anthony, 1, PLAYER-AST), (Roy Hibbert, 10, PLAYER-REB), (Roy Hibbert, 18, PLAYER-PT)

Gold The Indiana Pacers (17 - 31) throttled the New York Knicks (9 - 38) 103 - 82 Thursday at the Bankers Life Fieldhouse . The Knicks held a five - point deficit at halftime before having a complete meltdown in the third quarter . The Pacers outscored the Knicks 33 - 12 in the third to all but ice the game . Roy Hibbert scored 10 of his 18 points in the third and added 10 rebounds in the contest to notch his 12th double - double on the season . New York earned an impressive 100 - 92 win over Oklahoma City on Wednesday but was unable to carry that momentum into Thursday 's contest against Indiana . The Knicks were unable to get much of anything going in the second half . Carmelo Anthony had n't scored less than 20 points in any game this month , but was held to 18 points on 7 - of - 16 shooting against the Pacers . The Knicks have two days off before hosting the Lakers on Sunday . Indiana was able to rebound nicely after its 104 - 91 loss to Toronto on Tuesday . George Hill and C.J Miles were both able to play , but were on minutes restrictions and only saw 12 and 13 minutes respectively . Rodney Stuckey led the team in scoring , putting up 22 points on 8 - of - 13 shooting . The Pacers will look to secure their second straight win Saturday when they host the Kings .

Generated The Pacers (17 - 31) held on for a 103 - 82 win over the Knicks (9 - 38) on Friday . Carmelo Anthony led the way with a game - high 18 points on 7 - of - 16 shooting , to go along with one rebound , one assist and one block in 25 minutes . Jose Calderon was the only other starter to score in double figures , scoring 10 points on 4 - of - 7 shooting in 28 minutes off the bench . The Knicks shot 53 percent from the field and 33 percent from three - point range . The Knicks were led by Carmelo Anthony , who finished with 18 points (7 - 16 FG , 2 - 4 3Pt , 2 - 2 FT) , one rebound , one assist and one block in 25 minutes . Jose Calderon added 10 points (4 - 7 FG , 2 - 3 3Pt , 3 - 3 FT) , six assists , one rebound and one steal in 33 minutes . The Knicks shot 53 percent from the field and 33 percent from three - point range . Carmelo Anthony was the high - point man for the Knicks , with 18 points (7 - 16 FG , 2 - 4 3Pt , 2 - 2 FT) , to go with one rebound , one assist , one steal and one block . Jose Calderon chipped in 10 points (4 - 7 FG , 2 - 3 3Pt , 3 - 3 FT) , six assists , one rebound and one steal in 33 minutes . The Knicks shot 53 percent from the field and 31 percent from three - point range . The Knicks have now lost four of their last five games , as they continue to battle for the eighth seed in the Eastern Conference . They will look to get back on track as they play host to the Bucks on Sunday . The Pacers have now lost four of their last six games , as they continue to battle for a playoff spot in the Eastern Conference . They will look to bounce back as they play host to the Nets on Sunday .

Fig. 4. One example of our system. IE stands for triples extracted by the IE system, Refine stands for triples selected by the refining system, Gold stands for the reference text and Generated stands for the generated result.

The following text contains more fake information. This shows the limitations in generating long text for seq2seq models and some shortages of pre-selected refined text. For further improvement, we should improve the ability of the model to generate long text, and also consider dynamically giving information that the model needs instead of feeding fixed triples.

6 Conclusion

In this paper we propose a data-to-text generating model which can learn data selecting and ordering from an IE system. Different from previous methods, our model learns what to say and how to say from the supervision of an IE system. To achieve our goal, we propose to use a ranking unit to learn selecting and ordering content from the IE system and refine the input of the text generator. Experiments on the ROTOWIRE dataset verifies the effectiveness of our proposed method.

Acknowledgement. We thank the anonymous reviewers for their helpful comments on this paper. This work was partially supported by National Key Research and Development Project (2019YFB1704002) and National Natural Science Foundation of China (61876009 and 61572049). The corresponding author of this paper is Sujian Li.

References

1. Bosselut, A., Celikyilmaz, A., He, X., Gao, J., Huang, P.S., Choi, Y.: Discourse-aware neural rewards for coherent text generation. In: Proceedings of the 2018 Conference of the North American Chapter of the Association for Computational Linguistics: Human Language Technologies, Volume 1 (Long Papers), vol. 1, pp. 173–184 (2018)
2. Cao, Z., Qin, T., Liu, T.Y., Tsai, M.F., Li, H.: Learning to rank: from pairwise approach to listwise approach. In: Proceedings of the 24th International Conference on Machine Learning, pp. 129–136. ACM (2007)
3. Freund, Y., Iyer, R., Schapire, R.E., Singer, Y.: An efficient boosting algorithm for combining preferences. J. Mach. Learn. Res. **4**(Nov), 933–969 (2003)
4. Holmes-Higgin, P.: Text generation-using discourse strategies and focus constraints to generate natural language text by Kathleen R. Mckeown, Cambridge University Press, 1992, pp 246,£ 13.95, ISBN 0-521-43802-0. Knowl. Eng. Rev. **9**(4), 421–422 (1994)
5. Kukich, K.: Design of a knowledge-based report generator. In: 21st Annual Meeting of the Association for Computational Linguistics (1983). http://aclweb.org/anthology/P83-1022
6. Lebret, R., Grangier, D., Auli, M.: Neural text generation from structured data with application to the biography domain. In: Proceedings of the 2016 Conference on Empirical Methods in Natural Language Processing, EMNLP 2016, Austin, Texas, USA, 1–4 November 2016, pp. 1203–1213 (2016)
7. Mintz, M., Bills, S., Snow, R., Jurafsky, D.: Distant supervision for relation extraction without labeled data. In: Proceedings of the Joint Conference of the 47th Annual Meeting of the ACL and the 4th International Joint Conference on Natural Language Processing of the AFNLP, pp. 1003–1011. Association for Computational Linguistics (2009). http://aclweb.org/anthology/P09-1113
8. Novikova, J., Dušek, O., Rieser, V.: The E2E dataset: new challenges for end-to-end generation. In: Proceedings of the 18th Annual Meeting of the Special Interest Group on Discourse and Dialogue. Saarbrücken, Germany (2017). https://arxiv.org/abs/1706.09254. arXiv:1706.09254
9. Perez-Beltrachini, L., Lapata, M.: Bootstrapping generators from noisy data. In: Proceedings of the 2018 Conference of the North American Chapter of the Association for Computational Linguistics: Human Language Technologies, Volume 1 (Long Papers), vol. 1, pp. 1516–1527 (2018)
10. Puduppully, R., Dong, L., Lapata, M.: Data-to-text generation with content selection and planning. arXiv preprint arXiv:1809.00582 (2018)
11. Reiter, E., Dale, R.: Building applied natural language generation systems. Nat. Lang. Eng. **3**(1), 57–87 (1997)
12. dos Santos, C., Xiang, B., Zhou, B.: Classifying relations by ranking with convolutional neural networks. In: Proceedings of the 53rd Annual Meeting of the Association for Computational Linguistics and the 7th International Joint Conference on Natural Language Processing (Volume 1: Long Papers), vol. 1, pp. 626–634 (2015)

13. See, A., Liu, P.J., Manning, C.D.: Get to the point: summarization with pointer-generator networks. In: Proceedings of the 55th Annual Meeting of the Association for Computational Linguistics (Volume 1: Long Papers), vol. 1, pp. 1073–1083 (2017)
14. Sha, L., Mou, L., Liu, T., Poupart, P., Li, S., Chang, B., Sui, Z.: Order-planning neural text generation from structured data. arXiv preprint arXiv:1709.00155 (2017)
15. Sutskever, I., Vinyals, O., Le, Q.V.: Sequence to sequence learning with neural networks. In: Advances in Neural Information Processing Systems, pp. 3104–3112 (2014)
16. Wiseman, S., Shieber, S.M., Rush, A.M.: Challenges in data-to-document generation. In: Proceedings of the 2017 Conference on Empirical Methods in Natural Language Processing, EMNLP 2017, Copenhagen, Denmark, 9–11 September 2017, pp. 2253–2263 (2017)
17. Zhang, Z.: Weakly-supervised relation classification for information extraction. In: Proceedings of the Thirteenth ACM International Conference on Information and Knowledge Management, pp. 581–588. ACM (2004)

Plan-CVAE: A Planning-Based Conditional Variational Autoencoder for Story Generation

Lin Wang[1,2], Juntao Li[1,2], Dongyan Zhao[1,2], and Rui Yan[1,2(✉)]

[1] Center for Data Science, Academy for Advanced Interdisciplinary Studies, Peking University, Beijing, China
{wanglin,lijuntao,zhaody,ruiyan}@pku.edu.cn

[2] Wangxuan Institute of Computer Technology, Peking University, Beijing, China

Abstract. Story generation is a challenging task of automatically creating natural languages to describe a sequence of events, which requires outputting text with not only a consistent topic but also novel wordings. Although many approaches have been proposed and obvious progress has been made on this task, there is still a large room for improvement, especially for improving thematic consistency and wording diversity. To mitigate the gap between generated stories and those written by human writers, in this paper, we propose a planning-based conditional variational autoencoder, namely Plan-CVAE, which first plans a keyword sequence and then generates a story based on the keyword sequence. In our method, the keywords planning strategy is used to improve thematic consistency while the CVAE module allows enhancing wording diversity. Experimental results on a benchmark dataset confirm that our proposed method can generate stories with both thematic consistency and wording novelty, and outperforms state-of-the-art methods on both automatic metrics and human evaluations.

Keywords: Story generation · Planning-based method · Conditional variational autoencoder

1 Introduction

A narrative story is a sequence of sentences or words which describe a logically linked set of events [14]. Automatic story generation is a challenging task since it requires generating texts which satisfy not only thematic consistency but also wording diversity. Despite that considerable efforts have been made in the past decades, the requirement of thematic consistency and wording diversity is still one of the main problems in the task of story generation.

On the one hand, a well-composed story is supposed to contain sentences that are tightly connected with a given theme. To address this problem, most previous methods attempt to learn mid-level representations, such as events [12], prompts [3], keywords [26] or actions [4], to guide the sentences generation. Although

M. Sun et al. (Eds.): CCL 2020, LNAI 12522, pp. 95–109, 2020.
https://doi.org/10.1007/978-3-030-63031-7_8

these approaches have shown their encouraging effectiveness in improving the thematic consistency, most of them have no guarantee for the wording diversity. The main reason is that most of these methods are based on recurrent neural networks (RNNs), which tend to be entrapped within local word co-occurrences and cannot explicitly model holistic properties of sentences such as topic [2,9,10]. As a result, RNN tends to generate common words that appear frequently [29] and this will lead to both high inter- and intra-story content repetition rates.

On the other hand, a well-composed story also needs to contain vivid and diversified words. To address the issue of wording diversity, some studies have employed models based on variational autoencoder (VAE) [7] or conditional variational autoencoder (CVAE) as a possible solution. It has been proved that, through learning distributed latent representation of the entire sentences, VAE can capture global features such as topics and high-level syntactic properties, and thus can generate novel word sequences by preventing entrapping into local word co-occurrences [2]. As a modification of VAE, CVAE introduces an extra condition to supervise the generating process and has been used in multiple text generation tasks, e.g., dialogue response generation [29], Chinese poetry generation [25]. Recent researches [9,23] in story generation task have confirmed that CVAE can generate stories with novel wordings. Despite the promising progress, how to keep thematic consistency while improving wording diversity is still a challenging problem, since these two requirements are to some extent mutually exclusive [9]. Specifically, consistent stories may limit the choice of words, while diversified wordings may lead to the risk of inconsistent themes.

In this paper, we propose to conquer these two challenges simultaneously by leveraging the advantages of mid-level representations learning and the CVAE model in improving wording novelty. Specifically, we propose a planning-based CVAE model, targeting to generate stories with both thematic consistency and wording diversity. Our introduced method can be divided into two stages. In the *planning stage*, keyword extraction and expansion modules are used to generate keywords as sub-topics representations from the title, while in the *generation stage*, a CVAE neural network module is employed to generate stories under the guidance of previously generated keywords. In our method, the planning strategy aims to improve the thematic consistency while the CVAE module is expected to keep the wording diversity of the story. To evaluate our proposed method, we conduct experiments on a benchmark dataset, i.e., the Rocstories corpus [14]. Experimental results demonstrate that our introduced method can generate stories that are more preferable for human annotators in terms of thematic consistency and wording diversity, and meanwhile outperforms state-of-the-art methods on automatic metrics.

2 Related Work

2.1 Neural Story Generation

In recent years, neural network models have been demonstrated effective in natural language processing tasks [11,13,16,18,21,27]. In story generation, previous studies have employed neural networks for enhancing the quality of generated content. Jain et al. [5] explored generating coherent stories from independent short descriptions by using a sequence to sequence (S2S) architecture with a bidirectional RNN encoder and an RNN decoder. Since this model is insufficient for generating stories with consistent themes, to improve the thematic consistency of the generated stories, many other methods have been explored. Martin et al. [12] argued that using events representations as the guidance for story generation is able to improve the thematic consistency of generated content. Fan et al. [3] presented a hierarchical method that first generates a prompt from the title, and then a story is generated conditioned on the previously generated prompt. Following the idea of learning mid-level representations, Xu et al. [24] proposed a skeleton-based model that first extracts skeleton from previous sentences, and then generates new sentences under the guidance of the skeleton. Similarly, Yao et al. [26] explored using a storyline planning strategy for guiding the story generation process to ensure the output story can describe a consistent topic. Fan et al. [4] further adopted a structure-based strategy that first generates sequences of predicates and arguments, and then outputs a story by filling placeholder entities. Although these methods have achieved promising results, most of them are implemented with RNNs, which tend to encounter common words problem. In recent researches, the Conditional Variational Auto-Encoder model is regarded as a possible solution for improving the wording diversity in story generation [9].

2.2 Conditional Variational Autoencoder

The Variational Auto-Encoder (VAE) model is proposed in [7]. Through forcing the latent variables to follow a prior distribution, VAE is able to generate diverse text successfully by randomly sampling from the latent space [2]. Conditional Variational AutoEncoder (CVAE), as a variant of VAE, can generate specific outputs conditioned on a given input. CVAE has been used in many other related text generation tasks, such as machine translation [28], dialogue generation [19, 20, 29], and poem composing [10, 25]. Subsequently, in recent years, CVAE has begun to be applied in story generation task to tackle the common wording problem. Li et al. [9] explored adopting CVAE to generate stories with novel and diverse words, and Wang et al. [23] alter the RNN encoder and decoder of CVAE architecture with the Transformer encoder and decoder [22] for the story completing task. Although the CVAE model has achieved encouraging performance on improving wording diversity, it is a still challenging problem to generate stories with both thematic consistency and diverse wordings. To solve this problem, in this paper, we propose a Plan-CVAE, which leverages the advantages of CVAE to generate diverse sentences and keeps the thematic consistency of the whole generated stories by using a planning strategy.

3 Preliminary

3.1 VAE and CVAE

A VAE model consists of two parts, an encoder which is responsible for mapping the input x to a latent variable z, and a decoder which works by reconstructing the original input x from the latent variable z. In theory, VAE forces z to follow a prior distribution $p_\theta(z)$, generally a standard Gaussian distribution ($\mu = 0$, $\sigma = 1$). It first learns a posterior distribution of z conditioned on the input x via the encoder network, denoted as $q_\theta(z|x)$, and then applies the decoder network to computes another distribution of x conditioned on z, denoted as $p_\theta(x|z)$, where θ are the parameters of the network.

The training objective of VAE is to maximize the log-likelihood of reconstructing the input x, denoted as $log\, p_\theta(x)$, which involves an intractable marginalization [7]. To facilitate model parameters learning, VAE can be trained alternatively by maximizing the variational lower bound of the log-likelihood, and the true posterior distribution $q_\theta(z|x)$ is substituted by its variational approximation $q_\phi(z|x)$, where ϕ denotes the parameters of q. The objective can be written as

$$L(\theta, \phi; x) = -KL(q_\phi(z|x)||p_\theta(z)) + E_{q_\phi(z|x)}[log\, p_\theta(x|z)] \tag{1}$$

The objective mentioned above contains two terms, where the first term $KL(\cdot)$ represents the KL-divergence loss, which encourages the model to keep the posterior distribution $q_\phi(z|x)$ close to the prior $p_\theta(z)$. The second term $E[\cdot]$ is the reconstruction loss for guiding the decoder to reconstruct the original input x as much as possible.

CVAE is a modification version of VAE, it introduces an extra condition c to supervise the generative process. Correspondingly, the encoder computes a posterior distribution $q_\theta(z|x, c)$, representing the probability of generating z conditioned both on x and c. Similarly, the distribution computed by decoder is $p_\theta(x|z, c)$, and the prior distribution of z is $p_\theta(z|c)$. Accordingly, the objective of CVAE can be formulated as

$$L(\theta, \phi; x, c) = -KL(q_\phi(z|x, c)||p_\theta(z|c)) + E_{q_\phi(z|x,c)}[log\, p_\theta(x|z, c)] \tag{2}$$

3.2 Problem Formulation

We formulate the story generation task with the following necessary notations:

Input: A title $T = (t_1, t_2, ..., t_n)$ is given to the model to guide the story generation, where t_i refers the i-th word and n denotes the length of the given title.

Output: A story $S = \{S_1, S_2, ..., S_m\}$ should be generated by the model based on the given title, where S_i represents the i-th sentence and m denotes the total number of sentences in the story.

Keywords: A keywords sequence $K = (k_1, k_2, ..., k_m)$ is generated from title to enhance the process of story generation, where k_i is the i-th keyword which serves as the sub-topic or extra hint for S_i.

4 Planning-Based CVAE Method

4.1 Overview

The overview of our proposed method is shown in Fig. 1. Our method contains two stages: a planning stage and a generation stage. In the planning stage, a *Keywords-Extraction module* followed by a *Keywords-Expansion module* are used. In this stage, several keywords are first extracted from the title, and then the extracted keywords are expanded to match the number of sentences to be generated. In the generation stage, a *CVAE module* generates the story sentence-by-sentence conditioned on the keywords, i.e., keyword k_i is used as the sub-topic or hint of sentence S_i.

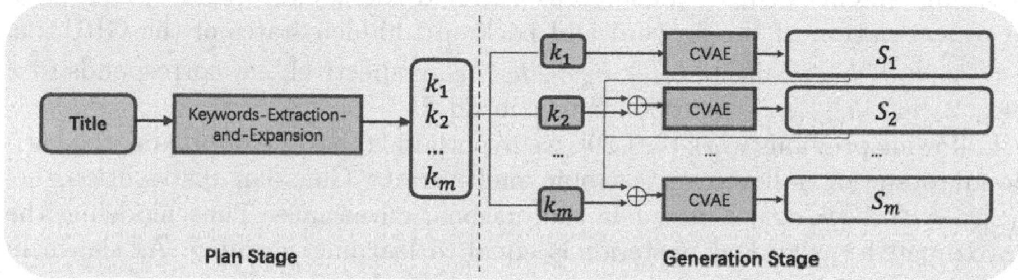

Fig. 1. An overview of our proposed method.

4.2 Planning Stage

In the planning stage, we first utilize RAKE algorithm [17] to extract keywords from the title. Since each sentence is to be generated under the guidance of a keyword, when the number of extracted keywords is not enough, we need to expand more keywords from existing ones. We adopt a language model with a long short-term memory network (LSTM) to predict the subsequent keywords based on the previously generated keywords.

To train the model, we collect training data from the story corpus. Specifically, for each story that contains m sentences in the corpus, we use RAKE to extract one keyword from one sentence. Then a keyword sequence $(k_1, k_2, ...k_m)$ corresponding to a story forms a sample in the training data. The language model is trained to maximize the log-likelihood of the subsequent keyword:

$$L(\theta) = log\, p_\theta(k_i|k_{1:i-1}) \tag{3}$$

where θ refers to the parameters of the language model, and $k_{1:i-1}$ denotes the preceding keywords.

Additionally, keywords can be directly generated by an RNN model from the title. Different from the straight-forward method, our method first extracts keywords from the title and then expands keywords to a sufficient number. Intuitively, the keywords extracted from the title possess a better consistency with the title. Thus, compared to the direct method, our method can lead to a better thematic consistency. To prove the superiority of our method, an ablation study is conducted to compare our method with the directed method, where the results are given in Table 2.

4.3 Generation Stage

We adopt the CVAE model for the generation stage. As demonstrated in Fig. 2, the CVAE model contains an encoder and a decoder. The encoder is implemented with a bidirectional GRU network to encode both the sentences and the keywords with shared parameters. At each step, the current sentence S_i, preceding sentences $S_{1:i-1}$ (denoted as ctx) and the keyword k_i are encoded as the concatenation of the forward and backward hidden states of the GRU, i.e. $h_i = [\overrightarrow{h_i}, \overleftarrow{h_i}]$[1], $h_{ctx} = [\overrightarrow{h_{ctx}}, \overleftarrow{h_{ctx}}]$, $h_k = [\overrightarrow{h_k}, \overleftarrow{h_k}]$, respectively. h_i corresponds to x in Eq. 2, and $[h_{ctx}, h_k]$ corresponds to c in Eq. 2.

Following previous work [7,9,29], we hypothesize that the approximated variational posterior follows an isotropic multivariate Gaussian distribution, i.e. $q_\phi(z|x,c) \sim \mathcal{N}(\mu, \sigma I)$, where I is the diagonal covariance. Thus modeling the approximated variational posterior is equal to learning μ and σ. As shown in Fig. 2, a recognition network is used to learn μ and σ. Specifically, we have

$$\begin{bmatrix} \mu \\ log\, \sigma \end{bmatrix} = W_r \begin{bmatrix} x \\ c \end{bmatrix} + b_r \tag{4}$$

where W_r and b_r are trainable parameters. Similarly, the prior is assumed to follow another multivariate Gaussian distribution, i.e. $p_\theta(z|c) \sim \mathcal{N}(\mu', \sigma' I)$, and μ' and σ' are learned by the prior network in Fig. 2, which is a one-layer fully-connected network (denoted as MLP) with $tanh(\cdot)$ as the activation function. Formally, it can be written as

$$\begin{bmatrix} \mu' \\ log\, \sigma' \end{bmatrix} = MLP_p(c) \tag{5}$$

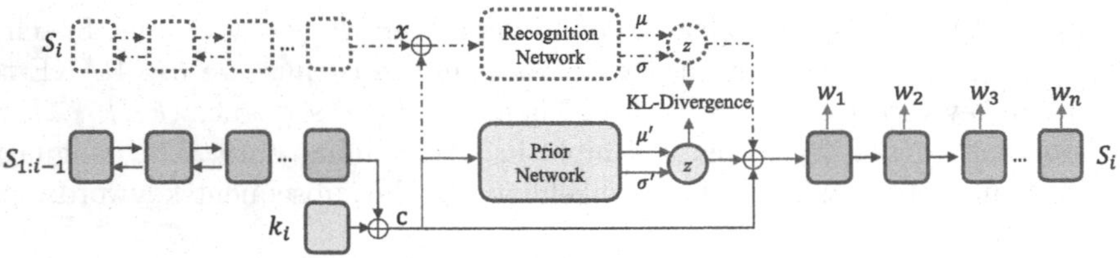

Fig. 2. The architecture of the CVAE module used in the generation stage. All components are used for training, while only the components with solid lines are for testing. ⊕ denotes the vector concatenation operation.

[1] → denotes forward and ← denotes backward.

The decoder is a one-layer GRU. The initial state of the decoder is computed as

$$S_{i,0} = W_d\left[z, c\right] + b_d \tag{6}$$

where W_d is a matrix for dimensional transformation, z is sampled from the recognition network during training and the prior network during testing. Meanwhile, a reparametrization trick [7] is used to sample z.

Moreover, previous researches proved that CVAE intends to encounter the latent variable vanishing problem in training [2]. Thus, in our implementation, KL cost annealing [2] and bag-of-word loss [29] are used to tackle the problem.

5 Experiments

5.1 Dataset

We conduct experiments on the ROCStories corpus [14], which contains 98159 stories. In our experiments, the corpus is randomly split into training, validation, and test datasets with 78527, 9816, 9816 stories. Every story in the dataset is comprised of one title and exactly five sentences, and the average word number of one story is 50.

5.2 Baselines

We utilize several strong and highly related methods of story generation as our baselines.

S2S, the sequence to sequence model [21] which has been widely used in multiple text generation tasks, such as machine translation and summarization. We implement it to generate stories in a sentence-by-sentence fashion, and the i-th sentence is generated by taking all the previous $i-1$ sentences as input.

AS2S, the sequence to sequence model enhanced by an attention mechanism [1], which is an improved version of S2S. It takes the same generation pipeline as S2S.

CVAE, the CVAE model without planning strategy. This pure CVAE model takes only the previous $i-1$ sentences as the condition c to generate the i-th sentence. This baseline is for demonstrating the performance of CVAE without planning strategy.

Plan-and-Write, the AS2S model with planning strategy proposed in [26]. Two different schema (static and dynamic) for keywords generation are proposed in the original paper. As the authors have proved that the static one is better, we implement the static scheme as our baseline.

5.3 Model Settings

We train our model with the following parameters and hyper-parameters. The word embedding size is set to 300, and the vocabulary is limited to the most frequent 30000 words. The hidden state size of encoder, decoder, and prior network are 500, 500, 600 respectively. And the size of the latent variable z is set to 300. To train our model, we adopt the Adam [6] optimization algorithm with an initial learning rate of 0.001 and gradient clipping of 5. All initial weights are sampled from a uniform distribution $[-0.08, 0.08]$. The batch size is 80.

5.4 Evaluation

We utilize both automatic and human metrics to evaluate the performance of our method.

BLUE Score. This metric is designed for calculating the word-overlap score between the golden texts and the generated ones [15], and has been used in many previous story generation works [9,26].

Distinct Score. To measure the diversity of the generated stories, we employ this metric to compute the proportion of distinct n-grams in the generated outputs [8]. Note that the final distinct scores are scaled to [0, 100].

Inter- and Intra-story Repetition. These two metrics are proposed in [26] and used for calculating the inter- and intra-story tri-grams[2] repetition rates by sentences and for the whole stories. The final results are also scaled to [0, 100].

Human Evaluation. We also employ three metrics for human evaluation, i.e., Readability, Consistency, and Creativity. Their descriptions are shown in Table 1. We randomly sample 100 generated stories from each baseline model and our method and then perform pairwise comparisons between our method and baselines. That is, for two stories with the same titles but generated by different two models, five well-educated human evaluators are asked to select the one they prefer on the three metrics. In comparison, no *equally good* option is given since the *equally good* option may leads to a careless comparison.

Table 1. Descriptions of human evaluation metrics.

Readability	Is the story formed with correct grammar?
Consistency	Does the story describe a consistent theme?
Creativity	Is the story narrated with diversified wordings?

[2] Results on four and five-grams have the same trends.

Table 2. Results of BLUE and Distinct scores. B-n and D-n represent the BLUE scores and Distinct scores on n-grams respectively. The final results are scaled to [0, 100]. The difference between Plan-CVAE* and Plan-CVAE is the former generates keywords directly from the title, while the latter generates keywords using our keywords-extraction and keyword-expansion module.

Model	Automatic evaluation							
	B-1	B-2	B-3	B-4	D-1	D-2	D-3	D-4
S2S	23.65	9.30	4.07	1.97	0.90	4.11	10.70	19.37
AS2S	24.70	9.68	4.27	2.07	0.93	4.53	11.13	19.41
CVAE	28.53	10.21	3.63	1.39	1.67	15.82	46.88	**76.64**
Plan-and-Write	27.39	11.78	**5.57**	**2.85**	0.84	5.15	14.67	28.28
Plan-CVAE*	29.57	11.32	4.43	1.85	1.52	14.13	42.30	71.42
Plan-CVAE	**30.25**	**12.05**	4.89	2.03	**1.75**	**16.38**	**46.98**	75.73
Human	–	–	–	–	2.87	26.74	62.92	86.67

6 Results and Analysis

Table 2 and Fig. 3 present the results of automatic evaluation, and Table 3 shows the results of human evaluation. Through analyzing these evaluation results, we have the following observations.

6.1 The Effect of the Planning Strategy

The Planning Strategy is Effective for Improving Thematic Consistency. As shown in Table 2, BLEU-[1–4] scores of Plan-CVAE are significantly higher than the pure CVAE model. Higher BLEU scores indicate that the planning strategy can improve the word-overlapping between the generated stories and the gold standard ones, which means the generated stories are more relevant with thematically consistent cases. For the subjective feelings of humans, as indicated by the human consistency evaluation in Table 3), Plan-CVAE can generate stories with better thematic consistency than the CVAE model. Meanwhile, Plan-CVAE outperforms all baselines on thematic consistency in human evaluation, this means the CVAE model gains a significant improvement on thematic consistency by using the planning strategy.

The Planning Strategy Does Not Affect the Wording Diversity. The planning strategy aims to enhance the CVAE model with better thematic consistency while preventing poor wording diversity. Plan-CVAE has a comparable performance with CVAE and outperforms other baselines on distinct scores in

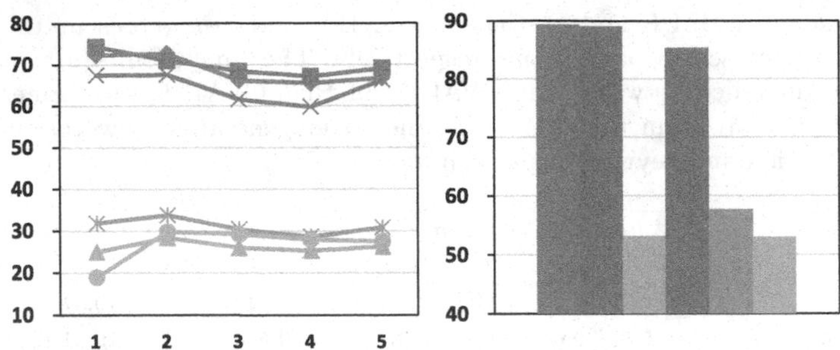

(a) Inter-story repetition curve by sentence.

(b) Inter-story aggregate repetition scores.

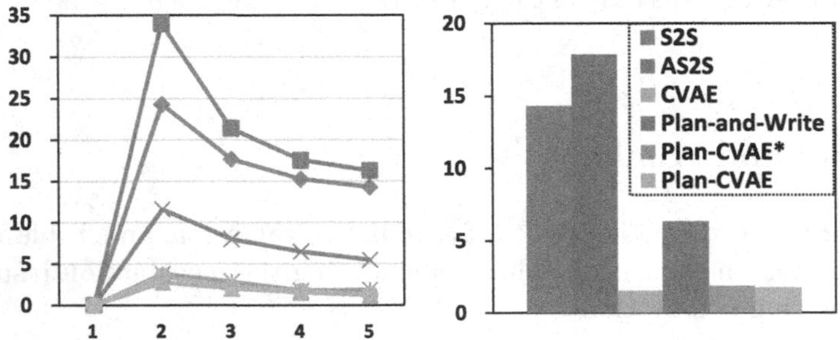

(c) Intra-story repetition curve by sentence.

(d) Intra-story aggregate repetition scores.

Fig. 3. Inter- and intra-story repetition rates by sentences (curves) and for the whole stories (bars). Final results are scaled to [0, 100], the lower the better.

Table 2 and the creativity metric in Table 3, which prove that the planning strategy does not affect the wording novelty. In Fig. 3, we also have a similar observation that both Plan-CVAE* and Plan-CVAE models achieve a quite low inter- and intra-story repetition rates, which means our proposed model can learn to create stories rather than copy and concatenate frequently occurred phrases in the training corpus.

Table 3. Results of human evaluation.

Readability			
Plan-CVAE	44%	**56%**	S2S
Plan-CVAE	**58%**	42%	AS2S
Plan-CVAE	**67%**	33%	CVAE
Plan-CVAE	47%	**53%**	Plan-and-Write
Consistency			
Plan-CVAE	**65%**	35%	S2S
Plan-CVAE	**61%**	39%	AS2S
Plan-CVAE	**84%**	16%	CVAE
Plan-CVAE	**58%**	42%	Plan-and-Write
Creativity			
Plan-CVAE	**93%**	7%	S2S
Plan-CVAE	**86%**	14%	AS2S
Plan-CVAE	**57%**	43%	CVAE
Plan-CVAE	**81%**	19%	Plan-and-Write

6.2 The Effect of CVAE

The CVAE Model can Effectively Improve the Wording Diversity. Plan-CVAE outperforms all baselines (excepts for CVAE) on automatic evaluations including distinct scores in Table 2 and inter- and intra-story repetition rates in Fig. 3, and on creativity score of human evaluation in Table 3. Specifically, all baselines based on RNNs, i.e., S2S, AS2S, and Plan-and-Write, achieve a quite low distinct score and high inter- and intra-story repetition rates, while Plan-CVAE significantly outperforms them by nearly doubling the distinct scores, reducing the repetition rates to about half of theirs, and achieving similar scores with the pure CVAE model. Results on the creativity metric in human evaluation also indicate the same conclusion. These results support the intuition that CVAE can address the poor wording diversity problem of RNN by randomly sampling from the latent space.

The Latent Variable in CVAE Reduces the Readability. CVAE improves the wording diversity by randomly sampling from the latent space. Thus, CVAE produces more uncertainty than RNNs and leads to inferior readability. This intuition is supported by the readability metric in human evaluation (Table 3).

6.3 Case Study

We present two groups of example stories in Table 4 to compare the performance of our proposed method with Plan-and-Write and CVAE since they perform well on either thematic consistency or wording novelty. When compared with

Table 4. Two group of example stories automatically generated by our proposed Plan-CVAE model and baseline methods.

Title: lake fun		
Plan-and-Write	Keywords	friends → lake → water → day → home
	Story	I went to the lake with my friends. We went to a lake. We went to the lake. We went to a water park. We had a great time
CVAE	Story	The kids rode outside of the neighborhood. The mom helped her to get away from the house. They played with the best and hung them out. Finally, they were a couple. Then, they played games
Plan-CVAE	Keywords	lake → friends → water → swim → shore
	Story	One summer day, my friends invited me to a different lake. While walking around the lake, we moved along. It was a great summer and my dad built a lot of stuff. Now the rest of the day I can swim. I was very late and went back to shore
Title: tax evasion		
Plan-and-Write	Keywords	money → taxes → pay → debt → bank
	Story	Bill had a lot of money. Bill had to pay for his taxes. Bill had to pay for his taxes. Bill was able to pay his taxes. Bill was able to pay his taxes
CVAE	Story	Sam and her husband wanted to be a lawyer. He worked hard to make it with his lawyer. He just needed to buy more and more things. He made a lot of money. He told her they both hurt more
Plan-CVAE	Keywords	tax evasion → money → taxes → pay → debt
	Story	Neil had recently moved to a tax preparer. He had applied for a few jobs before the loan. But he didn't notice that the bank was on his list! He was told he owed pay for tax fees. It was about to accept his taxes, but he had no interest

the pure CVAE model without planning strategy, we can observe that stories generated by CVAE are formed with novel words but without consistent topics, while stories generated by Plan-CVAE describe more consistent themes and are also narrated with novel wordings. On the other hand, when compared with the planning-based RNN method, we can see that stories generated by Plan-and-Write contain many repeated words and phrases, while stories composed by Plan-CVAE consist of more diverse words.

Although our proposed method has achieved a promising improvement in thematic consistency and wording diversity than baseline models, there is still a gap between stories generated by our method and humans. We also observed some bad cases generated by our method. These bad cases reflect three major problems, i.e., logical inconsistency, lacking sentiment, and weak readability. As for the logical inconsistency problem, one can introduce an extra control to dynamically adjust the keywords planning and content generation process, or establish a polishing mechanism to check and rewrite the generated content. To solve the sentiment problem, one can utilize a sentiment planning strategy to add sentimental information into the generated stories. In order to improve readability, it is worth to use pre-training strategy on larger related corpus for story generation. Our future work will focus on these issues.

7 Conclusion

In this paper, we proposed a planning-based conditional variational autoencoder model (Plan-CVAE) for story generation. Our proposed method involves two stages. In the planning stage, the keyword-extraction and keyword-expansion modules are used to generate keywords from the title. As for the generation stage, a CVAE neural network module is employed to generate stories under the guidance of keywords. In our method, the planning strategy aims to improve the thematic consistency while the CVAE module is expected to keep the wording diversity of the generated story. Experimental results of both automatic and human evaluations on a benchmark dataset, i.e., ROCStories corpus, show that our method performs better than existing methods on thematic consistency and wording diversity. The case study also confirms the effectiveness of our method.

Acknowledgements. We would like to thank the reviewers for their constructive comments. This work was supported by the National Key Research and Development Program of China (No. 2017YFC0804001), the National Science Foundation of China (NSFC No. 61876196 and NSFC No. 61672058) and the foundation of Key Laboratory of Artificial Intelligence, Ministry of Education, P.R. China. Rui Yan was sponsored by Beijing Academy of Artificial Intelligence (BAAI).

References

1. Bahdanau, D., Cho, K., Bengio, Y.: Neural machine translation by jointly learning to align and translate. CoRR abs/1409.0473 (2015)
2. Bowman, S.R., Vilnis, L., Vinyals, O., Dai, A., Jozefowicz, R., Bengio, S.: Generating sentences from a continuous space. In: Proceedings of The 20th SIGNLL Conference on Computational Natural Language Learning, pp. 10–21. Association for Computational Linguistics, Berlin, August 2016
3. Fan, A., Lewis, M., Dauphin, Y.: Hierarchical neural story generation. In: Proceedings of the 56th Annual Meeting of the Association for Computational Linguistics (Volume 1: Long Papers), pp. 889–898. Association for Computational Linguistics, Melbourne, July 2018
4. Fan, A., Lewis, M., Dauphin, Y.: Strategies for structuring story generation. In: Proceedings of the 57th Annual Meeting of the Association for Computational Linguistics, pp. 2650–2660. Association for Computational Linguistics, Florence, Italy, July 2019

5. Jain, P., Agrawal, P., Mishra, A., Sukhwani, M., Laha, A., Sankaranarayanan, K.: Story generation from sequence of independent short descriptions (2017)
6. Kingma, D.P., Ba, J.: Adam: a method for stochastic optimization. In: Bengio, Y., LeCun, Y. (eds.) 3rd International Conference on Learning Representations, ICLR 2015, San Diego, CA, USA, 7–9 May 2015, Conference Track Proceedings (2015)
7. Kingma, D.P., Welling, M.: Auto-encoding variational bayes. arXiv preprint arXiv:1312.6114 (2013)
8. Li, J., Galley, M., Brockett, C., Gao, J., Dolan, B.: A diversity-promoting objective function for neural conversation models. In: Proceedings of the 2016 Conference of the North American Chapter of the Association for Computational Linguistics: Human Language Technologies, pp. 110–119. Association for Computational Linguistics, San Diego, June 2016
9. Li, J., Bing, L., Qiu, L., Chen, D., Zhao, D., Yan, R.: Learning to write creative stories with thematic consistency. In: AAAI 2019: Thirty-Third AAAI Conference on Artificial Intelligence (2019)
10. Li, J., et al.: Generating classical Chinese poems via conditional variational autoencoder and adversarial training. In: Proceedings of the 2018 Conference on Empirical Methods in Natural Language Processing, pp. 3890–3900. Association for Computational Linguistics, Brussels, October-November 2018
11. Liu, D., et al.: A character-centric neural model for automated story generation. In: AAAI, pp. 1725–1732 (2020)
12. Martin, L.J., Ammanabrolu, P., Hancock, W., Singh, S., Harrison, B., Riedl, M.O.: Event representations for automated story generation with deep neural nets. ArXiv abs/1706.01331 (2018)
13. Mikolov, T., Karafiát, M., Burget, L., Černocký, J., Khudanpur, S.: Recurrent neural network based language model. In: Eleventh Annual Conference of the International Speech Communication Association (2010)
14. Mostafazadeh, N., et al.: A corpus and cloze evaluation for deeper understanding of commonsense stories. In: Proceedings of the 2016 Conference of the North American Chapter of the Association for Computational Linguistics: Human Language Technologies, pp. 839–849. Association for Computational Linguistics, San Diego, June 2016
15. Papineni, K., Roukos, S., Ward, T., Zhu, W.J.: BLEU: a method for automatic evaluation of machine translation. In: Proceedings of the 40th Annual Meeting of the Association for Computational Linguistics, pp. 311–318. Association for Computational Linguistics, Philadelphia, Pennsylvania, July 2002
16. Roemmele, M., Kobayashi, S., Inoue, N., Gordon, A.: An RNN-based binary classifier for the story cloze test. In: Proceedings of the 2nd Workshop on Linking Models of Lexical, Sentential and Discourse-level Semantics, pp. 74–80. Association for Computational Linguistics, Valencia, April 2017
17. Rose, S., Engel, D., Cramer, N., Cowley, W.: Automatic keyword extraction from individual documents, pp. 1–20 (2010)
18. Rush, A.M., Chopra, S., Weston, J.: A neural attention model for abstractive sentence summarization. arXiv preprint arXiv:1509.00685 (2015)
19. Serban, I.V., et al.: A hierarchical latent variable encoder-decoder model for generating dialogues. In: Proceedings of the Thirty-First AAAI Conference on Artificial Intelligence, AAAI 2017, pp. 3295–3301. AAAI Press (2017)
20. Shen, X., et al.: A conditional variational framework for dialog generation. In: Proceedings of the 55th Annual Meeting of the Association for Computational Linguistics (Volume 2: Short Papers), pp. 504–509. Association for Computational Linguistics, Vancouver, July 2017

21. Sutskever, I., Vinyals, O., Le, Q.V.: Sequence to sequence learning with neural networks. In: Advances in Neural Information Processing Systems, pp. 3104–3112 (2014)
22. Vaswani, A., et al.: Attention is all you need. In: Guyon, I., et al. (eds.) Advances in Neural Information Processing Systems, vol. 30, pp. 5998–6008. Curran Associates, Inc. (2017). http://papers.nips.cc/paper/7181-attention-is-all-you-need.pdf
23. Wang, T., Wan, X.: T-CVAE: transformer-based conditioned variational autoencoder for story completion. In: Proceedings of the Twenty-Eighth International Joint Conference on Artificial Intelligence, IJCAI-19, pp. 5233–5239. International Joint Conferences on Artificial Intelligence Organization (2019)
24. Xu, J., Ren, X., Zhang, Y., Zeng, Q., Cai, X., Sun, X.: A skeleton-based model for promoting coherence among sentences in narrative story generation. In: Proceedings of the 2018 Conference on Empirical Methods in Natural Language Processing, pp. 4306–4315. Association for Computational Linguistics, Brussels, October-November 2018
25. Yang, X., Lin, X., Suo, S., Li, M.: Generating thematic Chinese poetry using conditional variational autoencoders with hybrid decoders. In: Proceedings of the 27th International Joint Conference on Artificial Intelligence, IJCAI 2018, pp. 4539-4545. AAAI Press (2018)
26. Yao, L., Peng, N., Weischedel, R., Knight, K., Zhao, D., Yan, R.: Plan-and-write: towards better automatic storytelling. In: Proceedings of the AAAI Conference on Artificial Intelligence, vol. 33, pp. 7378–7385 (2019)
27. Yu, M.H., et al.: Draft and edit: automatic storytelling through multi-pass hierarchical conditional variational autoencoder. In: AAAI, pp. 1741–1748 (2020)
28. Zhang, B., Xiong, D., Su, J., Duan, H., Zhang, M.: Variational neural machine translation. In: Proceedings of the 2016 Conference on Empirical Methods in Natural Language Processing, pp. 521–530. Association for Computational Linguistics, Austin, November 2016
29. Zhao, T., Zhao, R., Eskenazi, M.: Learning discourse-level diversity for neural dialog models using conditional variational autoencoders. In: Proceedings of the 55th Annual Meeting of the Association for Computational Linguistics (Volume 1: Long Papers), pp. 654–664. Association for Computational Linguistics, Vancouver, July 2017

Knowledge Graph and Information Extraction

Towards Causal Explanation Detection with Pyramid Salient-Aware Network

Xinyu Zuo[1,2(✉)], Yubo Chen[1,2], Kang Liu[1,2], and Jun Zhao[1,2]

[1] National Laboratory of Pattern Recognition, Institute of Automation, Chinese Academy of Sciences, Beijing 100190, China
{xinyu.zuo,yubo.chen,kliu,jzhao}@nlpr.ia.ac.cn
[2] School of Artificial Intelligence, University of Chinese Academy of Sciences, Beijing 100049, China

Abstract. Causal explanation analysis (CEA) can assist us to understand the reasons behind daily events, which has been found very helpful for understanding the coherence of messages. In this paper, we focus on *Causal Explanation Detection*, an important subtask of causal explanation analysis, which determines whether a causal explanation exists in one message. We design a **P**yramid **S**alient-**A**ware **N**etwork (PSAN) to detect causal explanations on messages. PSAN can assist in causal explanation detection via capturing the salient semantics of discourses contained in their keywords with a bottom graph-based word-level salient network. Furthermore, PSAN can modify the dominance of discourses via a top attention-based discourse-level salient network to enhance explanatory semantics of messages. The experiments on the commonly used dataset of CEA shows that the PSAN outperforms the state-of-the-art method by 1.8% F1 value on the *Causal Explanation Detection* task.

Keywords: Causal explanation analysis · Causal semantic · Pyramid network.

1 Introduction

Causal explanation detection (CED) aims to detect whether there is a causal explanation in a given message (e.g. a group of sentences). Linguistically, there are coherence relations in messages which explain how the meaning of different textual units can combine to jointly build a discourse meaning for the larger unit. The explanation is an important relation of coherence which refers to the textual unit (e.g. discourse) in a message that expresses explanatory coherent semantics [12]. As shown in Fig. 1, M1 can be divided into three discourses, and D2 is the explanation that expresses the reason why it is advantageous for the equipment to operate at these temperatures. CED is important for tasks that require an understanding of textual expression [25]. For example, for question answering, the answers of questions are most likely to be in a group of sentences that contains causal explanations [22]. Furthermore, the summarization

M. Sun et al. (Eds.): CCL 2020, LNAI 12522, pp. 113–128, 2020.
https://doi.org/10.1007/978-3-030-63031-7_9

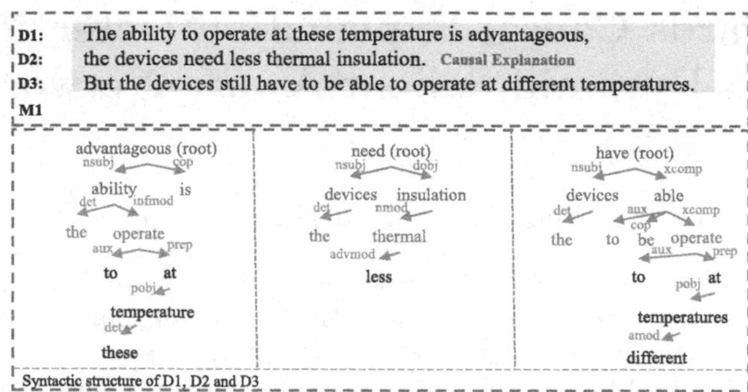

Fig. 1. Instance of causal explanation analysis (CEA). The top part is a message which contains its segmented discourses and a causal explanation. The bottom part is the syntactic dependency structures of three discourses divided from M1.

of event descriptions can be improved by selecting causally motivated sentences [9]. Therefore, CED is a problem worthy of further study.

The existing methods mostly regard this task as a classification problem [25]. At present, there are mainly two kinds of methods, feature-based methods and neural-based methods, for similar semantic understanding tasks in discourse granularity, such as opinion sentiment classification and discourse parsing [11,21,27]. The feature-based methods can extract the feature of the relation between discourses. However, these methods do not deal well with the implicit instances which lack explicit features. For CED, as shown in Fig. 1, D2 lacks explicit features such as *because of*, *due to*, or the features of tenses, which are not friendly for feature-based methods. The methods based on neural network are mainly Tree-LSTM model [30] and hierarchical Bi-LSTM model [25]. The Tree-LSTM models learn the relations between words to capture the semantics of discourses more accurately but lack further understanding of the semantics between discourses. The hierarchical Bi-LSTM models can employ sequence structure to implicitly learn the relations between words and discourses. However, previous work shows that compared with Tree-LSTM, Bi-LSTM lacks a direct understanding of the dependency relations between words. Therefore, the method of implicit learning of inter-word relations is not prominent in the tasks related to understanding the semantic relations of messages [16]. Therefore, how to directly learn the relations between words effectively and consider discourse-level correlation to further filter the key information is a valuable point worth studying.

Further analysis, why do the relations between words imply the semantics of the message and its discourses? From the view of computational semantics, the meaning of a text is not only the meaning of words but also the relation, order, and aggregation of the words. In other simple words is that the meaning of a text is partially based on its syntactic structure [12]. In detail, in CED, the core and subsidiary words of discourses contain their basic semantics. For example, as D1 shown in Fig. 1, according to the word order in syntactic structure, we can capture the *ability* of *temperature* is *advantageous*. We can understand the basic semantic of D1 which expresses some kind of *ability* is *advantageous* via root words *advantageous* and its affiliated words. Additionally, why the correlation and key information at the discourse level are so important to capture the causal explanatory semantics of the message? Through observation, the different discourse has a different status for the explanatory semantics of a message. For example, in M1, combined with D1, D2 expresses the explanatory semantics of *why the ability to work at these temperatures is advantageous*, while D3 expresses the semantic of transition. In detail, D1 and D2 are the keys to the explanatory semantics of M1, and if not treated D1, D2, and D3 differently, the transitional semantic of D3 can affect the understanding of the explanatory semantic of M1. Therefore, how to make better use of the information of keywords in the syntactic structure and pay more attention to the discourses that are key to explanatory semantics is a problem to be solved.

To this end, we propose a **P**yramid **S**alient-**A**ware **N**etworks (PSAN) which utilizes keywords on the syntactic structure of each discourse and focuses on the key discourses that are critical to explanatory semantics to detect causal explanation of messages. First, what are the keywords in a syntactic structure? From the perspective of syntactic dependency, the root word is the central element that dominates other words, while it is not be dominated by any of the other words, all of which are subordinate to the root word [33]. From that, the root and subsidiary words in the dependency structure are the keywords at the syntax level of each discourse. Specifically, we sample 100 positive sentences from training data to illuminate whether the keywords obtained through the syntactic dependency contain the causal explanatory semantics. And we find that the causal explanatory semantics of more than 80% sentences be captured by keywords in dependency structure[1]. Therefore, we extract the root word and its surrounding words on the syntactic dependency of each discourse as its keywords.

Next, we need to consider how to make better use of the information of keywords contained in the syntactic structure. To pay more attention to keywords, the common way is using attention mechanisms to increase the attention weight of them. However, this implicitly learned attention is not very interpretable. Inspired by previous researches [1,29], we propose a bottom graph-based word-level salient network which merges the syntactic dependency to capture the salient semantics of discourses contained in their keywords. Finally, how to consider the correlation at the discourse level and pay more attention to the discourses that are key to the explanatory semantics? Inspired by previous work [18], we propose a top attention-based discourse-level salient network to focus on the key discourses in terms of explanatory semantics.

In summary, the contributions of this paper are as follows:

- We design a **P**yramid **S**alient-**A**ware **N**etwork (PSAN) to detect causal explanations of messages which can effectively learn the pivotal relations between

[1] Five Ph.D. students majoring in NLP judge whether sentences could be identified as which containing causal explanatory semantics by the root word and its surrounding words in syntactic dependency, and the agreement consistency is 0.8.

keywords at word level and further filter the key information at discourse level in terms of explanatory semantics.
- PSAN can assist in causal explanation detection via capturing the salient semantics of discourses contained in their keywords with a bottom graph-based word-level salient network. Furthermore, PSAN can modify the dominance of discourses via a top attention-based discourse-level salient network to enhance explanatory semantics of messages.
- Experimental results on the open-accessed commonly used datasets show that our model achieves the best performance. Our experiments also prove the effectiveness of each module.

2 Related Works

Causal Semantic Detection: Recently, causality detection which detects specific causes and effects and the relations between them has received more attention, such as the researches proposed by Li [17], Zhang [35], Bekoulis [2], Do [5], Riaz [23], Dunietz [6] and Sharp [24]. Specifically, to extract the causal explanation semantics from the messages in a general level, some researches capture the causal semantics in messages from the perspective of discourse structure, such as capturing counterfactual conditionals from a social message with the PDTB discourse relation parsing [26], a pre-trained model with Rhetorical Structure Theory Discourse Treebank (RSTDT) for exploiting discourse structures on movie reviews [10], and a two-step interactive hierarchical Bi-LSTM framework [32] to extract emotion-cause pair in messages. Furthermore, Son [25] defines the causal explanation analysis task (CEA) to extract causal explanatory semantics in messages and annotates a dataset for other downstream tasks. In this paper, we focus on causal explanation detection (CED) which is the fundamental and important subtask of CEA.

Syntactic Dependency with Graph Network: Syntactic dependency is a vital linguistic feature for natural language processing (NLP). There are some researches employ syntactic dependency such as retrieving question answering passage assisted with syntactic dependency [4], mining opinion with syntactic dependency [31] and so on. For tasks related to causal semantics extraction from relevant texts, dependency syntactic information may evoke causal relations between discourse units in text [8]. And recently, there are some researches [20, 34] convert the syntactic dependency into a graph with graph convolutional network (GCN) [14] to effectively capture the syntactic dependency semantics between words in context, such as a semantic role model with GCN [20], a GCN-based model assisted with a syntactic dependency to improving relation extraction [34]. In this paper, we capture the salient explanatory semantics based on the syntactic-centric graph.

3 Methodology

The architecture of our proposed model is illustrated in Fig. 2. In this paper, the Pyramid Salient-Aware Network (PSAN) primarily involves the following three

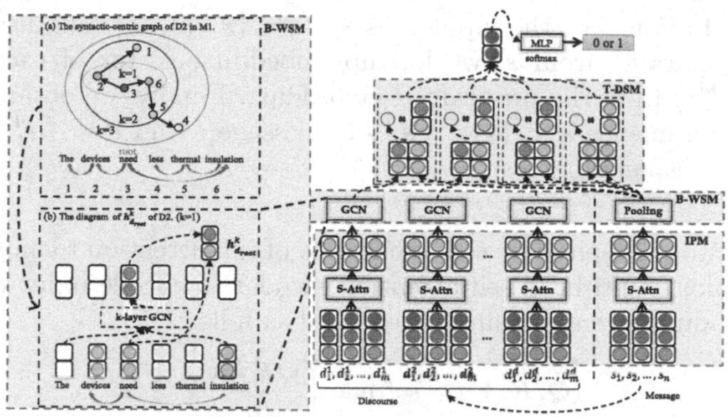

Fig. 2. The structure of PSAN. The left side is the detail of the bottom word-level salient-aware module (B-WSM), the top of right side is the top discourse-level salient-aware module (T-DSM) and the bottom of right side is the input processing module (IPM).

components: (i) **input processing module (IPM)**, which processes and encodes the input message and its discourses via self-attention module; (ii) **bottom word-level salient-aware module (B-WSM)**, which captures the salient semantics of discourses contained in their keywords based on the syntactic-centric graph; (iii) **top discourse-level salient-aware module (T-DSM)**, which modifies the dominance of different discourse based on the message-level constraint in terms of explanatory semantic via an attention mechanism, and obtain the final causal explanatory representation of input message m.

3.1 Input Processing Module

In this component, we split the input message m into discourses d. Specially, we utilize the self-attention encoder to encode input messages and their corresponding discourses.

Discourse Extraction. As shown in Fig. 1, we split the message into discourses with the same segmentation methods as Son [25] based on semantic coherence. In detail, first, we regard (','), ('.'), ('!'), ('?') tags and periods as discourse makers. Next, we also extract the discourse connectives set from PDTB2 as discourse makers. Specifically, we remove some simple connectives (e.g. I like running **and** basketball) from extracted discourse marks. Finally, we divide messages into discourses by the discourse makers.

Embedding Layer. For the input message $s = \{s_1, ..., s_n\}$ and discourse $d = \{d_1^d, ..., d_m^d\}$ separated from s, we lookup embedding vector of each word s_n (d_m^d) as $\boldsymbol{s_n}$ ($\boldsymbol{d_m^d}$) from the pre-trained embedding. Finally, we obtain the word representation sequence $\boldsymbol{s} = \{\boldsymbol{s_1}, ..., \boldsymbol{s_n}\}$ of message s and $\boldsymbol{d} = \{\boldsymbol{d_1^d}, ..., \boldsymbol{d_m^d}\}$ of discourse d corresponding to s.

Word Encoding. Inspired by the application of self-attention to multiple tasks [3,28], we exploit multi-head self-attention encoder to encode input words. The scaled dot-product attention can be described as follows:

$$(\boldsymbol{Q}, \boldsymbol{K}, \boldsymbol{V}) = \text{softmax}\left(\frac{\boldsymbol{Q}\boldsymbol{K}^T}{\sqrt{d}}\right)\boldsymbol{V} \tag{1}$$

where $\boldsymbol{Q} \in \mathbb{R}^{N \times 2dim_h}$, $\boldsymbol{K} \in \mathbb{R}^{N \times 2dim_h}$ and $\boldsymbol{V} \in \mathbb{R}^{N \times 2dim_h}$ are query matrices, keys matrices and value matrices, respectively. In our setting, $\boldsymbol{Q} = \boldsymbol{K} = \boldsymbol{V} = \boldsymbol{s}$ for encoding sentence, and $\boldsymbol{Q} = \boldsymbol{K} = \boldsymbol{V} = \boldsymbol{d}$ for encoding discourse.

Multi-head attention first projects the queries, keys, and values h times by using different linear projections. The results of attention are concatenated and once again projected to get the final representation. The formulas are as following:

$$head_i = \text{Attention}\left(\mathbf{Q}\mathbf{W}_i^Q, \mathbf{K}\mathbf{W}_i^K, \mathbf{V}\mathbf{W}_i^V\right) \tag{2}$$

$$\mathbf{H}' = (head_i \oplus \ldots \oplus head_h)\,\mathbf{W}_o \tag{3}$$

where, $\mathbf{W}_i^Q \in \mathbb{R}^{2dim_h \times dim_k}$, $\mathbf{W}_i^K \in \mathbb{R}^{2dim_h \times dim_k}$, $\mathbf{W}_i^V \in \mathbb{R}^{2dim_h \times dim_k}$ and $\mathbf{W}_o \in \mathbb{R}^{2dim_h \times 2dim_h}$ are projection parameters and $dim_k = 2dim_h/h$. And the output is the encoded message $\boldsymbol{H}_S^{ed} = \{\boldsymbol{h}_{s_1}^{ed}, ..., \boldsymbol{h}_{s_n}^{ed}\}$ and discourse $\boldsymbol{H}_{D^d}^{ed} = \{\boldsymbol{h}_{d_1^d}^{ed}, ..., \boldsymbol{h}_{d_m^d}^{ed}\}$.

3.2 Bottom Word-Level Salient-Aware Module

In this component, we aim to capture the salient semantics of discourses contained in their keywords based on syntactic-centric graphs. For each discourse, first, it extracts the syntactic dependency and constructs the syntactic-centric graph. Second, it collects the keywords and their inter-relations to capture the discourse-level salient semantic based on the syntactic-centric graph.

Syntactic-Centric Graph Construction. We construct a syntactic-centric graph of each discourse based on syntactic dependency to assist in capturing the semantics of discourses. We utilize Stanford CoreNLP tool[2] to extract the syntactic dependency of each discourse and convert them into syntactic-centric graphs. Specifically, in the syntactic-centric graph, the nodes represent words, and the edges represent whether there is a dependency relation between two words or not. As shown in the subplot (a) of Fig. 2, *need* is the root word in the syntactic dependency of *"the devices need less thermal insulation"* (D2 in S1), and words which are syntactically dependent on each other are connected with solid lines.

[2] https://stanfordnlp.github.io/CoreNLP/.

Keywords Collection and Salient Semantic Extraction. For each discourse, we collect the keywords based on the syntactic-centric graph and capture the salient semantic based on the syntactic-centric graph from its keywords. Firstly, as illustrated in Sect. 1, we combine the root word and the affiliated words that connected with the root word in k hops as the keywords. For example, as shown in Fig. 2, when $k = 1$, the keywords are {*need, devices and insulation*}, and the keywords are {*need, devices, insulation, the and thermal*} when $k = 2$. Secondly, inspired by previous works, we utilize k-layer graph convolutional network (GCN) [14] to encode the k hops connected keywords based on the syntactic-centric graph. For example, when $k = 1$, we encode 1-hop keywords with 1-layer GCN to capture the salient semantic. Specifically, we can capture different degrees of salient semantics by changing the value of k. However, it is not the larger the value of k, the deeper the salient semantics are captured. Conversely, the larger the k, the more noises are likely to be introduced. For example, when $k = 1$, *need*, *devices* and *insulation* are enough to express the salient semantic of D2 (working at these temperatures need less insulation). Finally, we select the representation of the root word in the final layer as the discourse-level representation which contains the salient semantic.

The graph convolutional network (GCN) [14] is a generalization of CNN [15] for encoding graphs. In detail, given a syntactic-centric graph with v nodes, we utilize an $v \times v$ adjacency matrix $\boldsymbol{A}$, where $A_{ij} = 1$ if there is an edge between node i and node j. In each layer of GCN, for each node, the input is the output $\boldsymbol{h}_i^{k-1}$ of the previous layer (the input of the first layer is the original encoded input words and features) and the output of node i at k-th layer is $\boldsymbol{h}_i^k$, the formula is as following:

$$\boldsymbol{h}_i^k = \sigma\left(\sum_{j=1}^{v} A_{ij} W^k \boldsymbol{h}_j^{k-1} + b^k\right) \tag{4}$$

where W^k is the matrice of linear transformation, b^k is a bias term and σ is a nonlinear function.

However, naively applying the graph convolution operation in Equation (3) could lead to node representations with drastically different magnitudes because the degree of a token varies a lot. This issue may cause the information in h_i^{k-1} is never carried over to h_i^k because nodes never connect to themselves in a dependency graph [34]. In order to resolve the issue that the information in h_i^{k-1} may be never carried over to h_i^k due to the disconnection between nodes in a dependency graph, we utilize the method raised by Zhang [34] which normalizes the activations in the GCN, and adds self-loops to each node in graph:

$$\boldsymbol{h}_i^k = \sigma\left(\sum_{j=1}^{v} \tilde{A}_{ij} W^k \boldsymbol{h}_j^{k-1} / d_i + b^k\right), \tag{5}$$

where $\tilde{\mathbf{A}} = \mathbf{A} + \mathbf{I}$, $\mathbf{I}$ is the $v \times v$ identify matrix and $d_i = \sum_{j=1}^{v} \tilde{A}_{ij}$ is the degree of word i in graph.

Finally, We select the representation $\boldsymbol{h}_{d_{root}}^k$ of the root word in final layer GCN as the salient representation of d-th discourse in message s. For example, as shown in the subplot (b) of Fig. 2, we choose the representation of *need* in the final layer as the salient representation of the discourse *"the devices need less thermal insulation"*.

3.3 Top Discourse-Level Salient-Aware Module

How to make better use of the relation between discourse and extract the message-level salient semantic? We modify the dominance of different discourse based on the message-level constraint in terms of explanatory semantic via an attention mechanism. First, we extract the global semantic of message s which contains its causal explanatory tendency. Next, we modify the dominance of different discourse based on global semantic. Finally, we combine the modified representation to obtain the final causal explanatory representation of input message s.

Global Semantic Extraction. Inspired by previous research [25], the average encoded word representation of all the words in message can represent its overall semantic simply and effectively. We utilize the average pooling on the encoded representation $\boldsymbol{H}_S^{ed}$ of message s to obtain the global representation which contains the global semantic of its causal explanatory tendency. The formula is as following:

$$\boldsymbol{h}_s^{glo} = \sum_{\boldsymbol{h}_s^{ed} \in \boldsymbol{H}_S^{ed}} \boldsymbol{h}_s^{ed}/n, \tag{6}$$

where $\boldsymbol{h}_s^{glo}$ is the global representation of message s via average pooling operation and n is the number of words.

Dominance Modification. We modify the dominance of different discourse based on the global semantic which contains its causal explanatory tendency via an attention mechanism. In detail, after obtaining the global representation $\boldsymbol{h}_s^{glo}$, we modify the salient representation $\boldsymbol{h}_{d_{root}}^k$ of discourses d constrained with $\boldsymbol{h}_s^{glo}$. Finally, we obtain final causal representation $\boldsymbol{h}_s^{caul}$ of message s via attention mechanism:

$$\alpha_{ss} = \boldsymbol{h}_s^{glo}\boldsymbol{W}_f(\boldsymbol{h}_s^{glo})^T \tag{7}$$

$$\alpha_{sd} = \boldsymbol{h}_s^{glo}\boldsymbol{W}_f(\boldsymbol{h}_{d_{root}}^k)^T \tag{8}$$

$$[\alpha_{ss}^{'}, \cdots, \alpha_{sd}^{'}] = softmax([\alpha_{ss}, ..., \alpha_{sd}]) \tag{9}$$

$$\boldsymbol{h}_s^{caul} = \alpha_{ss}^{'}\boldsymbol{h}_s^{glo} + ... + \alpha_{sd}^{'}\boldsymbol{h}_{d_{root}}^k, \tag{10}$$

where the $\boldsymbol{W}_f$ is matrice of linear transformation, $\alpha_{ss}^{'}$, $\alpha_{sd}^{'}$ are the attention weight. Finally, we mapping $\boldsymbol{h}_s^{caul}$ into a binary vector and get the output via a softmax operation.

4 Experiment

Dataset. We mainly evaluate our model on a unique dataset devoted to causal explanation analysis released by Son [25]. This dataset contains 3,268 messages consist of 1598 positive messages that contain a causal explanation and 1670 negative sentences randomly selected. Annotators annotate which messages contain causal explanations and which text spans are causal explanations (a discourse with a tendency to interpret something). We utilize the same 80% of the dataset for training, 10% for tuning, and 10% for evaluating as Son [25]. Additionally, to further prove the effectiveness of our proposed model, we regard sentences with causal semantic discourse relations in PDTB2 and sentences containing causal span pairs in BECauSE Corpus 2.0 [7] as supplemental messages with causal explanations to evaluate our model. In this paper, PDTB-CED and BECauSE-CED are used to represent the two supplementary datasets respectively.

Parameter Settings. We set the length of the sentence and discourse as 100 and 30 respectively. We set the batch size as 5 and the dimension of the output in each GCN layer as 50. Additionally, we utilize the 50-dimension word vector pre-trained with Glove. For optimization, we utilize Adam [13] with 0.001 learning rate. We set the maximum training epoch as 100 and adopt an early stop strategy based on the performance of the development set. All the results of different compared and ablated models are the average result of five independent experiments.

Compared Models. We compare our proposed model with feature-based and neural-based model: (1) **Lin et al.** [19]: an end-to-end discourse relation parser on PDTB, (2) **Linear SVM:** a linear designed feature based SVM classifier, (3) **RBF SVM**: a complex designed feature based SVM classifier, (4) **Random Forest:** a random forest classifier which relies on designed features, (5) **Son et al.** [25]: a hierarchical LSTM sequence model which is designed specifically for CEA. (6) **H-BiLSTM + BERT**[3,4]: a fine-tuned language model (BERT) which has been shown to improve the performance in some other classification tasks based on (5), (7) **H-Atten.:** a well-used Bi-LSTM model that captures hierarchical key information based on hierarchical attention mechanism, (8) **Our model:** our proposed pyramid salient-aware network (PSAN). Furthermore, we evaluate the performance of the model (5), (7), and (8) on the supplemental dataset to prove the effectiveness of our proposed model. Additionally, we design different ablation experiments to demonstrate the effectiveness of the bottom word-level salient-aware module (B-WSM), top discourse-level salient-aware module (T-DSM), and the influence of different depths in the syntactic-centric graph.

[3] https://github.com/huggingface/transformers.

[4] BERT can not be applied to the feature-based model suitably, so we deploy BERT on the latest neural model to make the comparison to prove the effectiveness of our proposed model.

4.1 Main Results

Table 1. Comparisons of the state-of-the-art methods on causal explanation detection.

Model	F1	F1	F1
	Facebook	PDTB-CED	BEcuasE-CED
Lin et al. [19]	63.8	–	–
Linear SVM [25]	79.1	–	–
RBF SVM [25]	77.7	–	–
Random Forest [25]	77.1	–	–
Son et al. [25]	75.8	63.6	69.6
H-Atten.	80.9	70.6	76.5
H-BiLSTM + BERT	85.0	–	–
Our model	**86.8**	**76.6**	**81.7**

Table 1 shows the comparison results on the Facebook dataset and two supplementary datasets. From the results, we have the following observations.

(1) Comparing with the current best feature-based and neural-based models on CED: **Lin et al.** [19], **Linear SVM** and **Son et al.** [25], **our model** improves the performance by 23.0, 7.7 and 11.0 points on F1, respectively. It illustrates that the pyramid salient-aware network (PSAN) can effectively extract and incorporate the word-level key relation and discourse-level key information in terms of explanatory semantics to detect causal explanation. Furthermore, comparing with the well-used hierarchical key information captured model (**H-Atten.**), **our model** improves the performance by 5.9 points on F1. This confirms the statement in Sect. 1 that directly employing the relation between words with syntactic structure is more effective than the implicit learning.

(2) Comparing the **Son et al.** [25] with pre-trained language model (**H-BiLSTM+BERT**), there is 9.2 points improvement on F1. It illustrates that the pre-trained language model (LM) can capture some causal explanatory semantics with the large-scale corpus. Furthermore, **our model** can further improve performance by 1.8 points compared with **H-BiLSTM+BERT**. We believe the reason is that the LM is pre-trained with large-scale regular sentences that do not contain causal semantics only, which is not specifically suitable for CED compared to the proposed model for explanatory semantic. Furthermore, the performance of **H-Atten.** is better than **Son et al.** [25] which indicates focusing on salient keywords and key discourses helps understand explanatory semantics.

(3) It is worth noting that, regardless of our proposed model, comparing the **Linear SVM** with **Son et al.** [25], the simple feature classifier is better than the simple deep learning model for CED on the Facebook dataset.

However, when combining the syntactic-centric features with deep learning, we could achieve a significant improvement. In other words, our model can effectively combine the *interpretable information* of the feature-based model with the *deep understanding* of the deep learning model.

(4) To further prove the effectiveness of the proposed model, we evaluate **our model** on supplemental messages with causal semantics in other datasets (PDTB-CED and BEcausE-CED). As shown in Table 1, the results show that the proposed model performs significantly better than the **Son et al.** [25] and **H-Atten.** on the other two datasets[5]. It further demonstrates the effectiveness of our proposed model.

(5) Moreover, **our model** is twice as fast as the **Son et al.** [25] during training because of the computation of self-attention and GCN is parallel. It illustrates that our model can consume less time and achieve significant improvement in causal explanation detection. Moreover, compared with the feature-based models, the neural-based models rely less on artificial design features.

4.2 Effectiveness of Bottom Word-Level Salient-Aware Module (B-WSM)

Table 2 tries to show the effectiveness of the salient information contained in the keywords of each discourse captured via the proposed B-WSM for causal explanation detection (Sect. 3.2). The results illustrate B-WSM can effectively capture the salient information which contains the most causal explanatory semantics. It is worth noting that when using the average encoded-word representation to represent each discourse (**w/o B-WSM + ave**), the model also achieves acceptable performance. This confirms the conclusion from Son [25] that the average word representation at word level contains certain causal explanatory semantic. Furthermore, only the root word of each discourse also contains some causal semantics (**w/o B-WSM + root**) which proves the effectiveness of capturing salient information via syntactic dependency from the keywords.

Table 2. Effectiveness of B-WSM. (**w/o** B-WSM denotes the models without B-WSM. + denotes repalcing the B-WSM with the module after +. **root** denotes using the encoded representation of the root word in each discourse to represent it. **ave** denotes using the average encoded representation of words in discourse to represent it.)

Dataset	Facebook		PDTB-CED		BEcausE-CED	
Model	F1	∇	F1	∇	**F1**	∇
Our model	**86.8**	–	**76.6**	–	**81.7**	–
w/o B-WSM + root	80.1	−6.7	69.9	−6.7	75.8	−5.9
w/o B-WSM + ave	84.7	−2.1	74.4	−2.2	79.8	−1.9

[5] We obtain the performance with the publicly released code by Son et al. [25]. The supplementary datasets are not specifically suitable for this task, and the architectural details of designed feature-based models are not public, so we only compare the performance of the latest model to prove the effectiveness of our proposed model.

4.3 Effectiveness of Top Discourse-Level Salient-Aware Module (T-DSM)

Table 3 tries to show the effectiveness of the salient information of the key discourses modified and incorporated via T-DSM for causal explanation detection (Sect. 3.3). The results compared with **w/o T-DSM + seq D** illustrate our T-DSM can effectively modify the dominance of different discourses based on the global semantic constraint via an attention mechanism to enhance the causal explanatory semantic. Specifically, the results of **w/o T-DSM + ave S/D** show that both discourse-level representation and global representation contain efficient causal explanatory semantics, which further proves the effectiveness of the proposed T-DSM.

Table 3. Effectiveness of T-DSM. (**w/o** T-DSM denotes models without T-DSM. + denotes replacing the T-DSM with the module after +. **seq D** denotes mapping the representation of discourses via a sequence LSTM to represent the whole message. **ave S/D** denotes using the average encoded representation of words in message and its discourses to represent the whole message.)

Dataset	Facebook		PDTB-CED		BEcausE-CED	
Model	F1	∇	F1	∇	F1	∇
Our model	**86.8**	–	**76.6**	–	**81.7**	–
w/o T-DSM + seq D	83.8	−3.0	72.9	−3.7	78.1	−3.6
w/o T-DSM + ave S/D	84.0	−2.8	73.5	−3.1	77.8	−3.9

4.4 Comparisons of Different Depths of Syntactic-Centric Semantic

To demonstrate the influence of the causal explanatory semantics contained in the syntactic-centric graph with different depths, we further compare the performance of our proposed model with a different number of GCN layers. As shown in Fig. 3, when the number of GCN layers is 2, the most efficient syntactic-centric information can be captured for causal explanation detection.

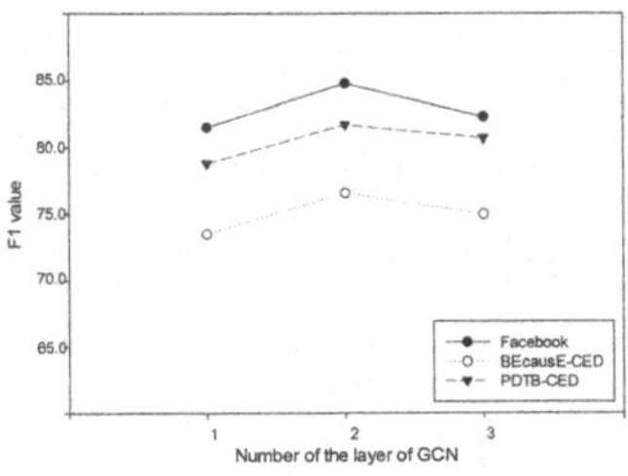

Fig. 3. Comparisons of different number of GCN layers.

4.5 Error Analysis

As shown in Fig. 4, we find the two main difficulties in this task:

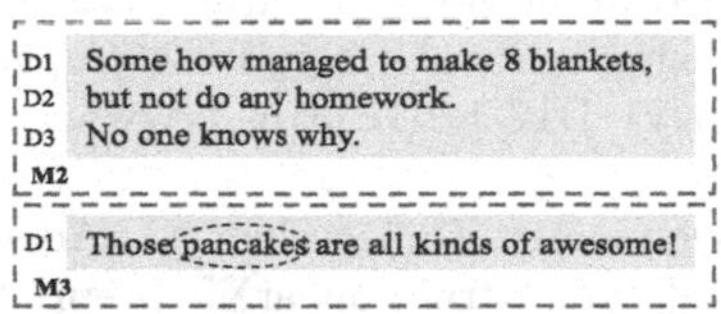

Fig. 4. Predictions of the proposed model.

(1) **Emotional tendency** The same expression can convey different semantic under different emotional tendencies, especially in this kind of colloquial expressions. As M2 shown in Fig. 4, *make 8 blankets* expresses *anger* over *not do any homework*, and our model wrongly predicts the *make 8 blankets* is the reason for *not do any homework*.

(2) **Excessive semantic parsing** Excessive parsing of causal intent by the model will lead to identifying messages that do not contain causal explanations as containing. As shown in Fig. 4, M3 means pancakes are awesome, but the model overstates the reason for *awesome* is a pancake.

5 Conclusion

In this paper, we devise a pyramid salient-aware network (PSAN) to detect causal explanations in messages. PSAN can effectively learn the key relation between words at the word level and further filter out the key information at the discourse level in terms of explanatory semantics. Specifically, we propose a bottom word-level salient-aware module to capture the salient semantics of discourses contained in their keywords based on a the syntactic-centric graph. We also propose a top discourse-level salient-aware module to modify the dominance of different discourses in terms of global explanatory semantic constraint via an attention mechanism. Experimental results on the open-accessed commonly used datasets show that our model achieves the best performance.

Acknowledgements. This work is supported by the Natural Key RD Program of China (No. 2018YFB1005100), the National Natural Science Foundation of China (No. 61533018, No. 61922085, No. 619- 76211, No. 61806201) and the Key Research Program of the Chinese Academy of Sciences (Grant NO. ZDBS-SSW-JSC006). This work is also supported by Beijing Academy of Artificial Intelligence (BAAI2019QN0301), CCF-Tencent Open Research Fund and independent research project of National Laboratory of Pattern Recognition.

References

1. Bastings, J., Titov, I., Aziz, W., Marcheggiani, D., Sima'an, K.: Graph convolutional encoders for syntax-aware neural machine translation. In: Proceedings of the 2017 Conference on Empirical Methods in Natural Language Processing, pp. 1957–1967. Association for Computational Linguistics, Copenhagen, September 2017. https://doi.org/10.18653/v1/D17-1209. https://www.aclweb.org/anthology/D17-1209
2. Bekoulis, G., Deleu, J., Demeester, T., Develder, C.: Adversarial training for multi-context joint entity and relation extraction. arXiv preprint arXiv:1808.06876 (2018)
3. Cao, P., Chen, Y., Liu, K., Zhao, J., Liu, S.: Adversarial transfer learning for Chinese named entity recognition with self-attention mechanism. In: Empirical Methods in Natural Language Processing, pp. 182–192. Association for Computational Linguistics, Brussels (2018)
4. Cui, H., Sun, R., Li, K., Kan, M.Y., Chua, T.S.: Question answering passage retrieval using dependency relations. In: ACM SIGIR, pp. 400–407. ACM (2005)
5. Do, Q.X., Chan, Y.S., Roth, D.: Minimally supervised event causality identification. In: Empirical Methods in Natural Language Processing, pp. 294–303. Association for Computational Linguistics (2011)
6. Dunietz, J., Levin, L., Carbonell, J.: Automatically tagging constructions of causation and their slot-fillers. TACL **5**, 117–133 (2017)
7. Dunietz, J., Levin, L., Carbonell, J.: The BECauSE corpus 2.0: annotating causality and overlapping relations. In: Proceedings of the 11th Linguistic Annotation Workshop, pp. 95–104. Association for Computational Linguistics, Valencia, April 2017. https://doi.org/10.18653/v1/W17-0812. https://www.aclweb.org/anthology/W17-0812
8. Gao, L., Choubey, P.K., Huang, R.: Modeling document-level causal structures for event causal relation identification. In: North American Chapter of the Association for Computational Linguistics, pp. 1808–1817. Association for Computational Linguistics, Minneapolis, June 2019
9. Hidey, C., McKeown, K.: Identifying causal relations using parallel Wikipedia articles. In: Proceedings of the 54th Annual Meeting of the Association for Computational Linguistics (Volume 1: Long Papers), pp. 1424–1433. Association for Computational Linguistics, Berlin, August 2016. https://doi.org/10.18653/v1/P16-1135. https://www.aclweb.org/anthology/P16-1135
10. Ji, Y., Smith, N.: Neural discourse structure for text categorization. arXiv preprint arXiv:1702.01829 (2017)
11. Jia, Y., Ye, Y., Feng, Y., Lai, Y., Yan, R., Zhao, D.: Modeling discourse cohesion for discourse parsing via memory network. In: Proceedings of the 56th Annual Meeting of the Association for Computational Linguistics (Volume 2: Short Papers), pp. 438–443. Association for Computational Linguistics, Melbourne, July 2018. https://doi.org/10.18653/v1/P18-2070. https://www.aclweb.org/anthology/P18-2070
12. Jurafsky, D.: Speech and language processing: an introduction to natural language (2010)
13. Kingma, D.P., Ba, J.: Adam: a method for stochastic optimization. arXiv preprint arXiv:1412.6980 (2014)
14. Kipf, T.N., Welling, M.: Semi-supervised classification with graph convolutional networks. arXiv preprint arXiv:1609.02907 (2016)

15. LeCun, Y., Bottou, L., Bengio, Y., Haffner, P., et al.: Gradient-based learning applied to document recognition. Proc. IEEE **86**(11), 2278–2324 (1998)
16. Li, J., Luong, T., Jurafsky, D., Hovy, E.: When are tree structures necessary for deep learning of representations? In: Proceedings of the 2015 Conference on Empirical Methods in Natural Language Processing, pp. 2304–2314. Association for Computational Linguistics, Lisbon, September 2015. https://doi.org/10.18653/v1/D15-1278. https://www.aclweb.org/anthology/D15-1278
17. Li, P., Mao, K.: Knowledge-oriented convolutional neural network for causal relation extraction from natural language texts. Expert Syst. Appl. **115**, 512–523 (2019)
18. Li, Q., Li, T., Chang, B.: Discourse parsing with attention-based hierarchical neural networks. In: Proceedings of the 2016 Conference on Empirical Methods in Natural Language Processing, pp. 362–371. Association for Computational Linguistics, Austin, November 2016. https://doi.org/10.18653/v1/D16-1035. https://www.aclweb.org/anthology/D16-1035
19. Lin, Z., Ng, H.T., Kan, M.Y.: A pdtb-styled end-to-end discourse parser. Nat. Lang. Eng. **20**(2), 151–184 (2014)
20. Marcheggiani, D., Titov, I.: Encoding sentences with graph convolutional networks for semantic role labeling. In: Empirical Methods in Natural Language Processing, pp. 1506–1515, September 2017
21. Nejat, B., Carenini, G., Ng, R.: Exploring joint neural model for sentence level discourse parsing and sentiment analysis. In: Proceedings of the 18th Annual SIGdial Meeting on Discourse and Dialogue, pp. 289–298. Association for Computational Linguistics, Saarbrücken, August 2017. https://doi.org/10.18653/v1/W17-5535. https://www.aclweb.org/anthology/W17-5535
22. Oh, J.H., Torisawa, K., Hashimoto, C., Sano, M., De Saeger, S., Ohtake, K.: Why-question answering using intra-and inter-sentential causal relations. In: Association for Computational Linguistics, vol. 1, pp. 1733–1743 (2013)
23. Riaz, M., Girju, R.: In-depth exploitation of noun and verb semantics to identify causation in verb-noun pairs. In: SIGDIAL, pp. 161–170 (2014)
24. Sharp, R., Surdeanu, M., Jansen, P., Clark, P., Hammond, M.: Creating causal embeddings for question answering with minimal supervision. arXiv preprint arXiv:1609.08097 (2016)
25. Son, Y., Bayas, N., Schwartz, H.A.: Causal explanation analysis on social media. In: Empirical Methods in Natural Language Processing (2018)
26. Son, Y., et al.: Recognizing counterfactual thinking in social media texts. In: Association for Computational Linguistics, pp. 654–658 (2017)
27. Soricut, R., Marcu, D.: Sentence level discourse parsing using syntactic and lexical information. In: Proceedings of the 2003 Human Language Technology Conference of the North American Chapter of the Association for Computational Linguistics, pp. 228–235 (2003). https://www.aclweb.org/anthology/N03-1030
28. Tan, Z., Wang, M., Xie, J., Chen, Y., Shi, X.: Deep semantic role labeling with self-attention. In: AAAI (2018)
29. Vashishth, S., Bhandari, M., Yadav, P., Rai, P., Bhattacharyya, C., Talukdar, P.: Incorporating syntactic and semantic information in word embeddings using graph convolutional networks. In: Proceedings of the 57th Annual Meeting of the Association for Computational Linguistics, pp. 3308–3318. Association for Computational Linguistics, Florence, July 2019. https://doi.org/10.18653/v1/P19-1320. https://www.aclweb.org/anthology/P19-1320

30. Wang, Y., Li, S., Yang, J., Sun, X., Wang, H.: Tag-enhanced tree-structured neural networks for implicit discourse relation classification. In: Proceedings of the Eighth International Joint Conference on Natural Language Processing (Volume 1: Long Papers), pp. 496–505. Asian Federation of Natural Language Processing, Taipei, Taiwan, November 2017. https://www.aclweb.org/anthology/I17-1050
31. Wu, Y., Zhang, Q., Huang, X., Wu, L.: Phrase dependency parsing for opinion mining. In: Empirical Methods in Natural Language Processing, pp. 1533–1541. Association for Computational Linguistics, Singapore, August 2009
32. Xia, R., Ding, Z.: Emotion-cause pair extraction: a new task to emotion analysis in texts. In: Association for Computational Linguistics, pp. 1003–1012. Association for Computational Linguistics, Florence, Italy, July 2019. https://doi.org/10.18653/v1/P19-1096. https://www.aclweb.org/anthology/P19-1096
33. Zhang, X., Zong, C.: Statistical natural language processing (second edition). Mach. Transl. **28**(2), 155–158 (2014)
34. Zhang, Y., Qi, P., Manning, C.D.: Graph convolution over pruned dependency trees improves relation extraction. In: Empirical Methods in Natural Language Processing, pp. 2205–2215 (2018)
35. Zhang, Y., Zhong, V., Chen, D., Angeli, G., Manning, C.D.: Position-aware attention and supervised data improve slot filling. In: Empirical Methods in Natural Language Processing, pp. 35–45 (2017)

Named Entity Recognition with Context-Aware Dictionary Knowledge

Chuhan Wu[1(✉)], Fangzhao Wu[2], Tao Qi[1], and Yongfeng Huang[1]

[1] Department of Electronic Engineering and BNRist, Tsinghua University, Beijing 100084, China
wuchuhan15@gmail.com, taoqi.qt@gmail.com, yfhuang@tsinghua.edu.cn
[2] Microsoft Research Asia, Beijing 100080, China
wufangzhao@gmail.com

Abstract. Named entity recognition (NER) is an important task in the natural language processing field. Existing NER methods heavily rely on labeled data for model training, and their performance on rare entities is usually unsatisfactory. Entity dictionaries can cover many entities including both popular ones and rare ones, and are useful for NER. However, many entity names are context-dependent and it is not optimal to directly apply dictionaries without considering the context. In this paper, we propose a neural NER approach which can exploit dictionary knowledge with contextual information. We propose to learn context-aware dictionary knowledge by modeling the interactions between the entities in dictionaries and their contexts via context-dictionary attention. In addition, we propose an auxiliary term classification task to predict the types of the matched entity names, and jointly train it with the NER model to fuse both contexts and dictionary knowledge into NER. Extensive experiments on the CoNLL-2003 benchmark dataset validate the effectiveness of our approach in exploiting entity dictionaries to improve the performance of various NER models.

Keywords: Named entity recognition · Dictionary knowledge · Context-aware

1 Introduction

Named entity recognition (NER) aims to extract entity names from texts and classify them into several pre-defined categories, such as person, location and organization [15]. It is an important task in natural language processing, and a prerequisite for many downstream applications such as entity linking [8] and relation extraction [19,21,39]. Thus, NER is a hot research topic. In this paper, we focus on the English NER task.

Many methods have been proposed for English NER, and most of them model this task as a word-level sequence labeling problem [4]. For example, Ma and

M. Sun et al. (Eds.): CCL 2020, LNAI 12522, pp. 129–143, 2020.
https://doi.org/10.1007/978-3-030-63031-7_10

Table 1. Two examples of context-dependent entities.

1	Jordan won against Houston He will give talks in **Jordan** and **Houston**	Red: PER Orange: ORG Blue: LOC
2	Brown is the former prime minister **Brown** shoes are my favourate	

Hovy [22] proposed a CNN-LSTM-CRF model for English NER. They used CNN to learn word representations from characters, LSTM to model the contexts of words, and CRF to decode labels. These existing NER methods usually rely on massive labeled data for model training, which is costly and time-consuming to annotate. When training data is scarce, their performance usually significantly declines [26]. In addition, their performance on recognizing entities that rarely or do not appear in training data is usually unsatisfactory [37].

Fortunately, many large-scale entity dictionaries such as Wikipedia [11] and Geonames[1] are off-the-shelf, and they can be easily derived from knowledge bases and webpages [24]. These entity dictionaries contain both popular and rare entity names, and can provide important information for NER models to identify these entity names. There are a few researches on incorporating entity dictionary into NER [20,23] and most of them are based on dictionary matching features. For example, Wang et al. [37] proposed to combine token matching features with token embeddings and LSTM outputs. However, in many cases entities are context-dependent. For instance, in Table 1, the word "Jordan" can be a person name or a location name in different contexts. Thus, it is not optimal to directly apply entity dictionaries to NER without considering the contexts.

In this paper, we propose a neural approach for named entity recognition with context-aware dictionary knowledge (CADK). We propose to exploit dictionary knowledge in a context-aware manner by modeling the relatedness between the entity names matched by entity dictionaries and their contexts. In addition, we propose an auxiliary term classification task to predict the types of the matched entity names in different contexts. Besides, we propose a unified framework to jointly train the NER model and the term classification model to incorporate entity dictionary knowledge and contextual information into the NER model. Extensive experiments show our approach can effectively exploit entity dictionaries to improve the performance of various NER models and reduce their dependence on labeled data.

2 Related Work

Named entity recognition is usually modeled as a sequence labeling problem [36]. Many traditional NER methods are based on statistical sequence modeling methods, such as Hidden Markov Models (HMM) and Conditional Random Fields

[1] https://www.geonames.org.

(CRF) [2,6,25,29]. Usually, a core problem in these methods is how to build the feature vector for each word, and these features are traditionally constructed via manual feature engineering [29]. For example, Ratinov and Roth [29] used many features such as word n-grams, gazetteers and prediction histories as the word features. Passos et al. [25] used features such as character n-grams, word types, capitalization pattern and lexicon matching features. They also incorporated lexicon embedding learned by skip-gram model to enhance the word representations. Designing these hand-crafted features usually needs a huge amount of domain knowledge. In addition, the feature vectors may be very sparse and their dimensions can be huge.

In recent years, many neural network based NER methods have been proposed [1,3–5,7,9,14,16,17,22,27,28,30,40]. For example, Lample et al. [14] proposed to use LSTM to learn the contextual representation of each token based on global context in sentences and use CRF for joint label decoding. Chiu and Nichols [4] proposed to use CNN to learn word representations from original characters and then learn contextual word representation using Bi-LSTM. Ma and Hovy [22] proposed to combine the CNN-LSTM framework with CRF for better performance. Peters et al. [27] proposed a semi-supervised approach named TagLM for NER by pre-training a language model on a large corpus to provide contextualized word representations. Devlin et al. [9] proposed a bidirectional pre-trained language model named BERT, which can empower downstream tasks like NER by using deep Transformers [35] to model contexts accurately. However, these neural network based methods heavily rely on labeled sentences to train NER models, which need heavy effort of manual annotation. In addition, their performance on recognizing entities which rarely or do not appear in labeled data is usually unsatisfactory [37].

There are several approaches on utilizing entity dictionaries for named entity recognition [6,18,20,25,31,32,37,38]. In traditional methods, dictionaries are often incorporated as additional features. For example, Cohen et al. [6] proposed to extract dictionary features based on entity matching and similarities, and they incorporated these features into an HMM based model. There are also a few methods to incorporate dictionary knowledge into neural NER models [4,20,37]. For example, Wang et al. [37] proposed to incorporate dictionaries into neural NER model for detecting clinical entities. They manually designed several features based on the matches with a clinical dictionary and then concatenated these features with the embedding vector as the input of the LSTM-CRF model. These methods rely on domain knowledge to design these dictionary based features, and these handcrafted features may not be optimal. Different from these methods, in our approach we introduce a term-level classification task to exploit the useful information in entity dictionary without manual feature engineering. We jointly train our model in both the NER and term classification tasks to enhance the performance of NER model in an end-to-end manner.

There are also a few methods that explore to incorporate dictionary knowledge into Chinese NER models in an end-to-end manner by using graph neural networks [10,33]. For example, Sui et al. [33] propose a character-based

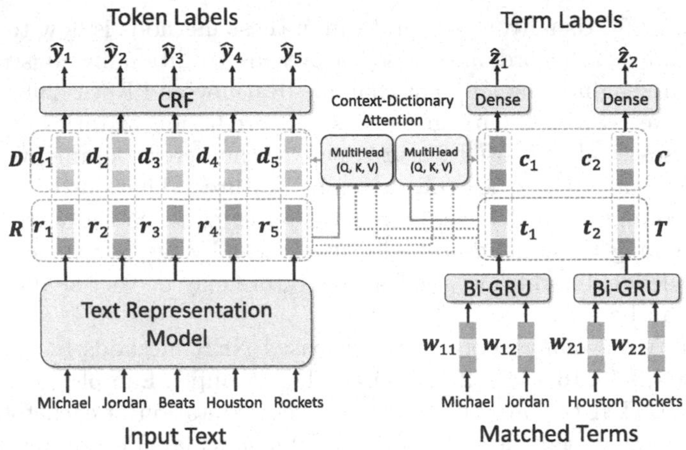

Fig. 1. The architecture of our CADK approach.

collaborative graph neural network to learn the representations of characters and words matched by dictionaries from three word-character graphs, i.e., a containing graph that describes the connection between characters and matched words, a transition graph that builds the connections between characters and the nearest contextual matched words, and a Lattice graph that connects each word with its boundary characters. However, these methods mainly model the interactions between matched entities and their local contexts, while ignore the relations between entities and their long-distance contexts. Different from these methods, our approach can model the interactions between the matched terms with the global contexts via entity-dictionary attention.

3 CADK Approach for NER

In this section, we introduce our NER approach with Context-Aware Dictionary Knowledge (CADK). The architecture of our approach is illustrated in Fig. 1. Our approach mainly contains five components, i.e., *text representation*, *term representation*, *context-dictionary attention*, *term classification* and *sequence tagging*. Next, we introduce the details of each module as follows.

3.1 Text Representation

The first module is a text representation model, which is used to learn the contextual representation of each word in an input text. It can be implemented by various neural text representation models, such as CNN [40], LSTM [12] and GRU [27] or pre-trained language models like ELMo [28] and BERT [9]. We denote the word sequence of the input text as $[w_1, w_2, ...w_N]$, where N is the number of words. The text representation model outputs a sequence that contains the contextual representation of each word, which is denoted as $\mathbf{R} = [\mathbf{r}_1, \mathbf{r}_2, ..., \mathbf{r}_N]$.

3.2 Term Representation

The second module is *term representation*, which is used to obtain the representations of the terms matched by the entity dictionaries. Usually, entity dictionaries contain both popular (e.g., America) and rare entity names (e.g., Chatham), and can help NER models recognize these entity names correctly. Thus, entity dictionaries have the potential to improve the performance of NER and reduce the dependence on labeled data. To incorporate useful information in entity dictionaries, we use them to match the input text and obtain a candidate list with M entity terms. We denote the word sequence of the i_{th} term as $[w_{i1}, w_{i2}, ...w_{iP}]$, where P represents the number of words in this term. In the *term representation* module, we first use a word embedding layer to convert the sequence of words in each term into a sequence of low-dimensional vectors. The word embedding parameters in this layer are shared with the *text representation* model. The word embedding sequence of the i_{th} term is denoted as $[\mathbf{w}_{i1}, \mathbf{w}_{i2}, ...\mathbf{w}_{iP}]$. Then, we apply a word-level Bi-GRU network to the word embedding sequence of each term to learn a hidden term representation. The GRU layer scans the word embedding sequence of each term in two directions, and combines the last hidden states in both directions as the representation of this term. For the i_{th} term, its representation is denoted as $\mathbf{t}_i$. We denote the sequence of the representations of the M matched terms as $\mathbf{T} = [\mathbf{t}_1, \mathbf{t}_2, ..., \mathbf{t}_M]$.

3.3 Context-Dictionary Attention

The third module is *context-dictionary attention*. Many entity names are context-dependent. For example, in the sentence "Jordan is a famous NBA player", the word "Jordan" is in a person name, while it is also frequently used as a location name. Thus, we propose to incorporate dictionary knowledge in a context-aware manner by modeling the relationships between the matched entity terms and their contexts. It is used to model the interactions between terms matched by dictionaries with the contexts in sentences. Usually, entity names may interact with other words in the same text, and such interactions are important for recognizing these entities. For example, in the sentence "Jordan is a basketball player", the interaction between the entity "Jordan" and the word "player" is very informative for identifying the type of this entity is "person". In addition, an entity may interact with multiple words. For instance, in the sentence "He travels from Houston to Seattle", the interactions between the entity "Houston" and its contexts like "travels" and "Seattle" are useful clues for recognizing this entity. Motivated by these observations, we propose a context-dictionary attention module to model the interactions between the terms matched by dictionaries with all words in texts. The context-dictionary attention network takes both the sequences of word representations $\mathbf{R} = [\mathbf{r}_1, \mathbf{r}_2, ..., \mathbf{r}_N]$ and term representations $\mathbf{T} = [\mathbf{t}_1, \mathbf{t}_2, ..., \mathbf{t}_M]$ (N and M are numbers of words and terms) as inputs, and outputs dictionary-aware representations of words in texts (denoted as **D**) and context-aware representations of terms (denoted as **C**). We use the multi-head productive attention mechanism [35] to model the interactions between terms and contexts. The dictionary-aware word representation sequence **D** is computed as follows:

$$\mathbf{D}^i = \text{Softmax}[\mathbf{W}_Q^i \mathbf{R}(\mathbf{W}_K^i \mathbf{T})^T](\mathbf{W}_V^i \mathbf{T}), \tag{1}$$

$$\mathbf{D} = \text{Concat}(\mathbf{D}^1, \mathbf{D}^2, ..., \mathbf{D}^h), \tag{2}$$

where $\mathbf{W}_Q^i$, $\mathbf{W}_K^i$, and $\mathbf{W}_V^i$ respectively stand for the parameters in the i_{th} head for transforming the query, key and value, h represents the number of parallel attention heads. The context-aware term representation sequence **C** is computed in a similar way as follows:

$$\mathbf{C}^i = \text{Softmax}[\mathbf{U}_Q^i \mathbf{T}(\mathbf{U}_K^i \mathbf{R})^T](\mathbf{U}_V^i \mathbf{R}), \tag{3}$$

$$\mathbf{C} = \text{Concat}(\mathbf{C}^1, \mathbf{C}^2, ..., \mathbf{C}^h), \tag{4}$$

where $\mathbf{U}_Q^i$, $\mathbf{U}_K^i$, and $\mathbf{U}_V^i$ are parameters. We concatenate **D** with the word representations **R**, and **C** with the term representations **T**, in a position-wise manner. In this way, entity dictionary with contextual information can be incorporated into a neural NER model.

3.4 Term Classification

The fourth module is *term classification*, which is used to classify the types of the terms matched by dictionaries based on the representations of terms and their interactions with the contexts. To fully exploit the useful information in the entity dictionary, we propose an auxiliary term classification task which predicts the type of the entity names matched by the entity dictionary. For example, in the sentence "Michael Jordan Beats Houston Rockets", if the terms "Michael Jordan" and "Houston Rockets" are matched by the dictionary, our model is required to classify the types of these terms in the context of this sentence. We use a dense layer with the softmax activation function to classify the type of each term as $\hat{z}_i = \text{softmax}(\mathbf{U}[\mathbf{c}_i; \mathbf{t}_i] + \mathbf{v})$, where $\mathbf{U}$ and $\mathbf{v}$ are parameters, $\mathbf{c}_i$ is the context-aware representation of the i_{th} term, and $\hat{z}_i$ is the predicted type label of this term. The gold type label of the matched term can be derived from the token labels of the input sentence. For example, if the label sequence of a sentence is "O-BLOC-ELOC-O", we can know that the gold type of the entity in this sentence is "location". The loss function of the term classification task is the cross-entropy of the gold and the predicted labels of all terms, which is evaluated as follows:

$$\mathcal{L}_{Term} = -\sum_{i=1}^{S}\sum_{j=1}^{M}\sum_{k=1}^{K} \hat{z}_{ijk} \log(z_{ijk}), \tag{5}$$

where S is the number of sentences for model training, K is the number of entity categories, z_{ijk} and $\hat{z}_{ijk}$ are the gold and predicted type labels of the j_{th} term from the i_{th} sentence in the k_{th} category.

3.5 Sequence Tagging

The last module is *sequence tagging*. Usually the label at each position may have relatedness with the previous ones. For example, in the *BIOES* tagging scheme, the label "I-LOC" can only appear after "B-LOC" and "I-LOC". Thus, a CRF layer is usually employed to jointly decode the label sequence. The loss function of the NER task is denoted as $\mathcal{L}_{NER}$. To incorporate the useful information in entity dictionary into NER models more effectively, we propose a unified framework based on multi-task learning to jointly train our model in both NER and term classification tasks. The final loss function is the weighted summation of the NER and term classification loss, which is formulated as follows:

$$\mathcal{L} = (1 - \lambda)\mathcal{L}_{NER} + \lambda\mathcal{L}_{Term}, \tag{6}$$

where $\mathcal{L}_{NER}$ is the loss of CRF model, $\lambda \in [0, 1]$ is a coefficient to control the relative importance of the term classification task.

4 Experiments

4.1 Dataset and Experimental Settings

Our experiments were conducted on the CoNLL-2003 dataset [34], which is a widely used benchmark dataset for NER. This dataset contains four different types of named entities, i.e., locations, persons, organizations, and miscellaneous entities that do not belong in the three previous categories. Following previous works [29], we used the BIOES labeling scheme. In our experiments, we used an entity dictionary provided by [11], which is derived from the WikiPedia database. This dictionary contains 297,073,139 entity names. The coefficient λ in Eq. (6) was 0.4. We used Adam [13] with gradient norms clipped at 5.0 as the optimizer for model training, and the learning rate was 0.001. The batch size was 64. These hyperparameters were tuned on the validation set. Each experiment was repeated 5 times independently, and the average performance in terms of precision, recall and Fscore were reported.

4.2 Comparison with Baseline Methods

To verify the effectiveness of the proposed *CADK* method, we compare several popular models and their variants using different methods for incorporating entity dictionaries. The methods to be compared including: (1) LSTM-CRF [12], a neural NER method that uses LSTM to learn word representations and CRF to decode labels; (2) TagLM [27], a neural NER model which uses GRU networks and a pre-trained language model to learn word representations, and uses CRF to decode labels; (3) ELMo [9], a pre-trained language model with bidirectional deep LSTM network. We apply an LSTM-CRF network based on the contextualized word embeddings generated by the ELMo model; (4) BERT [9], a pre-trained language model with bidirectional transformers. We fine-tune the

Table 2. Performance of different NER methods under different ratios of training data. P, R, F respectively stand for precision, recall and Fscore.

Model	10%			25%			100%		
	P	R	F	P	R	F	P	R	F
LSTM-CRF	84.23	88.22	86.18	87.75	87.86	87.81	90.75	90.14	90.36
LSTM-CRF+Feature	84.90	89.02	86.91	88.33	88.40	88.37	91.14	90.18	90.66
LSTM-CRF+GNN	85.54	88.74	87.11	88.53	88.56	88.54	90.99	90.51	90.75
LSTM-CRF+CADK	85.94	89.27	87.58	89.34	88.72	89.03	91.58	90.81	91.19
TagLM	85.63	88.70	87.14	88.64	89.05	88.85	92.01	91.40	91.71
TagLM+Feature	85.77	90.14	87.90	89.44	89.25	89.35	92.41	91.64	92.02
TagLM+GNN	86.27	90.02	88.10	89.79	89.34	89.56	92.62	91.91	92.26
TagLM+CADK	86.56	90.68	88.57	89.98	90.14	90.06	93.03	92.33	92.68
ELMo	85.34	89.24	87.25	88.76	89.13	88.95	92.42	92.23	92.30
ELMo+Feature	86.01	89.96	87.94	89.51	89.39	89.45	92.73	92.19	92.46
ELMo+GNN	86.71	89.97	88.31	89.70	89.65	89.68	92.92	92.28	92.60
ELMo+CADK	87.09	90.36	88.70	90.40	89.82	90.11	93.49	92.57	93.03
BERT	84.76	87.87	86.29	87.91	88.11	88.01	91.89	91.23	91.49
BERT+Feature	85.48	88.86	87.14	88.60	88.43	88.51	91.99	91.41	91.70
BERT+GNN	85.73	88.72	87.20	88.65	88.90	88.77	92.12	91.64	91.88
BERT+CADK	86.20	89.30	87.72	89.19	89.32	89.26	92.40	92.00	92.20

BERT-base version in the NER task; The methods for incorporating entity dictionaries including: (a) Feature [37], incorporating entity dictionaries using feature engineering. We combines the dictionary matching features with the hidden representations learned by the aforementioned methods; (b) GNN [33], using graph neural networks to incorporate entity dictionary knowledge; (c) CADK, our proposed method with context-aware dictionary knowledge.

We randomly sampled different ratios (i.e., 10%, 25% and 100%) of samples from the data for model training to evaluate these methods under different amounts of labeled data. The results are summarized in Table 2.[2] From Table 2, we find that when the training data is scarce, the performance of the methods without dictionary knowledge declines significantly. This is probably because these neural network based methods are data-intensive and require a large amount of labeled data for model training. When training data is scarce, many entities in the test set are unseen in the training data, making it difficult for existing NER methods to recognize them. Compared with methods without dictionaries, the methods that consider dictionary knowledge achieve better performance, and their advantage is larger when training data is more scarce. This is probably because incorporating dictionary knowledge can help recognize unseen or rare entities more effectively, which can reduce the dependency on labeled data. In addition, compared with the methods using dictionary matching features, the methods that can model the contexts of matched entities (*GNN* and *CADK*) perform better. This is probably because manually crafted features

[2] The performance of BERT is surprisingly unsatisfactory though we used the officially released model and carefully tuned hyperparameters.

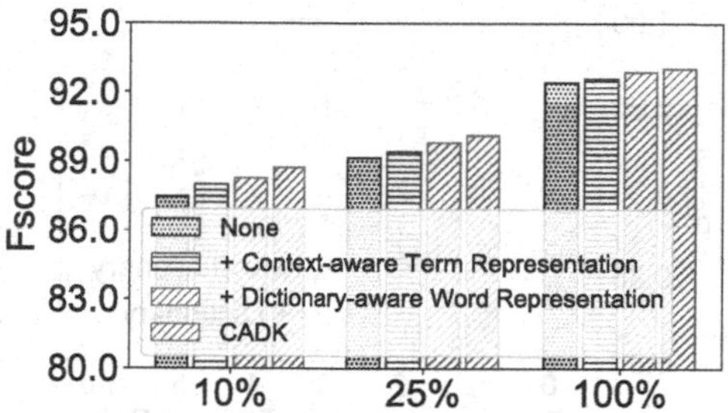

Fig. 2. Effectiveness of the context-dictionary attention module.

may be not optimal to utilize entity dictionaries, and the contexts of the matched entity names in different texts are not considered. Besides, our *CADK* method is better than the *GNN* method in exploiting dictionary knowledge for NER. Different from the *GNN* method that can only model the local contexts of matched entity names, in our approach we use the context-dictionary attention model to capture the global contexts of the matched terms, and we jointly train our model in both NER and term classification tasks to incorporate dictionary knowledge in a unified framework. Thus, our method can exploit dictionary information more accurately to improve neural NER model (Fig. 2).

4.3 Effectiveness of Context-Dictionary Attention

In this section, we conduct several ablation studies to validate the effectiveness of the context-dictionary attention module in our *CADK* method. Since it mainly aims to generate the dictionary-aware word representation and the context-aware term representation, we compare the performance of *ELMo-CADK* under different ratios of training data by removing one or both of them. The results are shown in Fig. 3. According to the results, we find that the dictionary-aware word representation can effectively improve the performance of our approach. This is because the dictionary-aware word representation encodes the information of the entities matched by dictionaries, which is helpful for recognizing them more accurately. In addition, incorporating the context-aware term representation can also improve the model performance. This is because many entities are context-dependent, and modeling their relations with the contexts is beneficial for NER. These results show the effectiveness of context-dictionary attention in injecting context-aware dictionary knowledge into neural NER models.

4.4 Performance on Rare Entities

In this section, we explore the influence of incorporating dictionary knowledge on recognizing the entities rarely appearing in the training data. We evaluate the recall of the entities in the test set with different appearance times in the training data. We conduct experiments under 25% of training data and the results of the

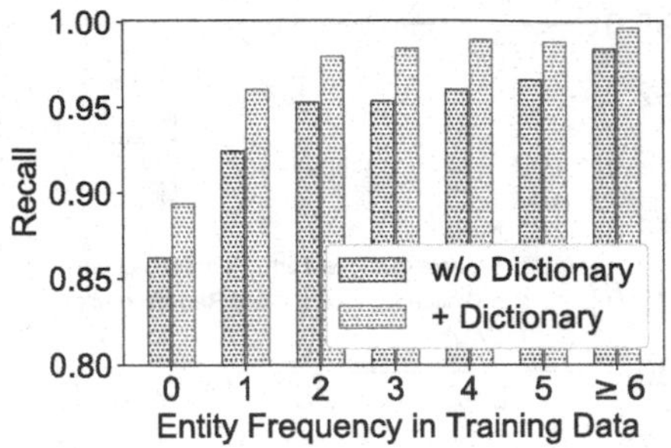

Fig. 3. Recall of entities in the test set with different frequencies in the training data.

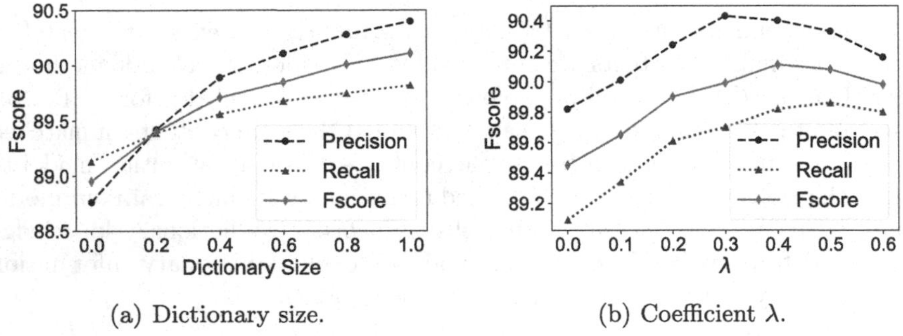

(a) Dictionary size.

(b) Coefficient λ.

Fig. 4. Model performance w.r.t. different dictionary sizes and different values of λ.

ELMo+CADK model are shown in Fig. 3, which reveals two findings. First, the performance on entities that do not or rarely appear in the training data is much lower than recognizing common entities. This result shows that rare entities are more difficult to recognize. Second, our approach can effectively improve the performance on entities that rarely appear in the training data. This is because our approach can utilize dictionary knowledge to help neural NER model recognize these rare entities more accurately.

4.5 Influence on Dictionary Size

In this section, we study the influence of the size of entity dictionaries. We randomly sampled different ratios of entities from the dictionary for entity matching and compare the performance of the *ELMo-CADK* model under 25% of training data. The results are shown in Fig. 4(a). We find that the model performance consistently improves when the dictionary size grows. This is because a larger dictionary usually has better entity coverage, and our approach can exploit richer information from the entity dictionary to help recognize entities more accurately.

Table 3. Several named entity recognition examples. Red, orange, and blue words represent the predicted person, location and organization entities respectively.

Example	Method	NER result
1	ELMo	Third one-day match: December 8, in Karachi
	ELMo+Feature	Third one-day match: December 8, in Karachi
	ELMo+CADK	Third one-day match: December 8, in Karachi
2	ELMo	Partizan - Dejan Koturovic 21
	ELMo+Feature	Partizan - Dejan Koturovic 21
	ELMo+CADK	Partizan - Dejan Koturovic 21
3	ELMo	Bolesy (Florida manager John Boles) told me ...
	ELMo+Feature	Bolesy (Florida manager John Boles) told me ...
	ELMo+CADK	Bolesy (Florida manager John Boles) told me ...

4.6 Influence of Hyper-parameters

In this section we explore the influence of an important hyper-parameter on our approach, i.e., λ in Eq. (6), which is used to control the relative importance of the term classification loss. The experimental results on λ using the *ELMo-CADK* model with 25% of training data are shown in Fig. 4(b). According to Fig. 4(b), the performance of our approach improves when λ increases. However, when λ becomes too large the performance declines. This is because when λ is too small, the useful information in the term classification task is not fully exploited. Thus, the performance is sub-optimal. When λ goes too large, the auxiliary task is dominant and the NER task is not fully respected. Thus, the performance is also sub-optimal. These results lead to a moderate selection of λ (e.g., 0.4).

4.7 Case Study

In this section, we conducted several case studies to better understand our approach in incorporating dictionary knowledge in a context-aware manner. Several representative samples are shown in Table 3. This experiment is conducted using 10% of training data. According to Table 3, incorporating entity dictionaries can help a NER model better recognize rare entities. For example, "Partizan" is a name of a football team, which only appears once in the training set. The basic NER model recognized it as a person name, while the approaches using dictionaries can make correct predictions. Our approach can also correctly recognize the context-dependent entities which the basic model and the model based on dictionary features fail to recognize. For example, the entity "Florida" is recognized as a location by *ELMo* and *ELMo+Feature*, since it is usually used as a location name. Our approach can recognize this entity correctly based on its contexts. These results show that our approach can effectively exploit the useful information in entity dictionaries with contextual information.

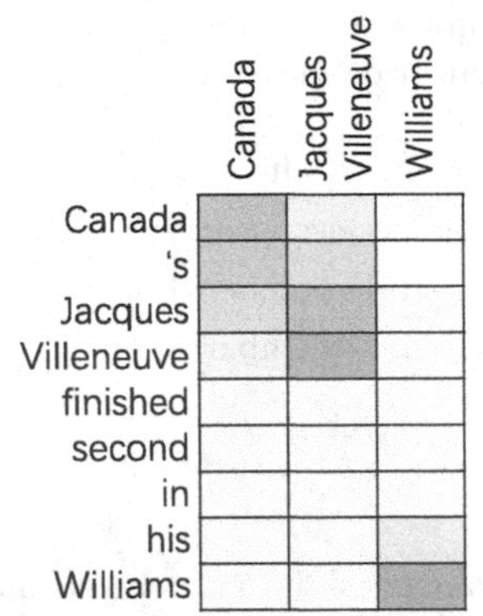

(a) Using word representations as attention query and term representations as key.

(b) Using term representations as attention query and word representations as key.

Fig. 5. Visualization of attention weights in the context-dictionary attention network.

Next, we visualize the attention weights in the context-dictionary attention to better understand the interactions between contexts and matched terms. The visualization results are shown in Fig. 5. According to the results, we can see that our approach can effectively model the interactions between entity terms and contexts. For example, in Fig. 5(a), the interaction between the word "Jacques" and the term "Jacques Villeneuve" is highlighted, which is important for identifying the word "Jacques" belongs to an entity name. In addition, in Fig. 5(b), the interaction between the term "Jacques Villeneuve" and the word "his" is also highlighted, which is an important clue for inferring the type of this entity is "person". These results indicate that our approach can effectively capture the relationships between the entity names matched by dictionaries and their contexts to learn context-aware dictionary knowledge.

5 Conclusion

In this paper we propose a neural NER approach which can incorporate entity dictionaries with contextual information. In our approach, we propose a context-dictionary attention network to model the interactions between entity names matched by dictionaries and their contexts in texts. In addition, we propose an auxiliary term classification task to classify the types of the terms matched by dictionaries based on contexts, and we jointly train our model in both NER and term classification tasks to incorporate the information of entity dictionaries and contexts into NER. Extensive experiments on the CoNLL-2003 benchmark dataset show that our approach can effectively improve the performance of NER especially when training data is insufficient.

Acknowledgments. Supported by the National Key Research and Development Program of China under Grant No. 2018YFC1604002, the National Natural Science Foundation of China under Grant Nos. U1936208, U1936216, U1836204 and U1705261.

References

1. Akbik, A., Blythe, D., Vollgraf, R.: Contextual string embeddings for sequence labeling. In: COLING, pp. 1638–1649 (2018)
2. Arora, R., Tsai, C.T., Tsereteli, K., Kambadur, P., Yang, Y.: A semi-Markov structured support vector machine model for high-precision named entity recognition. In: ACL (2019)
3. Chen, H., Lin, Z., Ding, G., Lou, J., Zhang, Y., Karlsson, B.: GRN: gated relation network to enhance convolutional neural network for named entity recognition. In: AAAI (2019)
4. Chiu, J., Nichols, E.: Named entity recognition with bidirectional LSTM-CNNs. TACL **4**(1), 357–370 (2016)
5. Clark, K., Luong, M.T., Manning, C.D., Le, Q.: Semi-supervised sequence modeling with cross-view training. In: EMNLP, pp. 1914–1925 (2018)
6. Cohen, W.W., Sarawagi, S.: Exploiting dictionaries in named entity extraction: combining semi-Markov extraction processes and data integration methods. In: KDD, pp. 89–98. ACM (2004)
7. Collobert, R., Weston, J., Bottou, L., Karlen, M., Kavukcuoglu, K., Kuksa, P.: Natural language processing (almost) from scratch. JMLR **12**(Aug), 2493–2537 (2011)
8. Derczynski, L., et al.: Analysis of named entity recognition and linking for tweets. Inf. Process. Manag. **51**(2), 32–49 (2015)
9. Devlin, J., Chang, M.W., Lee, K., Toutanova, K.: BERT: pre-training of deep bidirectional transformers for language understanding. In: NAACL-HLT, pp. 4171–4186 (2019)
10. Gui, T., et al.: A lexicon-based graph neural network for Chinese NER. In: EMNLP-IJCNLP, pp. 1039–1049 (2019)
11. Higashinaka, R., Sadamitsu, K., Saito, K., Makino, T., Matsuo, Y.: Creating an extended named entity dictionary from Wikipedia, In: COLING, pp. 1163–1178 (2012)
12. Huang, Z., Xu, W., Yu, K.: Bidirectional LSTM-CRF models for sequence tagging. arXiv preprint arXiv:1508.01991 (2015)
13. Kingma, D.P., Ba, J.: Adam: a method for stochastic optimization. arXiv preprint arXiv:1412.6980 (2014)
14. Lample, G., Ballesteros, M., Subramanian, S., Kawakami, K., Dyer, C.: Neural architectures for named entity recognition. In: NAACL-HLT, pp. 260–270 (2016)
15. Levow, G.A.: The third international Chinese language processing bakeoff: word segmentation and named entity recognition. In: Proceedings of the Fifth SIGHAN Workshop on Chinese Language Processing, pp. 108–117 (2006)

16. Li, P.H., Dong, R.P., Wang, Y.S., Chou, J.C., Ma, W.Y.: Leveraging linguistic structures for named entity recognition with bidirectional recursive neural networks. In: EMNLP, pp. 2664–2669 (2017)
17. Lin, B.Y., Lu, W.: Neural adaptation layers for cross-domain named entity recognition. In: EMNLP, pp. 2012–2022 (2018)
18. Lin, H., Li, Y., Yang, Z.: Incorporating dictionary features into conditional random fields for gene/protein named entity recognition. In: Washio, T., et al. (eds.) PAKDD 2007. LNCS (LNAI), vol. 4819, pp. 162–173. Springer, Heidelberg (2007). https://doi.org/10.1007/978-3-540-77018-3_18
19. Lin, Y., Shen, S., Liu, Z., Luan, H., Sun, M.: Neural relation extraction with selective attention over instances. In: ACL, pp. 2124–2133 (2016)
20. Liu, T., Yao, J.g., Lin, C.Y.: Towards improving neural named entity recognition with gazetteers. In: ACL, pp. 5301–5307 (2019)
21. Luo, X., Zhou, W., Wang, W., Zhu, Y., Deng, J.: Attention-based relation extraction with bidirectional gated recurrent unit and highway network in the analysis of geological data. IEEE Access **6**, 5705–5715 (2018)
22. Ma, X., Hovy, E.: End-to-end sequence labeling via bi-directional LSTM-CNNs-CRF. In: ACL, vol. 1, pp. 1064–1074 (2016)
23. Magnolini, S., Piccioni, V., Balaraman, V., Guerini, M., Magnini, B.: How to use gazetteers for entity recognition with neural models. In: Proceedings of the 5th Workshop on Semantic Deep Learning, pp. 40–49 (2019)
24. Neelakantan, A., Collins, M.: Learning dictionaries for named entity recognition using minimal supervision. In: EACL, pp. 452–461 (2014)
25. Passos, A., Kumar, V., McCallum, A.: Lexicon infused phrase embeddings for named entity resolution. CoNLL-2014, p. 78 (2014)
26. Peng, M., Xing, X., Zhang, Q., Fu, J., Huang, X.: Distantly supervised named entity recognition using positive-unlabeled learning. In: ACL, pp. 2409–2419 (2019)
27. Peters, M., Ammar, W., Bhagavatula, C., Power, R.: Semi-supervised sequence tagging with bidirectional language models. In: ACL, vol. 1, pp. 1756–1765 (2017)
28. Peters, M., et al.: Deep contextualized word representations. In: NAACL-HLT, pp. 2227–2237 (2018)
29. Ratinov, L., Roth, D.: Design challenges and misconceptions in named entity recognition. In: CoNLL, pp. 147–155 (2009)
30. Rei, M.: Semi-supervised multitask learning for sequence labeling. In: ACL, pp. 2121–2130 (2017)
31. Rocktäschel, T., Huber, T., Weidlich, M., Leser, U.: WBI-NER: the impact of domain-specific features on the performance of identifying and classifying mentions of drugs. In: SemEval 2013, vol. 2, pp. 356–363 (2013)
32. Song, M., Yu, H., Han, W.S.: Developing a hybrid dictionary-based bio-entity recognition technique. BMC Med. Inform. Decis. Mak. **15**(1), S9 (2015)
33. Sui, D., Chen, Y., Liu, K., Zhao, J., Liu, S.: Leverage lexical knowledge for Chinese named entity recognition via collaborative graph network. In: EMNLP-IJCNLP, pp. 3821–3831 (2019)
34. Tjong Kim Sang, E.F., De Meulder, F.: Introduction to the CoNLL-2003 shared task: language-independent named entity recognition. In: NAACL-HLT, pp. 142–147 (2003)
35. Vaswani, A., et al.: Attention is all you need. In: NIPS, pp. 5998–6008 (2017)
36. Wan, X., et al.: Named entity recognition in Chinese news comments on the web. In: IJCNLP, pp. 856–864 (2011)

37. Wang, Q., Zhou, Y., Ruan, T., Gao, D., Xia, Y., He, P.: Incorporating dictionaries into deep neural networks for the chinese clinical named entity recognition. J. Biomed. Inform. **92**, 103133 (2019)
38. Yu, X., Lam, W., Chan, S.K., Wu, Y.K., Chen, B.: Chinese NER using CRFs and logic for the fourth SIGHAN bakeoff. In: Proceedings of the Sixth SIGHAN Workshop on Chinese Language Processing (2008)
39. Zeng, W., Tang, J., Zhao, X.: Entity linking on Chinese microblogs via deep neural network. IEEE Access **6**, 25908–25920 (2018)
40. Zhu, Y., Wang, G.: CAN-NER: convolutional attention network for Chinese named entity recognition. In: NAACL-HLT, pp. 3384–3393 (2019)

Chinese Named Entity Recognition via Adaptive Multi-pass Memory Network with Hierarchical Tagging Mechanism

Pengfei Cao[1,2(✉)], Yubo Chen[1,2], Kang Liu[1,2], and Jun Zhao[1,2]

[1] National Laboratory of Pattern Recognition, Institute of Automation, Chinese Academy of Sciences, Beijing 100190, China
{pengfei.cao,yubo.chen,kliu,jzhao}@nlpr.ia.ac.cn
[2] School of Artificial Intelligence, University of Chinese Academy of Sciences, Beijing 100049, China

Abstract. Named entity recognition (NER) aims to identify text spans that mention named entities and classify them into pre-defined categories. For Chinese NER task, most of the existing methods are character-based sequence labeling models and achieve great success. However, these methods usually ignore lexical knowledge, which leads to false prediction of entity boundaries. Moreover, these methods have difficulties in capturing tag dependencies. In this paper, we propose an **A**daptive **M**ulti-pass **M**emory **N**etwork with **H**ierarchical **T**agging Mechanism (**AMMNHT**) to address all above problems. Specifically, to reduce the errors of predicting entity boundaries, we propose an adaptive multi-pass memory network to exploit lexical knowledge. In addition, we propose a hierarchical tagging layer to learn tag dependencies. Experimental results on three widely used Chinese NER datasets demonstrate that our proposed model outperforms other state-of-the-art methods.

Keywords: Named entity recognition · Memory network · Lexical knowledge

1 Introduction

The task of named entity recognition (NER) is to recognize the named entities from a plain text and classify them into pre-defined types. NER is a fundamental and preliminary task in natural language processing (NLP) area and is beneficial for many downstream NLP tasks such as relation extraction [1], event extraction [5] and question answering [28]. In recent years, numerous methods have been carefully studied for NER task, including Conditional Random Fields (CRFs) [13] and Support Vector Machines (SVMs) [11]. Currently, with the development of deep learning methods, neural networks have been introduced for the NER task. In particular, sequence labeling neural network models have achieved state-of-the-art performance [14,33].

M. Sun et al. (Eds.): CCL 2020, LNAI 12522, pp. 144–158, 2020.
https://doi.org/10.1007/978-3-030-63031-7_11

Though sequence labeling neural network methods have achieved great success for Chinese NER task, some challenging issues still have not been well addressed. One significant drawback is that previous methods usually **fail to correctly predict entity boundaries**. To conduct a quantitative analysis, we perform a BiLSTM+CRF model proposed by Huang et al. [10], which is the most representative Chinese NER sequence labeling system, on WeiboNER dataset [8,20], OntoNotes 4 dataset [27] and MSRA dataset [15]. The F1 scores are 55.84%, 63.17% and 89.13%, respectively. We do a further analysis and find that the errors of predicting entity boundaries are particularly serious. The average proportion of predicting entity boundaries errors is 82% on these three datasets. For example, the character-based BiLSTM+CRF model fails to predict the entity boundaries of "北海道 (Hokkaido)" in Fig. 1. To reduce the errors of predicting entity boundaries, some works [2,21] try to jointly perform Chinese NER with Chinese word segmentation (CWS) for using word boundaries information. However, the joint model requires additional annotated training data for CWS task.

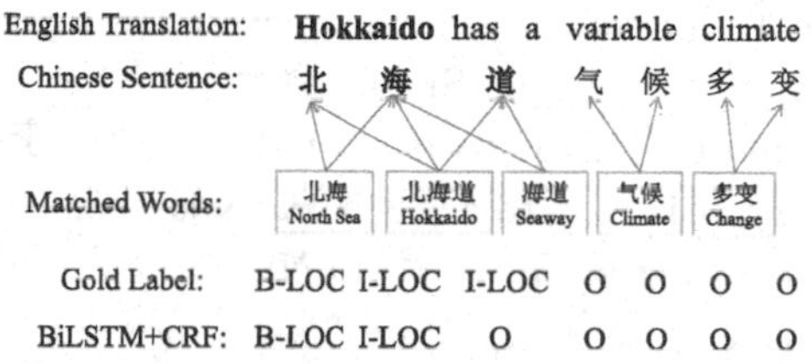

Fig. 1. An example of Chinese NER with wrong entity boundaries using the BiLSTM+CRF model. It also shows the matched words for each character.

Fortunately, existing lexicons can provide information on word boundaries and we refer to the information as lexical knowledge. In addition, the cost of obtaining lexicon is low and almost all fields have their lexicons, such as biomedical, social science fields and so on. Recently, Zhang et al. [33] propose a lattice LSTM model capable of leveraging lexicon for Chinese NER. Though effective, the lattice LSTM cannot exploit all matched words. When the candidate labeled character is within a matched word (i.e. the character is not the first or the last character of the matched word), the lattice model cannot explicitly and directly exploit the matched word. For example, for the candidate labeled character "海", it can match "北海 (North Sea)", "海道 (Seaway)" and "北海道 (Hokkaido)" in lexicon according to its context. When exploiting the matched words for character "海 (Sea)", the lattice model only considers "北海 (North Sea)" and "海道 (Seaway)", ignoring "北海道 (Hokkaido)" which can help determine that the character "海 (Sea)" is the middle of an entity rather than beginning or ending. Moreover, the lattice model only processes the matched words once, when learning the lexical knowledge for a character. However, it needs more reasoning passes on the matched words to better learn lexical knowledge in complex sentences intuitively. Take the sentence "南京市长江大桥 (Nanjing Yangtze River

Bridge)" for example, it is more complicated than the sentence in Fig. 1 because it is prone to be misunderstood as "南京市长/江大桥 (The mayor of Nanjing is Jiang Daqiao)". Thus, it needs more reasoning passes to learn the lexical knowledge for recognizing the entity "长江大桥 (Yangtze River Bridge)" than the entity "北海道 (Hokkaido)" in Fig. 1. However, if the reasoning passes are too many, the performance will decrease in word sense disambiguation task [17]. We argue that the problem also exists in Chinese NER task. Hence, how to exploit all matched words and perform flexible multi-pass reasoning according to the complexity of sentences should be well investigated.

Another issue is that most of the existing methods **cannot efficiently capture tag dependencies**. In sequence labeling neural network models, CRF is usually used as a decoding layer. Although the CRF decoder has achieved improvements, the transition matrix in CRF layer only learns the neighboring tag dependencies, which are typically first order dependencies [32]. Thus, CRF cannot well handle long-distance tag dependency problems. For example, in the sentence "耐克拥有比李宁更大的市场 (Nike has a larger market than Li Ning)", the tag of "李宁 (Li Ning)" is dependent on the tag of "耐克 (Nike)", as they should be the same entity type. Since "李宁 (Li Ning)" can be a person or an organization, it is more difficult to predict the tag of "李宁 (Li Ning)" than "耐克 (Nike)". However, it is easy to tag "耐克 (Nike)" as an organization. If we capture the dependencies between "李宁 (Li Ning)" and "耐克 (Nike)", we will have ample evidence to tag "李宁 (Li Ning)" as an organization. To address the issue, Zhang et al. [32] exploit the LSTM as decoder instead of CRF. However, the unidirectional LSTM decoder only leverages the past labels and ignores the future labels. In another sentence "李宁努力地同耐克竞争 (Li Ning strives to compete with Nike)", when predicting the tag of "李宁 (Li Ning)", the future tag of "耐克 (Nike)" can help us to determine the tag of "李宁 (Li Ning)". Thus, how to capture bidirectional (past and future) tag dependencies in the whole sentence is another challenging problem.

In this paper, we propose an **A**daptive **M**ulti-pass **M**emory **N**etwork with **H**ierarchical **T**agging Mechanism (**AMMNHT**) to address the aforementioned problems. To exploit all matched words and perform multi-pass reasoning across matched words for a character, memory network [24] can be utilized for Chinese NER. However, conventional memory network follows pre-defined passes to perform multi-pass reasoning and cannot perform adaptive and proper deliberation passes according to the change of input sentence. We utilize reinforcement learning [25] to adaptively determine the deliberation passes of memory network according to the complexity of sentences. Although we do not have explicit supervision for the reasoning passes of the memory network, we can obtain long-term feedback (or *reward*) from the final prediction, which inspires us to utilize reinforcement learning techniques. To capture bidirectional tag dependencies in the whole sentence, we propose a hierarchical tagging mechanism for Chinese NER task. In summary, the contributions of this paper are listed as follows:

- We propose a novel framework to integrate lexical knowledge from the lexicon for Chinese NER task, which can explicitly exploit all matched words and adaptively choose suitable reasoning passes for each sentence. To our best knowledge, this is the first work to automatically determine the reasoning passes of memory network via reinforcement learning techniques.
- We propose a hierarchical tagging mechanism for Chinese NER to capture bidirectional tag dependencies in the whole sentence.
- Experiments on three widely used Chinese NER datasets show that our proposed model outperforms previous state-of-the-art methods.

2 Related Work

In recent years, the NER task has attracted much research attention. Many methods have been proposed to perform the task. Early studies on NER often exploit CRFs [13] and SVMs [11]. These methods rely heavily on feature engineering. However, the designed features may be not appropriate for the task, which can lead to error propagation problem. Currently, neural network methods have been introduced into NER task and achieved state-of-the-art performance [14]. Huang et al. [10] use the bidirectional long short term memory (BiLSTM) for feature extraction and the CRF for decoding. The model is trained via the end-to-end paradigm. After that, the BiLSTM+CRF model is usually exploited as the baseline model for NER task. Ma et al. [18] use a character convolutional neural network (CNN) to represent spelling characteristic. Then the charcter representation vector is concatenated with word embedding as the input of the LSTM. Peters et al. [22] leverage a character language model to enhance the input of the model.

For Chinese NER, character-based methods have been the dominant approaches [6,16]. These methods only focus on character sequence information, ignoring word boundaries information, which can cause errors of predicting entity boundaries. Thus, how to better exploit lexical knowledge has received much research attention. Word segmentation information is used as extra features for Chinese NER task [8,20]. Peng et al. [21] and Cao et al. [2] propose a joint model for Chinese NER, which is jointly trained with CWS task. Zhang et al. [33] investigate a lattice LSTM to encode a sequence of input characters as well as words that match a lexicon. However, the lattice model cannot exploit all matched words and only processes the matched words once. Recently, graph-based models have been proposed for Chinese NER [7,23]. Based on the lattice structure, Sui et al. [23] propose a graph neural network to encode word information.

Tag dependencies is also a challenging problem, but few attention has been paid to tackling the problem. Zhang et al. [32] leverages LSTM as decoder for sequence labeling task. However, the unidirectional LSTM decoder only exploits the past predicted tags information, ignoring the future un-predicted tags. Hence, we propose a hierarchical tagging mechanism to capture bidirectional tag dependencies in the whole sentence. To our best knowledge, we are the first to introduce the hierarchical tagging mechanism to Chinese NER task. Moreover, to better capture the dependencies between tags, we try different hierarchical tagging mechanism.

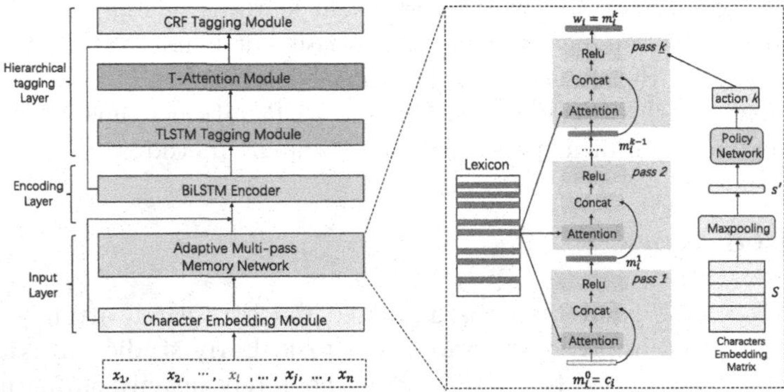

Fig. 2. The architecture of our proposed adaptive multi-pass memory network with hierarchical tagging mechanism. The right part is the adaptive multi-pass memory network (AMMN). For each character, the lexical knowledge (i.e., w_i in the figure) is obtained via the AMMN. We concatenate the character embeddings and lexical knowledge as the input of the encoding layer. In this figure, we use the character x_i as an example to illustrate the process.

3 Method

The architecture of our proposed model is shown in Fig. 2. The proposed model consists of three components: input layer, BiLSTM encoding layer and hierarchical tagging layer. In the following sections, we will describe the details of our model.

3.1 Input Layer

The inputs of our proposed model are character embeddings and lexical knowledge, which are obtained via character embedding module and adaptive multi-pass memory network, respectively.

Character Embedding Module. Similar to other methods using neural networks, the first step of our proposed model is to map discrete language symbols to distributed representations. Formally, given a Chinese sentence $s = \{x_1, x_2, \ldots, x_n\}$, each character x_i is represented by looking up embedding vector from a pre-trained character embedding table:

$$c_i = E^c(x_i) \tag{1}$$

where E^c is a pre-trained character embedding table and $c_i \in \mathbb{R}^{d_c}$. We obtain the characters embedding matrix, denoted as $S = \{c_1, c_2, \ldots, c_n\}$.

Adaptive Multi-pass Memory Network. The adaptive multi-pass memory network has three inputs: the candidate character embedding c_i as the initial query vector, the characters embedding matrix S and the matched words $\{w_{i1}, w_{i2}, \ldots, w_{iN_i}\}$ of the character x_i as the external memory, where N_i is the number of matched words. Since a candidate character may match multiple words in a lexicon and one-pass attention calculation may not accurately learn lexical knowledge, memory network is exploited to perform a deep reasoning process to highlight the correct lexical knowledge. After each pass, we need to update the query vector for the next pass. Therefore, the memory network contains two phases: ***attention calculation*** and ***update mechanism***.

Attention Calculation: During each pass, the query vector is the output of the former pass. We use attention to model the relationship between the query vector and the matched words. At pass k, the attention calculation can be formulated as follows:

$$\begin{aligned} e_{it}^k &= w_{it}^T m_i^{k-1} \\ \alpha_{it}^k &= \frac{\exp(e_{it}^k)}{\sum_{j=1}^{N_i} \exp(e_{ij}^k)} \end{aligned} \tag{2}$$

where m_i^{k-1} denotes the output of pass $k-1$. We treat the candidate character embedding c_i as m_i^0.

Update Mechanism: After calculating the attention, we can obtain the memory state at the current pass:

$$u_i^k = \sum_{t=1}^{N_i} \alpha_{it}^k w_{it} \tag{3}$$

We update the query vector by taking the former pass output and memory state of current pass into consideration for the next pass:

$$m_i^k = \text{Relu}(W_m[m_i^{k-1} : u_i^k] + b_m) \tag{4}$$

where $[:]$ is the concatenation operation. $W_m \in \mathbb{R}^{d_w \times 2d_w}$ and $b_m \in \mathbb{R}^{d_w}$ are trainable parameters. We use the output of the last pass as the lexical knowledge of the character x_i, denoted as w_i.

Empirically, different reasoning passes may obtain different performances [17]. We assume that less reasoning passes are enough to tackle simple sentences than complicated sentences. However, conventional memory network cannot perform adaptive and proper deliberation passes according to the complexity of the input sentence. Therefore, we utilize reinforcement learning to automatically control the reasoning passes of the memory network. We will introduce ***state***, ***action*** and ***reward*** as follows:

State: We use the sentence embedding s' as the state. After getting the characters embedding matrix S, we perform the max-pooling operation and treat the result as the sentence embedding:

$$s' = \text{Maxpooling}(S) \tag{5}$$

Action: We regard the reasoning pass as the action $a \in \{1, 2, \ldots, N\}$, where N is the maximal pass. We sample the value of a by a policy network $\pi_{\Theta}(a|s')$, which can be formulated as follows:

$$\pi_{\Theta}(a|s') = \text{Softmax}(W_p s' + b_p) \tag{6}$$

where $W_p \in \mathbb{R}^{N \times d_c}$ and $b_p \in \mathbb{R}^N$ are trainable parameters. $\Theta = \{W_p, b_p\}$.

Reward: We can obtain a terminal reward after finishing the final prediction. In this work, we use the F1 score of each sentence as the reward r.

Given T training instances, the objective function of policy network is defined as:

$$J_1 = \sum\nolimits_{i=1}^{T} \log \pi_{\Theta}(a_{(i)}|s'_{(i)}) r_{(i)} \tag{7}$$

where $a_{(i)}$, $s'_{(i)}$ and $r_{(i)}$ are the action, state and reward of the training instance i, respectively. We use the policy gradient method to learn the parameter set Θ.

3.2 BiLSTM Encoding Layer

After obtaining character embeddings and lexical knowledge, we concatenate them as the input of the encoding layer. Long short term memory (LSTM) is a variant of recurrent neural network (RNN), which is designed to address the gradient vanishing and exploding problems in RNN via introducing gate mechanism and memory cell. In order to incorporate information from both sides of sequence, we use BiLSTM to extract features. The hidden state of BiLSTM can be defined as follows:

$$h_i = [\overrightarrow{h}_i : \overleftarrow{h}_i] \tag{8}$$

where $\overrightarrow{h}_i \in \mathbb{R}^{d_h}$ and $\overleftarrow{h}_i \in \mathbb{R}^{d_h}$ are the hidden states at position i of the forward and backward LSTM, respectively.

3.3 Hierarchical Tagging Layer

In the hierarchical tagging layer, we exploit the LSTM as the first tagging module named as TLSTM and the CRF as the second tagging module.

The First Tagging Module: TLSTM. When detecting the tag of character x_i, the inputs of the first tagging module are: h_i from the BiLSTM encoding layer, former hidden state $\hat{h}_{i-1}$, and former predicted tag vector $\widehat{T}_{i-1}$. Formally, the TLSTM can be written precisely as follows:

$$\begin{aligned}
\begin{bmatrix} i_i \\ o_i \\ f_i \\ \tilde{c}_i \end{bmatrix} &= \begin{bmatrix} \sigma \\ \sigma \\ \sigma \\ \tanh \end{bmatrix} \left(W_d^T \begin{bmatrix} h_i \\ \hat{h}_{i-1} \\ \widehat{T}_{i-1} \end{bmatrix} + b_d \right) \\
\hat{c}_i &= \hat{c}_{i-1} \odot f_i + \tilde{c}_i \odot i_i \\
\hat{h}_i &= o_i \odot \tanh(\hat{c}_i) \\
\widehat{T}_i &= W_{td} \hat{h}_i + b_{td}
\end{aligned} \tag{9}$$

where i, f, o are the input gate, forget gate and output gate, respectively. $\widehat{T}$ is the predicted tagging vector.

Tagging Attention Module: T-Attention. Tagging attention aims to dynamically leverage the hidden states and preliminary predictions of the TLSTM. $\widehat{H} = \{\hat{h}_1, \hat{h}_2, \ldots, \hat{h}_n\}$ and $T_{raw} = \{\widehat{T}_1, \widehat{T}_2, \ldots, \widehat{T}_n\}$ denote the hidden states and preliminary predictions of the TLSTM, respectively. The attention is expressed as follows:

$$\begin{aligned} \hat{h}_{di} &= [\hat{h}_i : \widehat{T}_i] \\ m_i &= u_d^T \tanh(W_{da}\hat{h}_{di} + b_{da}) \\ \alpha_i &= \frac{\exp(m_i)}{\sum_{j=1}^{n} \exp(m_j)} \\ r_i &= \tanh(\sum\nolimits_{j=1}^{n} \alpha_j \hat{h}_{dj}) \end{aligned} \tag{10}$$

where $u_d \in \mathbb{R}^{d_{da}}$ is the context vector, which is randomly initialized and learned during the training process [30]. r_i denotes the representation of the hidden states and preliminary predictions of the TLSTM.

The Second Tagging Module: CRF. $H = \{h_1, h_2, \ldots, h_n\}$ and $R = \{r_1, r_2, \ldots, r_n\}$ denote the outputs of BiLSTM encoding layer and tagging attention module, respectively, which are concatenated as the input of the CRF module, denoted as $H_c = \{h_{c1}, h_{c2}, \ldots, h_{cn}\}$.

Given a sentence $s = \{x_1, x_2, \ldots, x_n\}$ with a final predicted tag sequence $y = \{y_1, y_2, \ldots, y_n\}$, the CRF tagging process is formalized as follows:

$$\begin{aligned} o_i &= W_o h_{ci} + b_o \\ s(s, y) &= \sum\nolimits_{i=1}^{n} (o_{i,y_i} + T_{y_{i-1},y_i}) \\ y^* &= \arg\max_{y \in Y_s} s(s, y) \end{aligned} \tag{11}$$

where o_{i,y_i} is the score of the y_i-th tag of the character x_i. T denotes the transition matrix which defines the scores of two successive labels. Y_s represents all candidate tag sequences for given sentence s. We use the Viterbi algorithm to get the final best-scoring tag sequence y^*.

3.4 Training

The probability of the ground-truth tag sequence $\bar{y}$ can be computed by:

$$p(\bar{y}|s) = \frac{\exp(s(s, \bar{y}))}{\sum_{\tilde{y} \in Y_s} \exp(s(s, \tilde{y}))} \tag{12}$$

Given a set of manually labeled training data $\{s^{(i)}, \bar{y}^{(i)}\}|_{i=1}^{T}$, the objective function of the tagging layer can be defined as follows:

$$J_2 = \sum\nolimits_{i=1}^{T} \log p(\bar{y}^{(i)} | s^{(i)}) \tag{13}$$

The objective function of the whole model is listed as follows:

$$J = \lambda J_1 + J_2 \tag{14}$$

As the adaptive multi-pass memory network and hierarchical tagging layer are correlated mutually, we train them jointly. We pre-train the model before the joint training process starts using the objective function J_2. Then, we jointly train the model using the objective function J.

4 Experiments

4.1 Datasets

We evaluate our proposed model on three widely used datasets, including MSRA [15], OntoNotes 4 [27] and WeiboNER [8,20]. The MSRA dataset contains three entity types (person, location and organization). The OntoNotes 4 dataset annotates 18 named entity types. In this work, we use the four most common named entity types (person, location, organization and geo-political), as same as previous studies [3,33]. The WeiboNER dataset is annotated with four entity types (person, location, organization and geo-political), including named entities and nominal mentions.

For MSRA dataset, we use the same data split as Dong et al. [6]. For OntoNotes 4 dataset, we take the same data split as Che et al. [3] and Zhang et al. [33]. For WeiboNER dataset, we use the same training, development and testing splits as Peng et al. [20] and He et al. [8].

4.2 Evaluation Metrics and Experimental Settings

For evaluation metrics, we use the Micro averaged Precision (P), Recall (R) and F1 score as metrics in our experiments, as the same as previous works [3,33], which are calculated per-span.

Hyper-parameters tuning is made through adjustments according to the performance on development sets. The size of character embedding d_c is 100. The size of word embedding d_w is 50. The hidden size of LSTM d_h is set to 300. The dropout rate is 0.3. The λ is set to 0.1. Adam [12] is used for optimization, with an initial learning rate of 0.001. The character embeddings are pre-trained on Chinese Wikipedia corpus by using word2vec toolkit [19]. We use the same lexicon as Zhang et al. [33].

4.3 Compared with State-of-the-art Methods

Evaluation on MSRA. We compare our proposed model with previous methods on MSRA dataset. The results are listed in Table 1. Zhang et al. [31] leverage rich handcrafted features for Chinese NER. The model gives very competitive performance. Dong et al. [6] incorporate radical features into neural LSTM+CRF model, achieving the F1 score of 90.95%. We can observe that our proposed

Table 1. Experimental results on MSRA dataset.

Models	P (%)	R (%)	F1 (%)
Chen et al. [4]	91.22	81.71	86.20
Zhou et al. [34]	88.94	84.20	86.51
Zhang et al. [31]*	92.20	90.18	91.18
Zhou et al. [35]	91.86	88.75	90.28
Dong et al. [6]	91.28	90.62	90.95
Zhang et al. [33]	93.57	92.79	93.18
Cao et al. [2]	91.73	89.58	90.64
AMMNHT	**93.62**	**92.96**	**93.29**

* in Table 1, 2 and 3 denotes that a model exploits additional labeled data.

model gets significant improvements over previous state-of-the-art methods. For example, compared with the latest model [2] which uses additional CWS training data, our proposed method improves the F1 score from 90.64% to 93.29%. Moreover, compared with Zhang et al. [33], our model also greatly improves the performance, which indicates that our method outperforms all of the compared methods.

Table 2. Experimental results on OntoNotes 4 dataset. The first and second blocks list word-based methods and character-based method, respectively.

Models	P (%)	R (%)	F1 (%)
Che et al. [3]*	77.71	72.51	75.02
Wang et al. [26]*	76.43	72.32	74.32
Yang et al. [29]	65.59	71.84	68.57
Yang et al. [29]*	72.98	80.15	76.40
Zhang et al. [33]	76.35	71.56	73.88
AMMNHT	**76.51**	**71.70**	**74.03**

* in Table 1, 2 and 3 denotes that a model exploits additional labeled data.

Evaluation on OntoNotes. We evaluate our proposed model on OntoNotes 4 dataset. Table 2 lists the results of our proposed model and previous state-of-the-art methods. In the first two blocks, we give the performance of word-based and character-based methods for Chinese NER, respectively. Based on the gold segmentation, Che et al. [3] propose an integer linear program based inference algorithm with bilingual constraints for NER. The model gives a 75.02% F1 score. With gold word segmentation, the word-based models achieve better performance than the character-based model. This demonstrates that word boundaries information is useful for Chinese NER task. Compared with the character-based method [33], our model improves the F1 score from 73.88% to 74.03%. Compared with the word-based method [26], our model also achieves better performance. The great improvements over previous state-of-the-art methods demonstrate the effectiveness of our proposed model.

Table 3. F1 scores (%) on WeiboNER dataset.

Models	NE	NM	Overall
Peng et al. [20]	51.96	61.05	56.05
Peng et al. [21]*	**55.28**	62.97	58.99
He et al. [8]	50.60	59.32	54.82
He et al. [9]*	54.50	62.17	58.23
Zhang et al. [33]	53.04	62.25	58.79
Cao et al. [2]	54.34	57.35	58.70
AMMNHT	54.09	62.43	**59.04**

* in Table 1, 2 and 3 denotes that a model exploits additional labeled data.

Evaluation on WeiboNER. We compare our proposed model with the latest models on WeiboNER dataset. The experimental results are shown in Table 3, where NE, NM and Overall denote F1 scores for named entities, nominal entities and both, respectively. Peng et al. [21] propose a model that jointly performs Chinese NER and CWS task, which achieves better results than Peng et al. [20] for named entity, nominal mention and overall. Recently, Zhang et al. [33] propose a lattice LSTM model to exploit word sequence information. The model gives a 58.79% F1 score on overall performance. It can be observed that our proposed model achieves great improvements compared with previous methods. For example, compared with the lattice LSTM model, our proposed model improves the F1 score from 53.04% to 54.09% for named entity. It proves the effectiveness of our proposed model.

Table 4. F1 score (%) of AMMNHT and its simplified models on MSRA, OntoNotes 4 and WeiboNER datasets, respectively.

Models	MSRA	OntoNotes	WeiboNER
BiLSTM+CRF	89.13	63.17	55.84
BiLSTM+CRF+AMMN	92.40	73.11	58.65
BiLSTM+HT	90.53	64.14	56.55
AMMNHT	**93.29**	**74.03**	**59.04**

4.4 Ablation Experiment

To investigate the effectiveness of adaptive multi-pass memory network and hierarchical tagging mechanism, we conduct the ablation studies. The baseline and simplified models of the proposed model are detailed as follows: (1) **BiLSTM+CRF**: The model is exploited as the strong baseline in our experiment. (2) **BiLSTM+CRF+AMMN**: The model integrates lexical knowledge from a lexicon via adaptive multi-pass memory network. (3) **BiLSTM+HT**: The model exploits the BiLSTM to extract features and uses the hierarchical tagging layer to predict labels.

From the results listed in Table 4, we have several important observations as follows:

- **Effectiveness of Adaptive Multi-pass Memory Network.** We observe that the BiLSTM+CRF+AMMN model outperforms the BiLSTM+CRF on these three datasets. For example, compared with the baseline, it improves the F1 score from 89.13% to 92.40% on MSRA dataset. Compared the AMMNHT with BiLSTM+HT, we can find similar phenomenon. The great improvements demonstrate the effectiveness of the adaptive multi-pass memory network.
- **Effectiveness of Hierarchical Tagging Mechanism.** Compared with the BiLSTM+CRF, the BiLSTM+HT model improves the performance, achieving 1.40% improvements of F1 score on MSRA dataset. Moreover, the AMMNHT also outperforms the BiLSTM+CRF+AMMN. The great improvements indicate the hierarchical tagging mechanism is very effective for Chinese NER task.
- **Effectiveness of Adaptive Multi-pass Memory Network and Hierarchical Tagging Mechanism.** We observe that the proposed model AMMNHT achieves better performance than its simplified models on the three datasets. For example, compared with BiLSTM+CRF, the AMMNHT model improves the F1 score from 89.13% to 93.29% on MSRA dataset. It indicates that simultaneously exploiting the adaptive multi-pass memory network and hierarchical tagging mechanism is also very effective.

Table 5. F1 score (%) of different passes from 1 to 5 and adaptive passes on the test sets. It shows suitable reasoning passes of memory network can boost the performance.

Pass	MSRA	OntoNotes	WeiboNER
1	92.64	72.87	58.52
2	92.96	73.50	58.83
3	93.14	73.77	58.74
4	93.12	73.85	58.34
5	93.03	73.46	58.13
Adaptive	**93.29**	**74.03**	**59.04**

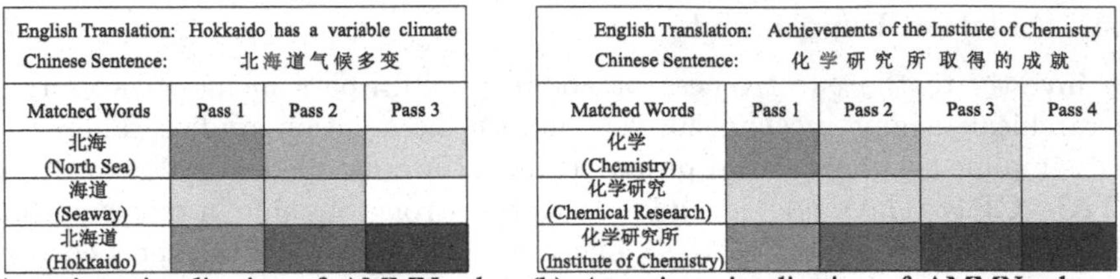

English Translation: Hokkaido has a variable climate Chinese Sentence: 北海道气候多变			
Matched Words	Pass 1	Pass 2	Pass 3
北海 (North Sea)			
海道 (Seaway)			
北海道 (Hokkaido)			

(a) Attention visualization of AMMN when learning lexical knowledge for the candidate character "海 (sea)".

English Translation: Achievements of the Institute of Chemistry Chinese Sentence: 化 学 研 究 所 取 得 的 成 就				
Matched Words	Pass 1	Pass 2	Pass 3	Pass 4
化学 (Chemistry)				
化学研究 (Chemical Research)				
化学研究所 (Institute of Chemistry)				

(b) Attention visualization of AMMN when learning lexical knowledge for the candidate character '学 (subject)".

Fig. 3. Two examples of attention weights in adaptive multi-pass memory network. The reasoning passes are 3 and 4, respectively. Darker colors mean that the attention weight is higher.

4.5 Adaptive Multiple Passes Analysis

To better illustrate the influence of multiple passes and adaptive multi-pass memory network, we give the results of fixed multiple passes and adaptive multi-pass memory network in Table 5. The results show that multiple passes operation performs better than one pass. The reason is that multiple passes reasoning can help to highlight the most appropriate matched words. The cases in Fig. 3 show that the deep deliberation can recognize the correct lexical knowledge by enlarging the attention gap between correct matched words and incorrect ones. When the number of passes is too large, the performance stops increasing or even decreases due to over-fitting. In contrast to the fixed multiple passes memory network, the adaptive multi-pass memory network has 0.21% improvements of F1 score on the WeiboNER dataset. Furthermore, the two examples in Fig. 3 show that adaptive multi-pass memory network can choose suitable reasoning passes according to the complexity of the input sentence, which also demonstrates the effectiveness of the adaptive multi-pass memory network.

5 Conclusion

In this paper, we propose an adaptive multi-pass memory network to incorporate lexical knowledge from a lexicon for Chinese NER task which can adaptively choose suitable reasoning passes according to the complexity of each sentence. Besides, we devise a hierarchical tagging layer to capture tag dependencies in the whole sentence. The adaptive memory network and hierarchical tagging mechanism can be easily applied to similar tasks involving multi-pass reasoning and decoding process, such as knowledge base question answering and machine translation. Experimental results on three widely used datasets demonstrate that our proposed model outperforms previous state-of-the-art methods.

Acknowledgments. This work is supported by the Natural Key R&D Program of China (No. 2017YFB1002101), the National Natural Science Foundation of China (No.

61533018 No. 61922085, No. 61976211, No. 61806201) and the Key Research Program of the Chinese Academy of Sciences (Grant NO. ZDBS-SSW-JSC006). This work isalso supported by Beijing Academy of Artificial Intelligence (BAAI2019QN0301), the CCF-Tencent Open Research Fund and independent research project of National Laboratory of Pattern Recognition.

References

1. Bunescu, R., Mooney, R.: A shortest path dependency kernel for relation extraction. In: Proceedings of EMNLP, pp. 724–731 (2005)
2. Cao, P., Chen, Y., Liu, K., Zhao, J., Liu, S.: Adversarial transfer learning for Chinese named entity recognition with self-attention mechanism. In: Proceedings of EMNLP (2018)
3. Che, W., Wang, M., Manning, C.D., Liu, T.: Named entity recognition with bilingual constraints. In: Proceedings of NAACL-HLT, pp. 52–62 (2013)
4. Chen, A., Peng, F., Shan, R., Sun, G.: Chinese named entity recognition with conditional probabilistic models. In: Proceedings of the Fifth SIGHAN Workshop on Chinese Language Processing, pp. 213–216 (2006)
5. Chen, Y., Xu, L., Liu, K., Zeng, D., Zhao, J.: Event extraction via dynamic multi-pooling convolutional neural networks. In: Proceedings of ACL, pp. 167–176 (2015)
6. Dong, C., Zhang, J., Zong, C., Hattori, M., Di, H.: Character-based LSTM-CRF with radical-level features for Chinese named entity recognition. In: Lin, C.-Y., Xue, N., Zhao, D., Huang, X., Feng, Y. (eds.) ICCPOL/NLPCC -2016. LNCS (LNAI), vol. 10102, pp. 239–250. Springer, Cham (2016). https://doi.org/10.1007/978-3-319-50496-4_20
7. Gui, T., et al.: A lexicon-based graph neural network for Chinese NER. In: EMNLP-IJCNLP (2019)
8. He, H., Sun, X.: F-score driven max margin neural network for named entity recognition in Chinese social media. arXiv preprint arXiv:1611.04234 (2016)
9. He, H., Sun, X.: A unified model for cross-domain and semi-supervised named entity recognition in Chinese social media. In: Proceedings of AAAI (2017)
10. Huang, Z., Xu, W., Yu, K.: Bidirectional LSTM-CRF models for sequence tagging. arXiv preprint arXiv:1508.01991 (2015)
11. Isozaki, H., Kazawa, H.: Efficient support vector classifiers for named entity recognition. In: Proceedings of the 19th International Conference on Computational Linguistics, vol. 1, pp. 1–7 (2002)
12. Kingma, D.P., Ba, J.: Adam: a method for stochastic optimization. arXiv preprint arXiv:1412.6980 (2014)
13. Lafferty, J., McCallum, A., Pereira, F.C.: Conditional random fields: probabilistic models for segmenting and labeling sequence data (2001)
14. Lample, G., Ballesteros, M., Subramanian, S., Kawakami, K., Dyer, C.: Neural architectures for named entity recognition. In: Proceedings of NAACL-HLT, pp. 260–270 (2016)
15. Levow, G.A.: The third international Chinese language processing bakeoff: word segmentation and named entity recognition. In: Proceedings of the Fifth SIGHAN Workshop on Chinese Language Processing, pp. 108–117 (2006)
16. Lu, Y., Zhang, Y., Ji, D.H.: Multi-prototype Chinese character embedding. In: Proceedings of LREC (2016)
17. Luo, F., Liu, T., Xia, Q., Chang, B., Sui, Z.: Incorporating glosses into neural word sense disambiguation. In: Proceedings of ACL, pp. 2473–2482 (2018)

18. Ma, X., Hovy, E.: End-to-end sequence labeling via bi-directional LSTM-CNNs-CRF. In: Proceedings of ACL, pp. 1064–1074 (2016)
19. Mikolov, T., Chen, K., Corrado, G., Dean, J.: Efficient estimation of word representations in vector space. arXiv preprint arXiv:1301.3781 (2013)
20. Peng, N., Dredze, M.: Named entity recognition for Chinese social media with jointly trained embeddings. In: Proceedings of EMNLP, pp. 548–554 (2015)
21. Peng, N., Dredze, M.: Improving named entity recognition for Chinese social media with word segmentation representation learning. In: Proceedings of ACL, pp. 149–155 (2016)
22. Peters, M., Ammar, W., Bhagavatula, C., Power, R.: Semi-supervised sequence tagging with bidirectional language models. In: Proceedings of ACL, pp. 1756–1765 (2017)
23. Sui, D., Chen, Y., Liu, K., Zhao, J., Liu, S.: Leverage lexical knowledge for Chinese named entity recognition via collaborative graph network. In: EMNLP-IJCNLP (2019)
24. Sukhbaatar, S., Weston, J., Fergus, R., et al.: End-to-end memory networks. In: Proceedings of NeurIPS, pp. 2440–2448 (2015)
25. Sutton, R.S., Barto, A.G., et al.: Introduction to reinforcement learning. MIT Press, Cambridge (1998)
26. Wang, M., Che, W., Manning, C.D.: Effective bilingual constraints for semi-supervised learning of named entity recognizers. In: Proceedings of AAAI (2013)
27. Weischedel, R., et al.: OntoNotes release 4.0. LDC2011T03, Philadelphia, Penn.: Linguistic Data Consortium (2011)
28. Yahya, M., Berberich, K., Elbassuoni, S., Weikum, G.: Robust question answering over the web of linked data. In: Proceedings of CIKM, pp. 1107–1116 (2013)
29. Yang, J., Teng, Z., Zhang, M., Zhang, Y.: Combining discrete and neural features for sequence labeling. In: International Conference on Intelligent Text Processing and Computational Linguistics, pp. 140–154 (2016)
30. Yang, Z., Yang, D., Dyer, C., He, X., Smola, A., Hovy, E.: Hierarchical attention networks for document classification. In: Proceedings of NAACL-HLT, pp. 1480–1489 (2016)
31. Zhang, S., Qin, Y., Wen, J., Wang, X.: Word segmentation and named entity recognition for SIGHAN bakeoff3. In: Proceedings of the Fifth SIGHAN Workshop on Chinese Language Processing, pp. 158–161 (2006)
32. Zhang, Y., Chen, H., Zhao, Y., Liu, Q., Yin, D.: Learning tag dependencies for sequence tagging. In: Proceedings of IJCAI, pp. 4581–4587 (2018)
33. Zhang, Y., Yang, J.: Chinese NER using lattice LSTM. In: Proceedings of ACL, pp. 1554–1564 (2018)
34. Zhou, J., He, L., Dai, X., Chen, J.: Chinese named entity recognition with a multi-phase model. In: Proceedings of the Fifth SIGHAN Workshop on Chinese Language Processing (2006)
35. Zhou, J., Qu, W., Zhang, F.: Chinese named entity recognition via joint identification and categorization. Chin. J. Electron. **22**, 225–230 (2013)

A Practice of Tourism Knowledge Graph Construction Based on Heterogeneous Information

Dinghe Xiao[1], Nannan Wang[2], Jiangang Yu[1], Chunhong Zhang[2](✉), and Jiaqi Wu[1]

[1] Hainan Sino-intelligent-Info Technology Ltd., Hainan, China
[2] School of Information and Communication Engineering, Beijing University of Posts and Telecommunications, Beijing, China
zhangch@bupt.edu.cn

Abstract. The increasing amount of semi-structured and unstructured data on tourism websites brings a need for information extraction (IE) so as to construct a Tourism-domain Knowledge Graph (TKG), which is helpful to manage tourism information and develop downstream applications such as tourism search engine, recommendation and Q & A. However, the existing TKG is deficient, and there are few open methods to promote the construction and widespread application of TKG. In this paper, we present a systematic framework to build a TKG for Hainan, collecting data from popular tourism websites and structuring it into triples. The data is multi-source and heterogeneous, which raises a great challenge for processing it. So we develop two pipelines of processing methods for semi-structured data and unstructured data respectively. We refer to tourism InfoBox for semi-structured knowledge extraction and leverage deep learning algorithms to extract entities and relations from unstructured travel notes, which are colloquial and high-noise, and then we fuse the extracted knowledge from two sources. Finally, a TKG with 13 entity types and 46 relation types is established, which totally contains 34,079 entities and 441,371 triples. The systematic procedure proposed by this paper can construct a TKG from tourism websites, which can further applied to many scenarios and provide detailed reference for the construction of other domain-specific knowledge graphs.

Keywords: Heterogeneous data · TKG · Systematic procedure

1 Introduction

Tourism has become increasingly popular in people's daily life. Before people set out to travel, they often need to make clear the travel guides and matters needing attention for their destinations. Nowadays, with the development of the Internet, many tourism websites have appeared and provide a variety of travel information, such as attractions, tickets, bus routes, travel guides, etc. However, there

M. Sun et al. (Eds.): CCL 2020, LNAI 12522, pp. 159–173, 2020.
https://doi.org/10.1007/978-3-030-63031-7_12

may be some errors in the miscellaneous information on the tourism websites, and information on different tourism websites may be inconsistent. As shown in screenshots of Sina Micro-Blog users' blogs in Fig. 1, there are still tourists who are worried about making travel strategies despite rich information on all kinds of tourism-related search engines. How to collect and integrate valuable tourism knowledge on websites is a very important issue.

Fig. 1. Screenshots of Sina Micro-Blog users' blogs. In the blogs, people with tourism intentions complain that it is difficult to formulate travel strategies.

Recently, Knowledge Graph (KG) has received much attention and research interest in industry and academia. The KG utilizes a set of subject-predicate-object triplets to represent the diverse entities and their relations in real-world scenes, which are respectively represented as nodes and edges in the graph. The KG is a graph-based large-scale knowledge representation and integration method, which has been applied in various scenarios such as enterprise [15], medical [17] and industry [29]. Naturally, we consider applying KG in the field of Tourism to integrate and organize relevant knowledge, so as to provide tourists with easier tools to develop travel strategies.

At present, several General Knowledge Graphs (GKGs) have been built both in Chinese and English [1,16,20,21]. The Domain-specific Knowledge Graph (DKG) in which the stored knowledge is limited to a certain field has also been implemented and put into use in many domains [30]. However, Tourism-domain Knowledge Graph (TKG) is still deficient, which undoubtedly hinders the development of intelligent tourism system. In this paper, we propose a systematic framework to construct a TKG under the background of Hainan Tourism. We combine the semi-structured knowledge crawled from the encyclopedia pages of tourism websites with the unstructured travel notes shared by tourists on the websites as the data source. Because of the lack of sufficient high-quality data and the difficulty of language processing, constructing a Chinese-based TKG still faces several challenges as follows:

Travel Notes are Colloquial and High-Noise. The writing style of travel notes is often arbitrary, and tourists tend to add various pictures, emoticons and special characters to travel notes, which will introduce much noise for unstructured data.

The Lack of Datasets Dedicated to Tourism. There is a serious lack of normative datasets in the tourism field, which are basis of model training.

Are the General Algorithms Suitable for Tourism? Entity extraction and relation extraction are the key steps in knowledge graph construction. Most of the existing algorithms for these two tasks are tested on the general datasets, we need to verify whether these algorithms are suitable for the tourism field.

How to Integrate Data from Different Sources? Data from different sources inevitably have some overlaps and ambiguities, which should be eliminated in the KG.

Facing this challenges, we put forward corresponding methods to deal with them. In detail, the contributions of our work are highlighted as follows:

- A specific method of collecting and processing tourism-domain data is described, and labeled datasets for information extraction in the field of tourism is constructed;
- The most suitable models for our tourism data are identified, and a tourism-domain knowledge graph is finally constructed.
- Experience in constructing the TKG can provide detailed reference for the construction of other domain-specific knowledge graphs.

2 Related Work

In recent years, the KG has been applied in many fields to complete knowledge storage, query, recommendation and other functions. In the tourism scene, experts and scholars have also begun to explore the application value of knowledge graphs. DBtravel [3] is an English tourism-oriented knowledge graph generated from the collaborative travel site Wikitravel. A Chinese TKG was also constructed by [27], which extracted tourism-related knowledge from existing Chinese general knowledge graph such as zhishi.me [16] and CN-DBpedia [21]. Unlike their Chinese TKG, we extensively obtain data and extract knowledge from popular tourism websites. In this way, the completeness of our knowledge graph does not depend on the existing knowledge graph, but on the amount of data we acquire. To construct the TKG, we need to extract triples form all kinds of information resources. The conversion process from semi-structured data to structured data is more standardized and has fewer errors, but semi-structured data often cannot contain all the knowledge. With the development of Natural Language Processing (NLP), more and more knowledge graphs are constructed based on unstructured corpus, using named entity recognition (NER) and relation extraction (RE) technologies.

As a hot research direction in the field of NLP, many Chinese NER models have been proposed over the years. The purpose of NER task is to identify mentions of named entities from text and match them to pre-defined categories.

As a classic branch of NER models, the dictionary-based methods recognize named entities by constructing a dictionary and matching text with it. For example, CMEL [14] built a synonym dictionary for Chinese entities from Microblog and adopts improved SVM to get textual similarity for entity disambiguation. Another line of related work is to apply traditional machine learning techniques to complete the NER task, just like the Conditional Random Fields (CRFs)-based NER System proposed by [7]. Recently, neural network-based (NN-based) models have shown great future prospects in improving the performance of NER systems, including bidirectional Long Short-Term Memory (LSTM) model [9], lattice-structured LSTM model [28], convolution neural network (CNN)-based model [6] and so on. In our work, we adopt the most mainstream NN-based NER algorithm at present, which combines BiLSTM and CRF.

Relation extraction (RE) is also one of the most important tasks in NLP. On the premise of pre-defined relation categories, RE is often transformed into a relation classification task. Similar to entity extraction, the mainstream algorithms for RE in recent years have also focused on NN-based ones. [23] utilized CNNs to classify relations and made representative progress. However, because CNN can not extract contextual semantic information well, recurrent neural network (RNN) [24], which is often used to process texts, is proposed for relation extraction. Since RNN is difficult to learn long-term dependencies, LSTM [26] was introduced into the RE task. To capture the most important information in a sentence, Attention-Based Bidirectional Long Short-Term Memory Networks (Att-BLSTM) [31] was come up and become a popular RE algorithm. The above supervise learning algorithms are time-consuming and costly to label data. In order to solve these problems, some distant supervision algorithms have also been developed [8,11,22]. Because the TKG only contains knowledge in the field of tourism, the corpus for training is not large, so we do not consider using distant supervision algorithms.

3 Implementation

In this paper, we crawl semi-structured and unstructured data related to Hainan Tourism from popular travel websites, and extract the structured knowledge from these two types of data in two pipelines. Figure 2 shows the overview of our method.

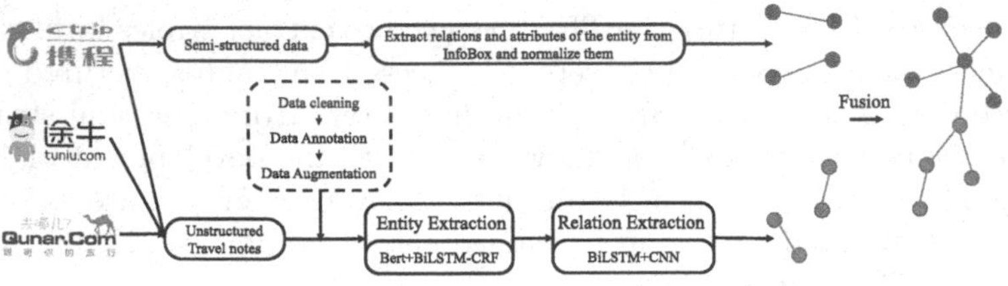

Fig. 2. The overview of our method.

3.1 Data Preparation

Tourism is an intelligent application market with great potential. Tourism data on the Internet has a large quantity but not effectively used, and standardized tourism datasets are not yet available. In this section, we will describe our data preparation process in detail, which is mainly divided into four steps including data acquisition, data cleaning, data annotation and data augmentation, and the last three steps are mainly aimed at unstructured data that is noisy and irregular.

Data Acquisition: This step aims to collect raw data in the field of tourism, which will be processed later to be used as input to the information extraction models. There are many popular Chinese tourism websites that cover numerous tourism-related knowledge on the Internet. We crawled semi-structured data on the Ctrip[1], where tourism-related entities (scenic areas, hotels, cities, etc.) have their corresponding descriptive pages. The Information Boxes (InfoBox) in these pages with clear structure contain a great number of named entities, relations and attributes, which can be used to fill the TKG. For example, the InfoBox of "Haikou Ublaya Inn" is shown in the Fig. 3(a). Meanwhile, we crawled tourists' travel notes related to Hainan on the three major Chinese travel websites, Ctrip[2], Tuniu[3] and Qunar[4]. Travel notes are rich in content and easy to obtain, which may supplement the information not contained in semi-structured data, and Fig. 3(b) shows an example of travel notes on the Tuniu.

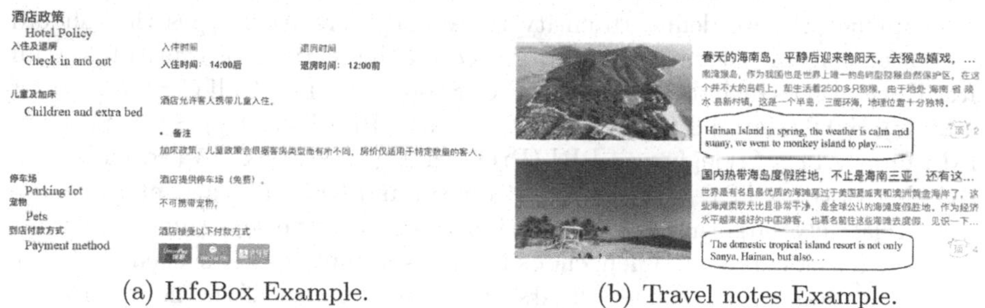

(a) InfoBox Example. (b) Travel notes Example.

Fig. 3. An example of (a) an InfoBox of "Haikou Ublaya Inn" and (b) travel notes related to Hainan on the Tuniu, which respectively correspond to the semi-structured data and unstructured data that we want to crawl on the travel websites.

We have crawled 33177 pages corresponding to Hainan-related entities from the Ctrip. In addition, a total of 19,023 travel notes are obtained after crawling the above three popular websites. The combination of semi-structured data and

[1] https://you.ctrip.com/place/100001.html.
[2] https://you.ctrip.com/travels/.
[3] https://trips.tuniu.com/.
[4] https://travel.qunar.com/.

unstructured data helps to provide a more complete source of information in the construction of TKG.

Data Cleaning: For unstructured data, due to the colloquial and casual nature, the travel notes crawled from the travel websites usually contain some noise that should be cleaned up, including some inconsistent Traditional Chinese characters, emoticons, Uniform Resource Locator (URL) links and some special characters like #, &, $, {, }, etc. We mainly delete these redundant contents through regular expressions. In view of the fact that some paragraphs in travel notes are relatively longer than the ideal length required by the models for entity extraction and relation extraction, we further perform paragraph segmentation to reduce the pressure of model training.

Data Annotation: For unstructured text, we should label it to build datasets that meet the training requirements for subsequent entity recognition and relation recognition algorithms. Before annotating data, we must first define the types of entities and relations that need to be extracted in the field of tourism. In order to truly understand the issues that users are concerned about, we crawl the text about the keyword "Hainan" in the QA modules of Ctrip and Tuniu, mainly including some users' questions and the answers given by other users, and then the word frequency in the Q & A data is analyzed through TF-IDF (Term Frequency-Inverse Document Frequency) algorithm. The statistical results of word frequency in our work are shown in Fig. 4(a). The results show that high-frequency words are mainly concentrated on types such as hotel, scenic spot, city, food, restaurant, etc. Referring to the definition of entities and relations in CN-DBpedia [21], we define 16 entity types and 51 relation types that should be extracted from unstructured data based on the features of tourism-domain data. Entity types include DFS (Duty Free Shop), GOLFC (Golf Course), FUNF (Funfair), HOT (Hotel), FOLKC (Folk Custom), SPE (Specialty), SNA (Snacks), TIM (Time), TEL (Telephone), PRI (Price), TIC (Ticket), SCEA (Scenic Area), PRO (Province), CITY (City), COU (County) and RES (Restaurant). Because of the relatively large number of relation types, we give an example to illustrate the relation types. When choosing a restaurant, tourists need to figure out the location, price, business hours and phone number of the hotel, and the location must be specific. So we define seven relations for RES type, including res_locatedin_scea, res_locatedin_pro, res_locatedin_city, res_locatedin_cou, res_open_time, res_phonenumber, res_PRI, where res_locatedin_scea means that the restaurant is in a certain scenic area, and the explanation of the remaining relations is similar.

After defining the entity & relation types to be extracted, for a sentence in travel notes, we should first label entity mentions in it, and then label the relation between entity pairs according to semantics. We adopt BRAT [19] as the main tool to label entities and relations in the text. There exist some problems when using BRAT to label entities and relations in the field of tourism. When labeling entities, 1) The travel notes are expressed by different people in a colloquial way, which makes it difficult to determine the boundary of the entities. We reasonably label the entity mentions with the boundary as large as possible, so as to make

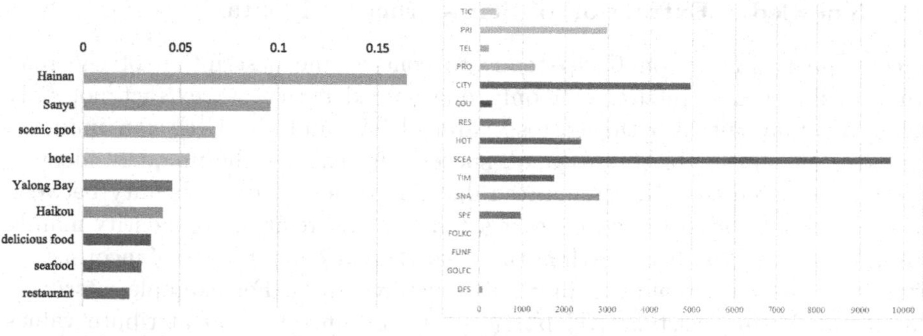

(a) Statistics of word frequency. (b) Statistics of entities of 16 types.

Fig. 4. (a) Word frequency statistics in the Q & A data, where high-frequency words need to be focused on; (b) Statistics of the number of 16 types of entities, it shows that the number of entities is unevenly distributed.

the entity mention more complete and specific; 2) In different contexts, entities with the same mention may belong to different types. So we label relations based on the semantics of the context. There are also some problems when labeling relations, 1) When multiple entities appear in a sentence, and there is more than one entity pair that has connections, we label as many entity-relation-entity combinations to obtain adequate relation annotated data; 2) A sentence may contain two entities, and there may be a connection between the two entities according to external knowledge, but the context of the sentence cannot reflect this connection. For this situation, we will not label the relation, so as not to have a negative impact on the subsequent training of the RE model.

After handling the above problems, 1902 travel notes are annotated. Because labeling relations needs to consider the context, which affects the speed of the labeling, we have not labeled all crawled travel notes, but only labeled the number enough to train the models. The details of the datasets will be shown in Sect. 4.1.

Data Augmentation: The number of entities in travel notes is not evenly distributed in categories. We make statistics on the number of entities of each entity type contained in the annotated dataset, as shown in the Fig. 4(b). We can see that there are a large number of labeled entities in SCEA and CITY types, and the proportion of other types is relatively small. In order to reduce the training error brought by data imbalance, we use substitution method to expand the types with a small amount of data. We take the DFC entities with a small proportion as example. First, select some sentences containing the DFC entity from the dataset, and then replace the DFC entity mentions in each sentence with other different DFC mentions. Although such replacement destroys the authenticity of the original data, the training for models is appropriate. We use this technology to augment a total of more than 8,000 pieces of sentence.

3.2 Knowledge Extraction of Semi-structured Data

Since a page crawled on Ctrip tends to contain the description of the relevant attributes and relations of only one named entity, we extract not only entity mention but also the corresponding URL, and the URL can uniquely represent the entity. In this way, we successfully extract the uniquely identifiable entities from the semi-structured data, and there is no ambiguity between these entities. In addition, we extract attributes and relations of a entity mainly through the InfoBox. It is worth noting that there are many cases of inconsistent attributes and value conflicts in Semi-structured data. For example, attribute names can be inconsistent (telephone, contact number), and attribute values can be inconsistent(086-6888-8888 and 68888888), so for the extracted semi-structured knowledge, we further refer to CN-DBpedia [21] for attribute normalization and value normalization to further obtain well-organized knowledge, and then we finally obtain about 370,000 triples from semi-structured data.

3.3 Knowledge Extraction of Unstructured Data

In this section, we mainly utilize mainstream deep learning algorithms to extract entities and relations in unstructured text.

Entity Extraction. For unstructured data from tourist travel notes, we take the method of Named Entity Recognition (NER) to extract entity mentions from the text. The main work of NER is sequence labeling, and Long Short-Term Memory (LSTM) networks have natural advantages in processing time series related tasks. The Conditional Random Field (CRF) model can effectively consider the mutual influence of output labels between characters. Therefore, the BiLSTM model and the CRF model are usually used together to become the mainstream model in the NER field.

Based on the classic BiLSTM-LSTM [10] model, We make further improvements in the embedding layer. After the Google BERT [5] model was proposed, the innovation of the pretrained language model has enabled many NLP tasks to achieve state-of-the-art performance, and large pretrained language models have become a hot tool. After BERT, other large pretrained language models like ALBERT [12] model have also been proposed. ALBERT is a simplified BERT version, and the number of parameters is much smaller than the traditional BERT architecture. In this paper, we utilize the pretrained BERT and ALBERT model to obtain the embedding matrix in embedding layer respectively, which is constant during the training process.

Relation Extraction. Relation Extraction (RE) is an important task of natural language processing (NLP) and also a key link in knowledge graph construction. After RE, a triple (s, r, o) is usually obtained, where s represents the head entity, o represents the tail entity, and r represents the relation between them. In our travel data, the number of relations is limited, so we can choose to transform the RE into a relation classification task, and we treat each relation type as a class. Comprehensively considering the advantages and disadvantages of the mainstream relationship classification model and characteristics of tourism data, we choose to adopt supervised algorithm, BiLSTM+CNN [25], for RE task in our work, whose framework can be shown in Fig. 5. CNN can extract local features of sentences, but it is not good at handling long dependencies among words, which can be made up by BiLSTM.

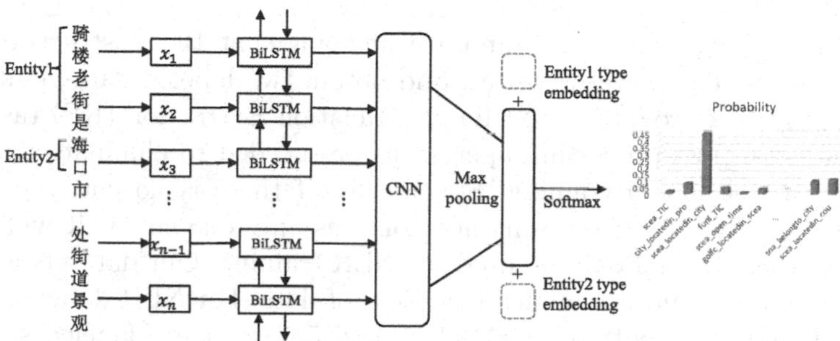

Fig. 5. The framework of the relation extraction model based on BiLSTM+CNN. The input in English is "Qilou Old Street is a street view of Haikou City", and from the bar graph we know that the output relation is scea_locatedin_city, then we can get the triple (Qilou Old Street, scea_locatedin_city, Haikou).

At the same time, considering that the entity category information may have an impact on the relation classification, the entity type information is introduced into the model [13]. Specifically, each entity type is represented as distributed embedding. As is shown in Fig. 5, after the CNN layer, we concatenate the entity type embedding of entity1 and entity2 with the output vector of Max pooling layer, and then feed it to the fully connected layer for subsequent label prediction.

Entity Alignment. There is often a situation where multiple mentions refer to the same entity. Entity alignment is to determine whether two entities with different mentions are the same entity by calculating and comparing their similarity. We observe the names of the entities that need to be aligned and find that the names of the two entities to be aligned are similar in most cases, just like "Nantian Ecological Grand View Garden" and "Nantian Grand View Garden". Therefore, basic distance measurement-based models are suitable enough for our entity alignment task, which is to calculate the distance between the names of the two entities. Common distance measurement algorithms include Jaccard coefficient, Euclidean distance, and editing distance. We weight and sum the distances measured under these three distance metrics, so as to discriminate whether entities with different names belong to the same entity. Although this method is simple, but it can solve most of the problems we encounter. Finally, we obtain about 220,000 triples from unstructured data.

In summary, we first construct independent knowledge graphs from two heterogeneous data sources respectively, and then we fuse the two sub-knowledge graphs to obtain a more complete knowledge graph, which is the Tourism-domain Knowledge Graph finally constructed.

4 Experiments

4.1 Datasets

In Sect. 3.1, we acquire, clean, annotate and augment the unstructured text crawled from popular travel websites, and obtain two labeled datasets suitable for Named Entity Recognition (NER) and Relation Extraction (RE) tasks. For labeled datasets, post-processing operations are needed to eliminate data that is meaningless for model training. Specifically, if there is no entity in a sentence, delete it directly. If the sentence contains only one entity, it will be cut to the proper length and only be used for NER training. Our datasets are both based on sentences, and a sentence is a piece of data. For NER dataset, we use train, valid, and test splits of 5490, 1178, and 591 sequence labeled sentences respectively. And train, valid, and test sets for RE task contain 6225, 1000 and 400 sentences respectively. Using the datasets we construct and divide, we next conduct comparative experiments to measure the model performance.

4.2 Model Training and Results

In order to obtain a named entity recognition model suitable for tourism-domain data, we compare several mainstream NER models including BERT [2], ALBERT [12], BiLSTM-CRF [10], BERT+BiLSTM-CRF [4], and BERT-CRF [18] on our NER dataset. For this task, we use Precision (P), Recall (R) and F1 score (F1) to evaluate the effect of NER model, which are standard information extraction metrics. The experimental results in Table 1 show that the BiLSTM-CRF algorithm based on the pretrained language model BERT has the best performance with F1-score 90.6%. BERT+BiLSTM-CRF practiced by [4] is used to complete the task of Chinese electronic medical records named entity recognition, and BERT+BiLSTM-CRF achieves approximately 75% F1 score and performs better than other models like BiLSTM-CRF and BiGRU-CRF in their work, which is consistent with our results. Both in their and our practice, the effectiveness of combining pretrained models with mainstream models is reflected. Meanwhile, we can see that baselines other than BERT+BiLSTM-CRF that have good performance on the general standard datasets can also achieve comparative results in the application of actual projects.

The NER models share the same divided NER dataset and training environment, and all models are trained with 15 epochs.

Similar to entity extraction, we also compare three mainstream models in relation extraction task, including BiLSTM+ATT [31], CNN [23] and BiLSTM-CNN [25]. The evaluation metrics applied in RE models are also P, R and F1. Among these models, as is shown in Table 1, BiLSTM-CNN shows the relatively better performance than BiLSTM+ATT and CNN on our RE dataset.

Table 1. Comparison of experimental results with NER baselines and RE baselines on our datasets.

	Model	P	R	F1
5*NER	BiLSTM-CRF [10]	**0.890**	0.876	0.883
	BERT-CRF [18]	0.862	0.904	0.882
	BERT [2]	0.822	0.867	0.839
	ALBERT [12]	0.837	0.829	0.828
	BERT+BiLSTM-CRF [4]	0.887	**0.926**	**0.906**
4*RE	BiLSTM+ATT [31]	0.766	0.681	0.702
	CNN [23]	0.803	0.651	0.701
	BiLSTM-CNN [25]	**0.941**	0.791	0.842
	BiLSTM-CNN(with types)	0.918	**0.914**	**0.909**

In order to further verify the validity of adding entity type embedding in RE, comparative experiments are carried out on the model BiLSTM-CNN. Table 1 shows that by introducing entity type information, the F1 score of BiLSTM-CNN is improved by 6.7%, which is the highest among our experimental models. The main reason may be that by introducing entity type information into the model, the scope of classification is narrowed, that is to say, entity type information restricts the classification to a certain extent, so as to significantly improve the effect of relation classification. The above RE models share the same divided RE dataset and training environment, and all models are trained with 64 epochs.

To sum up, based on the above analysis of the experimental results of each model, BERT+BiLSTM-CRF is selected as NER model and BiLSTM + CNN model with entity type information introduced is selected as the RE model in our work.

4.3 Knowledge Construction

We fuse the two sub-knowledge graphs obtained from semi-structured data and unstructured data to get the complete TKG. The final TKG with a total of 441,371 triples contains 13 entity types and 46 relation types. In Fig. 6, a knowledge graph composed of partial triples is depicted. The central node is Sansha that belongs to CITY type, and we show a part of the nodes around it and the adjacent relations and attributes.

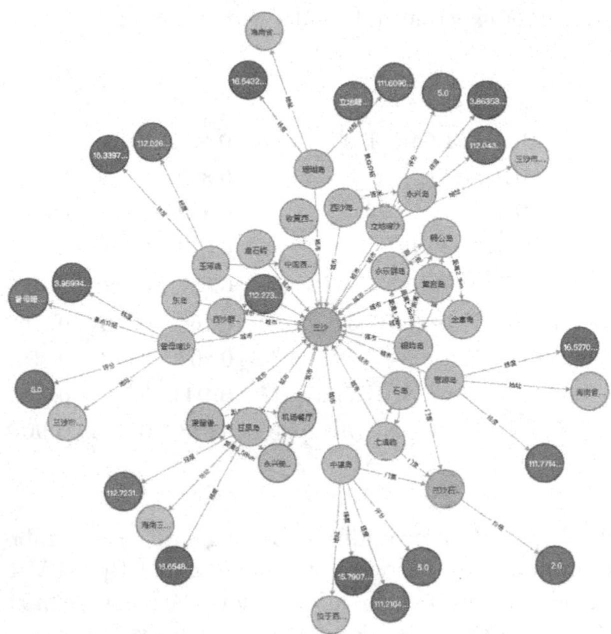

Fig. 6. Partial triples in tourism knowledge graph, which shows the part of the tourism-domain knowledge graph with CITY Sansha as the central node.

5 Conclusions

With the development of tourism, information management and utilization in the field of tourism is a very important task. We proposed a systematic approach to construct the Chinese tourism knowledge graph, using the information on the tourism websites. We leveraged semi-structured data and unstructured data to extract entities and relations synchronously, and they can be combined to obtain more complete sets of entities and relations than only one of them. Due to the lack of standardized datasets in the field of tourism, we first proposed a strategy for constructing datasets to facilitate the extraction of entities and relations from the complex network text data. In addition, we used several algorithms to complete the named entity recognition (NER) task and relation extraction (RE) task on the datasets we created, and compare the results. We found that BERT+BILSTM-CRF has the best performance for NER task and BiLSTM+CNN with entity type information introduced performs best on RE task.

We have implemented a relatively complete information extraction system on the tourism knowledge graph. In the future work, we want to solve the problem of how to update the knowledge in real time, because the knowledge on the tourism websites is always increasing and changing. In addition, we intend to explore some domain-adaptive techniques to make our model can be used widely.

Acknowledgements. This work is supported by the Science and Technology Department of Hainan Province, "Intelligent analysis platform of Hainan tourist's behavior and accurate service mining prediction" project (ZDKJ201808).

References

1. Auer, S., Bizer, C., Kobilarov, G., Lehmann, J., Cyganiak, R., Ives, Z.: DBpedia: a nucleus for a web of open data. In: Aberer, K., et al. (eds.) ASWC/ISWC -2007. LNCS, vol. 4825, pp. 722–735. Springer, Heidelberg (2007). https://doi.org/10.1007/978-3-540-76298-0_52
2. Cai, Q.: Research on Chinese naming recognition model based on BERT embedding. In: 2019 IEEE 10th International Conference on Software Engineering and Service Science (ICSESS), pp. 1–4. IEEE (2019)
3. Calleja, P., Priyatna, F., Mihindukulasooriya, N., Rico, M.: DBtravel: a tourism-oriented semantic graph. In: Pautasso, C., Sánchez-Figueroa, F., Systä, K., Murillo Rodríguez, J.M. (eds.) ICWE 2018. LNCS, vol. 11153, pp. 206–212. Springer, Cham (2018). https://doi.org/10.1007/978-3-030-03056-8_19
4. Dai, Z., Wang, X., Ni, P., Li, Y., Li, G., Bai, X.: Named entity recognition using BERT BiLSTM CRF for Chinese electronic health records. In: 2019 12th International Congress on Image and Signal Processing, BioMedical Engineering and Informatics (CISP-BMEI), pp. 1–5. IEEE (2019)
5. Devlin, J., Chang, M.W., Lee, K., Toutanova, K.: BERT: pre-training of deep bidirectional transformers for language understanding. arXiv preprint arXiv:1810.04805 (2018)
6. Gui, T., Ma, R., Zhang, Q., Zhao, L., Jiang, Y.G., Huang, X.: CNN-based Chinese NER with lexicon rethinking. In: Proceedings of the 28th International Joint Conference on Artificial Intelligence, pp. 4982–4988. AAAI Press (2019)
7. Han, A.L.-F., Wong, D.F., Chao, L.S.: Chinese named entity recognition with conditional random fields in the light of chinese characteristics. In: Kłopotek, M.A., Koronacki, J., Marciniak, M., Mykowiecka, A., Wierzchoń, S.T. (eds.) IIS 2013. LNCS, vol. 7912, pp. 57–68. Springer, Heidelberg (2013). https://doi.org/10.1007/978-3-642-38634-3_8
8. Han, X., Sun, L.: Global distant supervision for relation extraction. In: Thirtieth AAAI Conference on Artificial Intelligence (2016)
9. He, Z., Zhou, Z., Gan, L., Huang, J., Zeng, Y.: Chinese entity attributes extraction based on bidirectional LSTM networks. Int. J. Comput. Sci. Eng. **18**(1), 65–71 (2019)
10. Huang, Z., Xu, W., Yu, K.: Bidirectional LSTM-CRF models for sequence tagging. arXiv preprint arXiv:1508.01991 (2015)
11. Ji, G., Liu, K., He, S., Zhao, J.: Distant supervision for relation extraction with sentence-level attention and entity descriptions. In: Thirty-First AAAI Conference on Artificial Intelligence (2017)

12. Lan, Z., Chen, M., Goodman, S., Gimpel, K., Sharma, P., Soricut, R.: ALBERT: a lite BERT for self-supervised learning of language representations. arXiv preprint arXiv:1909.11942 (2019)
13. Lee, J., Seo, S., Choi, Y.S.: Semantic relation classification via bidirectional lstm networks with entity-aware attention using latent entity typing. Symmetry **11**(6), 785 (2019)
14. Meng, Z., Yu, D., Xun, E.: Chinese microblog entity linking system combining wikipedia and search engine retrieval results. In: Zong, C., Nie, J.-Y., Zhao, D., Feng, Y. (eds.) NLPCC 2014. CCIS, vol. 496, pp. 449–456. Springer, Heidelberg (2014). https://doi.org/10.1007/978-3-662-45924-9_41
15. Miao, Q., Meng, Y., Zhang, B.: Chinese enterprise knowledge graph construction based on linked data. In: Proceedings of the 2015 IEEE 9th International Conference on Semantic Computing (IEEE ICSC 2015), pp. 153–154. IEEE (2015)
16. Niu, X., Sun, X., Wang, H., Rong, S., Qi, G., Yu, Y.: Zhishi.me - weaving chinese linking open data. In: Aroyo, L., et al. (eds.) ISWC 2011. LNCS, vol. 7032, pp. 205–220. Springer, Heidelberg (2011). https://doi.org/10.1007/978-3-642-25093-4_14
17. Rotmensch, M., Halpern, Y., Tlimat, A., Horng, S., Sontag, D.: Learning a health knowledge graph from electronic medical records. Sci. Rep. **7**(1), 1–11 (2017)
18. Souza, F., Nogueira, R., Lotufo, R.: Portuguese named entity recognition using BERT-CRF. arXiv preprint arXiv:1909.10649 (2019)
19. Stenetorp, P., Pyysalo, S., Topić, G., Ohta, T., Ananiadou, S., Tsujii, J.: BRAT: a web-based tool for NLP-assisted text annotation. In: Proceedings of the Demonstrations at the 13th Conference of the European Chapter of the Association for Computational Linguistics, pp. 102–107. Association for Computational Linguistics (2012)
20. Suchanek, F.M., Kasneci, G., Weikum, G.: YAGO: a core of semantic knowledge. In: Proceedings of the 16th International Conference on World Wide Web, pp. 697–706 (2007)
21. Xu, B., et al.: CN-DBpedia: a never-ending Chinese knowledge extraction system. In: Benferhat, S., Tabia, K., Ali, M. (eds.) IEA/AIE 2017. LNCS (LNAI), vol. 10351, pp. 428–438. Springer, Cham (2017). https://doi.org/10.1007/978-3-319-60045-1_44
22. Zeng, D., Liu, K., Chen, Y., Zhao, J.: Distant supervision for relation extraction via piecewise convolutional neural networks. In: Proceedings of the 2015 Conference on Empirical Methods in Natural Language Processing, pp. 1753–1762 (2015)
23. Zeng, D., Liu, K., Lai, S., Zhou, G., Zhao, J., et al.: Relation classification via convolutional deep neural network (2014)
24. Zhang, D., Wang, D.: Relation classification via recurrent neural network. arXiv preprint arXiv:1508.01006 (2015)
25. Zhang, L., Xiang, F.: Relation classification via BiLSTM-CNN. In: Tan, Y., Shi, Y., Tang, Q. (eds.) DMBD 2018. LNCS, vol. 10943, pp. 373–382. Springer, Heidelberg (2018). https://doi.org/10.1007/978-3-319-93803-5_35
26. Zhang, S., Zheng, D., Hu, X., Yang, M.: Bidirectional long short-term memory networks for relation classification. In: Proceedings of the 29th Pacific Asia Conference on Language, Information and Computation, pp. 73–78 (2015)
27. Zhang, W., Cao, H., Hao, F., Yang, L., Ahmad, M., Li, Y.: The chinese knowledge graph on domain-tourism. In: Park, J.J., Yang, L.T., Jeong, Y.-S., Hao, F. (eds.) MUE/FutureTech -2019. LNEE, vol. 590, pp. 20–27. Springer, Singapore (2020). https://doi.org/10.1007/978-981-32-9244-4_3

28. Zhang, Y., Yang, J.: Chinese NER using lattice LSTM. arXiv preprint arXiv:1805.02023 (2018)
29. Zhao, M., et al.: Construction of an industrial knowledge graph for unstructured chinese text learning. Appl. Sci. **9**(13), 2720 (2019)
30. Zhao, Z., Han, S.K., So, I.M.: Architecture of knowledge graph construction techniques. Int. J. Pure Appl. Math. **118**(19), 1869–1883 (2018)
31. Zhou, P., et al.: Attention-based bidirectional long short-term memory networks for relation classification. In: Proceedings of the 54th Annual Meeting of the Association for Computational Linguistics (Volume 2: Short Papers), pp. 207–212 (2016)

A Novel Joint Framework for Multiple Chinese Events Extraction

Nuo Xu[1,2(✉)], Haihua Xie[2], and Dongyan Zhao[1]

[1] Wangxuan Institute of Computer Technology, Peking University, Beijing 100080, China
{xunuo2019,zhaodongyan}@pku.edu.cn

[2] State Key Laboratory of Digital Publishing Technology, Peking University Founder Group Co., Ltd., Beijing 100871, China
xiehh@founder.com

Abstract. Event extraction is an essential yet challenging task in information extraction. Previous approaches have paid little attention to the problem of roles overlap which is a common phenomenon in practice. To solve this problem, this paper defines event relation triple to explicitly represent relations among triggers, arguments and roles which are incorporated into the model to learn their inter-dependencies. A novel joint framework for multiple Chinese events extraction is proposed which jointly performs predictions for event triggers and arguments based on shared feature representations from pre-trained language model. Experimental comparison with state-of-the-art baselines on ACE 2005 dataset shows the superiority of the proposed method in both trigger classification and argument classification.

Keywords: Chinese multiple event extraction · Pre-trained language models · Roles overlap problem · Event relation triple

1 Introduction

Event extraction (EE) is of utility and challenge task in natural language processing (NLP). It aims to identify event triggers of specified types and their arguments in text. As defined in Automatic Content Extraction (ACE) program, the event extraction task is divided into two subtasks, i.e., trigger extraction (identifying and classifying event triggers) and argument extraction (identifying arguments and labeling their roles).

Chinese event extraction is a more difficult task because of language specific issue in Chinese [1]. Since Chinese does not have delimiters between words, segmentation is usually a necessary step for further processing, leading to word-trigger mismatch problem [2]. The approaches based on word-wise classification paradigm commonly suffer from this. It is hard to extract accurately when

Supported by National Key Research and Development Program (No. 2019YFB1406302), China Postdoctoral Science Foundation (NO. 2020M670057) and Beijing Postdoctoral Research Foundation (No. ZZ2019-92).

M. Sun et al. (Eds.): CCL 2020, LNAI 12522, pp. 174–183, 2020.
https://doi.org/10.1007/978-3-030-63031-7_13

a trigger is part of a word or cross multiple words. To avoid this issue, we formulate Chinese event extraction as a character-based classification task. In addition, another interesting issue in event extraction which is rarely followed requires more efforts. It is the roles overlap problem that we concern in this paper, including the problems of either roles sharing the same argument or arguments overlapping on some words. There are multiple events existing in the one sentence, which commonly causes the roles overlap problem and is easy to overlook [3]. Figure 1(a) shows example of roles sharing the same argument in ACE 2005 dataset. "控" (accuse) triggers a Charge-Indict event and "杀害" (kill) triggers an Attack event, while argument "他们" (them) plays the role "Defendant" as well as the role "Attacker" at the same time. Figure 1(b) shows example of arguments overlapping on some words in ACE 2005 dataset. "来往" (traveled between) triggers a Transport event, while argument "中国" (China) plays not only the role "Origin" but "Destination" and argument "来往于中国和澳大利亚之间的乘客" (passengers who traveled between China and Australia) plays the role "Artifact". We observe that the above two arguments overlap on word "中国" (China), which is more challenging for traditional methods to simultaneously identify these two arguments, especially for those being long noun phrases. Research shows that there exist about 10% events in ACE 2005 dataset [4] having the roles overlap problem [3]. Moreover, the results of event extraction could affect the effectiveness of many other NLP tasks, such as the construction of knowledge graph. Therefore, the roles overlap problem is of great importance and needs to be seriously addressed.

It is thus appealing to design a single architecture to solve the problem. In this paper, we propose a single framework to jointly extract triggers and arguments. Inspired by the effectiveness of pre-trained language models, we adopt bidirectional encoder representation from transformer (BERT) as the encoder to obtain the shared feature representations. Specifically, the relations among triggers (t), arguments (a) and roles (r) are defined as event relation triples $<t, r, a>$ where r represents the dependencies of a on t in the event triggered by t. The event sentence of Fig. 1(b) could be represented by event relation triples as <来往, , Origin, 中国>, <来往, Destination, 中国>, <来往, Origin, 澳大利亚>, <来往, Destination, 澳大利亚>, <来往, Artifact, 来往于中国和澳大利亚之间的乘客> . As is seen, event relation triples could explicitly describe relations among the three items. The task of argument classification is converted to relation extraction. Specially, to extract multiple events and relation triples, we utilize multiple sets of binary classifiers to determine the spans (each span includes a start and an end). By this approach, not only roles overlap problem but also word-trigger mismatch and word boundary problems in Chinese language are solved. Our framework avoids human involvements and elaborate engineering features in event extraction, but yields better performance over prior works.

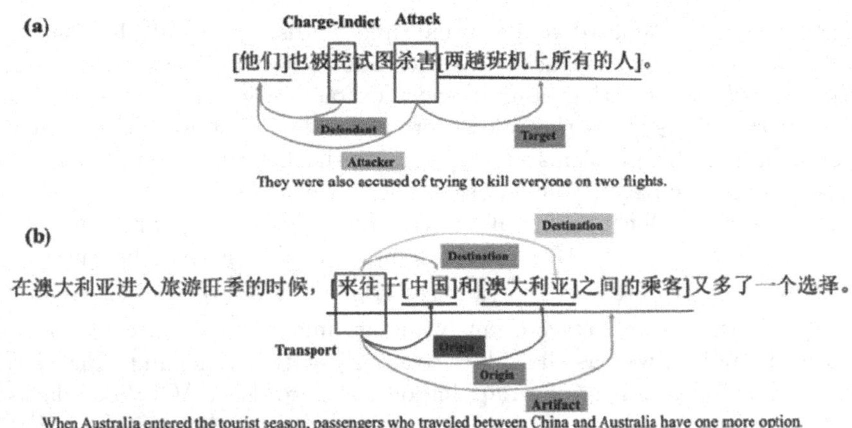

Fig. 1. Examples of roles overlap problem

2 Related Work

EE is an important task which has attracted many attentions. There are two main paradigms for EE: a) the joint approach that predicts event triggers and arguments jointly [5,6], and b) the pipelined approach that first identifies trigger and then identifies arguments in separate stages [7]. The advantages of such a joint system are twofold: (1) mitigating the error propagation from the upstream component (trigger extraction) to the downstream classifier (argument extraction), and (2) benefiting from the inter-dependencies among event triggers and argument roles [8]. Traditional methods that rely heavily on hand-craft features are hard to transfer among languages and annotation standards [9–11]. The neural network based methods that are able to learn features automatically [12,13] have achieved significant progress. Most of them have followed the pipelined approach. Some improvements have been made by jointly predicting triggers and arguments [6–8] and introducing more complicated architectures to capture larger scale of contexts. These methods have achieved promising results in EE.

Although roles overlap problem has been put forward [3,5,6], there are only few works in the literature to study this. He and Duan construct a multi-task learning with CRF enhanced model to jointly learn sub-events [5]. However, their method relies on hand-crafted features and patterns, which makes them difficult to be integrated into recent neural models. Yang et al. adopt a two-stage event extraction by adding multiple sets of binary classifiers to solve roles overlap problem which suffers from error propagation [3]. It does not employ shared feature representations as we do in this work.

In recent years, pre-trained language models are successful in capturing words semantic information dynamically by considering their context. McCann et al.(2017) pre-trained a deep LSTM encoder for machine translation (MT) to contextualize word vectors [14]. ELMo (Embeddings from Language Models) improved 6 challenging NLP problems by learning the internal states of the stacked bidirectional LSTM (Long Short-Term Memory) [15]. Open AI GPT (Generative Pre-Training) improved the state-of-the-art in 9 of 12 tasks [16]. BERT obtained new state-of-the-art results on 11 NLP tasks [17].

3 Extraction Model

This section describes our approach that is designed to extract events. We now define the scope of our work. The task of argument extraction is defined as automatically extracting event relation triples defined. In our model, instead of treating entity mentions as being provided by human annotators, only event label types and argument role types are utilized as training data for both trigger and argument extraction.

We propose a pre-trained language model based joint multiple Chinese event extractor (JMCEE). Let $s = \{c_1, c_2, \&, c_n\}$ be annotated sentence s with n as the number of characters and c_i as the ith character. Given the set of event relation triples $E = \{<t, r, a>\}$ in s, the goal of our framework is to perform the task of trigger extraction T and argument extraction A jointly:

$$P(A, T|s) = P(A|T, s) \times P(T|s) = \prod_{(r,a)\in E|t} p((r, a)|t, s) \prod_{t\in E} p(l, t|s) \quad (1)$$

Here $(r, a) \in E|t$ denotes an argument and role pair (r, a) in the event triples E triggered by t and l denotes the event label type. Based on Eq. (1), we first predict all possible triggers and their label types in a sentence; then for each trigger, we integrate information of predicted trigger word to extract event relation triple $<t, r, a>$ by simultaneously predicting all possible roles and arguments, as illustrated in Fig. 2. We employ a pre-trained BERT encoder to learn the representation for each character in one sentence, then feed it into downstream modules. Token [CLS] and [SEP] are placed at the start and end of the sentence. Multiple sets of binary classifiers are added on the top of the BERT encoder to implement predictions for multiple events and relation triples. For trigger extraction, we need to predict the start and end of event type l for $c_i \in s$ (l could be "Other" type to indicate that there is no word triggering any event) with each set of binary classifiers severing for an event type to determine the starts and ends of all triggers. For argument extraction, we need to extract event relation triple $<t, r, a>$ by predicting the start and end of role type r for c_i in sentence s based on predicted triggers (r is set to "Other" if there is no word triggering any event as well) with each set of binary classifiers severing for a role to determine the starts and ends of all arguments that play it. The roles overlap problem could be solved since the prediction could belong to different arguments and roles. Besides, our JMCEE enables to identify those arguments being long noun phrases like "来往于中国和澳大利亚之间的乘客" (passengers who traveled between China and Australia), which tackles the word boundary problem often encountered in Chinese. Compared with sentence-level sequential modeling methods, our approach also avoids suffering low ef ciency in capturing very long-range dependencies in previous works [18].

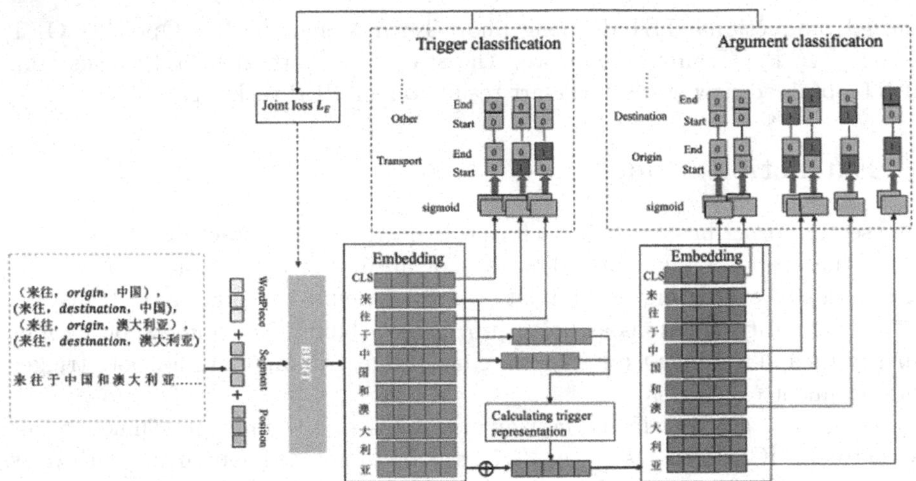

Fig. 2. The framework of JMCEE, including the trigger extract component and the argument extract component. The extraction procedure of the event instance is shown.

3.1 Trigger Extraction

Trigger extraction aims to predict whether a token is a start or an end of a trigger for type label l. A token c_i is predicted as the start of a trigger with probability for type label l through feeding it into a fully-connected layer with sigmoid activation function:

$$P_{Ts}^{l}(c_i) = \sigma(W_{Ts}^{l}\beta(c_i) + b_{Ts}^{l}) \tag{2}$$

while as the end with probability:

$$P_{Te}^{l}(c_i) = \sigma(W_{Te}^{l}\beta(c_i) + b_{Te}^{l}) \tag{3}$$

where we utilize subscript "s" to denote "start" and subscript "e" to denote "end". W_{Ts} and b_{Ts} are respectively the trainable weights and bias of binary classifier that targets to detect starts of triggers' labels, while W_{Te} and b_{Te} are respectively the trainable weights and bias of another binary classifier that targets to detect ends of triggers' labels. β is the BERT embedding. Set thresholds of detecting starts and ends as $\delta^l = \{\delta_s^l, \delta_e^l\}$, δ_s^l and δ_e^l are respectively the thresholds of binary classifiers that targets to detect starts and ends of triggers' labels. If $P_{Ts}^{l}(c_i) > \delta_s^l$, token c_i is identified as the start of type label l. If $P_{Te}^{l}(c_i) > \delta_e^l$, token c_i is identified as end of type label l.

3.2 Argument Extraction

Once the triggers and their type labels have been identified, we come to the argument extraction component. Argument classification is converted to event relation extraction for triple $<t, r, a>$. Note that when the sentence is identified as "Other" type, we simply skip the following operation for argument role extraction. To better learn the inter-dependencies among the multiple events appearing in one sentence, we randomly pick one of predicted triggers in a sentence during the training phase, while in the evaluation phase, all the predicted triggers are picked in turn to predict corresponding arguments and roles played in the triggering events. We integrate information of predicted trigger word to argument extraction component. In ACE corpus, more than 98.5% triggers contain no more than 3 characters, so we simply pick the embedding vectors of start $\beta_s(c_i)$ and end $\beta_e(c_j)$ of one predicted trigger word t, and then generate representation of trigger word $\beta(t)$ by averaging these two vectors.

$$\beta(t) = \frac{(\beta_s(c_i) + \beta_e(c_j))}{2} \tag{4}$$

When obtain representations of trigger words $\beta(t)$, we add original embedding generated by BERT and $\beta(t)$ together:

$$\beta^{'}(s) = \beta(s) + \beta(t) \tag{5}$$

After integrate information of predicted trigger word to BERT sentence encoding, feed $\beta^{'}(s)$ into a full-connected layer with sigmoid activation function. A token c_k is predicted as the start of an argument triggered by word t which plays role r with probability:

$$P_{As}(c_k, r|t) = \sigma(W^r_{As}\beta^{'}(c_k) + b^r_{As}) \tag{6}$$

while as the end triggered by word t with probability:

$$P_{Ae}(c_k, r|t) = \sigma(W^r_{Ae}\beta^{'}(c_k) + b^r_{Ae}) \tag{7}$$

where W_{Ae} and b_{As} are respectively the trainable weights and bias of binary classifier that targets to detect starts of arguments' roles, while W_{Ae} and b_{Ae} are respectively the trainable weights of the other binary classifier that detects ends of arguments' roles. Set thresholds of detecting starts and ends as $\varepsilon^r = \{\varepsilon^r_s, \varepsilon^r_e\}$, ε^r_s and ε^r_e are respectively the thresholds of binary classifiers that target to detect starts and ends of triggers' labels. If $P_{As}(c_k, r|t) > \varepsilon^r_s$,token c_k is identified as the start of argument role r. If $P_{Ae}(c_k, r|t) > \varepsilon^r_e$, token c_k is identified as the end of argument role r.

3.3 Model Training

We train the joint model and define L_T as the loss function of all binary classifiers that are responsible for detecting triggers, shown as follows:

$$L_T = \frac{1}{m \times n}(\sum_{l=0}^{m}\sum_{i=0}^{n} -logP^l_{Ts}(c_i) + \sum_{l=0}^{m}\sum_{i=0}^{n} -logP^l_{Te}(c_i)) \tag{8}$$

L_T denotes the average of cross entropy of output probabilities of all binary classifiers which detect starts and ends of triggers on each type label. In the same way, we define L_A as the loss function of all binary classifiers that are responsible for detecting event relation triples:

$$L_A = \frac{1}{m \times n}(\sum_{r=0}^{m}\sum_{i=0}^{n} -logP_{As}(c_k, r|t) + \sum_{r=0}^{m}\sum_{i=0}^{n} -logP_{Ae}(c_k, r|t)) \quad (9)$$

where m denotes the sum of event label types and argument role types. L_A denotes the average of cross entropy of output probabilities of all binary classifiers which detect starts and ends of arguments on each role. The final loss function $L_E = L_T + L_A$. We minimize the final loss function to optimize the parameters of the model.

4 Experiments

We evaluate JMCEE framework on the ACE 2005 dataset that contains 633 Chinese documents. We follow the same setup as [1,2,13], in which 549/20/64 documents are used for training/development/test set. The proposed model is compared with the following state-of-the-art methods:

1) DMCNN [12] adopts dynamic multi-pooling CNN to extract sentence-level features automatically.
2) Rich-C [9] is a joint-learning, knowledge-rich approach including character-based features and discourse consistency features.
3) C-BiLSTM [13] designs a convolutional Bi-LSTM model which conduct Chinese event extraction from perspective of a character-level sequential labeling paradigm.
4) NPNs [1] performs event extraction in a character-wise paradigm, where a hybrid representation is learned to capture both structural and semantic information from both characters and words.

ACE 2005 dataset annotates 33 event subtypes and 35 role classes. The tasks of event trigger classification and argument classification in this paper are combined into a 70-category task along with "None" word and "Other" type. In order to evaluate the effectiveness of our proposed model, we evaluate models by micro-averaged Precision (P), Recall (R) and F1-score followed the computation measures of Chen and Ji (2009). It is worth noting that all the predicted roles for an argument are required to match with the golden labels, instead of just one of them. We take a further step to see the impacts of pipelined model and joint model. The pipelined model called MCEE which identifies triggers and arguments in two separate stages based our classification algorithm. The highest F-score parameters on the development set are picked and listed in Table 1.

Table 2 shows the results of trigger extraction on ACE 2005. The performances of Rich-C and C-BiLSTM are reported in their papers. As is seen, our JMCEE framework achieves the best F1 scores for trigger classi cation among

Table 1. Hyper-parameters for experiments.

Hyper-parameter	Trigger classification	Argument classification
Character embedding	768	768
Maximum length	510	510
Batch size	8	8
Learning rate of Adam	0.0005	0.0005
Classification thresholds	[0.5, 0.5, 0.5, 0.5]	[0.5, 0.4, 0.5, 0.4]

all the compared methods. Our JMCEE gains at least 8% F1-score improvements on trigger classification task on ACE 2005, which steadily outperforms all baselines. The improvement on the trigger extraction is quite significant, with a sharp increase of near 10% on the F1 score compared with these conventional methods. Table 3 shows results of argument extraction. Compared with these baselines, our JMCEE is at least 3% higher over other models on F1-score on argument classification task. While the improvement in argument extraction is not so obvious comparing to trigger extraction. This is probably due to the rigorous evaluation metric we have taken and the difficulty of argument extraction. Note that by our approach we identify 89% overlap roles in test set. Moreover, results show that our joint model substantially outperforms the pipelined model whether on trigger classification or argument classification. It is seen that joint model enables to capture the dependencies and interactions between the two subtasks and communicate deeper information between them, and thus improves the overall performance.

Table 2. Comparison of different methods on Chinese trigger extraction on ACE 2005 test set. Bold denotes the best result.

Model	Trigger identification			Trigger classification		
	P	R	F1	P	R	F1
DMCNN	66.6	63.6	65.1	61.6	58.8	60.2
Rich-C	62.2	71.9	66.7	58.9	68.1	63.2
C-BiLSTM	65.6	66.7	66.1	60.0	60.9	60.4
NPNs	75.9	61.2	67.8	73.8	59.6	65.9
MCEE (BERT-Pipeline)	82.5	78.0	80.2	72.6	68.2	70.3
JMCEE (BERT-Joint)	**84.3**	**80.4**	**82.3**	**76.4**	**71.7**	**74.0**

Table 3. Comparison of different methods on Chinese argument extraction on ACE 2005 test set. Bold denotes the best result.

Model	Argument identification			Argument classification		
	P	R	F1	P	R	F1
Rich-C	43.6	**57.3**	49.5	39.2	**51.6**	44.6
C-BiLSTM	53.0	52.2	52.6	47.3	46.6	46.9
MCEE (BERT-Pipeline)	59.5	40.4	48.1	51.9	37.5	43.6
JMCEE (BERT-Joint)	**66.3**	45.2	**53.7**	**53.7**	46.7	**50.0**

5 Conclusions

In this paper, we propose a simple yet effective joint Chinese multiple events extraction framework which jointly extracts triggers and arguments. Our contribution in this work is as follows:

1) Event relation triple is defined and incorporated into our framework to learn inter-dependencies among event triggers, arguments and arguments roles, which solves the roles overlap problem.
2) Our framework performs event extraction in a character-wise paradigm by utilizing multiple sets of binary classifiers to determine the spans, which allows to extract multiple events and relation triples and avoids Chinese language specific issues.

Our future work will focus on data generation to enrich training data and try to extend our framework to the open domain.

References

1. Chen, Z., Ji, H.: Language specific issue and feature exploration in Chinese event extraction. In: Proceedings of NAACL-HLT 2009, pp. 209–212 (2009)
2. Lin, H.Y., et al.: Nugget proposal networks for Chinese event detection. In: Proceedings of the 56th Annual Meeting of the Association for Computational Linguistics, pp. 1565–1574 (2018)
3. Yang, S., et al.: Exploring pre-trained language models for event extraction and generation. In: Proceedings of the 57th Annual Meeting of the Association for Computational Linguistics, pp. 5284–5294 (2019)
4. Doddington, G.R., et al.: The automatic content extraction (ACE) program-tasks, data, and evaluation. In: LREC (2004)
5. He, R.F., Duan, S.Y.: Joint Chinese event extraction based multi-task learning. J. Softw. **30**(4), 1015–1030 (2019)
6. Liu, X., Luo, Z., Huang, H.: Jointly multiple events extraction via attention-based graph information aggregation. arXiv preprint arXiv:1809.09078 (2018)
7. Nguyen, T.H., Cho, K., Grishman, R.: Joint event extraction via recurrent neural networks. In: Proceedings of NAACL-HLT 2016, pp. 300–309 (2016)

8. Nguyen, T.M., Nguyen, T.H.: One for all: neural joint modeling of entities and events. In: Proceedings of the AAAI Conference on Artificial Intelligence, vol. 33, pp. 6851–6858 (2019)
9. Chen, C., Ng, V.: Joint modeling for Chinese event extraction with rich linguistic features. In: Proceedings of COLING 2012, pp. 529–544 (2012)
10. Liao, S., Grishman, R.: Using document level cross-event inference to improve event extraction. In: Proceedings of the 48th Annual Meeting of the Association for Computational Linguistics, pp. 789–797 (2010)
11. Li, Qi., Ji, H., Huang, L.: Joint event extraction via structured prediction with global features. In: Proceedings of the 51st Annual Meeting of the Association for Computational Linguistics, pp. 73–82 (2013)
12. Chen, Y.B., et al.: Event extraction via dynamic multi-pooling convolutional neural networks. In: Proceedings of the 53rd Annual Meeting of the Association for Computational Linguistics and the 7th International Joint Conference on Natural Language Processing, vol.1, pp. 167–176 (2015)
13. Zeng, Y., Yang, H., Feng, Y., Wang, Z., Zhao, D.: A convolution BiLSTM neural network model for Chinese event extraction. In: Lin, C.-Y., Xue, N., Zhao, D., Huang, X., Feng, Y. (eds.) ICCPOL/NLPCC -2016. LNCS (LNAI), vol. 10102, pp. 275–287. Springer, Cham (2016). https://doi.org/10.1007/978-3-319-50496-4_23
14. McCann, B., et al.: Learned in translation: contextualized word vectors. In Advances in Neural Information Processing Systems, pp. 6294–6305 (2017)
15. Peters, M., et al.: Deep contextualized word representations. In: Proceedings of the 2018 Conference of the North American Chapter of the Association for Computational Linguistics, vol. 1, pp. 2227–2237 (2018)
16. Radford, A., et al.: Improving language understanding by generative pre-training (2018). https://www.cs.ubc.ca/amuham01/LING530/papers/radford2018 improving.pdf
17. Devlin, J., et al.: BERT: pre-training of deep bidirectional transformers for language understanding. arXiv preprint arXiv:1810.04805 (2018)
18. Sha, L., et al.: Jointly extracting event triggers and arguments by dependency-bridge RNN and tensor-based argument interaction. In: Proceedings of the 32nd AAAI Conference on Artificial Intelligence, pp. 5916–5923 (2018)

Entity Relative Position Representation Based Multi-head Selection for Joint Entity and Relation Extraction

Tianyang Zhao[1(✉)], Zhao Yan[2], Yunbo Cao[2], and Zhoujun Li[1]

[1] State Key Lab of Software Development Environment, Beihang University, Beijing, China
{tyzhao,lizj}@buaa.edu.cn

[2] Tencent Cloud Xiaowei, Beijing, China
{zhaoyan,yunbocao}@tencent.com

Abstract. Joint entity and relation extraction has received increasing interests recently, due to the capability of utilizing the interactions between both steps. Among existing studies, the Multi-Head Selection (MHS) framework is efficient in extracting entities and relations simultaneously. However, the method is weak for its limited performance. In this paper, we propose several effective insights to address this problem. First, we propose an entity-specific Relative Position Representation (eRPR) to allow the model to fully leverage the distance information between entities and context tokens. Second, we introduce an auxiliary Global Relation Classification (GRC) to enhance the learning of local contextual features. Moreover, we improve the semantic representation by adopting a pre-trained language model BERT as the feature encoder. Finally, these new keypoints are closely integrated with the multi-head selection framework and optimized jointly. Extensive experiments on two benchmark datasets demonstrate that our approach overwhelmingly outperforms previous works in terms of all evaluation metrics, achieving significant improvements for relation F1 by +2.40% on CoNLL04 and +1.90% on ACE05, respectively.

1 Introduction

The entity-relation extraction task aims to recognize the entity spans from a sentence and detect the relations holds between two entities. Generally, it can be formed as extracting triplets (e_1, r, e_2), which denotes that the relation r holds between the head entity e_1 and the tail entity e_2, i.e., (*John Smith*, `Live-In`, *Atlanta*). It plays a vital role in the information extraction area and has attracted increasing attention in recent years.

Traditional pipelined methods divide the task into two phases, named entity recognition (NER) and relation extraction (RE) [3,15,18]. As such methods neglect the underlying correlations between the two phases and suffer from the

T. Zhao—Work done during an internship at Tencent.

M. Sun et al. (Eds.): CCL 2020, LNAI 12522, pp. 184–198, 2020.
https://doi.org/10.1007/978-3-030-63031-7_14

error propagation issue, recent works propose to extract entities and relations jointly. These joint models fall into two paradigms. The first paradigm can be denoted as $(e_1, e_2) \rightarrow r$, which first recognizes all entities in the sentence, then classifies the relation depend on each extracted entity pairs. However, these methods require enumerating all possible entity pairs and the relation classification may be affected by the redundant ones. While another paradigm is referred as $e1 \rightarrow (r, e2)$, which detects head entities first and then predicts the corresponding relations and tail entities [2,14,26]. Comparing with the first paradigm, the second one can jointly identify entities and all the possible relations between them at once. A typical approach is the Multi-Head Selection (MHS) framework [2]. It first recognizes head entities using the BiLSTM-CRF structure and then performs tail entity extraction and relation extraction in one pass based on multiclass classification. The advantage of the MHS framework is obvious - it is efficient to work with the scenario, that one entity can involve several relational triplets, making this solution suitable for large scale practical applications. In this paper, we focus on the second paradigm of the joint models, especially on the MHS framework.

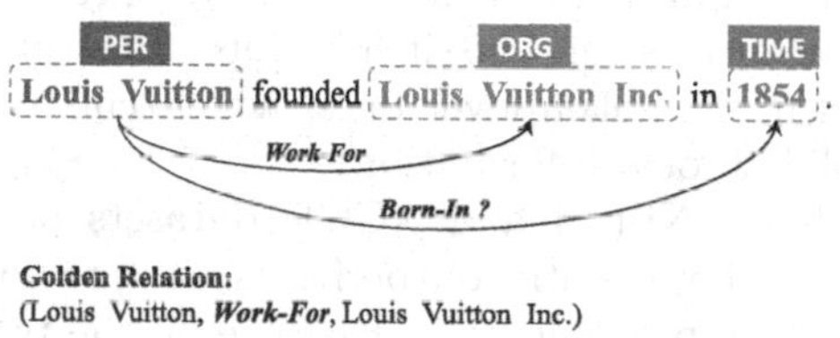

Fig. 1. An example to show the impact of entity-specific relative position.

Despite the efficiency of the MHS framework, it is weak for the limited performance compared with other complex models. Intuitively, the distance between entities and other context tokens provide important evidence for entity and relation extraction. Meanwhile, the distance information of non-entity words is less important. As shown in the sentence of Fig. 1, the "Louis Vuitton" that is far from the word "Inc." is a person entity, while the one adjacent to "Inc." denotes an organization. Such an entity-specific relative position can be a useful indicator to differentiate entity tokens and non-entity tokens and enhance interactions between entities. While the existing model pays equal attention to each context tokens and ignores the relative distance information of entities. As a result, the entity-specific features may become less obscure and mislead the relation selection. Second, the existing model predicts the relations and tail entities merely based on the local contextual features of the head entity, and the incomplete local information may confuse the predictor. While the semantic of the whole sentence always has a significant impact on relation prediction. For example, in Fig. 1, the relation between "Louis Vuitton" and "1854" may easily be mislabeled as "`Born-In`" without considering the meaning of the whole sentence. Therefore, the global semantics should also be taken into account.

To address the aforementioned limitations, we present several new key points to improve the existing multi-head selection framework. First, we propose an entity-specific Relative Position Representation (eRPR) to leverage the distance information between entities and their contextual tokens, which provides important positional information for each entity. Then, in order to better consider the sentence-level semantic during relation prediction, we add up an auxiliary Global Relational Classification (GRC) to guide the optimization of local context features. In addition, different from the original MHS structure, we adopt the pre-trained transformer-based encoder (BERT) to enhance the ability of semantic representations. Notably, the proposed method can address the entity and multiple-relation extraction simultaneously and without relying on any external parsing tools or hand-crafted features. We conduct extensive experiments on two widely-used datasets CoNLL04 and ACE05, and demonstrate the effectiveness of the proposed framework.

To summarize, the contributions of this paper are as follows:

- We propose an entity-specific relative position representation to allow the model aware of the distance information of entities, which provides the model with richer semantics and handles the issue of obscure entity features.
- We introduce a global relation classifier to integrate the essential sentence-level semantics with the token-level ones, which can remedy the problem caused by incompleted local information.
- Experiments on the CoNLL04 and ACE05 datasets demonstrate that the proposed framework significantly outperforms the previous work, achieving +2.40% and +1.90% improvements in F1-score on the two datasets.

2 Related Work

In this section, we introduce the related studies for this work, entity and relation extraction as well as the positional representation.

Entity and Relation Extracion. As a crucial content of information extraction, the entity-relation extraction task has always been widely concerned. Previous studies [3,15,18] mainly focus on pipelined structure, which divides the task into two independent phases, all entities are extracted first by an entity recognizer, and then relations between every entity pairs are predicted by a relation classifier. The pipelined methods suffer from error propagation issue and they ignore the interactions between the two phrases. To ease these problems, many joint models have been proposed to extract the relational triplets (e_1, r, e_2), simultaneously. According to different extraction order, the joint models can be categorized into two paradigms. The first paradigm identifies all entities in the sentence first, then traverses each pair of entities and determines their potential relation. Various models have achieved promising results by exploiting recurrent neural network [16,17], graph convolutional network [9,23] and transformer-based structure [7, 25]. Though effective, these models need to examine every possible entity pairs, which inevitably contains a lot of redundant pairs. In the second paradigm, the head entities are detected first and the corresponding relations and tail entities are extracted later. Bekoulis et al. [2] present the multi-head selection framework to automatically extract multiple entities and relations at once. Huang et al. [12] improve the MHS framework by using NER pretraining and soft label embedding features. Recently, Li et al. [14] cast the task as a question answering problem and identify entities based on a machine reading comprehension model. Different from the first one, the second paradigm is able to extract entities and all the relations at once without enumerating every entity pair each time, which reduces redundant prediction and improves work efficiency.

Positional Representation. Many works design representations to encode positional information for non-recurrent models, which can fall into three categories. The first one designs the position encodings as a deterministic function of position or learned parameters [10,22]. These encodings are combined with input elements to expose position information to the model. For example, the convolutional neural networks inherently capture the relative positions within each convolutional kernels. The second category is the absolute position representation. The Transformer structure [24] contains neither recurrence nor convolution, in order to inject the positional information to the model, it defines the sine and consine functions of different frequencies to encode absolute positions. However, such absolute positions cannot model the interaction information between any two input tokens explicitly. Therefore, the third category extends the self-attention mechanism to consider the relative positions between sequential elements [4,21]. Differently, we propose the relative positions especially for entities to enhance the interactions between them. While we do not consider the relative positions for non-entity tokens to alleviate the unnecessary noise.

Our work is inspired by the multi-head selection framework but enjoys new points as follows. 1) We propose an entity-specific relative position representation to better encode the distance between entities and context tokens. 2) We incorporate the sentence-level information for relation classification to revise the learning of local features. 3) We enhance the original MHS framework with a pre-trained self-attentive encoder. Together these improvements contribute to the extraction performance remarkably.

3 Method

In this section, we briefly present the details of the relative position representation based multi-head selection framework. The concept of multi-head means that any head entity may be relevant to multiple relations and tail entities [2].

Formally, denote $\mathcal{E}$ and $\mathcal{R}$ as the set of pre-defined entity types and relation categories, respectively. Given an input sentence with N tokens $\boldsymbol{s} = \{s_1, s_2, \ldots, s_N\}$, the entity-relation extraction task aims at extracting a set of named entities $\boldsymbol{e} = \{e_1, e_2, \ldots, e_M\}$ with specific types $\boldsymbol{y} = \{y_1, y_2, \ldots, y_M\}$, and predict the relation r_{ij} for each entity pair (e_i, e_j), where $y_i \in \mathcal{E}$ and $r_{ij} \in \mathcal{R}$. Triplets such as (e_i, r_{ij}, e_j) are formulated as the output, where e_i is the head entity and e_j is the tail entity, e.g., (*John Smith*, `Live-In`, *Atlanta*).

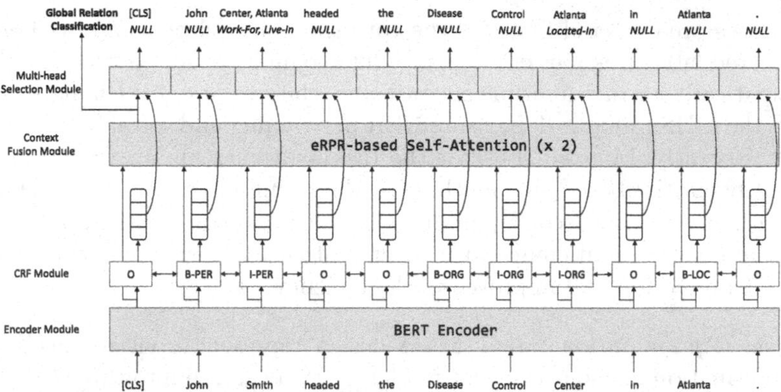

Fig. 2. The overview of the relative position representation based multi-head selection framework. We take a sentence from CoNLL04 dataset as an example. In this sentence, the golden relational triplets are: (*John Smith*, `Live-In`, *Atlanta*), (*John Smith*, `Work-For`, *Disease Control Center*) and (*Disease Control Center*, `Located-In`, *Atlanta*). The *NULL* label denotes a case of no relation.

As illustrated in Fig. 2, our framework consists of four modules as follows: the encoder module, the CRF module, the context fusion module and the multi-head selection module. The token sequence is taken as the input of the framework and is fed into the BERT encoder to capture contextual representations. The CRF module is applied afterward to extract potential head entities (i.e., boundaries and types). Then, the hidden states of BERT and the entity information are fed into the context fusion module to encoder the entity position-based features. Finally, a multi-head selection module is employed to simultaneously extract tuples of relation and tail entity for the input token (e.g., (`Work-For`, *Center*) and (`Live-In`, *Atlanta*) for the head entity *Simth*). Additionally, we present the strategy of global relation classification. We will elaborate on each of the modules in the following subsections.

3.1 Encoder Module

The encoder module aims at mapping discrete tokens into distributed semantic representations. Bidirectioal Encoder Representations from Transformers (BERT) [5] is a pre-trained language representations built on the bidirectional self-attentive models. It is known for its powerful feature representative ability and recently breaks through the leaderboards of a wide range of natural language processing tasks, such as named entity recognition, word segmentation and question answering. Different from the previous work [2] which uses the BiLSTM as the feature encoder, we use the BERT instead to better represent contextual features.

As illustrated in Fig. 2, given a N-token sentence $\boldsymbol{s} = \{s_1, s_2, \ldots, s_N\}$, a special classification token ([CLS]) is introduced as the first token of the input sequence as $\{[\text{CLS}], s_1, s_2, \ldots, s_N\}$. The sequence is encoded by the multi-layer bidirectional attention structure. The output of the BERT layer is the contextual representation of each token as $\boldsymbol{h} = \{h_0, h_1, \ldots, h_N\}$ where $h_i \in \mathbb{R}^{d_h}$, where d_h denotes the dimension of the hidden state of BERT.

3.2 CRF Module

The conditional random field is a probabilistic method that jointly models interactions between entity labels, which is widely used in named Entity recognition task. Similarly, we employ a linear-chain CRF over the BERT layer to obtain the most possible entity label for each token, e.g., B-PER.

Given the BERT outputs $\boldsymbol{h} = \{h_0, h_1, \ldots, h_N\}$, the corresponding entity label sequence is denoted as $\boldsymbol{y} = \{y_0, y_1, \ldots, y_N\}$. Specifically, we use the BIO (Begin, Inside, Non-Entity) tagging scheme. For example, B-PER denotes the beginning token of a person entity. The probability of using $\boldsymbol{y}$ as the label prediction for the input context is calculated as

$$p(\boldsymbol{y}|\boldsymbol{h}) = \frac{\prod_{i=1}^{N} \phi_i(y_{i-1}, y_i, \boldsymbol{h})}{\sum_{y' \in \mathcal{Y}(\boldsymbol{h})} \prod_{i=1}^{N} \phi_i(y'_{i-1}, y'_i, \boldsymbol{h})}. \tag{1}$$

$\mathcal{Y}(\boldsymbol{h})$ is the set of all possible label predictions. And $\phi_i(y_{i-1}, y_i, \boldsymbol{h}) = \exp(\mathbf{W}_{\text{CRF}}^{y_i} h_i + \mathbf{b}_{\text{CRF}}^{y_{i-1} \rightarrow y_i})$, where $\mathbf{W}_{\text{CRF}} \in \mathbb{R}^{d_h \times d_l}$, $\mathbf{b}_{\text{CRF}} \in \mathbb{R}^{d_l \times d_l}$ with d_l denoting the size of the entity label set. $\mathbf{W}_{\text{CRF}}^{y_i}$ is the column corresponding to label y_i, and $\mathbf{b}_{\text{CRF}}^{y_{i-1} \rightarrow y_i}$ is the transition probability from label y_{i-1} to y_i.

During training, the NER loss function is defined as the negative log-likelihood:

$$\mathcal{L}_{\text{NER}} = -\sum_{\boldsymbol{h}} \log p(\boldsymbol{y}|\boldsymbol{h}). \tag{2}$$

During decoding, the most possible label sequence y^* is the sequence with maximal likelihood of the prediction probability:

$$y^* = \arg\max_{\boldsymbol{y} \in \mathcal{Y}(\mathbf{h})} p(\boldsymbol{y}|\boldsymbol{h}). \tag{3}$$

The final labels can be efficiently addressed by the Viterbi algorithm [8].

3.3 Context Fusion Module

The context fusion module focuses on injecting the entity-specific relative position representation into the semantic feature of entities to capture the distance information between entities and other context tokens. The self-attention structure in BERT introduces sine and cosine functions of varying frequency to represent the absolute position representation (APR) of tokens. However, such absolute position representation neglects the relative distance information between entities and other tokens, while such distance plays a crucial role in entity-relation prediction. Hence, we introduce an entity-specific relative position representation to efficiently encode the relative distance.

Formally, for the output states of BERT encoder $\boldsymbol{h} = \{h_0, h_1, \ldots, h_N\}$ where $h_i \in \mathbb{R}^{d_h}$, the relative position layer outputs a transformed sequence $\boldsymbol{p} = \{p_0, p_1, \ldots, p_N\}$ where $p_i \in d_p$ with d_p as the hidden dimension of self-attention structure.

Consider two input states h_i and h_j, where h_i denotes an entity and h_j denotes a contextual token, $i, j \in 0, 1, \ldots, N$. In order to inject the relative position information into x_i, we define $a_{ij}^K \in d_p, a_{ij}^V \in d_p$ as two different relative distances between h_i and h_j. Suppose that the impacts of tokens beyond a maximum distance on the current token are negligible. Therefore, we clip the relative position within a maximum distance δ and only consider the position information of δ tokens on the left and δ tokens on the right. We define $\boldsymbol{\omega}^K = (\omega_{-\delta}^K, \ldots, \omega_{\delta}^K)$ and $\boldsymbol{\omega}^V = (\omega_{-\delta}^V, \ldots, \omega_{\delta}^V)$ as two relative position representations, where $\omega_i^K, \omega_i^V \in \mathbb{R}^{d_p}$ are initialized randomly and will be learned during training. Figure 3 illustrates an example of the relative position representations. Then, a_{ij}^K and a_{ij}^V are assigned as:

$$\begin{aligned} a_{ij}^K &= \omega_{\text{clip}(j-i,\delta)}^K \\ a_{ij}^V &= \omega_{\text{clip}(j-i,\delta)}^V \\ clip(x,\delta) &= \max(-\delta, \min(x,\delta)). \end{aligned} \tag{4}$$

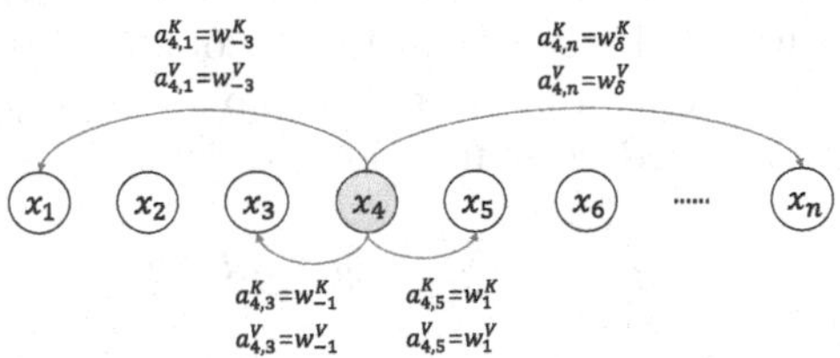

Fig. 3. An example to illustrate the entity relative position representation. x_4 is considered as an entity, we show the eRPR between x_4 and the context tokens within the clipped distance δ. Assuming $3 <= \delta <= n - 4$ in this example.

Based on the relative position representations a_{ij}^K, a_{ij}^V, the attention matrix between h_i and h_j is calculated as:

$$\alpha_{ij} = \text{softmax}\Big(\frac{(h_i W^Q)(h_j W^K + a_{ij}^K)^T}{\sqrt{d_p}}\Big), \tag{5}$$

where $W^Q \in \mathbb{R}^{d_h \times d_p}, W^K \in \mathbb{R}^{d_h \times d_p}$ are parameter matrices for multi-head projections. The attentional output of h_i is the weighted sum of h_j which also consider the relative position:

$$p_i = \sum_{j=1}^{n} \alpha_{ij}(h_j W^V + a_{ij}^V). \tag{6}$$

Specifically, we only consider the relative position of named entities rather than every tokens in the sentence. So $\boldsymbol{\omega}^{\boldsymbol{K}}$ and $\boldsymbol{\omega}^{\boldsymbol{V}}$ are set as 0 for non-entity tokens. the This entity-only RPR approach comes with the following key advantages: 1) it encodes unique features for entities and thus can better differentiate entities from other plain tokens; 2) it provides entity-specific information and helps the relation and tail entity prediction.

3.4 Multi-head Selection Module

The multi-head selection module aims to predict the possible relations and tail entities simultaneously for each head entity [2]. Given a sequence of entity labels $\boldsymbol{y} = \{y_0, y_1, \ldots, y_N\}$ predicted by the CRF module, we map each label to a distributed label embedding as $\boldsymbol{l} = \{l_0, l_1, \ldots, l_N\}$, $l_i \in \mathbb{R}^{d_l}$, where d_l is the label embedding size. The mapping dictionary is randomly initialized and fine-tuned during training. During training, we use the golden entity labels.

As shown in Fig. 2, the input to the multi-head selection layer are the concatenation of label embedding and the outputs of relative position layer as:

$$z_i = [l_i; p_i], i = 0, 1, \ldots, N. \tag{7}$$

For each input state z_i, we compute the score between z_i and z_j given a relation $r_k, r_k \in \mathcal{R}$ as:

$$g(z_i, z_j, r_k) = V^r f(U^r z_j + W^r z_i + b^r), \tag{8}$$

where $V^r \in \mathbb{R}^{d_r}$, $U^r, W^r \in \mathbb{R}^{d_r \times (d_h + d_l)}$, $b^r \in \mathbb{R}^{d_r}$, $f(\cdot)$ is the element-wise RELU function. The most probable tail entity s_j with the relation r_k corresponding to the head entity s_i is predicted as:

$$\Pr(\text{tail} = s_j, \text{relation} = r_k | \text{head} = s_i) = \sigma(g(z_i, z_j, r_k)), \tag{9}$$

where $\sigma(\cdot)$ denotes the sigmoid function.

During training, we optimize the cross-entropy loss $\mathcal{L}_{\text{MHS}}$ for the candidate tail entity s_{ij} and relation r_{ij} given the head entity s_i as:

$$\mathcal{L}_{\text{MHS}} = \sum_{i=0}^{N}\sum_{j=0}^{M} -\log \Pr(tail = s_j, relation = r_j | head = s_i), \tag{10}$$

where M is the number of golden relations for s_i. During testing, we select the tuple of the relation and tail entity $(\hat{r_k}, \hat{s_j})$ with a score exceeding the confidence threshold η. In this way, multiple tail entities and relations for the head entity s_i can be predicted simultaneously.

3.5 Global Relation Classification

Generally, detecting the relation between entities needs to consider the theme of the sentence. The previous work only use the local context information for relation and entity prediction, which may lead to the deviation of global semantics.

We introduce the global relation classification strategy to guide the training of local semantic features. As illustrated in Fig. 2, the first output of the relative position layer corresponding to the hidden state of [CLS] token p_0, which can be considered as the aggregate representation of the sentence. Therefore, we use the [CLS] token to predict the relations relevant to the whole sentence $\boldsymbol{s}$ as:

$$\Pr(\text{relation} = \boldsymbol{r}|\boldsymbol{s}) = \sigma(W^g p_0 + b^g), \tag{11}$$

where $\boldsymbol{r} \subseteq \mathcal{R}$, $W^g \in \mathbb{R}^{d_h \times |\mathcal{R}|}$, $b^r \in \mathbb{R}^{|R|}$, $\sigma(\cdot)$ is the sigmoid function. During training, we minimize the binary cross-entropy loss for the global classification as:

$$\mathcal{L}_{\text{GRC}} = \sum_{i=0}^{T} \Pr(\text{relation} = \boldsymbol{r}|\boldsymbol{s}), \tag{12}$$

where T denotes the number of golden relations in the sentence.

3.6 Joint Training

To train the model jointly, we optimize the final combined objective function as:

$$\mathcal{L} = \mathcal{L}_{\text{NER}} + \lambda \mathcal{L}_{\text{GRC}} + \mathcal{L}_{\text{MHS}}, \tag{13}$$

where $\mathcal{L}_{\text{NER}}$, $\mathcal{L}_{\text{GRC}}$, and $\mathcal{L}_{\text{MHS}}$ denote the loss function for head entity recognition, global relation classification and multi-head selection, respectively (Eq. 2, 12, 10), $\lambda \in [0, 1]$ is the weight controlling the trade-off of the global relation classification. $\mathcal{L}$ is averaged over samples for each batch.

4 Experiment

In this section, we conduct extensive experiments to verify the effectiveness of our framework, and make detailed analyses to show its advantages.

4.1 Dataset

We conduct evaluation on two widely-used benchmarks for entity and relation extraction: CoNLL04 and ACE05.

CoNLL04 [20] defines 4 entity types as Location (`LOC`), Organization (`ORG`), Person (`PER`) and `Other` and 5 relation categories as `Located-In`, `OrgBased-In`, `Live-In`, `Kill` and `Work-For`. It consists of news articles from the Wall Street Journal and Associated Press. We use the data split by Gupta et al. [11] (910 instances for training, 243 for validation and 288 for testing).

ACE05 [6] provides 7 entity types: Location (`LOC`), Organization (`ORG`), Person (`PER`), Geopolitical Entity (`GPE`), Vehicle (`VEH`), Facility (`FAC`), Weapon (`WEA`) and 6 relation types: `ORG-AFF`, `PER-SOC`, `ART`, `PART-WHOLE`, `GEN-AFF`, `PHYS`. It contains documents from different domains such as newswire and online forums. We adopt the same data splits as the previous work [17] (351 documents for training, 80 for validation and 80 for testing).

4.2 Implemental Details

Following previous works, we use the standard precision (P), recall (R), and micro-F1 score (F1) as the evaluation metrics. A relation is correct if the arguments of triplet (e_1, r, e_2) are correct. Other experimental settings are as follows. We initialize the BERT encoder layer using the pre-trained BERT-Base-Cased checkpoint[1] which has 12 layers, a hidden size of 768. We use Adam optimizer with an initial learning rate of 5×10^{-5}. During training, we do warm-up startup first and employ a linearly decrease with 0.05 as the decay rate. For the model structure, we adopt 2-layer eRPR-based self-attention after the BERT encoder layer. The self-attention layer has an identical structure as the layer in BERT. The relative position representations $\boldsymbol{\omega}^{\boldsymbol{K}}, \boldsymbol{\omega}^{\boldsymbol{V}}$ are initiaized randomly with a uniform distribution. The maximum relative distance is set as $\delta = 4$. The GRC loss weight is set as $\lambda = 1$. The size of entity label embedding is set as $d_l = 50$. The threshold for multi-head selection $\eta = 0.5$.

Specifically, we use both the *relaxed* and the *strict* evaluation settings for comparison. In the *relaxed* setting, a multi-token entity is correct if at least one of its comprising token types is correct; a relation is correct if the two argument entities are correct and the relation type is correct. In the *strict* setting, we consider an entity is correct if the entity type and the boundaries are both correct; a relation is correct if the relation type and the argument entities are both correct.

4.3 Results and Analyses

Comparison Baseline. As shown in Table 1, we list the following baselines for comparison. Gupta et al. [11] propose a table-filling based method that relies on hand-crafted features and external NLP tools. Adel and Schütze [1] use a global normalized convolutional neural networks to extract entities and relations. Miwa and Bansal [1] adopt a BiLSTM to extract entities and a Tree-LSTM to model the dependency relations between entities. Bekoulis et al. [2] propose the multi-head selection structure, which adopts BiLSTM as the feature encoder and uses CRF for entity recognition and can extract the relational triplet simultaneously. The results on CoNLL04 and ACE05 are directly copied from the published paper.

Main Results. Table 1 presents the performance comparisons on CoNLL04 and ACE05 datasets. eRPR MHS is the proposed full model, which uses the BERT at encoder module, and follows by two eRPR self-attention layers and adopts the GRC strategy. As we can see, our eRPR MHS overwhelmingly outperforms all the baseline models in terms of all three evaluation metrics on the two datasets. by a large margin for both entity and relation extraction. Especially, comparing with the model by Bekoulis et al. [2], our model achieves significant boosts by 2.40% and 1.90% for relation F1 on CoNLL04 and ACE05, respectively. These

[1] BERT checkpoints are available at https://github.com/google-research/bert.

Table 1. Performance comparison with baselines on CoNLL04 and ACE05. eRPR denotes models adopt the self-attention with entity-specific relative position representation at the context fusion module. The ✓ and ✗ marks stand for whether or not the model builds on hand-crafted features or NLP tools. eRPR MHS is the proposed full model.

Model	Pre-calculated features	Evaluation	Entity			Relation		
			P	R	F1	P	R	F1
		CoNLL04						
Gupta et al. [11]	✓	*relaxed*	92.50	92.10	92.40	78.50	63.00	69.90
Gupta et al. [11]	✗	*relaxed*	88.50	88.90	88.80	64.60	53.10	58.30
Adel and Schütze [1]	✗	*relaxed*	-	-	82.10	-	-	62.50
Bekoulis et al. [2]	✗	*relaxed*	93.41	93.15	93.26	72.99	63.37	67.01
eRPR MHS	✗	*relaxed*	94.32	93.81	**94.06**	73.85	64.41	**68.81**
Miwa and Sasaki [19]	✓	*strict*	81.20	80.20	80.70	76.00	50.90	61.00
Bekoulis et al. [2]	✗	*strict*	83.75	84.06	83.90	63.75	60.43	62.04
eRPR MHS	✗	*strict*	86.85	85.62	**86.23**	64.20	64.69	**64.44**
		ACE05						
Miwa and Bansal [17]	✓	*strict*	80.80	82.90	81.80	48.70	48.10	48.40
Katiyar and Cardie [13]	✗	*strict*	81.20	78.10	79.60	46.40	45.53	45.70
eRPR MHS	✗	*strict*	86.26	84.66	**85.45**	60.60	60.84	**60.72**

results show that, with our enhanced components, i.e., the eRPR layers, the global relation classification and the BERT encoder, the model performance can be significantly improved. Such improvements highlight the effectiveness of our proposed framework.

Ablation Study. As shown in Table 2, we list variant models (Model 1–5) for each component in our framework. Model 1 stands for the original MHS framework proposed by Bekoulis et al. [2]. By comparison, we come to the following conclusions. 1) Replacing the BiLSTM with pre-trained BERT can improve the performance obviously (Model 2 v.s. Model 1). 2) Adding the context fusion module after the encoder module can enhance the semantic representation, leading to higher results (Model 3 v.s. Model 2). 3) Comparing Model 4 with the above variations, incorporating eRPR into the self-attention structure can significantly increase the precision of models and thus contribute to better overall F1 scores. For example, it increases the relation F1 from 63.96% to 64.21% on CoNLL04. We attribute it to that, the eRPR injects distance information into entity features, which can provide useful information to the multi-head selection. 4) Comparing Model 5 and Model 4, the GRC strategy can further improve model performance. Therefore, global information is instructive for learning local features. Finally, combining all these components, we achieve significant improvements over the original MHS.

Table 2. Ablation study on CoNLL04 and ACE05. APR denotes models adopt the general self-attention with absolute position representation at the context fusion module. eRPR denotes models adopt the self-attention with entity-specific relative position representation at the context fusion module. The ✓mark refers to the model including the global relation classification. We use the *strict* evaluation setting here.

Model	Encoder	Context fusion	GRC	Entity			Relation		
				P	R	F1	P	R	F1
				CoNLL04					
1	BiLSTM	-	-	83.75	84.06	83.90	63.75	60.43	62.04
2	BERT	-	-	85.75	86.28	86.00	65.15	62.56	63.83
3	BERT	APR Layer ×2	-	86.32	85.68	86.00	64.53	63.40	63.96
4	BERT	eRPR Layer ×2	-	86.75	85.56	86.15	63.93	64.50	64.21
5	BERT	APR Layer ×2	✓	86.78	85.66	86.22	64.18	64.30	64.24
6	BERT	eRPR Layer ×2	✓	86.85	85.62	**86.23**	64.20	64.69	**64.44**
				ACE05					
1	BiLSTM	-	-	84.88	84.10	84.49	57.40	60.32	58.82
2	BERT	-	-	85.70	84.25	84.96	59.92	60.06	59.99
3	BERT	APR Layer ×2	-	86.18	84.55	85.36	60.23	60.82	60.52
4	BERT	eRPR Layer ×2	-	86.24	84.60	85.41	60.57	60.76	60.66
5	BERT	APR Layer ×2	✓	86.22	84.57	85.39	60.46	60.76	60.61
6	BERT	eRPR Layer ×2	✓	86.26	84.66	**85.45**	60.60	60.84	**60.72**

4.4 Effect of the Maximum Relative Distance

In this subsection, we evaluate the effect of varying the maximum relative distance δ. Following previous studies [21], we conduct experiments on CoNLL04 with different maximum relative distance δ, increases exponentially from 0 to 64. Figure 4a shows the experimental results. We observe that when $\delta = 8$, the entity F1 has the best result, and when $\delta = 4$, the relation F1 has the best result. Meanwhile, the larger value of δ (i.e., $\delta = 64$) is meaningless for both entity and relation extraction, which verifies that the impacts of tokens beyond a maximum distance can be negligible. Therefore, to ensure a better performance for relation extraction, we set $\delta = 4$ for all the experiments.

4.5 Effect of the GRC Loss Weight

In this subsection, we evaluate the effect of different GRC loss weight λ to the model performance. We keep the maximum relative distance δ as 4 and conduct the experiments on the CoNLL04 dataset with λ from 0 to 1 at the interval of 0.2. As shown in Fig. 4a, the setting with $\lambda = 0$ denotes the GRC is not used in

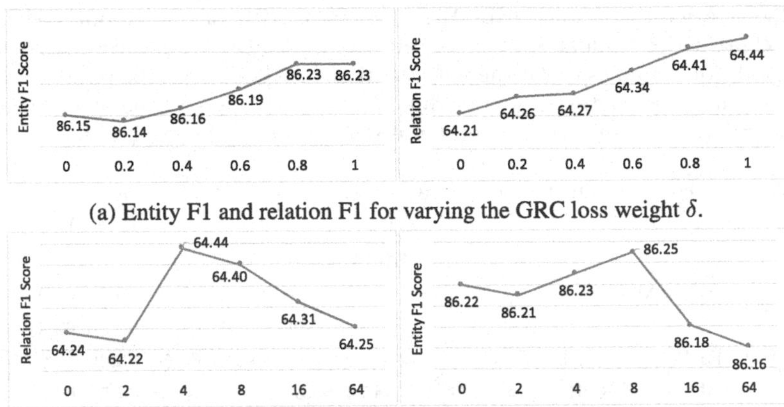

(a) Entity F1 and relation F1 for varying the GRC loss weight δ.

(b) Entity F1 and relation F1 for varying the maximum relative distance λ.

Fig. 4. Results for varying the GRC loss weight δ and the maximum relative distance λ.

the framework and its performance is much lower than settings with larger λ In addition, with the growth of λ, both entity and relation F1 scores are increased continuously. As such, we keep $\lambda = 1$ for all the above experiments. These comparison results further demonstrate the effectiveness of GRC. Therefore, the sentence-level information can be utilized fruitfully for multi-head selection and helps improve the overall performance.

5 Conclusion

In this paper, we propose a relative position representation based multi-head selection framework for joint entity and relation extraction. Different with the existing multi-head selection method, we introduce the relative position representation to capture the distance information of entities. We then propose a global relation classification to guide the learning of local features. Additionally, BERT is incorporated in the framework for semantic representation. Experimental results on CoNLL04 and ACE05 datasets show that our framework significantly outperforms all the baseline models for both entity and relation extraction.

Acknowledgements. This work was supported in part by the National Natural Science Foundation of China (Grant Nos. U1636211, 61672081,61370126), the Beijing Advanced Innovation Center for Imaging Technology (Grant No. BAICIT-2016001), and the Fund of the State Key Laboratory of Software Development Environment (Grant No. SKLSDE-2019ZX-17).

References

1. Adel, H., Schütze, H.: Global normalization of convolutional neural networks for joint entity and relation classification. In: Proceedings of the 2017 Conference on Empirical Methods in Natural Language Processing, pp. 1723–1729 (2017)
2. Bekoulis, G., Deleu, J., Demeester, T., Develder, C.: Joint entity recognition and relation extraction as a multi-head selection problem. Expert Syst. Appl. **114**, 34–45 (2018)
3. Chan, Y.S., Roth, D.: Exploiting syntactico-semantic structures for relation extraction. In: Proceedings of the 49th Annual Meeting of the Association for Computational Linguistics: Human Language Technologies, pp. 551–560 (2011)
4. Dai, Z., Yang, Z., Yang, Y., Carbonell, J.G., Le, Q., Salakhutdinov, R.: Transformer-xl: attentive language models beyond a fixed-length context. In: Proceedings of the 57th Annual Meeting of the Association for Computational Linguistics, pp. 2978–2988 (2019)
5. Devlin, J., Chang, M.W., Lee, K., Toutanova, K.: Bert: pre-training of deep bidirectional transformers for language understanding. In: Proceedings of the 2019 Conference of the North American Chapter of the Association for Computational Linguistics: Human Language Technologies, vol. 1 (Long and Short Papers), pp. 4171–4186 (2019)
6. Doddington, G.R., Mitchell, A., Przybocki, M.A., Ramshaw, L.A., Strassel, S.M., Weischedel, R.M.: The automatic content extraction (ACE) program-tasks, data, and evaluation. In: LREC, pp. 837–840 (2004)
7. Eberts, M., Ulges, A.: Span-based joint entity and relation extraction with transformer pre-training. arXiv preprint arXiv:1909.07755 (2019)
8. Forney, G.D.: The viterbi algorithm. In: Proceedings of the IEEE, pp. 268–278 (1973)
9. Fu, T.J., Li, P.H., Ma, W.Y.: Graphrel: modeling text as relational graphs for joint entity and relation extraction. In: Proceedings of the 57th Annual Meeting of the Association for Computational Linguistics, pp. 1409–1418 (2019)
10. Gehring, J., Auli, M., Grangier, D., Yarats, D., Dauphin, Y.N.: Convolutional sequence to sequence learning. In: Proceedings of the 34th International Conference on Machine Learning, vol. 70, pp. 1243–1252 (2017)
11. Gupta, P., Schütze, H., Andrassy, B.: Table filling multi-task recurrent neural network for joint entity and relation extraction. In: Proceedings of COLING 2016, the 26th International Conference on Computational Linguistics: Technical Papers, pp. 2537–2547 (2016)
12. Huang, W., Cheng, X., Wang, T., Chu, W.: BERT-based multi-head selection for joint entity-relation extraction. In: Tang, J., Kan, M.-Y., Zhao, D., Li, S., Zan, H. (eds.) NLPCC 2019. LNCS (LNAI), vol. 11839, pp. 713–723. Springer, Cham (2019). https://doi.org/10.1007/978-3-030-32236-6_65
13. Katiyar, A., Cardie, C.: Going out on a limb: joint extraction of entity mentions and relations without dependency trees. In: Proceedings of the 55th Annual Meeting of the Association for Computational Linguistics (vol. 1: Long Papers), pp. 917–928 (2017)
14. Li, X., et al.: Entity-relation extraction as multi-turn question answering. In: Proceedings of the 57th Annual Meeting of the Association for Computational Linguistics, pp. 1340–1350 (2019)

15. Lin, Y., Shen, S., Liu, Z., Luan, H., Sun, M.: Neural relation extraction with selective attention over instances. In: Proceedings of the 54th Annual Meeting of the Association for Computational Linguistics (vol. 1: Long Papers), pp. 2124–2133 (2016)
16. Luan, Y., Wadden, D., He, L., Shah, A., Ostendorf, M., Hajishirzi, H.: A general framework for information extraction using dynamic span graphs. In: Proceedings of the 2019 Conference of the North American Chapter of the Association for Computational Linguistics: Human Language Technologies, vol. 1 (Long and Short Papers), pp. 3036–3046 (2019)
17. Miwa, M., Bansal, M.: End-to-end relation extraction using LSTMS on sequences and tree structures. In: Proceedings of the 54th Annual Meeting of the Association for Computational Linguistics (vol. 1: Long Papers), pp. 1105–1116 (2016)
18. Miwa, M., Sætre, R., Miyao, Y., Tsujii, J.: A rich feature vector for protein-protein interaction extraction from multiple corpora. In: Proceedings of the 2009 Conference on Empirical Methods in Natural Language Processing, pp. 121–130 (2009)
19. Miwa, M., Sasaki, Y.: Modeling joint entity and relation extraction with table representation. In: Proceedings of the 2014 Conference on Empirical Methods in Natural Language Processing, pp. 1858–1869 (2014)
20. Roth, D., Yih, W.T.: A linear programming formulation for global inference in natural language tasks. Technical report, Illinois Univ At Urbana-Champaign Department of Computer Science (2004)
21. Shaw, P., Uszkoreit, J., Vaswani, A.: Self-attention with relative position representations. In: Proceedings of the 2018 Conference of the North American Chapter of the Association for Computational Linguistics: Human Language Technologies, vol. 2 (Short Papers), pp. 464–468 (2018)
22. Sukhbaatar, S., Weston, J., Fergus, R., et al.: End-to-end memory networks. In: Advances in Neural Information Processing Systems, pp. 2440–2448 (2015)
23. Sun, C., et al.: Joint type inference on entities and relations via graph convolutional networks. In: Proceedings of the 57th Annual Meeting of the Association for Computational Linguistics, pp. 1361–1370 (2019)
24. Vaswani, A., et al.: Attention is all you need. In: Advances in Neural Information Processing Systems, pp. 5998–6008 (2017)
25. Wang, H., et al.: Extracting multiple-relations in one-pass with pre-trained transformers. In: Proceedings of the 57th Annual Meeting of the Association for Computational Linguistics, pp. 1371–1377 (2019)
26. Zhao, T., Yan, Z., Cao, Y., Li, Z.: Asking effective and diverse questions: a machine reading comprehension based framework for joint entity-relation extraction. In: Proceedings of the Twenty-Ninth International Joint Conference on Artificial Intelligence, pp. 3948–3954 (2020)

Machine Translation and Multilingual Information Processing

A Mixed Learning Objective for Neural Machine Translation

Wenjie Lu, Leiying Zhou, Gongshen Liu(✉), and Quanhai Zhang(✉)

School of Electronic Information and Electrical Engineering, Shanghai Jiao Tong University, Shanghai, China
{jonsey,zhouleiying,lgshen,qhzhang}@sjtu.edu.cn

Abstract. Evaluation discrepancy and overcorrection phenomenon are two common problems in neural machine translation (NMT). NMT models are generally trained with word-level learning objective, but evaluated by sentence-level metrics. Moreover, the cross-entropy loss function discourages model to generate synonymous predictions and overcorrect them to ground truth words. To address these two drawbacks, we adopt multi-task learning and propose a mixed learning objective (MLO) which combines the strength of word-level and sentence-level evaluation without modifying model structure. At word-level, it calculates semantic similarity between predicted and ground truth words. At sentence-level, it computes probabilistic n-gram matching scores of generated translations. We also combine a loss-sensitive scheduled sampling decoding strategy with MLO to explore its extensibility. Experimental results on IWSLT 2016 German-English and WMT 2019 English-Chinese datasets demonstrate that our methodology can significantly promote translation quality. The ablation study shows that both word-level and sentence-level learning objectives can improve BLEU scores. Furthermore, MLO is consistent with state-of-the-art scheduled sampling methods and can achieve further promotion.

Keywords: Neural machine translation · Evaluation discrepancy · Overcorrection phenomenon

1 Introduction

In recent years, tremendous progresses have been made in the field of neural machine translation (NMT) [12,23]. A typical NMT model can be formulated as an encoder-decoder-attention architecture [1,5] with maximum likelihood estimation (MLE) objective. Given sufficient parallel corpora, NMT models can achieve promising performance.

Despite much success, NMT models suffer from two major drawbacks. First, there exists a discrepancy between training objectives and evaluation metrics. Most NMT models are trained with MLE objective under the teacher forcing algorithm [28], i.e., models calculate and accumulate cross-entropy loss between predicted and ground truth sentences word by word. A lower cross-entropy value

M. Sun et al. (Eds.): CCL 2020, LNAI 12522, pp. 201–213, 2020.
https://doi.org/10.1007/978-3-030-63031-7_15

means the predictions are closer to ground truth at word level. Model parameters are updated through backpropagation to minimize the value of loss function. However, translation quality is measured by sentence-level metrics such as BLEU [21], ROUGE [10], etc. This way of word-level optimization mismatches sentence-level evaluation metrics, which may mislead the promotion of translation performance. Second, the MLE training objective brings about overcorrection phenomenon [30]. To be specific, models are trained to learn absolutely correct translations and overcorrect synonymous words and phrases. Once the model predicts a word different from the ground truth word, the cross-entropy loss will immediately punish it and lead the model to the correct direction. As for synonymous phrases, it may result in translating wrong phrases while reducing the diversity of translation.

In this paper, we present a novel approach to solve the above problems. Instead of training NMT models with word-level cross-entropy loss, we propose to train models with a mixed learning objective (MLO), which can combine the strength of word-level and sentence-level training. At word level, MLO estimates semantic similarity between the predicted and the ground truth words. Synonymous words will be encouraged rather than overcorrected. At sequence level, MLO calculates probabilistic n-gram matching score between the predicted and the ground truth sentences. The differentiable property of MLO enables NMT models to be trained flexibly without modifying structure. Most important of all, it can relieve the problem of evaluation discrepancy and overcorrection phenomenon.

The major contributions of this paper are summarized as follows:

- We present a novel mixed learning objective for training NMT models, aiming at alleviating evaluation discrepancy and overcorrection phenomenon. The mixed learning objective can encourage word-level semantic similarity and balance sequence-level n-gram precision of the translation.
- We explore the extensibility of mixed learning objective and adopt a novel loss-sensitive scheduled sampling instead of teacher forcing algorithm. The proposed objective is more consistent with state-of-the-art scheduled sampling methods and can achieve better performance.
- We demonstrate the effectiveness of our approach on IWSLT 2016 German-English and WMT 2019 English-Chinese datasets, and achieve significant improvements. Moreover, the mixed learning objective can be flexibly applied by various model structures and algorithms.

2 Related Work

2.1 Evaluation Discrepancy

To tackle the problem of discrepancy between word-level MLE objective and sentence-level evaluation metrics, some researches utilize techniques like generative adversarial network (GAN) [6] or reinforcement learning (RL) [24]. Borrowed idea from DAD [26] and beam search [17,23], reference [16] proposed

Mixed Incremental Cross-Entropy Reinforce (MIXER) to directly optimized model parameters with respect to the metric used at inference time. Further, reference [20] presented minimum risk training (MRT) to minimize the expected loss (i.e., risk) on the training data. Reference [27] proposed to train NMT models with semantic similarity based on MRT. Reference [29] introduced beam-search optimization schedule for model to learn global sequence scores. Moreover, reference [11] proposed RankGAN which can analyze and rank sentences by giving a reference group, and thus achieve high-quality language descriptions.

2.2 Overcorrection Phenomenon

As for overcorrection phenomenon, especially synonymous phrases, one solution is to utilize the model's previous predictions as input in training. The generation inconsistency between training and inference which called exposure bias [30] causes models to overcorrect from synonymous translations and generate wrong phrases. Reference [2] firstly proposed a scheduled sampling strategy based on an algorithm called Data As Demonstrator (DAD) [26]. At every decoding step, a dynamic probability p is used to decide whether to sample from ground truth or the previous word predicted by the model itself. Inspired by their method, reference [30] came up with sampling from ground truth and inferred sentences word by word through force decoding.

3 Methodology

3.1 Model Overview

Without loss of generality, we utilize a common RNN attention model [1] as baseline to demonstrate our approach. Suppose that the source sentence $X = (x_1, x_2, ..., x_{T_x})$ and the target sentence $Y = (y_1, y_1, ..., y_{T_y})$. The RNN model encodes the source sentence as follows:

$$h_t = \phi(h_{t-1}, x_t) \tag{1}$$

where h_0 is an initial vector and ϕ is a nonlinear function. Then context vector $c_i, i = 1, 2, ..., T_y$ is calculated by:

$$c_i = \sum_{j=1}^{T_x} \alpha_{ij} \cdot h_j \tag{2}$$

where α_{ij} is the attention weight between c_i and h_j.

When the decoder receives the context c_t, it calculates the hidden layer vector s_t by:

$$s_t = f(s_{t-1}, y_{t-1}, c_t) \tag{3}$$

where s_0 is an initial vector, f is a nonlinear function of hidden layers, y_{t-1} is the historical output at time $t-1$ in inference and ground truth word in training, and y_0 is the end flag of source sentence X.

According to the hidden layer state s_t, the probability of inferring the word y_t can be computed by:

$$P(y_t) = softmax(W_o \cdot p(y_t \mid y_1, ..., y_{t-1}, x)) \quad (4)$$

$$p(y_t \mid y_1, ..., y_{t-1}, x) = g(y_{t-1}, s_t, c_t) \quad (5)$$

where g is a nonlinear function and W_o is a mapping matrix.

Finally, given a set of sequence pairs $(X_i, Y_i), i = 1, 2, ..., N$ in the parallel corpora, the training objective is to maximize the likelihood as follows:

$$\hat{\theta}_{MLE} = argmax\{L(\theta)\} \quad (6)$$

where $L(\theta)$ is the loss function computed by:

$$L(\theta) = \sum_{i=1}^{N} logP(Y_i \mid X_i, \theta) = \sum_{i=1}^{N}\sum_{t=1}^{T_y} logP(y_t) \quad (7)$$

3.2 Word-Level Semantic Similarity Objective

The original cross-entropy loss measures the probability of predicting right translation for each word, which means it only cares about how to generate ground truth words with maximum likelihood. This may cause two problems. First, generating any other words is discouraged. Although synonymous translations are right in the subjective sense, they will be punished and corrected to ground truth words. Second, suppose that the word with maximum probability is not ground truth word, and the model will choose it as predicted translation. The calculation of cross-entropy loss does not take into consideration of what exactly that word is, which is important for evaluating the model.

Therefore, we design the word-level learning objective in order to measure the semantic similarity between the generated translations and ground truth sentences. There have been lots of complex researches on semantic similarity [8,15]. In order not to include additional models, we adopt the cosine similarity method for measurement. Mathematically speaking, cosine similarity calculates the semantic similarity between two non-zero vectors, which is suitable for word embeddings.

Given the predicted translation $Y^* = (y_1^*, y_2^*, \cdots, y_{T_{Y^*}}^*)$, the semantic similarity between sentence Y and Y^* can be calculated by:

$$Sim(Y, Y^*) = \sum_{i=1}^{T_y} \frac{emb(y_i) \cdot emb(y_i^*)}{\|emb(y_i)\| \times \|emb(y_i^*)\|} \quad (8)$$

where $emb(\cdot)$ refers to the word embedding of each word.

Therefore, we can calculate semantic similarity between every translation and corresponding ground truth sentence. During training, the word-level training objective is defined as followings:

$$L_{word} = -\sum_{j=1}^{N} Sim(Y_j, Y_j^*) \quad (9)$$

3.3 Sentence-Level Probabilistic N-gram Objective

The word-level semantic similarity objective helps to foster translation diversity and relieve the problem of overcorrection, which can improve word-level translation accuracy. As for another important standard fluency in machine translation, we design a sentence-level probabilistic n-gram objective which is consistent with evaluation metrics.

The calculation of n-gram matching is widely used in machine translation evaluation metrics. Take BLEU for example, firstly n-grams in source sentence Y and Y^* are extracted and counted, denoted as $C(n-gram)$. Next, n-gram matches between Y and Y^* are computed and denoted as $C_{clip}(n-gram)$. The precision score can be calculated by their ratio.

However, the non-differentiable property of BLEU makes it unable to be adopted as loss function. Therefore, inspired by [19], we modified the calculation of n-gram matches as follows. Supposing that $(g_1, g_2, ..., g_n)$ is an n-gram sequence, then its occurrences in Y can be computed by:

$$\widetilde{C}_Y(n-gram) = \sum_{i=0}^{T_Y-n} \prod_{j=1}^{n} 1\{g_j = y_{i+j}\} \cdot P(y_{i+j}) \tag{10}$$

where $1\{\cdot\}$ denotes an indicator function and $P(\cdot)$ is calculated by Eq. (5). Then, the clip n-gram matches between two sentences and the precision score of translation Y can be computed as follows:

$$C_{clip}(n-gram) = min\{\widetilde{C}_Y(n-gram), C_{Y^*}(n-gram)\} \tag{11}$$

$$\widetilde{p_n} = \frac{\sum_{n-gram \in Y} C_{clip}(n-gram)}{\sum_{n-gram' \in Y} \widetilde{C}_Y(n-gram')} \tag{12}$$

Finally, to punish very long or short translations, BLEU is modified based on $\widetilde{p_n}$ and defined as follows:

$$\widetilde{BLEU}(Y, Y^*) = BP \cdot exp(\sum_{n=1}^{N} w_n log\widetilde{p_n}) \tag{13}$$

where BP is brevity penalty, w_n is positive weights and N is the maximum length of n-gram.

Therefore, we can calculate probabilistic n-gram matching score between every translation and corresponding ground truth sentence. During training, the sentence-level training objective is defined as followings:

$$L_{sent} = -\sum_{j=1}^{N} \widetilde{BLEU}(Y_j, Y_j^*) \tag{14}$$

3.4 Mixed Learning Objective

In order to alleviate the problem of evaluation discrepancy and overcorrection phenomenon, we propose the mixed learning objective. At word level, it can calculate semantic similarity for training evaluation and promote translation diversity. At sentence level, it can compute probabilistic n-gram precision of predicted sentence and promote translation fluency. The mixed learning objective is defined as follows:

$$L_{total} = L_{ce} + \alpha_{word} \cdot L_{word} + \alpha_{sent} \cdot L_{sent} \tag{15}$$

where L_{ce} refers to cross-entropy loss, α_{word} is the weight of word-level loss function and α_{sent} is the weight of sentence-level loss function.

Similar to reference [30], we adopt the Gumbel-Max technique [7,13] for generating more robust outputs. To be specific, the Gumbel noise is defined as follows:

$$G = -log(-logU) \tag{16}$$

where $U \sim Unif[0,1]$.

Then Eq. (5) is modified to:

$$P(y_t) = softmax(\frac{W_o \cdot p(y_t \mid y_1, ..., y_{t-1}, x) + G}{\tau}) \tag{17}$$

where τ is a temperature parameter controlling the generated distribution.

During training, we adopt a scheduled sampling strategy instead of teacher forcing algorithms. At every decoding step, a probability p is used to decide whether to sample from ground truth or inferred words. Specifically, assuming w_t is the input at each decoding step t and y'_{t-1} is the word obtained from inferred words, then $Pr(w_t = y_{t-1}) = p$ and $Pr(w_t = y'_{t-1}) = 1 - p$. We hope the probability p to decay from 1 to 0, so that the training process can gradually learned to deal with simulated inference situation.

Borrowing idea from the decay schedule in learning rate, sample probability can be defined as an inverse sigmoid curve with variable training epochs. Considering that a loss function intuitively reflects how well the model is trained, we define loss-sensitive sample probability as follows:

$$p = \frac{k}{k + exp(\frac{e}{k})} \cdot \sigma(L) \tag{18}$$

where k is a hyper-parameter, e is the current index of epoch, L is the average loss function value of epoch e, and σ is a non-linear function. For practice, we choose tanh function.

4 Experiments

4.1 Experimental Setup

We conduct our experiments comparable with previous work by using the following two datasets:

German-English. The German-English dataset is chosen from IWSLT 2016 [3]. We use official testset2013 as validation set. The training and validation data consists of 196,884 and 992 sentences respectively. As for evaluation, we use the testset dataset from 2010 to 2014 and tokenized BLEU scores as computed by the multi-bleu.perl script[1].

English-Chinese. The English-Chinese dataset is chosen from the casia2015 parallel corpus in WMT 2019 shared task. It consists of approximately 1.05M sentences. We use official newsdev2017 as validation set and evaluate on the newstest dataset from 2017 to 2019.

For all training data, we perform tokenization and truecasing using standard Moses tools. For Chinese corpora, we use jieba[2] for segmentation. Then, we employ byte pair encoding (BPE) [18] with 50,000 operations to alleviate Out-of-Vocabulary problem. To accelerate training and save cost, we discard sentences with more than 50 tokens. The dimension of word embeddings is set to 512.

We first pretrain the baseline model by MLE. Then, we replace the cross-entropy loss function with MLO. The model is trained with a batch size of 60. We use Adam [9] optimizer to tune the parameters. Besides, we use dropout regularization with a drop probability 0.5. During decoding, the beam size is set to 3. The hyper-parameter of sample probability k and temperature τ are set to 12 and 0.5 respectively. The weights of word-level and sentence-level loss function are set to 0.8 and 150 respectively.

4.2 Baseline Systems

We compare our method with existing common NMT systems including Transformer [25], Evolved Transformer [22] and DTMT [14]. Moreover, to explore the extensibility of MLO, we experiment on several state-of-the-art scheduled sampling works. These baseline systems are included as follows:

RNNsearch. A vanilla attention-based recurrent neural network which consists of 2-layer bidirectional GRU units [4]. The dimension of hidden layer is 512.

SS-NMT. A word-level scheduled sampling method [2] which utilizes an inverse sigmoid decay schedule to sample from previous predicted word and ground truth word.

OR-NMT. A sentence-level scheduled sampling method [30] which utilizes inverse sigmoid decay schedule to sample from predicted sentence and ground truth sentence. Predicted sentence is generated by beam search and force decoding.

---

1 https://github.com/moses-smt/mosesdecoder/blob/master/scripts/generic/multi-bleu.perl.

2 https://github.com/fxsjy/jieba.

4.3 Main Results

Table 1. Results of the proposed method on German-English dataset (BLEU).

Systems	testset2010	testset2011	testset2012	testset2014	Average
Transformer	25.17	30.03	26.20	24.24	26.41
Evolved transformer	26.33	31.45	27.28	25.36	27.61
DTMT	26.51	31.66	27.64	26.02	27.96
RNNsearch	24.46	28.06	24.92	22.94	25.10
+ SS-NMT	26.46	30.14	26.60	24.31	26.88
+ OR-NMT	27.37	30.72	27.54	25.20	27.71
+ MLO	25.84	29.85	26.32	23.73	26.44
+ SS-NMT + MLO	26.78	30.34	26.99	24.81	27.23
+ OR-NMT + MLO	**27.44**	**31.89**	**27.65**	**25.92**	**28.22**

Table 1 and Table 2 reports the results of the proposed method in comparison to other NMT systems on German-English and English-Chinese datasets respectively. As it can be seen, training with sentence-level scheduled sampling and mixed learning objective (AT-NMT + MLO) obtains the best published results on all testsets.

On German-English dataset, our full system can outperform RNNsearch by +3.11 BLEU averagely. On English-Chinese dataset, our full system can have an improvement of +2.68 BLEU on three testsets.

To validate the effectiveness of mixed learning objective, we carry out ablation study to evaluate the performance of word-level and sentence-level learning objective respectively. The mixed learning objective is proposed to encourage word-level semantic similarity and balance sequence-level n-gram precision of the translation. Meanwhile, it aims at relieving the problem of evaluation discrepancy and overcorrection. We will display and analyze the effect of mixed learning objective in detail in Sect. 4.4.

Another point of focus lies in the extensibility of mixed learning objective. As shown in Table 1 and Table 2, combining scheduled sampling strategy with mixed learning objective can achieve better translation performance. We will discuss the effect of scheduled sampling from two aspects in Sect. 4.5. Besides, the loss-sensitive sample probability is defined to sense the speed of converge and make adjustment on sample probability. We will analyse its effect on scheduled sampling methods to explore how to achieve better performance.

4.4 Effect of Mixed Learning Objective

Aiming to alleviate evaluation discrepancy and overcorrection phenomenon, we propose the mixed learning objective which can promote word-level semantic similarity and sequence-level n-gram precision. To explore the effect of

Table 2. Results of the proposed method on English-Chinese dataset (BLEU).

Systems	newstest2017	newstest2018	newstest2019	Average
Transformer	26.37	25.09	25.76	25.74
Evolved transformer	27.84	25.98	27.25	27.02
DTMT	28.07	26.10	27.34	27.17
RNNsearch	24.92	24.17	24.20	24.63
+ SS-NMT	25.89	25.12	25.43	25.48
+ OR-NMT	28.03	26.10	26.66	26.93
+ MLO	25.83	24.74	25.32	25.29
+ SS-NMT + MLO	26.60	25.42	25.63	25.88
+ OR-NMT + MLO	**28.18**	**26.63**	**27.13**	**27.31**

mixed learning objective, we conduct experiments on word-level and sentence-level learning objective respectively without scheduled sampling strategy on RNNsearch under the same conditions.

The experimental results are listed in Table 3 and Table 4. As it can be seen, only using word-level or sentence-level learning objective rather than cross-entropy loss can help achieve higher BLEU scores on two datasets. To be specific, word-level learning objective can get a promotion of +0.16–+0.73 BLEU averagely over RNNsearch on German-English and English-Chinese datasets. Sentence-level learning objective can outperform RNNsearch by $+0.43 \sim +0.95$ BLEU score on two datasets averagely.

For the experimental results, we make some simple analysis. The word-level learning objective takes into account semantic similarity between predicted and ground truth words, so that it can avoid forcing model to generate the only one correct translation. The promotion in BLEU scores verifies that discouraging and punishing other synonymous words is disadvantageous for NMT models. Therefore, the word-level learning objective can to some extent solve this problem and encourage translation diversity.

The sentence-level learning objective calculates probabilistic n-gram matching scores between predicted and ground truth sentence, which is coordinate with general evaluation metrics. On the one hand, it contributes to alleviate the problem of evaluation discrepancy without importing additional complex model. On the other hand, the objective can naturally promote translation performance on BLEU scores.

Furthermore, MLO which combines word-level and sentence-level learning objective can obtain best translation performance in BLEU scores. Specifically, MLO can outperform RNNsearch by +0.66–+1.34 BLEU score averagely on German-English and English-Chinese datasets.

Table 3. BLEU scores on German-English dataset.

Systems	testset2010	testset2011	testset2012	testset2014	Average
RNNsearch	24.46	28.06	24.92	22.94	25.10
+ L_{word}	25.18	28.57	25.74	23.83	25.83
+ L_{sent}	25.40	28.72	26.01	24.07	26.05
+ MLO	25.84	29.85	26.32	23.73	26.44

Table 4. BLEU scores on English-Chinese dataset.

Systems	newstest2017	newstest2018	newstest2019	Average
RNNsearch	24.92	24.17	24.20	24.63
+ L_{word}	25.26	24.45	24.67	24.79
+ L_{sent}	25.59	24.51	25.10	25.06
+ MLO	25.83	24.74	25.32	25.29

4.5 Effect of Loss-Sensitive Scheduled Sampling

To validate the extensibility of MLO, we conduct various experiments which combine MLO with state-of-the-art scheduled sampling methods. Moreover, we defined a novel loss-sensitive sample probability for model to be flexibly adapted to scheduled sampling strategy. Under the same experimental settings, we conduct experiments on German-English and English-Chinese datasets to validate the effectiveness of loss-sensitive scheduled sampling and analyze in two aspects.

Figure 1 gives the training loss curves of MLO, SS-NMT+MLO and OR-NMT+MLO during training. As the training epoch increases, MLO continues to decrease at the lowest value and gradually tends to be flat. Due to different sampling strategies, SS-NMT and OR-NMT gradually converge to a certain training loss value. Moreover, Fig. 2 gives the BLEU score curves of three methods. It can be seen that SS-NMT+MLO and OR-NMT+MLO can achieve better BLEU scores compared to RNNsearch on validation set. We can also conclude from Table 1 and Table 2 that SS-NMT+MLO can achieve a promotion of +0.35–+0.4 BLEU scores over SS-NMT and OR-NMT+MLO can outperform OR-NMT by +0.38–+0.51 BLEU scores.

Since the starting point of scheduled sampling is to solve the problem of exposure bias and overcorrection phenomenon, the original cross-entropy loss function may be hard to score the inference results and guide the training process. However, the MLO is proposed for alleviating these problems as well. Therefore, the idea of combining MLO with scheduled sampling is natural and proved to be effective.

Besides the mutual promotion of MLO and scheduled sampling, the last thing we want to point out is the necessity and effectiveness of loss-sensitive sample probability. We define $\sigma(L) = 1$ as non-sensitive sample probability and perform parallel tests. By observing their decay curves during training, we find that

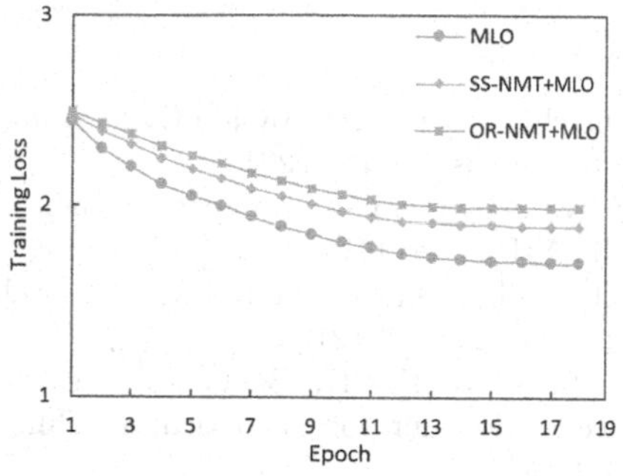

Fig. 1. The training loss curves of three baseline systems on the IWSLT 2016 German-English translation task.

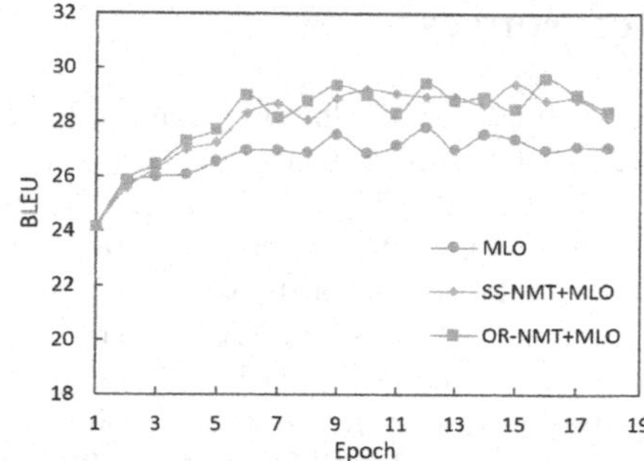

Fig. 2. Trends of BLEU scores of three baseline systems on the validation set on the IWSLT 2016 German-English translation task.

loss-sensitive sample probability is more flexible and helpful in adjusting a proper probability for different training scenes. Since $tanh(L) < 1$, the loss-sensitive probability is calculated to be lower than non-sensitive probability. From the perspective of feeding as input inferred rather than ground truth words, we make it harder for model to learn and correct mistakes. Meanwhile, the experimental results show promotion on translation quality.

5 Conclusion

In this paper, we propose a mixed learning objective for NMT so as to alleviate the problem of evaluation discrepancy and overcorrection phenomenon. At word-level, the objective measures semantic similarity between the generated and ground truth words. At sentence-level, the objective calculates probabilistic n-gram matching scores of the translations. Moreover, we combine loss-sensitive scheduled sampling methods with mixed learning objective for mutual promotion. Experimental results show that our proposed method can achieve significant improvement on BLEU scores compared to previous works.

Acknowledgment. This research work has been funded by the National Natural Science Foundation of China (Grant No. 61772337), the National Key Research and Development Program of China No. 2018YFC0830803.

References

1. Bahdanau, D., Cho, K., Bengio, Y.: Neural machine translation by jointly learning to align and translate. In: Bengio, Y., LeCun, Y. (eds.) ICLR (2015)
2. Bengio, S., Vinyals, O., Jaitly, N., Shazeer, N.: Scheduled sampling for sequence prediction with recurrent neural networks. In: NIPS, pp. 1171–1179 (2015)
3. Cettolo, M., Girardi, C., Federico, M.: Wit^3: web inventory of transcribed and translated talks. In: EAMT, Trento, Italy, pp. 261–268 (2012)
4. Cho, K., Gulcehre, B.V.M.C., Bahdanau, D., Schwenk, F.B.H., Bengio, Y.: Learning phrase representations using RNN encoder-decoder for statistical machine translation. In: EMNLP, pp. 1724–1734 (2014)
5. Forcada, M.L., Ñeco, R.P.: Recursive hetero-associative memories for translation. In: Mira, J., Moreno-Díaz, R., Cabestany, J. (eds.) IWANN 1997. LNCS, vol. 1240, pp. 453–462. Springer, Heidelberg (1997). https://doi.org/10.1007/BFb0032504
6. Goodfellow, I., et al.: Generative adversarial nets. In: NIPS, pp. 2672–2680 (2014)
7. Gumbel, E.J.: Statistical theory of extreme values and some practical applications. NBS Applied Mathematics Series, vol. 33 (1954)
8. Kenter, T., De Rijke, M.: Short text similarity with word embeddings. In: CIKM, pp. 1411–1420 (2015)
9. Kingma, D.P., Ba, J.: Adam: a method for stochastic optimization. arXiv preprint arXiv:1412.6980 (2014)
10. Lin, C.Y.: Rouge: a package for automatic evaluation of summaries. In: Text Summarization Branches Out, pp. 74–81 (2004)
11. Lin, K., Li, D., He, X., Zhang, Z., Sun, M.T.: Adversarial ranking for language generation. In: NIPS, pp. 3155–3165 (2017)
12. Luong, M.T., Pham, H., Manning, C.D.: Effective approaches to attention-based neural machine translation. In: EMNLP, pp. 1412–1421 (2015)
13. Maddison, C., Tarlow, D., Minka, T.: A* sampling. NIPS (2014)
14. Meng, F., Zhang, J.: DTMT: a novel deep transition architecture for neural machine translation. In: AAAI, vol. 33, pp. 224–231 (2019)
15. Pradhan, N., Gyanchandani, M., Wadhvani, R.: A review on text similarity technique used in IR and its application. Int. J. Comput. Appl. **120**(9), 29–34 (2015)
16. Ranzato, M., Chopra, S., Auli, M., Zaremba, W.: Sequence level training with recurrent neural networks. arXiv preprint arXiv:1511.06732 (2015)
17. Rush, A.M., Chopra, S., Weston, J.: A neural attention model for abstractive sentence summarization. In: EMNLP, pp. 379–389 (2015)
18. Sennrich, R., Haddow, B., Birch, A.: Neural machine translation of rare words with subword units. In: ACL, pp. 1715–1725 (2016)
19. Shao, C., Feng, Y., Chen, X.: Greedy search with probabilistic n-gram matching for neural machine translation. arXiv preprint arXiv:1809.03132 (2018)
20. Shen, S., et al.: Minimum risk training for neural machine translation. In: ACL, pp. 1683–1692 (2016)
21. Shterionov, D., Nagle, P., Casanellas, L., Superbo, R., O'Dowd, T.: Empirical evaluation of NMT and PBSMT quality for large-scale translation production. In: EAMT: User Track (2017)
22. So, D., Le, Q., Liang, C.: The evolved transformer. In: ICML, pp. 5877–5886 (2019)
23. Sutskever, I., Vinyals, O., Le, Q.V.: Sequence to sequence learning with neural networks. In: NIPS, pp. 3104–3112 (2014)
24. Sutton, R.S., Barto, A.G., et al.: Introduction to reinforcement learning, vol. 2. MIT Press, Cambridge (1998)

25. Vaswani, A., et al.: Attention is all you need. In: NIPS, pp. 5998–6008 (2017)
26. Venkatraman, A., Hebert, M., Bagnell, J.A.: Improving multi-step prediction of learned time series models. In: AAAI, pp. 3024–3030 (2015)
27. Wieting, J., Berg-Kirkpatrick, T., Gimpel, K., Neubig, G.: Beyond bleu: training neural machine translation with semantic similarity. In: ACL, pp. 4344–4355 (2019)
28. Williams, R.J., Zipser, D.: A learning algorithm for continually running fully recurrent neural networks. Neural Comput. **1**(2), 270–280 (1989)
29. Wiseman, S., Rush, A.M.: Sequence-to-sequence learning as beam-search optimization. In: EMNLP 2016, pp. 1296–1306 (2016)
30. Zhang, W., Feng, Y., Meng, F., You, D., Liu, Q.: Bridging the gap between training and inference for neural machine translation. In: ACL, pp. 4334–4343 (2019)

Multi-reward Based Reinforcement Learning for Neural Machine Translation

Shuo Sun, Hongxu Hou(✉), Nier Wu, Ziyue Guo, and Chaowei Zhang

College of Computer Science-college of Software,
Inner Mongolia University, Hohhot, China
sunshuo07@126.com, wunier04@126.com, guoziyue08@126.com,
zhangchaowei08@126.com, cshhx@imu.edu.cn

Abstract. Reinforcement learning (RL) has made remarkable progress in neural machine translation (NMT). However, it exists the problems with uneven sampling distribution, sparse rewards and high variance in training phase. Therefore, we propose a multi-reward reinforcement learning training strategy to decouple action selection and value estimation. Meanwhile, our method combines with language model rewards to jointly optimize model parameters. In addition, we add Gumbel noise in sampling to obtain more effective semantic information. To verify the robustness of our method, we not only conducted experiments on large corpora, but also performed on low-resource languages. Experimental results show that our work is superior to the baselines in WMT14 English-German, LDC2014 Chinese-English and CWMT2018 Mongolian-Chinese tasks, which fully certificates the effectiveness of our method.

Keywords: Reinforcement learning · Multi-reward · Gumbel noise

1 Introduction

Neural machine translation (NMT) [2,15,17] has drawn universal attention without the demand of numerous manual work. In training phase, generic NMT models employ maximum likelihood estimation (MLE) [4], which is the token-level objective function. However, it is inconsistent with sequence-level evaluation metrics such as BLEU [7]. Reinforcement learning (RL) are leveraged for sequence generation tasks including NMT to optimize sequence-level objectives, such as Actor-Critic [1] and Minimum Risk Training (MRT) [10]. In machine translation community, Wu el at. [15] provide the first comprehensive study of different aspects of RL training, they set a single reward to mitigate the inconsistency, and combine MLE with RL to stabilize the training process. Nevertheless, NMT based on reinforcement learning (RL) is unable to guarantee that the machine-translated sentences are as natural, sufficient and accurate as reference. To obtain smoother translation results, generative adversarial network (GAN) and deep reinforcement learning (DRL) [16] are employed to NMT. And [17] utilizes sentence-level BLEU Q as a reinforcement target based on the work of [16] to enhance the capability of the generator.

M. Sun et al. (Eds.): CCL 2020, LNAI 12522, pp. 214–227, 2020.
https://doi.org/10.1007/978-3-030-63031-7_16

Although this nova machine translation learning paradigm based on GAN and DRL reveals excellent manifestation, there are still some limitations: (1) when calculating rewards, the overestimation of Q value will give rise to a sub-optimal strategy update. (2) during training phase, it exists the problems with uneven sampling distribution, sparse rewards and high variance. What's more, the generator uses Monte Carlo to simulate the entire sentence, but it usually requires more calculation steps, resulting in too many parameters. (3) traditional NMT usually utilizes deterministic algorithms such as Beam Search or Greedy Decoding when predicting the next token. These methods lacks randomness, which may cause the potential best solution to be discarded.

In this paper, we propose some measures to address the above problems. Foremost, we adopt a novel multi-reward reinforcement learning method. That is, we weighted sum the actual reward of the discriminator, the language model reward and the sentence BLEU to obtain the total reward. Among them, we adopt reward shaping to alleviate the sparse reward when calculating sentence rewards. Nextly, our method employ Temporal-Difference Learning (TD) [11] to simulate the entire sentence. It effectively speeds up training and relieves the problem of error accumulation. Finally, we adopt Gumbel-Top-K Stochastic Beam Search [5] to predict the next token. The method trains model more efficiently by adding the noise obeying Gumbel distribution to control random sampled noise. Experiments on the datasets of the English-German, Chinese-English and Mongolian-Chinese translation tasks reveal our approach outperforms the best published results. In summary, we mainly made the following contributions:

- It is the first time that duel reward has been applied to neural machine translation. This method is applicable to arbitrary end-to-end NMT system.
- Our generator to optimize reward by using Gumbel-Top-K Stochastic Beam Search to sample different samples and Temporal-Difference Learning to simulate sentences.
- In English-German and Chinese-English translation tasks, we tested two different NMT models: RNNSearch and Transformer. Experimental results reveal that our proposed method performs well.

2 Background and Related Work

Common NMT models are based on an encoder decoder architecture. The encoder reads and encodes the source language sequence $X = (x_1, ..., x_n)$ into the context vector representation, and the decoder generates the corresponding target language sequence $\hat{Y} = (\hat{y}_1, ..., \hat{y}_m)$. Given H training sentence pairs $\{x^i, y^i\}_{i=1}^{H}$, at each timestep t, NMT is trained by maximum likelihood estimation(MLE) and generates the target words $\hat{y}_t$ by maximum the probability of translation conditioned on the source sentence X. The training goal is to maximize:

$$L_{MLE} = \sum_{i=1}^{H} logp(\hat{y}^i|x^i) = \sum_{i=1}^{H}\sum_{t=1}^{m} logp(\hat{y}_t^i|\hat{y}_1^i...\hat{y}_{t-1}^i, x^i) \quad (1)$$

where m is the length of sentence $\hat{y}^i$.

According [14], reinforcement learning enables NMT to optimize evaluation during training and usually estimates the overall expectation by sampling $\hat{y}$ with policy $p(\hat{y}|x)$. The training objective of RL is to maximize the expected reward:

$$L_{RL} = \sum_{i=1}^{H} R\left(\hat{y}^i, y^i\right), \hat{y}^i \sim p\left(\hat{y}|x^i\right), \forall_i \in [H]. \tag{2}$$

where $R(\hat{y}, y)$ is the final reward calculated by BLEU after generating the complete sentence $\hat{y}$. To increase stationarity, we combine the two simple linearly:

$$L_{COM} = \mu \times L_{MLE} + (1 - \mu) \times L_{RL} \tag{3}$$

where μ is the hyperparameter to control the balance between MLE and RL. L_{COM} is the strategy to stabilize RL training progress.

[17] proposed the BLEU reinforced conditional sequence generative adversarial net (BR-CSGAN) on the basis of reinforcement learning. The specific process is that generator G generates the target sentence based on the source sentence, and discriminator D detects whether the given sentence is groundtruth. During training status, G attempts to deceive discriminator D into believing that the generated sentence is groundtruth. The D strives to improve its anti-spoofing ability to distinguish machine-translated sentences from groundtruth. When G and D reach the Nash balance, the training results achieve the optimal state, and utilize BLEU to guide the learning of the generator.

3 Approach

In this section, we describe the multi-reward of reinforcement learning evaluation paradigm based on GAN model. The overall architecture is shown in Fig. 1. We introduce the generator G, discriminator D, sampling with Gumbel-Top-K Stochastic Beam Search, calculating final reward and training the entire model in detail.

3.1 GAN-Based NMT

The Generative Adversarial Net comprises of two adversarial sub models, a generator and a discriminator. The generator G is similar to the NMT model. Based on the source language sentence X, G aims to generate a target sentence $\hat{Y}$ which is indistinguishable from the reference Y. We take two different architectures for the generator, the traditional RNNSearch [2] and the state-of-the-art Transformer [12].

We utilize CNN [18] that performs better in classification tasks to construct the discriminator D. It aims to identify machine-generated sentences from a set of sentences containing machine translation $\hat{Y}$ and reference Y. To be specific, the generator's output $\hat{Y}$ or reference Y is spliced with the source language sentence

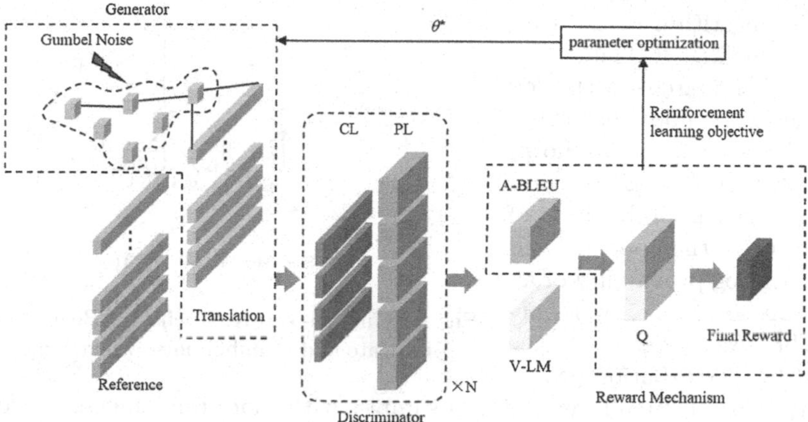

Fig. 1. The Illustration of the proposed multi-reward generative adversarial net(referred to as MR-GAN). The discriminator D is trained over the reference sentence pairs translated by the human and the generated sentence pairs (sampled with Gumbel noise) by G. Extract feature information via convolution (CL) and pooling (PL) operations, and adopt global features to jointly calculate rewards. Lastly, the generator G is trained by policy gradient where the final reward R is provided by D, V and A.

X to form a two-dimensional matrix, and the similarity between Y and X is measured by a convolution network. The optimization goal of the discriminator is minimize the coss-entropy loss of the binary classification:

$$L = -\left[a * log(p) + (1-a) * log(1-p)\right], p = \delta\left(V\left[r_X; r_{\hat{Y}}\right]\right) \tag{4}$$

where p is the probability that the target-language sentence is being real. r_X is the sentence representation of source language, which consists of extracting different features through different numbers of kernels with different window sizes. Similarly, $r_{\hat{Y}}$ is the target language sentence representation extracted from the target matrix $\hat{Y}_{1:T} = \hat{y}_1; \hat{y}_2; ...; \hat{y}_T$. V indicates the matrix which is used to merge r_X and $r_{\hat{Y}}$ into a low dimensional vector space. δ denote as the logistic function and a is a variable, which is correctly 1, otherwise 0.

3.2 Sampling

General NMT adopt Beam Search to generate the next token to reduce search space and speed up decoding. However, in many training methods such as RL or MRT, it is necessary to randomly collect multiple different samples from the model to calculate the sentence-level loss when decoding, but traditional methods can only produce similar results and loss of randomness. For this purpose, this work adopts an efficient and stable sampling method based on Gumbel-Top-K Stochastic Beam Search [5] to predict next token.

This algorithm uses Top-Down sampling [6] and performs Beam Search on the log probability of random perturbations. The structure is shown in Fig. 2 with $k = 2$. We first perturb the log probability of the root node, then perturb and correct the log probability of all candidate sequences, and only keep the two nodes with the highest log perturbation probability to expand. Finally we get two samples with more randomness as well as each sample is subject to the original distribution.

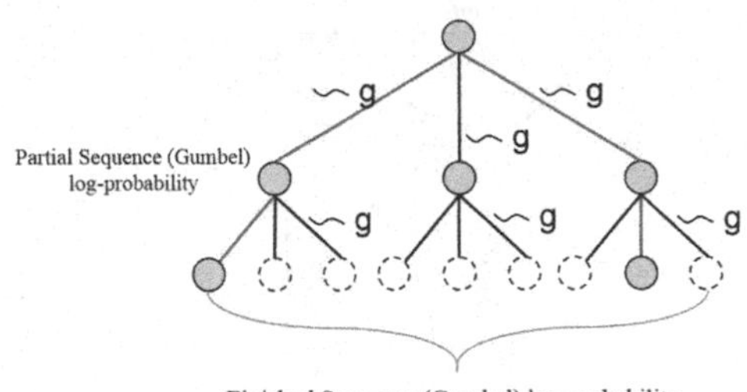

Fig. 2. Gumbel-Top-K Stochastic Beam Search. $\sim g$ indicate add Gumbel noise when sampling.

Specifically, for a category distribution I with N categories $I \sim Categorical\left(\frac{exp\phi_i}{\sum_{j\in N} exp\phi_i}\right)$, where ϕ_i is the log-probability of the i-th category and $i \in N$. If we take the logarithm of each category of I and add the noise g that obeys the Gumbel distribution, then take the Top-K from this slightly disturbed sentence with the Top-K probability (i.e. largest K categories after logarithmic calculation). The equations are as follows:

$$G \sim Gumbel\left(\phi\right) = \phi - log\left(-logU\right) = Gumbel\left(0\right) + \phi \tag{5}$$

$$I_{1,\ldots,k} = argtopK_{i\in N}Gumbel\left(\phi_i\right) \tag{6}$$

where $U \sim Uniform\left(0,1\right)$ and $G_i \sim Gumbel\left(0\right)$. It can be guaranteed that these K categories are subject to I and are different simultaneously, meanwhile the noise is controlled by the Gumbel distribution. With this method, we can train the model more efficiently and alleviate the problems of overtranslation and undertranslation in NMT.

3.3 Multi-reward of Reinforcement Learning

As shown in Fig. 1. Distinct with [17], which directly apply smoothed sentence-level BLEU as the specific objective Q for the generator. We aim to alleviate the overestimation of the reward, meanwhile, consider the fluency of machine translation and loyalty to the real translation. Therefore, our method is inspired by [13], given the generated sentence $\hat{Y}_{1:t}$ and the reference Y, the objective Q calculates a reward $Q(\hat{Y}_{1:t}, Y)$, which measures the fluency and loyalty of the generated sentence $\hat{Y}_{1:t}$, the equation is computed as:

$$Q(\hat{Y}_{1:t}, Y) = \lambda V\left(\hat{Y}_{1:t}\right) + (1-\lambda)\, A\left(\hat{Y}_{1:t}, Y\right) \tag{7}$$

where we set the independently generated language model reward as the value function $V\left(\hat{Y}_{1:t}\right)$, and the sentence reward as the advantage function $A\left(\hat{Y}_{1:t}, Y\right)$. λ is a hyper-parameter.

Value Function. For the sake of receiving smoother translation, we utilize the language model scores to participate in the calculation of rewards in reinforcement learning so that the NMT can consider the contextual and positional information of the corpus when translating. $V\left(\hat{Y}_{1:t}\right)$ represents the fluency score of sequence $\hat{y}_{1:t}$ including the current word, which guide the NMT to translate a sufficiently accurate and smooth result.

Typical language models have problems such as zero probability or statistical inadequacies. Good and Turing [3] proposed a new probabilistic formula to ease the "unsmoothness" problem. As shown in Fig. 3. They solved the zero-probability problem by down-regulating the frequency of words below the threshold and giving the out-of-vocabulary (OOV) a small non-zero value, where the sum of the down-regulated frequencies equals the probability of the OOV. The equation for 3-gram is as follow:

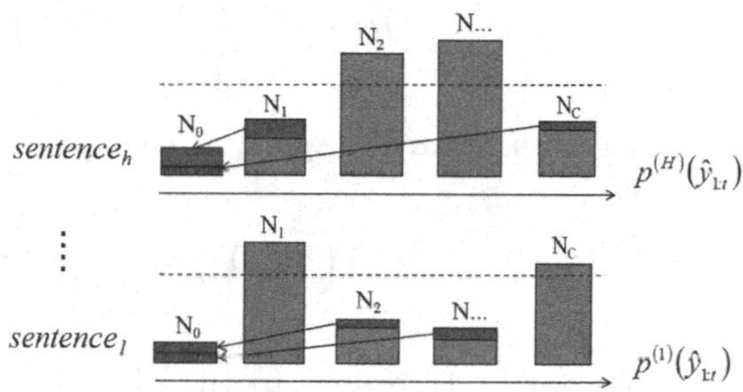

Fig. 3. Good-Turing smoothing algorithm. "...." represent threshold, decrease the frequency of words which number of occurrences is lower than the threshold, and give the sum of the resulting frequencies to the words that do not appear.

$$P_{GT}(\omega_i|\omega_{i-2},\omega_{i-1}) = \begin{cases} f(\omega_i|\omega_{i-2},\omega_{i-1}), c(\omega_{i-2},\omega_{i-1},\omega_i) \geq U \\ f_{gt}(\omega_i|\omega_{i-2},\omega_{i-1}), 0 < c(\omega_{i-2},\omega_{i-1},\omega_i) < U \\ Q(\omega_{i-2},\omega_{i-1}) \cdot P(\omega_i|\omega_{i-1}), otherwise \end{cases} \tag{8}$$

where we set $U = 9$, which is a threshold, and the function $f_{gt}(.)$ represents the relative frequency after Good-Turing estimation. Therefore, the probability is normalized to get $V\left(\hat{Y}_{1:t}\right)$, the equation is as follow:

$$V\left(\hat{Y}_{1:t}\right) = \sum_{s=1}^{S} c_s P_{GT}^{s} \tag{9}$$

where c_s represents the number of words that occur s times, and P_{GT}^{s} represents the probability of s occurrences obtained from Good-Turing smooth algorithm.

Advantage Function. From Eq. (2), the reward $R(\hat{y}, y)$ is only obtained after generate a complete sentence $\hat{y}$, it indicate only one reward is available for all actions(sample $\hat{y}_1 ... \hat{y}_T$). Consequently, RL training is inefficient due to the sparsity of rewards, and the model updates each token in the training sentence with the same reward without distinction. Following [15], we employ reward shaping to overcome the shortcoming. The current reward with reward shaping is defined as:

$$r_t(\hat{y}_t, y) = R(\hat{y}_{1...t}, y) - R(\hat{y}_{1...t-1}, y) \tag{10}$$

where $R(\hat{y}_{1...t}, y)$ is defined as the BLEU score of $\hat{y}_{1...t}$ respect to y. Reinforce algorithm has high variance because it use a single sample $\hat{y}$ to estimate the expectation. To improve the stability of the algorithm, we add an estimate of the average reward at each step t, and then subtract it from future cumulative reward. The cumulative reward are obtained from (11):

$$R(\hat{y}, y) = \sum_{i=1}^{m} r_t(\hat{y}, y), R(\hat{y}, y) - \hat{r}_t \tag{11}$$

Combined with reward shaping, at each step t the Advantage function is computed as:

$$A\left(\hat{Y}_{1:t}, Y\right) = \sum_{T=t}^{m} r_T(\hat{y}_T, y) - \hat{r}_t \tag{12}$$

Final Reward. According to the objective of the generator model (policy) $G_{\theta^*}(\hat{y}_t|\hat{Y}_{1:t-1})$ [19], to estimate $R^{G_{\theta^*}}_{D,V,A}$,which is the action-value function of a target sentence. Following Eq. (6), we consider the actual estimated probability of the discriminator D, the language model scores V and the sentence reward A as the final reward that update and optimize the generator G:

$$R^{G_{\theta^*}}_{D,V,A}\left(\hat{Y}_{1:T-1}, X, \hat{y}_T, Y\right) = \alpha\left(D\left(X, \hat{Y}_{1:T}\right) - b\left(X, \hat{Y}_{1:T}\right)\right) + \beta V\left(\hat{Y}_{1:t}\right) + \gamma A\left(\hat{Y}_{1:t}, Y\right) \tag{13}$$

where $b\left(X, \hat{Y}\right)$ represents the baseline value for reducing the variance estimation of rewards. We set $b\left(X, \hat{Y}\right) = 0.5$ based on experience. $\hat{Y}_{1:T}$ represents the generated target sentence and Y indicates the reference. α, β, γ are hyper-parameters.

However, D only provides a reward value for a entire generated target sequence. If $\hat{Y}_{1:T}$ is not the completed target sequence, the value of $D\left(X, \hat{Y}_{1:T}\right)$ is meaningless. Therefore, we cannot obtain the action-value of the intermediate state directly. Due to the large variance and parameters of Monte Carlo search, our work utilize Temporal-Difference (TD)[1] to sample the last $T - t$ tokens, it does not stop until the end of the sentence is sampled or the sampled sentence attains the maximum length. We implement the H TD emulation process as:

$$(\hat{Y}^1_{1:T_1}, ..., \hat{Y}^H_{1:T_H}) = TD^{G_{\theta^*}}\left(\left(\hat{Y}_{1:t}, X\right), H\right) \tag{14}$$

where $\left(\hat{Y}_{1:t}, X\right) = (\hat{y}_1 ... \hat{y}_t, X)$ is the current state, and $\hat{Y}^H_{t+1:T_H}$ is sampling based on G_{θ^*}. The discriminator rewards the sampled sentences separately and the

[1] Monte Carlo search is updated after sampled the complete sentence $\hat{y}$. It causes too many parameters and slower update speed when sentence length is longer. Temporal-Difference (TD) algorithm is an iterative way of calculating value function, which is updated once per sampling, accelerates the convergence speed and reduces variance.

Table 1. BLEU scores of different NMT systems on Chinese-English and English-German.

Model	Zh-En				En-De
	MT14	MT15	MT16	AVE	Newstest2014
Representative end-to-end NMT systems					
RNNSearch [2]	33.76	34.08	33.98	33.94	21.20
RNNSearch+BR-CSGAN [17]	35.47	35.71	36.14	35.77	22.89
Transformer [12]	41.82	41.67	41.92	41.80	27.30
Transformer+RL [15]	41.96	42.13	41.97	42.02	27.25
Transformer+BR-CSGAN [17]	42.46	42.54	42.83	42.61	27.92
Our work					
RNNSearch+MR-GAN	36.93	37.04	36.89	**36.95**	**24.61**
Transformer+MR-GAN	43.23	43.66	43.98	**43.62**	**28.69**

discriminator output is calculated as the average of the H rewards. Therefore, for a target sentence of length T, we calculate the reward for $\hat{y}_t$ as:

$$R_{D,V,A}^{G_{\theta^*}}\left(\hat{Y}_{1:t-1}, X, \hat{y}_T, Y\right) = \begin{cases} \frac{1}{H}\sum_{J-1}^{H} \alpha\left(D\left(X, \hat{Y}_{1:T_h}^h\right) - b\left(X, \hat{Y}_{1:T_h}^h\right)\right) + \beta V\left(\hat{Y}_{1:T_h}\right) + \gamma A\left(\hat{Y}_{1:T_h}, Y\right) t < T \\ \alpha D((X, \hat{Y}_{1:t}) - b(X, \hat{Y}_{1:t})) + \beta V\left(\hat{Y}_{1:t}\right) + \gamma A\left(\hat{Y}_{1:t}, Y\right) t - T \end{cases} \quad (15)$$

3.4 Training

The training goal is to train G from the initial state to achieve maximum expectations end rewards. The objective equation is as follows:

$$J(\theta^*) = \sum_{\hat{Y}_{1:T}} G_{\theta^*}\left(\hat{Y}_{1:T}|X\right) \cdot R_{D,V,A}^{G_{\theta^*}}\left(\hat{Y}_{1:T-1}, X, \hat{y}_T, Y\right) \quad (16)$$

where R is Eq. (16). Using sentence overall rewards to dynamically update the discriminator and then the generator.

$$min - E_{X,\hat{Y}\in P_{data}}\left[logD\left(X, \hat{Y}\right)\right] - E_{X,\hat{Y}\in G}\left[log\left(1 - D\left(X, \hat{Y}\right)\right)\right] \quad (17)$$

After completing the above operations, we adopt gradient descent to retrain the generator:

$$\nabla J(\theta^*) = \frac{1}{T}\sum_{t=1}^{T} E_{\hat{y}_t \in G_{\theta^*}}\left[R_{D,V,A}^{G_{\theta^*}}\left(\hat{Y}_{1:t-1}, X, \hat{y}_T, Y\right) \cdot \nabla_{\theta^*} logp\left(\hat{y}_t|\hat{Y}_{1:t-1}, X\right)\right] \quad (18)$$

4 Experiment and Analysis

We evaluate Chinese-English (Zh-En), English-German (En-De) and Mongolian-Chinese(Mo-Zh) tasks to verify the effectiveness of our MR-GAN.

4.1 Datasets and Preprocessing

For En-De translation, we conduct our experiments on WMT14 En-De dataset, which contains 4.5 million bilingual pairs. Sentences are encoded using byte-pair encoding (BPE) [9]. Newstest2012/2013 are chosen for development set, Newstest2014 as the test set. For the Zh-En translation, LDC2014 corpus as training set with a total of 1.6 million bilingual pairs. Both the source and target sentences are encoded with BPE. MT2013 is used as a development set and MT2014/2015/2016 as a test set. For Mo-Zh translation, the dataset adopts 261643 sentence pair Mongolian-Chinese bilingual aligned corpus provided by CWMT2018, we utilize 220000 sentence pairs as training set, 20822 as validation set, and the rest as test set. We perform word segmentation processing on the Chinese. On the Mongolian end, due to its own natural separator, so we encode it with BPE.

4.2 Setting

For Transformer-Big, following [12], we set $dropout = 0.1$ and set the dimension of the word embedding as 1024. We employ the Gumbel-Top-K Stochastic Beam Search to sample the target token with beam size $K = 4$. A single model obtained by averaging the last 20 checkpoints and we use adaptive methods to adjust the learning rate. For RNNSearch [2], it is an RNN-based encoder decoder framework with attention mechanism. We set the hidden layer nodes and word embedding dimensions of the encoder and decoder to 512 and $dropout = 0$. The learning rate and checkpoint settings are consistent with Transformer-Big.

For D, CNN consists of one input layer, three convolution + pool layer pairs, one MLP layer and softmax layer. When the model is down-sampling, we use a 3×3 convolution window to perform convolution calculations on the internal corpus, and the output size is 2×2 pooling window. In addition, we set the feature map and MLP hidden layer size as 20. The word embedding dimension and the number of nodes are consistent with G.

Considering the computational complexity of model and the hardware environment of experiment, we adopt ELMO[2] [8] to construct and train the language model, which fully consider contextual information in semantic learning. Furthermore, we adopt BLEU [7] to evaluate these tasks. All models are implemented in T2T tool and trained on two Titan XP GPUs. We stop training when the model does not improve on the tenth evaluation of the development set.

[2] http://allennlp.org/elmo/.
ELMO, which fully consider contextual information has shown certain potential in semantic learning. It has strong modeling capabilities, meanwhile, the parameters and complexity are relatively small, which is convenient for model construction and training.

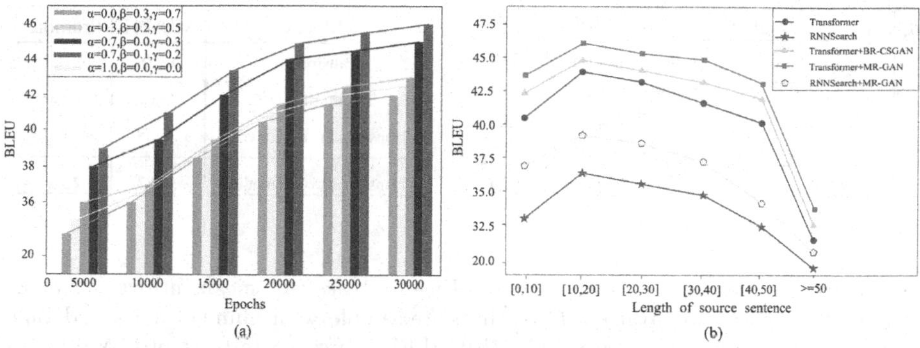

Fig. 4. (a): Training line charts with different hyper-parameters weights. (b): BLEU scores on test set of LDC2014 Zh→En over different length of source sentences.

4.3 The Pre-training of Model

When the generator and discriminator achieve the synchronization and coordination effect, the performance of the model will be optimal. Therefore, we need to pre-train the model. The first step is pre-train the generator G on bilingual training set until the best translation performance is achieved and we employ the traditional maximum likelihood estimation during the process. Then, generate the sentences(machine translations) by using the generator to decode the training data. The next step is pre-trian the discriminator on the combination of true bilingual data and machine translation data until the classification accuracy achieves at ξ. Finally, according to the study of Yang et al. [17], the method of jointly training the generator and discriminator and using the policy gradient to train the generator will lead to unstableness. Therefore, following [17], we adopt the teacher forcing approach to solve this problem. The parameter setting is exactly similar to [17], but the difference is that we employ the Temporal-Difference instead of Monte Carlo.

4.4 Main Results and Analysis

For RNNSearch, it is optimized with the mini-batch of 64 examples. For Transformer, each training batch contains a set of sentence pairs contains approximately 25000 source tokens and 25000 target tokens. Table 1 shows the comparison between existing NMT system and our work. It can be seen that on Transformer, our approach outperforms the beat performance model and achieves improvement up to +1.01 BLEU points averagely on Chinese-English test sets an +0.77 BLEU points on English-German test set. It is profit from the novel method we have adopted to calculate rewards. Compared with traditional reinforcement learning, the scope of reward calculation is wider and making translation results more accurate and fluent. Furthermore, our method adds Gumbel noise when sampling, which makes the sampling more random and alleviates the problem of overtranslation and undertranslation. Experiments on the

ID	V-LM	A-BLEU	G-N	ZH-EN	EN-DE
1	×	×	×	42.61	27.92
2	×	√	×	42.82	27.97
3	√	√	×	43.07	28.23
4	√	×	√	43.21	28.42
5	√	√	√	**43.62**	**28.69**

(a)

Model	MO-ZH	Promote
Transformer	34.56	–
Transformer+RL	35.02	0.46
Transformer+BR-CSGAN	35.63	1.07
Transformer+Our method	**36.82**	**2.26**

(b)

Fig. 5. (a): Ablation study on Zh→En and En→De tasks. "o" means utilize this module and "×" means not utilize. "G-N" indicate sample with Gumbel noise and Line 1 represent the result of BR-CSGAN. (b): BLEU scores on test set of CWMT2018 MO→ZH over different length of source sentences.

RNNSearch model shows the same trends, our approach still achieves 36.95 and 24.61 BLEU points on Chinese-English and English-German translations respectively.

4.5 Effect of Hyper-Parameters and Sentence Length

We conduct a set of typical experiments using Transformer on the Chinese-English task to verify the influence of hyper-parameters (Eq. 15) on experimental results. As shown in Fig. 4(a), the worst result is obtained when $\alpha = 0$. The effect of the model continues to improve as the value of α increases. In the case of $\alpha = 0.7$, $\beta = 0.1$. and $\gamma = 0.2$, it achieves the best performance in several groups of experiments, and when $\alpha = 0.7$, $\beta = 0$. and $\gamma = 0.3$, the effect is not satisfactory. It indicates that the integration of language model rewards to evaluate the fluency of translation can effectively improve the quality of the model. When $\alpha = 1.0$, the effect is very poor but better than $\alpha = 0$, which explains that the multiple rewards proposed in this paper are effectively.

To verify the performance of this method on long sentences, following [2], we divided the development set data and test set data of the Chinese-English task according to the sentence length. Figure 4(b) shows the BLEU scores for different sentence lengths. No matter on RNNSearch or Transformer, compared with baseline and the best performing BR-CSGANS [17], our work have outstanding behaviors continuously. It is due to our method not only calculates the single-step reward, but also adds a smoothing restriction, which makes our method perform better on both long and short sentences.

4.6 Ablation Study

Figure 5(a) shows the results of ablation study. Line 1 represent the result of BR-CSGAN and line 2 represent that reward shaping is used to calculate BLEU on the basis of BR-CSGAN. It is clear that language model reward plays a critical role since removing it impairs translation performance (line 3). As shown in line 4, sampling with Gumbel noise is also an essential part of our approach.

The sentence reward with each token is also shown to be benefit for improving performance (line 2) but seem to have relatively smaller contributions than the above two parts.

4.7 Result of Mongolian-Chinese

To verify the robustness of the proposed method, we conducted a low-resource language Mongolian-Chinese experiment on Transformer. The experimental results are shown in Fig. 5(b). Compared with the traditional Transformer, our approach improves 2.26 BLEU scores, meanwhile, it also increases 1.09 on the current best performance BR-CSGAN. It is fully proved that our method is also helpful for low-resource translation tasks.

5 Conclusion

In this paper, we propose a novel multi-reward reinforcement learning training paradigm to guide the optimization of model parameters, which makes the reward calculation more extensive. In addition, we employ Gumbel method instead of traditional beam search to selectively sample more random datas in the target space, and combining TD to calculate real-time reward. We validate the effectiveness of our method on the RNNSearch and the Transformer. A large number of experiments clearly show that our approach achieves significant improvements.

References

1. Bahdanau, D., et al.: An actor-critic algorithm for sequence prediction. In: 5th International Conference on Learning Representations, ICLR 2017, Toulon, France, 24–26 April 2017, Conference Track Proceedings (2017). https://openreview.net/forum?id=SJDaqqveg
2. Bahdanau, D., Cho, K., Bengio, Y.: Neural machine translation by jointly learning to align and translate. In: 3rd International Conference on Learning Representations, ICLR 2015, San Diego, CA, USA, 7–9 May 2015, Conference Track Proceedings (2015). http://arxiv.org/abs/1409.0473
3. Gale, W.A., Sampson, G.: Good-turing frequency estimation without tears. J. Quant. Linguistics **2**(3), 217–237 (1995). https://doi.org/10.1080/09296179508590051
4. Harris, C.M., Mandelbaum, J.: A note on convergence requirements for nonlinear maximum-likelihood estimation of parameters from mixture models. Comput. OR **12**(2), 237–240 (1985). https://doi.org/10.1016/0305-0548(85)90048-6
5. Kool, W., van Hoof, H., Welling, M.: Stochastic beams and where to find them: the gumbel-top-k trick for sampling sequences without replacement. In: Proceedings of the 36th International Conference on Machine Learning, ICML 2019, Long Beach, California, USA, 9–15 June 2019, pp. 3499–3508 (2019). http://proceedings.mlr.press/v97/kool19a.html

6. Maddison, C.J., Tarlow, D., Minka, T.: A* sampling. In: Advances in Neural Information Processing Systems 27: Annual Conference on Neural Information Processing Systems 2014, Montreal, Quebec, Canada, 8–13 December 2014, pp. 3086–3094 (2014). http://papers.nips.cc/paper/5449-a-sampling
7. Papineni, K., Roukos, S., Ward, T., Zhu, W.: Bleu: a method for automatic evaluation of machine translation. In: Proceedings of the 40th Annual Meeting of the Association for Computational Linguistics, Philadelphia, PA, USA, 6–12 July 2002, pp. 311–318 (2002). https://www.aclweb.org/anthology/P02-1040/
8. Peters, M.E., et al.: Deep contextualized word representations. In: Proceedings of the 2018 Conference of the North American Chapter of the Association for Computational Linguistics: Human Language Technologies, NAACL-HLT 2018, New Orleans, Louisiana, USA, 1–6 June 2018, vol. 1 (Long Papers), pp. 2227–2237 (2018). https://www.aclweb.org/anthology/N18-1202/
9. Sennrich, R., Haddow, B., Birch, A.: Neural machine translation of rare words with subword units. In: Proceedings of the 54th Annual Meeting of the Association for Computational Linguistics, ACL 2016, Berlin, Germany, 7–12 August 2016, vol. 1: Long Papers (2016). https://doi.org/10.18653/v1/p16-1162
10. Shen, S., et al.: Minimum risk training for neural machine translation. In: Proceedings of the 54th Annual Meeting of the Association for Computational Linguistics, ACL 2016, Berlin, Germany, 7–12 August 2016, vol. 1: Long Papers (2016). https://www.aclweb.org/anthology/P16-1159/
11. Sutton, R.S.: Learning to predict by the methods of temporal differences. Mach. Learn. **3**, 9–44 (1988). https://doi.org/10.1007/BF00115009
12. Vaswani, A., et al.: Attention is all you need. In: Advances in Neural Information Processing Systems 30: Annual Conference on Neural Information Processing Systems 2017, Long Beach, CA, USA, 4–9 December 2017, pp. 5998–6008 (2017). http://papers.nips.cc/paper/7181-attention-is-all-you-need
13. Wang, Z., Schaul, T., Hessel, M., van Hasselt, H., Lanctot, M., de Freitas, N.: Dueling network architectures for deep reinforcement learning. In: Proceedings of the 33nd International Conference on Machine Learning, ICML 2016, New York City, NY, USA, 19–24 June 2016, pp. 1995–2003 (2016). http://proceedings.mlr.press/v48/wangf16.html
14. Williams, R.J.: Simple statistical gradient-following algorithms for connectionist reinforcement learning. Mach. Learn. **8**, 229–256 (1992). https://doi.org/10.1007/BF00992696
15. Wu, L., Tian, F., Qin, T., Lai, J., Liu, T.: A study of reinforcement learning for neural machine translation. In: Proceedings of the 2018 Conference on Empirical Methods in Natural Language Processing, Brussels, Belgium, 31 October –4 November 2018, pp. 3612–3621 (2018). https://www.aclweb.org/anthology/D18-1397/
16. Wu, L., et al.: Adversarial neural machine translation. In: Proceedings of The 10th Asian Conference on Machine Learning, ACML 2018, Beijing, China, 14–16 November 2018, pp. 534–549 (2018). http://proceedings.mlr.press/v95/wu18a.html
17. Yang, Z., Chen, W., Wang, F., Xu, B.: Improving neural machine translation with conditional sequence generative adversarial nets. In: Proceedings of the 2018 Conference of the North American Chapter of the Association for Computational Linguistics: Human Language Technologies, NAACL-HLT 2018, New Orleans, Louisiana, USA, 1–6 June 2018, vol. 1 (Long Papers), pp. 1346–1355 (2018). https://www.aclweb.org/anthology/N18-1122/

18. Yin, W., Schütze, H., Xiang, B., Zhou, B.: ABCNN: attention-based convolutional neural network for modeling sentence pairs. TACL **4**, 259–272 (2016). https://tacl2013.cs.columbia.edu/ojs/index.php/tacl/article/view/831
19. Yu, L., Zhang, W., Wang, J., Yu, Y.: Seqgan: sequence generative adversarial nets with policy gradient. In: Proceedings of the Thirty-First AAAI Conference on Artificial Intelligence, San Francisco, California, USA, 4–9 February 2017, pp. 2852–2858 (2017). http://aaai.org/ocs/index.php/AAAI/AAAI17/paper/view/14344

Minority Language Information Processing

Low-Resource Text Classification via Cross-Lingual Language Model Fine-Tuning

Xiuhong Li[1], Zhe Li[2(✉)], Jiabao Sheng[2], and Wushour Slamu[3]

[1] College of Information Science and Engineering, Xinjiang University, Ürümqi, China

[2] College of Software, Xinjiang Laboratory of Multi-Language Information Technology, Xinjiang Multilingual Information Technology Research Center, Xinjiang University, Ürümqi, China
lizhe@stu.xju.edu.cn

[3] College of Information Science and Engineering, Xinjiang Laboratory of Multi-Language Information Technology, Xinjiang Multilingual Information Technology Research Center, Xinjiang University, Ürümqi, China

Abstract. Text classification tends to be difficult when data are inadequate considering the amount of manually labeled text corpora. For low-resource agglutinative languages including Uyghur, Kazakh, and Kyrgyz (UKK languages), in which words are manufactured via stems concatenated with several suffixes and stems are used as the representation of text content, this feature allows infinite derivatives vocabulary that leads to high uncertainty of writing forms and huge redundant features. There are major challenges of low-resource agglutinative text classification the lack of labeled data in a target domain and morphologic diversity of derivations in language structures. It is an effective solution which fine-tuning a pre-trained language model to provide meaningful and favorable-to-use feature extractors for downstream text classification tasks. To this end, we propose a low-resource agglutinative language model fine-tuning *AgglutiFiT*, specifically, we build a low-noise fine-tuning dataset by morphological analysis and stem extraction, then fine-tune the cross-lingual pre-training model on this dataset. Moreover, we propose an attention-based fine-tuning strategy that better selects relevant semantic and syntactic information from the pre-trained language model and uses those features on downstream text classification tasks. We evaluate our methods on nine Uyghur, Kazakh, and Kyrgyz classification datasets, where they have significantly better performance compared with several strong baselines.

Keywords: Transfer learning · Pre-training · Low-resources text classification · Fine-tuning

M. Sun et al. (Eds.): CCL 2020, LNAI 12522, pp. 231–246, 2020.
https://doi.org/10.1007/978-3-030-63031-7_17

1 Introduction

Text classification is the backbone of most natural language processing tasks such as sentiment analysis, classification of news topics, and intent recognition. Although deep learning models have reached the most advanced level on many Natural Language Processing (NLP) tasks, these models are trained from scratch, which makes them require larger datasets. Still, many low-resource languages lack rich annotated resources that support various tasks in text classification. For UKK languages, words are derived from stem affixes, so there is a huge vocabulary. Stems represent of text content and affixes provide semantic and grammatical functions. Diversity of morphological structure leads to transcribe speech as they pronounce while writing and suffer from high uncertainty of writing forms on these languages which causes the personalized spelling of words especially less frequent words and terms [2]. Data collected from the Internet are noisy and uncertain in terms of coding and spelling [1]. The main problems in NLP tasks for UKK languages are uncertainty in terms of spelling and coding and annotated datasets inadequate poses a big challenge for classifying short and noisy text data.

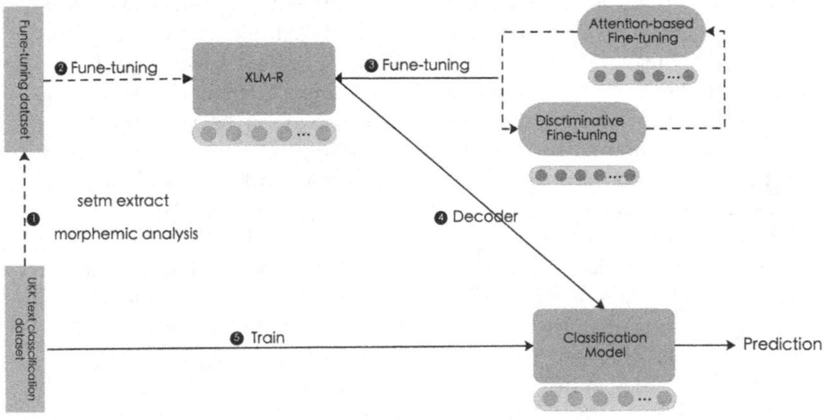

Fig. 1. High-level illustration of AgglutiFiT

Data augmentation can effectively solve the problem of insufficient marker corpus in low-resource language datasets. [15] present two simple text augmentation techniques using "crop" sentences by removing dependency links, and "rotates" sentences by moving the tree fragments around the root. However, this may not be sufficient for several other tasks such as cross-language text classification due to irregularities across UKK languages in these kinds of scenarios. Pre-trained language models such as *BERT* [7] or *XLM* [6] have become an effective way in NLP and yields state-of-the-art results on many downstream tasks. These models require only unmarked data for training, so they are especially useful when there is very little market data. Fully exploring fine-tuning can go a long way toward solving this problem [21]. [17] conduct an empirical study on fine-tuning, although these methods achieve better performance, they did not perform well on UKK low-resource agglutinative languages due to the morphologic diversity of derivations.

The significant challenge of using language model fine-tuning on low-resource agglutinative languages is how to capture feature information. To apprehend rich semantic patterns from plain text, [23] incorporating knowledge graphs (KGs), which provide rich structured knowledge facts for better language understanding. [24] propose to incorporate explicit contextual semantics from pre-trained semantic role labeling (SemBERT) which can provide rich semantics for language representation to promote natural language understanding. UKK languages are a kind of morphologically rich agglutinative languages, in which words are formed by a root (stem) followed by suffixes. These methods are difficult to capture the semantic information of UKK languages. As the stems are the notionally independent word particles with a practical meaning, and affixes provide grammatical functions in UKK languages, morpheme segmentation can enable us to separate stems and remove syntactic suffixes as stop words, and reduce noise and capture rich feature in UKK languages texts in the classification task.

In this paper, as depict in Fig. 1, we propose a low-resource agglutinative language model fine-tuning model: $AgglutiFiT$ that is capable of addressing these issues. First, we use $XLM-R$ pre-train a language model on a large cross-lingual corpus. Then we build a fine-tuning dataset by stem extraction and morphological analysis as the target task dataset to fine-tune the cross-lingual pre-training model. Moreover, we introduce an attention-based fine-tuning strategy that selects relevant semantic and syntactic information from the pre-trained language model and uses discriminative fine-tuning to capture different types of information on different layers. To evaluate our model, we collect and annotate nine corpora for text classification of UKK low-resource agglutinative language, including topic classification, sentiment analysis, intention classification. The experimental results show $AgglutiFiT$ can significantly improve the performance with a small number of labeled examples.

The contributions of this paper are summarized as follows:

- We collect three low-resource agglutinative languages including Uyghur, Kazakh, and Kyrgyz nine datasets, each of languages datasets contains topic classification, sentiment analysis, and intention classification three common text classification tasks.
- We propose a fine-tuning strategy on low-resource agglutinative language that builds a low-noise fine-tuning dataset by stem extraction and morphological analysis to fine-tune the cross-lingual pre-training model.
- We propose an attention-based fine-tuning method that better select relevant semantic and syntactic information from the pre-trained language model and uses discriminative fine-tuning capture different types of information different layers.

2 Related Work

In the field of natural language processing, low-resource text processing tasks receives increasing attention. We briefly reviewed three related directions: data augmentation, language model pre-training, and fine-tuning.

Data Augmentation. Data Augmentation is that solves the problem of insufficient data by creating composite examples that are generated from but not identical to the original document. [20] present EDA, easy data augmentation techniques to improve the performance of text classification task. For a given sentence in the training set, EDA randomly chooses and performs one of the following operations: synonym replacement, random insertion, random swap, random deletion. UKK languages have few synonyms for a certain word, so the substitution of synonyms cannot add much data. Its words are formed by a root (stem) followed by suffixes, and as the powerful suffixes can reflect semantically and syntactically, random insertion, random swap, random deletion may change the meaning of a sentence and cause the original tags to become invalid. In the text classification, training documents are translated into another language by using an external system and then converted back to the original language to generate composite training examples, this technology known as *backtranslation*. [16] work experiments with *backtranslation* as data augmentation strategies for text classification. The translation service quality of Uyghur is not good, and Kazakh and Kyrgyz do not have mature and robust translation service, so it is difficult to use the three languages in *backtranslation*. [15] propose an easily adaptable, multilingual text augmentation technique based on dependency trees. It augments the training sets of these low-resource languages which are known to have extensive morphological case-marking systems and relatively free word order including Uralic, Turkic, Slavic, and Baltic language families.

Cross-Lingual Pre-trained Language Model. Recently, Pre-training language models such as BERT [8] and GPT-2 [13] have achieved enormous success in various tasks of natural language processing such as text classification, machine translation, question answering, summarization, etc. The early work in the field of cross-language understanding has proven the effectiveness of cross-language pre-trained models on cross-language understanding. The multilingual $BERT$ model is pre-trained on Wikipedia in 104 languages using a shared vocabulary of word blocks. LASER [3] is trained on parallel data of 93 languages and those languages share BPE vocabulary. [6] also use parallel data to pre-train $BERT$. These models can achieve zero distance migration, but the effect is poor compared with the monolingual model. The $XLM-R$ [5] uses filtered commoncrawled data over 2TB to demonstrate that using a large-scale multilingual pre-training model can significantly improve the performance of cross-language migration tasks.

Fine-Tuning. When we adapt the pre-training model to NLP tasks in a target domain, a proper fine-tuning strategy is desired. [11] proposes the universal language model fine-tuning ($ULMFiT$) with several novel fine-tuning techniques. ULMFiT consists of three steps, namely general-domain LM pre-training, target task LM fine-tuning, and target task classifier fine-tuning. [9] combines the *ULMFiT* with the quasi-recurrent neural network (*QRNN*) [4] and subword tokenization [12] to propose multi-lingual language model fine-tuning (*MultiFit*) to enable practitioners to train and fine-tune language models efficiently. The *MultiFiT* language model consists of one subword embedding layer, four *QRNN* layers, one aggregation layer, and two linear layers. Moreover, a bootstrapping method [14] is applied to reduce the complexity of training. Although those approaches are general enough and have achieved state-of-the-art results on various classification datasets, the method is considered can not solve the problem of morphologic diversity of derivations in language structures on low-resource agglutinative language. [18] proposes an attention-based fine-tuning algorithm. With this algorithm, the customers can use the given language model and fine-tune the target model by their own data, but that does not capture different levels of syntactic and semantic information on different layers of a neural network. In this paper, we use a new fine-tuning strategy that provides a feature extractor to extract features and use these features for downstream text classification tasks.

3 Methodology

In this section, we will explain our methodology, which is also shown in Fig. 1. Our training consists of four stages. We first pre-train a language model on a large scale cross-lingual text corpus. Then the pre-trained model is fine-tuned by the fine-tuning dataset on unsupervised language modeling tasks. The fine-tuning dataset is constructed by means of stem extraction and morpheme analysis on the downstream classification datasets. Moreover, we use an attention-based fine-tuning to build our classification model and uses discriminative fine-tuning to capture different types of information on different layers. Finally, train the classifier using target task datasets.

3.1 LM Fine-Tuning Based on UKK Characteristics

When we apply the pre-training model to text classification tasks in a target domain, a proper fine-tuning strategy is desired. In this paper, we employ two fine-tuning methods as below.

Fine-Tuning Datasets Based on Morphemic Analysis. UKK languages are agglutinative languages, meaning that words are formed by a stem augmented by an unlimited number of suffixes. The stem is an independent semantic unit while the suffixes are auxiliary functional units. Both stems and suffixes are called morphemes. Morphemes are the smallest functional units in agglutinative languages. Because of this agglutinative nature, the number of words of these languages can be almost infinite, and most of the words appear very rarely in the text corpus. Modeling based on a smaller unit like morpheme can provide

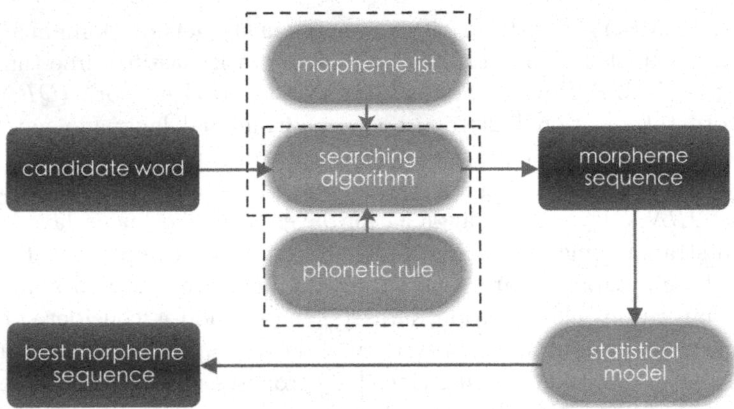

Fig. 2. Morpheme segmentation flow chart

stronger statistics hence robust models. The total number of suffixes in each of UKK languages is around 120. New suffixes may be created, but this is the typical case.

As shown in Fig. 2, we use a semi-supervised morpheme segmenter based on the suffix set [2]. For a candidate word, this tool designs an iterative searching algorithm to produce all possible segmentation results by matching the stem-set and the suffix set. The phonemes on the boundaries change their surface forms according to the phonetic harmony rules when the morphemes are merged into a word. Morphemes will harmonize each other, and appeal to the pronunciation of each other. When the pronunciation is precisely represented, the phonetic harmony can be clearly observed in the text. An independent statistical model can be adopted to pick the best result from N-best results in the UKK text classification task.

We adopt this tool to train a statistical model using word-morpheme parallel training corpus, extraction and greatly improved the UKK text classification task. Which included 10,000 Uyghur sentences, 5000 Kazakhh sentences, and 5000 Kyrgyz sentences. We selected 80% of them as the training corpus. The remainder is used as the testing corpus to execute morpheme segmentation and stem extraction experiments. We can collect necessary terms compose a less noise fine-tuning datasets by extracting stems in the UKK languages classification task. Then fine-tuning with XLM-R on this fine-tuning datasets for better performance. For example in Table 1, a stem can grasp the features of other words, and the feature will be greatly reduced.

Table 1. Examples of Uyghur word variants.

of Uyghur.pdf

Stem	Words	Affixes
ئىش work	worker ئىش+چى = ئىشچى	چى
	office ئىش+خانا = ئىشخانا	خانا
	position ئىش+ تات = ئىشتات	تات
ئوقۇ read	go to school ئوقۇ+ ش = ئوقۇش	ش
	student ئوقۇ+غۇچى = ئوقۇغۇچى	غۇچى
	teach ئوقۇ+ ت = ئوقۇت	ت

Discriminative Fine-Tuning. Different layers of a neural network can capture different levels of syntactic and semantic information [11,22]. Naturally, the lower layers of the $XLM - R$ model may contain more general information. Therefore, we can fine-tune them with assorted learning rates. Following [11], we use the discriminative fine-tuning method. We separate the parameters θ into $\{\theta^1, ..., \theta^L\}$, where θ^l contains the parameters of the l-th layer. Then the parameters are updated as follows:

$$\theta_t^l = \theta_{t-1}^l - \eta^l \cdot \nabla_{\theta^l} J(\theta), \tag{1}$$

where η^l represents the learning rate of the $l - th$ layer and t denotes the update step. Following [17], we set the base learning rate to η_L and use $\eta^{k-1} = \xi \cdot \eta_k$, where ξ is a decay factor and less than or equal to 1. When $\xi < 1$, the lower layer has a slower learning rate than the higher layer. When $\xi = 1$, all layers have the same learning rate, which is equivalent to the regular stochastic gradient descent (SGD).

Attention-Based Fine-Tuning. For classification tasks, we adopt an attention-based encoder-decoder structure. As the encoder, our pre-trained model learns the contextualized features from inputs of the dataset. Then the hidden states over time steps, denoted as $H = h_1, h_2, ..., h_T$, can be viewed as the representation of the data to be classified, which are also the input of the attention layer. Since we do not have any additional information from the decoder, we use the self-attention to extract the relevant aspects from the input states. Specifically, the alignment is computed as

$$u_t = \tanh(W_u h_t + bu) \tag{2}$$

for $t = 1, 2, ..., T$, where W_u and b_u are the weight matrix and bias term to be learned. Then the alignment scores are given by the following soft-max function:

$$\alpha_t = \frac{\exp(W_\alpha u_t)}{\sum_{i=1}^{T} \exp(W_\alpha u_t)} \tag{3}$$

The final context vector, which is also the input of the classifier, is computed by

$$c = \sum_{i=1}^{T} \alpha_t u_t \tag{4}$$

3.2 Text Classifier

For the classifier, we add two linear blocks with batch normalization and dropout, and ReLU activations for the intermediate layer and a Softmax activation for the output layer that calculates a probability distribution over target classes. Consider the output of the last linear block is S_o. Further, denote by $C = c_1, c_2, ..., c_M = XxY$ the target classification data, where $c_i = (x_i, y_i)$, x_i is the input sequence of tokens and y_i is the corresponding label. The classification loss we use to train the model can be computed by:

$$L_2(C) = \sum_{(x,y)\in C} \log p(y|x) \tag{5}$$

where

$$p(y|x) = p(y|x_1, x_2, ..., x_m) := softmax(W_{s_o}) \tag{6}$$

4 Datasets

4.1 Data Collection

We construct nine low-resource agglutinative language datasets including Uyghur, Kazakh, and Kyrgyz, these datasets cover common text classification tasks: topic classification, sentiment analysis, and intention classification. We use the web crawler technology to collect our text data, and download from the Uyghur, Kazakh and Kyrgyz's official websites as well as other main websites.[1]

4.2 Corpus Statistics

In this section, we introduce the detailed information of the corpus. We divided them into morpheme sequences and used morpheme segmentation tools to extract word stems. The method of subword extraction based on stem affix has achieved a good performance on the reduction of feature space. As a result, the vocabulary of morpheme is greatly reduced to about 30%, as shown in Table 2, Table 3 and Table 4. In addition, when the types and numbers of corpora increase, the accumulation of morphemes is only one-third of the accumulation of words.

[1] www.uyghur.people.com.cn, uy.ts.cn, Kazakhh.ts.cn, www.hawar.cn, Sina Weibo, Baidu Tieba and WeChat.

Topic Classification. The corpus for the Uyghur language cover 9 topics: law, finance, sports, culture, health, tourism, education, science, and entertainment. Each category has 1,200 texts, resulting in a total of 10,800 texts. We name this corpus as `ug-topic`. The corpus for the Kazakh language cover 8 topics: law, finance, sports, culture, tourism, education, science, and entertainment. Each of them contains 1,200 texts, so there are 9,600 texts totally. We name this corpus as `kz-topic`. The corpus for the Kyrgyz language cover 7 topics: law, finance, sports, culture, tourism, education. Each category contains 1,200 texts (totally 8,400 texts). We name this corpus as `ky-topics`. The details are shown in Table 2.

Sentiment Analysis. We constructed 3 sentiment analysis datasets for three-category classification, namely positive, negative, and neutral. Each language is related to 900 texts and each category contains 300 texts. We name these datasets as `ug-sen`, `kz-sen` and `ky-sen` as shown in Table 3.

Intention Classification. We construct 3 datasets of five-class user intent identification: news, life, travel, entertainment, and sports. Each language contains 200 texts. We name these datasets as `ug-intent`, `kz-intent` and `ky-intent` as shown in Table 4.

Table 2. Statistics of the topic classification dataset.

Corpus	of Class	Average text length	Word vocabulary	Morpheme vocabulary	Morpheme-word vocabulary ratio (%)
ug-topic	9	148.3	79,126	23,364	29.5%
kz-topic	8	130.9	68,334	20,600	30.1%
ky-topic	7	145.7	58,137	18,487	31.7%

Table 3. Statistics of the sentiment analysis datasets.

Corpus	of Class	Average text length	Word vocabulary	Morpheme vocabulary	Morpheme-word vocabulary ratio (%)
ug-sen	3	23.6	8,791	2,794	31.1%
kz-sen	3	20.7	7,933	2,403	30.3%
ky-sen	3	21.3	7,385	2,274	30.8%

Table 4. Statistics of the intention classification datasets.

Corpus	of Class	Average text length	Word vocabulary	Morpheme vocabulary	Morpheme-word vocabulary ratio (%)
ug-intent	5	18.9	12,651	3,997	31.6%
kz-intent	5	16.0	10,368	3,182	30.7%
ky-intent	5	15.4	11,343	3,720	32.8%

4.3 Corpus Examples

In this section, we present some examples of various language categorization tasks. Different from Kazakhstan and Kyrgyzstan, in China, the Kazakh language used by the Kazakh people and the Kyrgyz language borrowed from the Arabic alphabet. The red keywords indicate the words that have the same meaning. The blue keywords represent their meaning in English.

5 Experiment

5.1 Datasets and Tasks

We evaluate our method on nine agglutinative language datasets which we construct of three common text classification tasks: topic classification, sentiment analysis, and intention classification. We use 75% of the data as the training set, 10% as the validation set, and 15% as the test set.

5.2 Baselines

We compare our method with the cross-lingual classification model $ULMFiT$ [11], which introduces key techniques for fine-tuning language models, and $SemBERT$ [24], which is capable of explicitly absorbing contextual semantics over a BERT backbone. Moreover, we compare against the cross-lingual embedding model, namely $LASER$ [3], which uses a large parallel corpus. We also compare against $BWEs$ [10], a cross-lingual domain adaptation method for classification text. For cross-lingual pre-training language models, the $XLM-R$ model used in this paper is loaded from the torch.Hub. $XLM-R$ shows the possibility of training one model for many languages while not sacrificing per-language performance. It is trained on $2.5TB$ of CommonCrawl data, in 100 languages and uses a large vocabulary size of 250K. For the $ULMFiT$ and $BWEs$ model, we use English as the source language. $XLM-R$ and $ULMFiT$ are fine-tuned on target task datasets rather than the fine-tuning datasets that we built.

5.3 Hyperparameters

In our experiment, we use the $XLM-R_{Base}$ model, which uses a $BERT_{Base}$ architecture [19] with a hidden size of 768, 12 Transformer blocks and 12 self-attention heads. We fine-tune the $XLM-R_{Base}$ model on 4 T K80 GPUs and set the batch size to 24 to ensure that the GPU memory is fully utilized. The dropout probability is always 0.1. We use Adam with $\beta_1 = 0.9$ and $\beta_2 = 0.999$. Following [17], we use the discriminative fine-tuning method [11], where the base learning rate is $2e-5$, and the warm-up proportion is 0.1. We empirically set the max number of the epoch to 20 and save the best model on the validation set for testing.

Table 5. Example from the UKK datasets

Topic	Law	Uyghur	دۆلەتنى قانۇن بويىچە ئىدارە قىلىشتا چىڭ تۇرۇش
		Kazakh	مەملەكەتتى زاڭمەن باسقارۋعا تاباندى بولۋ
		Kyrgyz	ماملەكەتتى زاكون بويۇنچا جونگو سالۇۇ
		English	Ensuring every dimension of governance is law-based
	Finance	Uyghur	COVID-19 ئامېرىكا ئىقتىسادىغا تەسىر كۆرسىتەمدۇ؟
		Kazakh	جاڭا ءتيپتى وكپە ايدارشا ۆيرۋسى امەريكا ەكونوميكاسىنا ىقپال ەتەمە؟
		Kyrgyz	جاڭىچا تاجىسامان ۋىرۇس امەرىكا ىقتىسادىناتااسىر كوتسوتوبۇ
		English	Will the COVID-19 pandemic affect the US economy?
	Sports	Uyghur	كوبى بىر ئۇلۇغ ۋاسكېتبول تەنھەرىكەتچىسى.
		Kazakh	كوبە ۇلى باسكەتبول سپورتشىسى
		Kyrgyz	گوبى دەگەن بىر ۇلۇۇ ۋاسكىتبول چەبەرى
		English	Kobe is a great basketball player.
Sentiment	Positive	Uyghur	شىنجاڭنىڭ مەنزىرىسى سۈرەتتەك گۈزەل
		Kazakh	شينجياڭنىڭ كورىنىسى سۋرەتتەي كوركەم
		Kyrgyz	شىنجاڭدىن كورۇنۇشتورۇ سۇروتتوي كوركوم
		English	Xinjiang is a picturesque landscape
	Neutral	Uyghur	بىز ئىلمىي ماقالە يېزىۋاتىمىز.
		Kazakh	ءبىز علمي ماقالا جازىپ جاتىرمىز
		Kyrgyz	بىز ماقالاجازىپ جاتابىز
		English	We are writing a paper
	Negative	Uyghur	سىز نېمىشقا بويسۇنمايسىز؟
		Kazakh	سەن نەگە بويسىنبايسىڭ؟
		Kyrgyz	سىز نەگە مويۇن سۇنبايسىز
		English	Why are you disobedient?

5.4 Results and Analysis

In this section, we demonstrate the effectiveness of our low-resource agglutinative language fine-tuning model. Our approach significantly outperforms the previous work on cross-lingual classification. Separately, the best results in the metric are bold, respectively.

As given in Table 6, Table 7, and Table 8, We show results for topic classification, sentiment analysis, and intention classification. Our *AgglutiFiT* outperform their cross-lingual and domain adaptation method. Pre-training is most beneficial for tasks with low-resource datasets and enables generalization even

Table 6. Results on topic classification accuracy.

Model	ug-topic	kz-topic	ky-topic
ULMFiT	92.99%	92.93%	92.34%
LASER	83.19%	82.32%	82.13%
SemBERT	91.53%	90.12%	90.24%
BWEs	59.24%	59.12%	58.89%
AgglutiFiT	**96.45%**	**95.39%**	**94.89%**

Table 7. Results on sentiment analysis accuracy.

Model	ug-sen	kz-sen	ky-sen
ULMFiT	90.49%	90.39%	90.38%
LASER	74.32%	73.99%	72.13%
SemBERT	86.37%	88.47%	86.94%
BWEs	56.59%	56.39%	56.03%
AgglutiFiT	**92.81%**	**92.89%**	**92.23%**

with 100 labeled examples when fine-tuning with fine-tuning dataset, our approach has a greater performance boost.

Compared with *ULMFiT*, we perform better on all three tasks, although *ULMFiT* introduces techniques that are key for fine-tuning a language model including discriminative fine-tuning and target task classifier fine-tuning. The reason can be partly explained as we adopt a less noisy datasets in the fine-tuning phase and attention-based fine-tuning which makes it possible to obtain a closer distribution of data in the general domain to the target domain. *LASER* obtain strong results in multilingual similarity search for low-resource languages, but we work better than *LASER* contribute to we use attention-based fine-tuning and different learning rates at a different layer, which allows us to capture more syntactic and semantic information at each layer, moreover, *LASER* has no learn joint multilingual sentence representations for UKK languages. Experimental results on methods *SemBERT* are lower than *AgglutiFiT* on account of lack of the necessary semantic role labels to embedding in the parallel lead to does not capture more accurate semantic information. *BWEs* is significantly lower than other models, we conjecture is that the source language of method *BWEs* is English, which is quite different from the UKK languages in data distribution, more importantly, the datasets of UKK languages are too inadequacy to create good *BWEs*. Our three task experiments also show that using more high-quality datasets to fine-tune the results would be better.

5.5 Ablation Study

To evaluate the contributions of key factors in our method, we perform an ablation study as shown in Fig. 3. We run experiments on nine corpora that are representative of different tasks, genres, and sizes.

The Effect of Morphemic Analysis. In order to gauge the impact of fine-tuning datasets quality, we compare the fine-tuning on the constructed fine-tuning datasets with the target task datasets without stem-word extraction. The experimental results show that the performance of all tasks is greatly improved by using our fine-tuning datasets. Stem is a practical unit of vocabulary. Stem extraction enables us to capture effective and meaningful features and greatly reduce the repetition rate of features.

Table 8. Results on intention classification accuracy.

Model	ug-intent	kz-intent	ky-intent
ULMFiT	90.97%	91.23%	91.13%
LASER	77.21%	77.89%	77.33%
SemBERT	89.79%	87.28%	89.13%
BWEs	57.50%	57.48%	57.39%
AgglutiFiT	**93.47%**	**93.81%**	**93.28%**

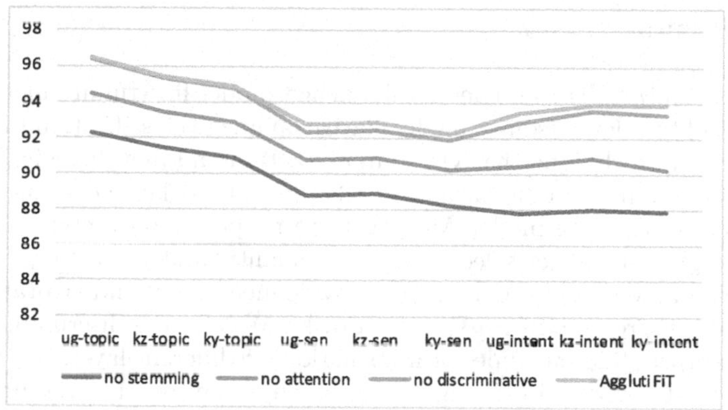

Fig. 3. Explore the influence of important factors on accuracy

The Effect of Attention-Based Fine-Tuning. As given in Fig. 3, we can observe that by adding an attention fine-tuning, our model advances accuracies. Attention-based fine tuning relies on a semantic between words that would influence the overall model performance. In order to see the effectiveness of the attention-based fine-tuning more clearly, we visualize the attention scores with respect to the input texts on Uyghur. The randomly chosen examples of visualization with respect to different classes are given in Fig. 4, where darker color means higher attention scores.

The Effect of Discriminative Fine-Tuning. We compare with and without discriminative fine-tuning on the model. Discriminative fine-tuning improve performance across all three tasks, however, the role of improvement is limited, we still need a better optimization method to explore how discriminative fine-tuning can be better applied in the model.

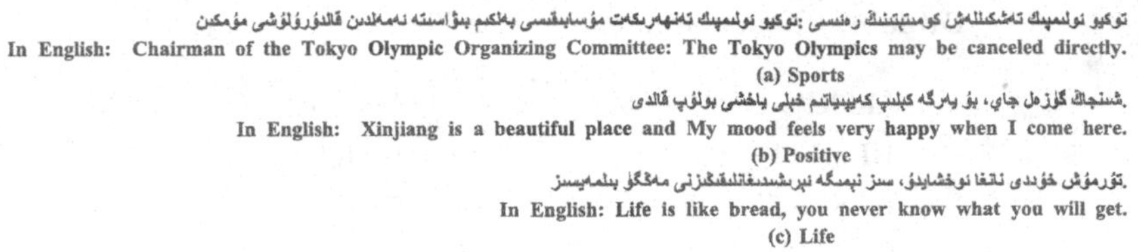

توكيو ئولىمپىك تەشكىللەش كومىتېتىنىڭ رەئىسى: توكيو ئولىمپىك تەنھەرىكەت مۇسابىقىسى بەلكىم بىۋاستە ئەمەلدىن قالدۇرۇلۇشى مۇمكىن

In English: Chairman of the Tokyo Olympic Organizing Committee: The Tokyo Olympics may be canceled directly.

(a) Sports

.شىنجاڭ گۈزەل جاي، بۇ يەرگە كېلىپ كەيپىياتىم خېلى ياخشى بولۇپ قالدى

In English: Xinjiang is a beautiful place and My mood feels very happy when I come here.

(b) Positive

.تۇرمۇش خۇددى نانغا ئوخشايدۇ، سىز نېمىگە ئېرىشىدىغانلىقىڭىزنى مەڭگۈ بىلمەيسىز

In English: Life is like bread, you never know what you will get.

(c) Life

Fig. 4. Examples of attention visualization on Uyghur with respect to different classes

6 Conclusion

We propose *AgglutiFiT*, an effective language model fine-tuning method that can be applied to a low-resource agglutinative language classification tasks. This novel fine-tuning technique that via stem extraction and morphological analysis builds a low-noise fine-tuning dataset as the target task dataset to fine-tune the cross-lingual pre-training model. Moreover, we propose an attention-based fine-tuning strategy that better selects relevant semantic and syntactic information from the pre-trained language model to provide meaningful and favorable-to-use feature for downstream text classification tasks. We also use discriminative fine-tuning to capture different types of information on different layers. Our method significantly outperformed existing strong baselines on nine low-resource agglutinative language datasets of three representative low-resource agglutinative text classification tasks. We hope that our results will catalyze new developments in low-resource agglutinative languages task for NLP.

Acknowledgments. This paper support by Xinjiang University Ph.D. Foundation Initiated Project Grant Number 620312343, Xinjiang Uygur Autonomous Region Graduate Research and Innovation Project Grant Number XJ2020G071, Dark Web Intelligence Analysis and User Identification Technology Grant Number 2017YFC0820702-3, National Language Commission Research Project Grant Number ZDI135-96, and funded by National Engineering Laboratory for Public Safety Risk Perception and Control by Big Data (PSRPC).

References

1. Ablimit, M., Kawahara, T., Pattar, A., Hamdulla, A.: Stem-affix based Uyghur morphological analyzer. Int. J. Future Gener. Commun. Netw. **9**(2), 59–72 (2016)
2. Ablimit, M., Parhat, S., Hamdulla, A., Zheng, T.F.: A multilingual language processing tool for Uyghur, Kazak and Kirghiz. In: 2017 Asia-Pacific Signal and Information Processing Association Annual Summit and Conference (APSIPA ASC), pp. 737–740. IEEE (2017)
3. Artetxe, M., Schwenk, H.: Massively multilingual sentence embeddings for zero-shot cross-lingual transfer and beyond. Trans. Assoc. Comput. Linguist. **7**, 597–610 (2019)
4. Bradbury, J., Merity, S.J., Xiong, C., Socher, R.: Quasi-recurrent neural network, US Patent App. 15/420,710, 10 May 2018

5. Conneau, A., et al.: Unsupervised cross-lingual representation learning at scale. arXiv preprint arXiv:1911.02116 (2019)
6. Conneau, A., Lample, G.: Cross-lingual language model pretraining. In: Advances in Neural Information Processing Systems, pp. 7057–7067 (2019)
7. Devlin, J., Chang, M.W., Lee, K., Toutanova, K.: Bert: pre-training of deep bidirectional transformers for language understanding. arXiv preprint arXiv:1810.04805 (2018)
8. Devlin, J., Chang, M.W., Lee, K., Toutanova, K.: Bert: pre-training of deep bidirectional transformers for language understanding. In: NAACL 2019, pp. 4171–4186 (2019)
9. Eisenschlos, J., Ruder, S., Czapla, P., Kadras, M., Gugger, S., Howard, J.: Multifit: efficient multi-lingual language model fine-tuning. In: Proceedings of the 2019 Conference on Empirical Methods in Natural Language Processing and the 9th International Joint Conference on Natural Language Processing (EMNLP-IJCNLP), pp. 5706–5711 (2019)
10. Hangya, V., Braune, F., Fraser, A., Schütze, H.: Two methods for domain adaptation of bilingual tasks: delightfully simple and broadly applicable. In: Proceedings of the 56th Annual Meeting of the Association for Computational Linguistics (vol. 1: Long Papers), pp. 810–820. Association for Computational Linguistics (2018). http://aclweb.org/anthology/P18-1075
11. Howard, J., Ruder, S.: Universal language model fine-tuning for text classification. arXiv:1801.06146 (2018)
12. Kudo, T.: Subword regularization: improving neural network translation models with multiple subword candidates. In: Proceedings of the 56th Annual Meeting of the Association for Computational Linguistics (vol. 1: Long Papers), pp. 66–75 (2018)
13. Radford, A., et al.: Better language models and their implications. OpenAI Blog (2019). https://openai.com/blog/better-language-models
14. Ruder, S., Plank, B.: Strong baselines for neural semi-supervised learning under domain shift. In: Proceedings of the 56th Annual Meeting of the Association for Computational Linguistics (vol. 1: Long Papers), pp. 1044–1054 (2018)
15. Şahin, G.G., Steedman, M.: Data augmentation via dependency tree morphing for low-resource languages. arXiv preprint arXiv:1903.09460 (2019)
16. Shleifer, S.: Low resource text classification with ulmfit and backtranslation. arXiv preprint arXiv:1903.09244 (2019)
17. Sun, C., Qiu, X., Xu, Y., Huang, X.: How to fine-tune BERT for text classification? In: Sun, M., Huang, X., Ji, H., Liu, Z., Liu, Y. (eds.) CCL 2019. LNCS (LNAI), vol. 11856, pp. 194–206. Springer, Cham (2019). https://doi.org/10.1007/978-3-030-32381-3_16
18. Tao, Y., Gupta, S., Krishna, S., Zhou, X., Majumder, O., Khare, V.: Finetext: text classification via attention-based language model fine-tuning. arXiv preprint arXiv:1910.11959 (2019)
19. Vaswani, A., et al.: Attention is all you need. In: Advances in Neural Information Processing Systems, pp. 5998–6008 (2017)
20. Wei, J.W., Zou, K.: Eda: easy data augmentation techniques for boosting performance on text classification tasks. arXiv preprint arXiv:1901.11196 (2019)
21. Xu, Y., Qiu, X., Zhou, L., Huang, X.: Improving Bert fine-tuning via self-ensemble and self-distillation. arXiv preprint arXiv:2002.10345 (2020)
22. Yosinski, J., Clune, J., Bengio, Y., Lipson, H.: How transferable are features in deep neural networks? In: Advances in Neural Information Processing Systems, pp. 3320–3328 (2014)

23. Zhang, Z., Han, X., Liu, Z., Jiang, X., Sun, M., Liu, Q.: Ernie: enhanced language representation with informative entities. In: Proceedings of the 57th Annual Meeting of the Association for Computational Linguistics, pp. 1441–1451 (2019)
24. Zhang, Z., et al.: Semantics-aware Bert for language understanding. arXiv preprint arXiv:1909.02209 (2019)

Constructing Uyghur Named Entity Recognition System Using Neural Machine Translation Tag Projection

Azmat Anwar[1,2,3], Xiao Li[1,2,3], Yating Yang[1,2,3](✉), Rui Dong[1,2,3], and Turghun Osman[1,2,3]

[1] Xinjiang Technical Institute of Physics and Chemistry, Chinese Academy of Sciences, Urumqi, China
{azmat,xiaoli,yangyt,dongrui,turghun}@ms.xjb.ac.cn

[2] University of Chinese Academy of Sciences, Beijing, China

[3] Xinjiang Laboratory of Minority Speech and Language Information Processing, Urumqi, China

Abstract. Although named entity recognition achieved great success by introducing the neural networks, it is challenging to apply these models to low resource languages including Uyghur while it depends on a large amount of annotated training data. Constructing a well-annotated named entity corpus manually is very time-consuming and labor-intensive. Most existing methods based on the parallel corpus combined with the word alignment tools. However, word alignment methods introduce alignment errors inevitably. In this paper, we address this problem by a named entity tag transfer method based on the common neural machine translation. The proposed method marks the entity boundaries in Chinese sentence and translates the sentences to Uyghur by neural machine translation system, hope that neural machine translation will align the source and target entity by the self-attention mechanism. The experimental results show that the Uyghur named entity recognition system trained by the constructed corpus achieve good performance on the test set, with 73.80% F1 score (3.79% improvement by baseline).

Keywords: Named Entity Recognition · Low resource · Cross-lingual

1 Introduction

Named Entity Recognition (NER) is a task of identifying named entities (NEs), especially person names (PER), location names (LOC), organization names (ORG), and classifying them into some pre-defined target entity classes [14]. NER is essential to many natural language processing (NLP) tasks such as relation extraction [8], event detection [4], knowledge graph construction [3] and so on. Although the NER achieves great success by the introduction of the advanced neural networks [7,9,16,18,21,23,33,34,41], these methods are highly dependent on a large amount of annotated training data, and thus challenging

M. Sun et al. (Eds.): CCL 2020, LNAI 12522, pp. 247–260, 2020.
https://doi.org/10.1007/978-3-030-63031-7_18

to apply these models to low resource languages including Uyghur. Constructing a well-annotated NE corpus manually is very time-consuming and labor-intensive. Instead, Cross-lingual transfer is an effective solution, which addresses this challenge by transferring knowledge from a high-resource source language with abundant entity labels to a low-resource target language with few or no labels. According to the resource availability of the target language, different types of NER methods are proposed, such as bilingual parallel corpus based tag projection [11,12,27,40,42], cross-lingual word embedding [13,15,39], cross-lingual Wikification [17,28,31,36] or multi-task learning [20,41].

As a low resource language, Uyghur has no well-annotated corpus available for NER, but it is easy to get Uyghur-Chinese bilingual parallel corpus as Uyghur-Chinese machine translation is an important task of China Conference on Machine translation (CCMT). A common way of constructing NER corpus for the language which has a bilingual parallel corpus is using off-the-shelf NER tool in the source language to get entity annotations and transfer them to target language combing with the automatic word alignment. Although some researchers have also applied this method to transfer NE annotations from Chinese to Uyghur and achieved remarkable results [24], these pipeline methods inevitably introduce errors from the source language, including errors from NER tools and automatic word alignment. Figure 1 illustrates an Example of NER corpus construction based on the bilingual parallel corpus and automatic word alignment.

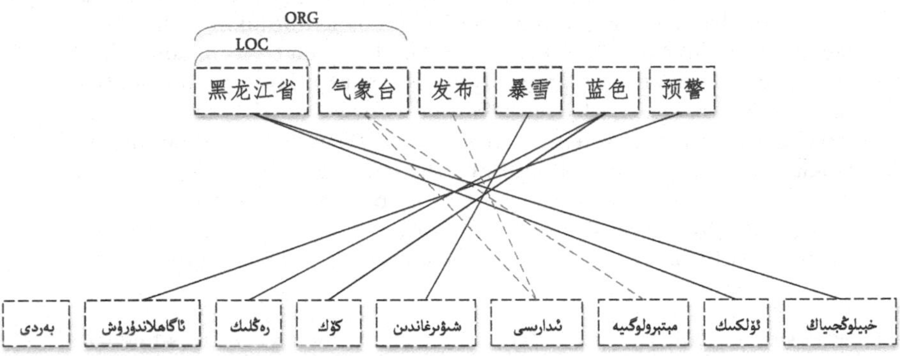

Fig. 1. Example of NER corpus construction based on the bilingual parallel corpus and automatic word alignment. Errors from NER tools and automatic word alignments remarked in red color while blue indicates correct. (Color figure online)

In this paper, we address these challenges by a NE annotation transfer method based on neural machine translation (NMT). Given an Uyghur-Chinese parallel corpus, first, we train a general-purpose Chinese-Uyghur NMT system using the parallel corpus. Then, add the NE boundary information directly to the source Chinese sentence by multiple off-the-shelf NER tools. Finally, translate the Chinese sentences with entity boundary to Uyghur language using the pre-trained NMT system, we hope that NMT will align the source and target entity by the self-attention mechanism. Our method can be illustrated by the following example provided in Fig. 2.

The main advantages of our method are used multi NER tools in the source language to minimize annotation errors and use general-purpose NMT without adding new tokens to indicate NE boundary in the parallel corpus, thus no need to annotate any training data manually.

Fig. 2. Example of transferring NE tags from Chinese to Uyghur using NER tools and NMT

2 Related Work

Named Entity Recognition: NER is typically framed as a task of sequence labeling which aims at automatic detection of NEs in free text [25]. CRF, SVM, and perceptron models with hand-crafted features are applied in early works [22,32,35]. With the great advantages of deep neural networks, research focuses on the neural network-based methods that need less feature engineering and domain knowledge [18,43,44]. Collebert [9] proposed a feed-forward neural network with a fixed-sized window for each word, which failed in considering useful relations between long-distance words. To overcome this limitation, Chiu et al. [7] presented a bidirectional LSTM-CNNs architecture that automatically detects word and character-level features. Ma et al. [23] further extended it into bidirectional LSTM-CNNs-CRF architecture, where the CRF module was added to optimize the output label sequence.

Transfer Learning for NER: Low-resource languages often suffer from a lack of annotated corpora to estimate high-performing neural network models for many NLP tasks. Transfer learning is an efficient way to bridge the gap across languages. Transfer learning methods for NER can be divided into two types: parallel corpora based and shared representation based transfer. Early works mainly focus on parallel corpora to projecting information from high-resource languages to low-resource languages [11,12,27,40,42]. Chen et al. [6] and Wang et al. [40] proposed to jointly identify and align bilingual named entities. Kim et al. [17], Nothman et al. [28] and Tsai et al. [36] using the Wikipedia information to improve low-resource NER. Mayhew et al. [26] created a cross-language NER system by translating annotated data of high-resource to low-resource which works well for very minimal resource languages. On the other hand, the shared representation methods do not require parallel corpora. Fang et al. [13] proposed cross-lingual word embeddings to transfer knowledge across resources. Pan et al. [31] proposes a large-scale cross-lingual named entity dataset which contains 282 languages for evaluation. Yang et al. [41], Wang et al. [39], Lin et al. [20] and Liu et al. [21] shows that jointly training on multiple tasks or languages helps improve performance. Different from transfer learning methods, multi-task learning aims at improving the performance of all the resources instead of low resource only.

Token Added Machine Translation (TAMT): The researchers proposed TAMT methods to solve the different types of problems. Ugawa et al. [37] add the entity tags to the source language sentences to disambiguate the multi-meaning entities in the target language. Li et al. [19] use NE tags to indicate the NE boundary information in the source language sentences to get better customized entity translation. Bai et al. [2] use some special tokens to mark the segmentation boundary for the slot value in the source sentence and transfer the source language spoken language understanding corpus to the target language.

3 Methodology

3.1 General-Purpose NMT System

Machine translation (MT) translates text sentences from a source language to a target language and the Transformer model is the first NMT model relying entirely on self-attention to compute representations of its input and output without using recurrent neural networks (RNN) or convolutional neural networks (CNN). Our general-purpose NMT system is based on the Transformer model.

The Transformer model is an encoder-decoder structure like most competitive neural sequence transduction models, as shown in Fig. 3. The encoder is including three steps, in the first step, the input words are projected into an embedding vector space, position embedding is also added to input vectors to capture the notion of token position within the sequence. The second step is a multi-head self-attention. This is an extension of the previous attention scheme. Instead of using a single attention function, this step computes multiple attention blocks over the source, concatenates them and projects them linearly back onto a space with the original dimensionality. The scaled dot-product attention with different linear projections is computed over attention blocks individually. Finally, a position-wise fully connected feed-forward network is used, which consists of two linear transformations with a ReLU activation.

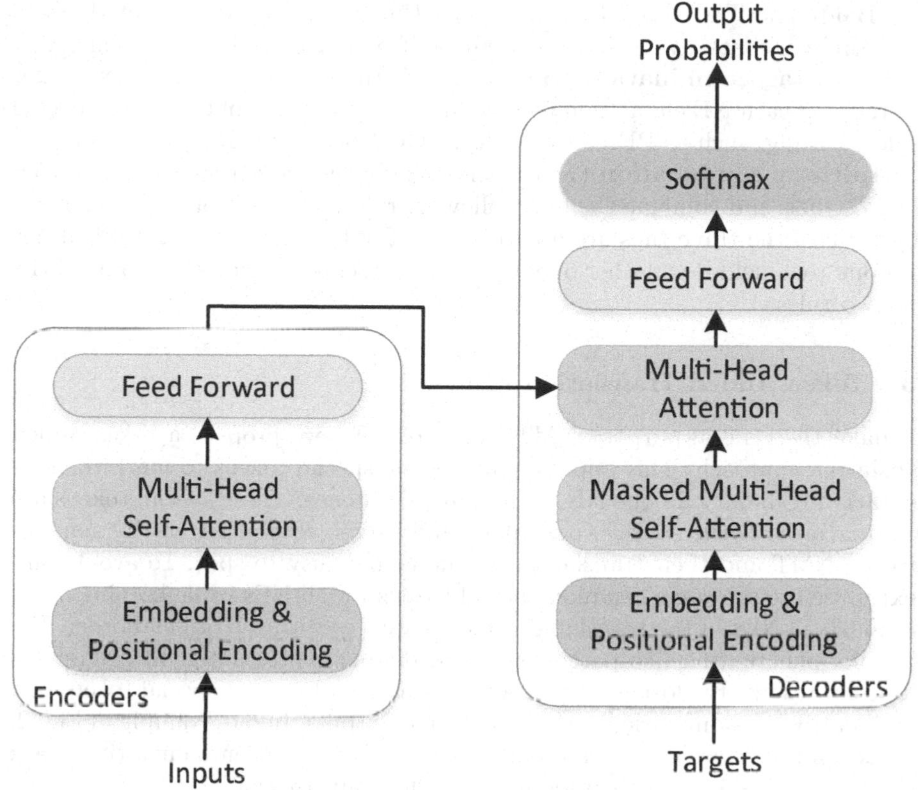

Fig. 3. Simplified diagram of the transformer model

The decoder works similarity, from left to right with generates one word at a time. It including five steps. The first step: embedding and position encoding, is similar to the encoder. The second step is masked multi-head attention, which masks future words forces to attend only to past words. The third step is a multi-head attention that not only attends to these past words, but also to the final representations generated by the encoder. The fourth step is another feed-forward network. Finally, a softmax layer applied to map target word scores into target word probabilities. More details about the model are found in the original paper [38].

3.2 Source Language Named Entity Tags

We consider three NE classes in this paper (PER, LOC, ORG). For every NE in source sentence, we generate the candidate NE class tags using three types of third-party NER tools: Pyltp[1] from Harbin Institute of Technology, PaddleHub[2]

[1] https://github.com/HIT-SCIR/pyltp.
[2] https://github.com/PaddlePaddle/PaddleHub.

from Baidu and THULAC[3] from Tsinghua University. In order to get the best tags from candidates, we will try two kinds of strategies described as follows:

Single tag combination (STC): Check these tools on a test set to get accuracy for each NE class, then use the highest accuracy tool to get the specific single class tag, such as PER from Pyltp, LOC from PaddleHub and so on.

Multi-tag combination (MTC): For single sentences, tags are comes from all three tools and combine them by following rules:(1) Tag kept for a single NE only if all of the three tags are identical. (2) Tag kept for the longer NE if NE from one tool includes another one. (3) Drop the sentences not satisfy any of the first two rules.

3.3 Token-Added Translation

To make the general-purpose NMT aware of NEs, we propose a token added translation approach. This approach uses some special tokens to mark the segmentation boundary for the NE in the source sentence. These special tokens are common in both the source vocabulary and target vocabulary of the general-purpose NMT and their translation is unique and easy to spot. To avoid complexity, we use the same common special tokens for all NEs while keeping order and mark all NEs in the translated target sentences with the original order. For example, punctuation like parentheses and double quotes are good candidates as special tokens. Enclosing NEs in source sentences by these special tokens can help identify NE boundaries in the translation outputs. In our example in Fig. 2, the special tokens we choose is a pair of Chinese punctuation named title mark (《》), which translated to corresponding Uyghur punctuation («»).

In token-added translation, no additional word alignment process is required. However, such an approach relies heavily on the NMT general training data where the special tokens (e.g. parentheses or double quotation marks) are kept in both source and target data. For different language pairs, different special tokens might be chosen for the best translation quality. Empirically we find that title marks are highly effective for Chinese to Uyghur NE translation.

3.4 NER Model

The hierarchical CRF model consists of three components: a character-level neural network, either an RNN or a CNN, that allows the model to capture subword information, such as morphological variations and capitalization patterns; a word-level neural network, usually an RNN, that consumes word representations and produces context-sensitive hidden representations for each word; and a linear-chain CRF layer that models the dependency between labels and performs inference.

In this paper, we closely follow the architecture proposed by Lample et al. [18], and use bidirectional LSTMs for both the character level and word level neural networks. Specifically, given an input sequence of words $(w_1, w_2, ..., w_n)$,

[3] https://github.com/thunlp/THULAC-Python.

and each word's corresponding character sequence, the model first produces a representation for each word, x_i, by concatenating its character representation with its word embedding. Subsequently, the word representations of the input sequence $(x_1, x_2, ..., x_n)$ are fed into a word level Bi-LSTM, which models the contextual dependency within each sentence and outputs a sequence of context sensitive hidden representations $(h_1, h_2, ..., h_n)$. A CRF layer is then applied on top of the word level LSTM and takes in as its input the sequence of hidden representations $(h_1, h_2, ..., h_n)$, and defines the joint distribution of all possible output label sequences. The Viterbi algorithm is used during decoding.

4 Experiment

4.1 Data

The CCMT 2017 Chinese-Uyghur corpus[4] is used to train the general-purpose Chinese-Uyghur NMT system and the MSRA dataset from international Chinese language processing Bakeoff 2006[5] is used to evaluate the performance of Chinese NER tools. As no publicly available test set to evaluate the performance of Uyghur NER, we will randomly choose 2000 sentences from Uyghur named entity relation corpus [1], in which tagged entity tags and relation types, checked the entity tags manually and used 1000 sentences as our Uyghur NER test set and another 1000 sentences as development set. The 1,500,000 Uyghur sentences crawled from the Tianshan website[6] is used to train the Uyghur word embeddings and the BIO tag schema is used where the B, I, O refer to the beginning, inside and outside of an entity, respectively.

4.2 Setup

General-Purpose NMT: We use the Transformer model [38] implemented in PyTorch in the fairseq-py [30] toolkit and all experiments are based on the "base" transformer model. We use word representations of size 512, feed-forward layers with inner dimension 2048, and multi-headed attention with 8 attention heads. We apply dropout with probability 0.3. Models are optimized with Adam using β_1= 0.9, β_2 = 0.98, and ε = 1e-8. We use the same learning rate schedule as Vaswani et al. [38] , i.e., the learning rate increases linearly for 4,000 steps to 5e-4 (or 1e-3 in experiments that specify 2x *lr*), after which it is decayed proportionally to the inverse square root of the number of steps. We use label smoothing with 0.1 weight for the uniform prior distribution over the vocabulary. All experiments are run on 2 NVIDIA V100 GPUs interconnected by Infiniband.

NER Model: We use the 300-dimensional word embeddings pretrained by Word2Vec, FastText, and Glove respectively. We set the character embedding

[4] http://ee.dlut.edu.cn/CWMT2017/index_en.html.
[5] http://sighan.cs.uchicago.edu/bakeoff2006/.
[6] http://uy.ts.cn/.

size to be 100, character level LSTM hidden size to be 25, and word-level LSTM hidden size to be 100. For OOV words, we initialize an unknown embedding by uniformly sampling from range $[-\sqrt{\frac{3}{emb}}, +\sqrt{\frac{3}{emb}}]$ where emb is the size of embedding, 300 in our case. We train the model for 100 epochs and optimize the parameters by Stochastic Gradient Descent (SGD) with momentum, gradient clipping, and learning rate decay. We set the learning rate (lr) and the decay rate (dr) as 0.01 and 0.05 respectively. To prevent overfitting, we apply dropout with a rate of 0.5 on outputs of the two Bi-LSTMs.

4.3 Results and Analysis

Comparison of tag combination strategy

1) Result of the STC Strategy

To obtain the accuracy of the three named entity recognition systems for the recognition of each entity type, we conducted experiments on the MSRA data set, and the experimental results are shown in Table 1.

The experimental results show that PaddleHub has the best recognition for ORG while Pyltp for LOC and THULAC for PER. Therefore, the results of three NER systems are fused according to the STC strategy, and the fusion results are shown in Table 2. It can be seen that the recognition performance of the single and all entity is higher than the original system.

2) Result of the MTC Strategy

The result of the MTC is strategy shown in Table 3. Comparing the results of Table 2 and Table 3, it can be seen that the STC strategy is better than the MTC strategy for the recognition of Chinese named entities, and the following experiments are based on the STC strategy.

Table 1. The results of three Chinese NER system.

NER system	Entity type	Accuracy	Recall	F1
PaddleHub	LOC	81.09	66.77	73.24
	PER	83.16	80.08	81.59
	ORG	70.31	61.38	**65.54**
	ALL	79.51	69.86	74.37
Pyltp	LOC	86.26	71.81	**78.38**
	PER	90.73	61.53	73.33
	ORG	82.21	48.61	61.10
	ALL	86.88	63.53	73.40
THULAC	LOC	73.58	65.73	69.43
	PER	86.93	85.25	**86.08**
	ORG	78.06	16.30	26.97
	ALL	79.24	61.32	69.14

Table 2. The results of the STC.

Strategy	Entity type	Accuracy	Recall	F1
STC	LOC	90.10	70.56	79.14
	PER	86.93	85.25	86.08
	ORG	70.47	61.31	65.57
	ALL	84.70	73.26	**78.56**

Baseline. To show the effectiveness of the proposed method, a strong baseline system is needed. In this paper, we will gradually explore the impact of different word alignment tools and different word vector models on cross-lingual entity migration, and finally, build a cross-language entity migration baseline system based on the parallel corpus and word alignment tools.

1) Comparison of Word Alignment Tools

Word alignment accuracy is very important for word alignment based cross-lingual NER system and GIZA++ [5], fast_align [10], and efmaral[29] are currently popular word alignment tools. We will construct an Uyghur NER system using these three types of word alignment tools with the STC strategy based on the Uyghur-Chinese parallel corpus and The performance is shown in Table 4. It can be seen that efmaral word alignment tool has the best performance for our task.

2) Comparison of Word Embeddings

Word embeddings can provide rich semantic information and allow the system to better capture the semantic relevance between words. we will use the static word embeddings generated from Word2Vec, Glove, and FastText separately to initialize the network input and explore the effect of different word vectors on Uyghur NER construction. The Experimental results are shown in Table 5 and it can be seen that Word2Vec generated embeddings have good performance for our task.

Analysis of Token-added Translation Method. We use the STC strategy to get named entity tags in Chinese and use the proposed token-added translation method to translated the entities to Uyghur to construct tagged NER corpus. Finally, train an Uyghur NER system using this data and the performance is shown in Table 6.

Table 3. The results of MTC.

Strategy	Entity type	Accuracy	Recall	F1
MTC	LOC	88.35	67.23	76.37
	PER	81.56	71.89	75.31
	ORG	69.67	54.38	61.08
	ALL	79.86	64.50	71.36

Table 4. Comparison of three word alignment tools.

Tools	Entity type	Accuracy	Recall	F1
GIZA++	LOC	82.36	43.22	56.69
	PER	95.15	32.24	48.16
	ORG	73.07	41.11	52.62
	ALL	82.04	39.92	**53.71**
fast_align	LOC	80.61	56.36	66.43
	PER	96.93	36.35	52.87
	ORG	65.30	44.25	52.75
	ALL	79.08	48.37	**60.03**
efmaral	LOC	80.17	69.83	74.65
	PER	87.54	40.46	55.34
	ORG	66.88	53.48	59.44
	ALL	77.93	58.44	**66.69**

Table 5. Comparison of three types of word embedding.

Word embedding	Accuracy	Recall	F1
Random	77.93	58.44	66.69
Glove	78.50	61.37	68.89
FastText	78.58	61.50	69.00
Word2Vec	79.17	62.75	**70.01**

Table 6. Uyghur NER based on token-added translation method.

Method	Entity type	Accuracy	Recall	F1
Token-add translation	LOC	66.45	50.91	57.65
	PER	79.96	69.57	74.41
	ORG	47.28	28.75	35.75
	ALL	66.70	50.33	**57.37**

From Table 6, it can be seen that the token-add translation method has worse performance compared with baseline. After analyzing the data, we found that only Uyghur stems are included in the special token while most of the affixes appended by the stem are being excluded. As an agglutinative language, Uyghur has rich affixes to express grammatical information in the sentence. For example, as shown in Fig. 4, the original Chinese entity "新疆" is included in the Chinese bookmark (《》) and translated to Uyghur by MT, it can be found that the translated Uyghur entity also included in Uyghur bookmark («») while appended affix is excluded.

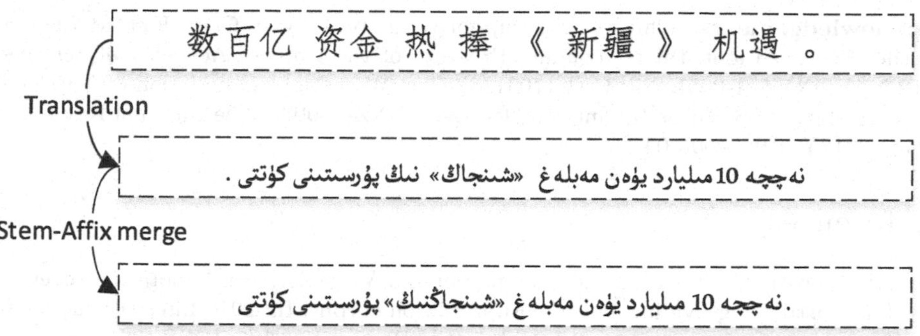

Fig. 4. The example of entity boundary characters based entity translation

Table 7. Result of Stem-Affix merged method

Method	Entity type	Accuracy	Recall	F1
Stem-Affix merged	LOC	78.57	70.91	74.54
	PER	80.78	81.58	81.18
	ORG	67.05	61.32	64.06
	ALL	76.47	71.32	**73.80**

To prevent the problem, we apply a stem-affix merge method for translated Uyghur sentences and merge the stem with the followed word if it is affix. We train a new Uyghur NER system using handled corpus and the result is shown in Table 7. It can be seen that the combination of stem and affixes can effectively avoid the affix as a separate word in the corpus, thereby greatly improving the quality of the corpus and the performance of trained Uyghur NER system significantly, the f1 score is 3.79% higher than the baseline.

5 Conclusion

Aiming at the lack of Uyghur named entity recognition training corpus, this paper proposes a cross-language named entity tag transfer method based on general machine translation and entity boundary token. First obtains the named entity tags of Chinese sentences in Chinese-Uyghur parallel corpus through a variety of Chinese named entity recognition tools and uses tag fusion strategies to fuse multi-source tags, then select appropriate special symbols to surround the entities and uses Chinese-Uyghur neural machine translation system to translate the Chinese sentences to Uyghur. Finally, the Uyghur stems and affixes merge method is used to obtain a high-quality Uyghur named entity recognition corpus. The Uyghur NER system trained with this corpus achieved good performance, which was 3.79% points higher than the baseline system.

Acknowledgements. This work is supported in part by A Class Funded Project of the Western Light Talent Training Program of the Chinese Academy of Sciences (2017-XBQNXZ-A-005), NSFC (U1703133), The West Light Foundation of The Chinese Academy of Sciences (Grant No. 2019-XBQNXZ-B-008), The National Key R&D Plan (2017YFC0822505-04).

References

1. Abiderexiti, K., Maimaiti, M., Yibulayin, T., Wumaier, A.: Annotation schemes for constructing uyghur named entity relation corpus. In: 2016 International Conference on Asian Language Processing (IALP), pp. 103–107. IEEE (2016)
2. Bai, H., Zhou, Y., Zhang, J., Zhao, L., Hwang, M.Y., Zong, C.: Source-critical reinforcement learning for transferring spoken language understanding to a new language. arXiv preprint arXiv:1808.06167 (2018)
3. Bosselut, A., Rashkin, H., Sap, M., Malaviya, C., Celikyilmaz, A., Choi, Y.: Comet: commonsense transformers for automatic knowledge graph construction. arXiv preprint arXiv:1906.05317 (2019)
4. Cakır, E., Virtanen, T.: Convolutional recurrent neural networks for rare sound event detection. Deep Neural Networks for Sound Event Detection, vol. 12 (2019)
5. Casacuberta, F., Vidal, E.: Giza++: training of statistical translation models (2007). Retrieved 29 October 2019
6. Chen, Y., Zong, C., Su, K.Y.: On jointly recognizing and aligning bilingual named entities. In: Proceedings of the 48th Annual Meeting of the Association for Computational Linguistics. pp. 631–639. Association for Computational Linguistics (2010)
7. Chiu, J.P., Nichols, E.: Named entity recognition with bidirectional LSTM-CNNs. Trans. Assoc. Comput. Linguist. **4**, 357–370 (2016)
8. Christopoulou, F., Miwa, M., Ananiadou, S.: A walk-based model on entity graphs for relation extraction. arXiv preprint arXiv:1902.07023 (2019)
9. Collobert, R., Weston, J., Bottou, L., Karlen, M., Kavukcuoglu, K., Kuksa, P.: Natural language processing (almost) from scratch. J. Mach. Learn. Res. **12**(Aug), 2493–2537 (2011)
10. Dyer, C., Chahuneau, V., Smith, N.A.: A simple, fast, and effective reparameterization of IBM model 2. In: Proceedings of the 2013 Conference of the North American Chapter of the Association for Computational Linguistics: Human Language Technologies, pp. 644–648 (2013)
11. Ehrmann, M., Turchi, M., Steinberger, R.: Building a multilingual named entity-annotated corpus using annotation projection. In: Proceedings of the International Conference Recent Advances in Natural Language Processing 2011, pp. 118–124 (2011)
12. Fang, M., Cohn, T.: Learning when to trust distant supervision: An application to low-resource pos tagging using cross-lingual projection. arXiv preprint arXiv:1607.01133 (2016)
13. Fang, M., Cohn, T.: Model transfer for tagging low-resource languages using a bilingual dictionary. arXiv preprint arXiv:1705.00424 (2017)
14. Hobbs, J.R., et al.: FASTUS: a cascaded finite-state transducer for extracting information from natural-language text. In: Finite-State Language Processing, pp. 383–406 (1997)
15. Huang, L., Cho, K., Zhang, B., Ji, H., Knight, K.: Multi-lingual common semantic space construction via cluster-consistent word embedding. arXiv preprint arXiv:1804.07875 (2018)

16. Huang, Z., Xu, W., Yu, K.: Bidirectional LSTM-CRF models for sequence tagging. arXiv preprint arXiv:1508.01991 (2015)
17. Kim, S., Toutanova, K., Yu, H.: Multilingual named entity recognition using parallel data and metadata from Wikipedia. In: Proceedings of the 50th Annual Meeting of the Association for Computational Linguistics: Long Papers, vol. 1, pp. 694–702. Association for Computational Linguistics (2012)
18. Lample, G., Ballesteros, M., Subramanian, S., Kawakami, K., Dyer, C.: Neural architectures for named entity recognition. arXiv preprint arXiv:1603.01360 (2016)
19. Li, Z., Wang, X., Ai, A.T., Chng, E.S., Li, H.: Named-entity tagging and domain adaptation for better customized translation (2018)
20. Lin, Y., Yang, S., Stoyanov, V., Ji, H.: A multi-lingual multi-task architecture for low-resource sequence labeling. In: Proceedings of the 56th Annual Meeting of the Association for Computational Linguistics (vol. 1: Long Papers), pp. 799–809 (2018)
21. Liu, L., et al.: Empower sequence labeling with task-aware neural language model. In: Thirty-Second AAAI Conference on Artificial Intelligence (2018)
22. Luo, G., Huang, X., Lin, C.Y., Nie, Z.: Joint entity recognition and disambiguation. In: Proceedings of the 2015 Conference on Empirical Methods in Natural Language Processing, pp. 879–888 (2015)
23. Ma, X., Hovy, E.: End-to-end sequence labeling via bi-directional LSTM-CNNs-CRF. arXiv preprint arXiv:1603.01354 (2016)
24. Maimaiti, M., Wumaier, A., Abiderexiti, K.: Construction of Uyghur named entity corpus. Belt & Road: Language Resources and Evaluation, p. 2 (2018)
25. Marrero, M., Urbano, J., Sánchez-Cuadrado, S., Morato, J., Gómez-Berbís, J.M.: Named entity recognition: fallacies, challenges and opportunities. Comput. Standards Interfaces **35**(5), 482–489 (2013)
26. Mayhew, S., Tsai, C.T., Roth, D.: Cheap translation for cross-lingual named entity recognition. In: Proceedings of the 2017 Conference on Empirical Methods in Natural Language Processing, pp. 2536–2545 (2017)
27. Ni, J., Dinu, G., Florian, R.: Weakly supervised cross-lingual named entity recognition via effective annotation and representation projection. arXiv preprint arXiv:1707.02483 (2017)
28. Nothman, J., Ringland, N., Radford, W., Murphy, T., Curran, J.R.: Learning multilingual named entity recognition from Wikipedia. Artif. Intell. **194**, 151–175 (2013)
29. Östling, R., Tiedemann, J.: Efficient word alignment with Markov chain monte Carlo. Prague Bull. Math. Linguist. **106**(1), 125–146 (2016)
30. Ott, M., et al.: fairseq: a fast, extensible toolkit for sequence modeling. arXiv preprint arXiv:1904.01038 (2019)
31. Pan, X., Zhang, B., May, J., Nothman, J., Knight, K., Ji, H.: Cross-lingual name tagging and linking for 282 languages. In: Proceedings of the 55th Annual Meeting of the Association for Computational Linguistics (vol. 1: Long Papers), pp. 1946–1958 (2017)
32. Passos, A., Kumar, V., McCallum, A.: Lexicon infused phrase embeddings for named entity resolution. arXiv preprint arXiv:1404.5367 (2014)
33. Peters, M.E., Ammar, W., Bhagavatula, C., Power, R.: Semi-supervised sequence tagging with bidirectional language models. arXiv preprint arXiv:1705.00108 (2017)
34. Peters, M.E., et al.: Deep contextualized word representations. arXiv preprint arXiv:1802.05365 (2018)

35. Ratinov, L., Roth, D.: Design challenges and misconceptions in named entity recognition. In: Proceedings of the Thirteenth Conference on Computational Natural Language Learning (CoNLL-2009), pp. 147–155 (2009)
36. Tsai, C.T., Mayhew, S., Roth, D.: Cross-lingual named entity recognition via wikification. In: Proceedings of the 20th SIGNLL Conference on Computational Natural Language Learning, pp. 219–228 (2016)
37. Ugawa, A., Tamura, A., Ninomiya, T., Takamura, H., Okumura, M.: Neural machine translation incorporating named entity. In: Proceedings of the 27th International Conference on Computational Linguistics, pp. 3240–3250 (2018)
38. Vaswani, A., et al.: Attention is all you need. In: Advances in Neural Information Processing Systems, pp. 5998–6008 (2017)
39. Wang, D., Peng, N., Duh, K.: A multi-task learning approach to adapting bilingual word embeddings for cross-lingual named entity recognition. In: Proceedings of the Eighth International Joint Conference on Natural Language Processing (vol. 2: Short Papers), pp. 383–388 (2017)
40. Wang, M., Che, W., Manning, C.D.: Joint word alignment and bilingual named entity recognition using dual decomposition. In: Proceedings of the 51st Annual Meeting of the Association for Computational Linguistics (vol. 1: Long Papers), pp. 1073–1082 (2013)
41. Yang, Z., Salakhutdinov, R., Cohen, W.: Multi-task cross-lingual sequence tagging from scratch. arXiv preprint arXiv:1603.06270 (2016)
42. Yarowsky, D., Ngai, G., Wicentowski, R.: Inducing multilingual text analysis tools via robust projection across aligned corpora. In: Proceedings of the First International Conference on Human Language Technology Research, pp. 1–8. Association for Computational Linguistics (2001)
43. Zhou, J.T., Zhang, H., Jin, D., Peng, X., Xiao, Y., Cao, Z.: Roseq: robust sequence labeling. IEEE Trans. Neural Netw. Learn. Syst. (2019)
44. Žukov-Gregorič, A., Bachrach, Y., Coope, S.: Named entity recognition with parallel recurrent neural networks. In: Proceedings of the 56th Annual Meeting of the Association for Computational Linguistics (vol. 2: Short Papers), pp. 69–74 (2018)

Recognition Method of Important Words in Korean Text Based on Reinforcement Learning

Feiyang Yang, Yahui Zhao(✉), and Rongyi Cui

Department of Computer Science and Technology, Yanbian University, 977 Gongyuan Road, Yanji 133002, China
903873610@qq.com

Abstract. The manual labeling work for constructing the Korean corpus is too time-consuming and laborious. It is difficult for low-minority languages to integrate resources. As a result, the research progress of Korean language information processing is slow. From the perspective of representation learning, reinforcement learning was combined with traditional deep learning methods. Based on the Korean text classification effect as a benchmark, and studied how to extract important Korean words in sentences. A structured model Information Distilled of Korean (IDK) was proposed. The model recognizes the words in Korean sentences and retains important words and deletes non-important words. Thereby transforming the reconstruction of the sentence into a sequential decision problem. So you can introduce the Policy Gradient method in reinforcement learning to solve the conversion problem. The results show that the model can identify the important words in Korean instead of manual annotation for representation learning. Furthermore, compared with traditional text classification methods, the model also improves the effect of Korean text classification.

Keywords: Reinforcement learning · Attention mechanism · Korean natural language processing · Structure discovery

1 Introduction

The languages of ethnic minorities have created the diversity of Chinese characters and are an important part of Chinese characters, providing important support for the development of national culture. However, the research on Korean natural language processing in my country is still in the development stage, and the related research is still relatively lagging behind South Korea and North Korea [2]. For manual annotation of Korean sentences, the structure division requires a lot of energy and time. So for this problem, we associate the method of representation learning. Representation learning has been widely used in text classification, sentiment analysis, language reasoning and other fields in recent

M. Sun et al. (Eds.): CCL 2020, LNAI 12522, pp. 261–272, 2020.
https://doi.org/10.1007/978-3-030-63031-7_19

years. It is a basic problem in the field of artificial intelligence, and it is particularly important in the field of natural language processing. Therefore, we use this method as the core logic of the model, aiming at Korean text, identifying important words and performing sentence classification tasks on newly constructed sentences. The resulting structural representation does not require manual annotation, greatly reducing manpower and scientific research resources.

In order to find important Korean words in sentences, we use the effect of text classification as feedback in reinforcement learning. The current mainstream text classification models are roughly divided into four types: bag-of-words model, sequence model, structure representation model, and attention model. The bag-of-words representation model often ignores the order of words, such as deep average networks, self-encoders [5]; the sequence representation model often only considers the words themselves, but ignores the phrase structure, such as CNN, RNN and other neural network models [14]; structural representation models often rely on pre-specified parse trees to construct structured representations, such as Tree-LSTM, recursive autoencoders [17]; representation models based on attention mechanisms need to use input words or sentences The attention scoring function is used to construct a representation, such as Self-Attention [15], and the effect is very dependent on the reliability of scoring. In the existing structured representation model, the structure can be provided as input, or it can be predicted using the supervised method of explicit tree annotations, but few studies have studied the representation with automatically optimized structure. *Yogatama et al.* proposed to construct a binary tree structure for sentence representation only under the supervision of downstream tasks, but this structure is very complicated and the depth is too large, resulting in unsatisfactory classification performance [16]. *Chung et al.* proposed a hierarchical representation model to capture the latent structure of sequences with latent semantics, but the structure can only be found in the hidden space [3]. *Tianyang et al.* proposed a method that combines the strategy gradient method in reinforcement learning with the LSTM model in deep learning. The effect of text classification is used as the baseline for reinforcement learning to carry out unsupervised structuring, and its structuring effect is closer to Human, and the classification effect is significantly better than other mainstream models [12].

Inspired by *Tianyang et al.*, we propose a method that incorporates reinforcement learning. By identifying the structure related to the task, it does not require explicit structural annotations to construct a sentence representation. Among them, the structure discovery problem is transformed into a sequence decision problem. Using the policy gradient method in reinforcement learning (Policy Gradient), the value of the delayed reward function is used to guide the self-discovery of the structure. The definition of the reward function is expressed in the same text according to the structure. The classification effect of the classifier is derived, Each time the structured representation obtained needs to be used after all sequential decisions have been made, The model incorporates an attention mechanism and a baseline of reinforcement learning convergence on the basis of predecessors to optimize it. The main purpose of the model IDK we designed is to delete the unimportant words in the sentence, retain the words most relevant to the task, and construct a new sentence representation, in which the strategy network, the structured representation, and the classification network are seamlessly integrated. The strategy network defines the strategy used to discover the structure. The classification network calculates the classification accuracy based on the structured sentence representation, and passes the value of the reward function to the strategy network to promote the self-optimization of the entire network model.

2 Strategy Network Based on Reinforcement Learning Combined with Attention Mechanism

The core idea of the strategy network is the Policy Gradient method, which is different from the traditional method. Instead of backpropagating through errors, the observation reward value is used to enhance or weaken the possibility of selecting actions. That is, the probability that a good action will be selected next time will increase; the probability that a bad action will be selected next time will decrease. A complete strategy represents a sequence of actions taken in each state in a round, The cumulative sum of the revenue generated by each action represents the round reward value. We use a random strategy $\pi\left(a_t|\mathbf{s}_t;\Theta\right)$. And use the reward generated by each round delay to guide strategy learning, For each state of the structured representation model generated each time, different actions are sampled. First, all the words of the entire sentence must be sampled for action, so as to determine the actions of all states corresponding to a sentence, Secondly, the determined action sequence is passed into the representation model to generate a new structured representation; then the generated structured representation is passed into the classification network, and this representation model is used to calculate the classification accuracy $P(y|X)$, Finally, the calculated reward is used for strategy learning. In the loop iteration, a better strategy is found, so that a better structured representation is obtained, and then a better classification effect is obtained. The strategy is defined as follows:

$$\pi\left(a_t|\mathbf{s_t};\Theta\right)=\sigma\left(\mathbf{W}*\mathbf{s}_t+\mathbf{b}\right) \tag{1}$$

Which a_t represents the probability of choosing at; σ represents the sigmoid function; Θ represents the parameters of the strategy network. During the training, actions are sampled according to the probability in Eq. 1. During the test, the action with the highest probability will be selected to achieve a better classification effect.

$$a_t^*=\mathrm{argmax}_a\pi(a\,|\mathrm{s_t};\Theta) \tag{2}$$

When all actions are sampled by the strategy network, the structured representation of the sentence is determined by the representation model, and the determined representation model is passed to the classification network to obtain, where y is the classification label, reward will be calculated from the predicted distribution, and There are also factors for thinking about the trend of structural choices. Therefore, the Policy Gradient method in the reinforcement learning algorithm is used to optimize the parameters of the strategy network [6], so as to maximize the expected return, as shown in Eq. 3.

$$\begin{aligned} J(\Theta) &= \sum_{\mathbf{s_1}a_1\cdots\mathbf{s_L}a_L} P_\Theta\left(\mathbf{s_1}a_1\cdots\mathbf{s_L}a_L\right)R_L \\ &= \sum_{\mathbf{s_1}a_1\cdots\mathbf{s_L}a_L} p\left(\mathbf{s_1}\right)\prod_t \pi_\Theta\left(a_t|\mathbf{s_t}\right)p\left(\mathbf{s_{t+1}}|\mathbf{s_t},a_t\right)R_L \\ &= \sum_{\mathbf{s_1}a_1\cdots\mathbf{s_L}a_L} \prod_t \pi_\Theta\left(a_t|\mathbf{s_t}\right)R_L \end{aligned} \tag{3}$$

The reward calculation is only for one round, because the state at step $t+1$ is completely determined by the state at step t, so the probability sum is 1. Through the likelihood ratio technique, the following gradient update strategy network [11] is finally used, where N represents the round number. In the iterative process of reinforcement learning, the variance is generally large. If the loss value is always positive, the direction of the iteration is easy to move toward. It has been proceeding in the wrong direction, so the introduction of b as a baseline can accelerate convergence, as shown in Eq. 4.

$$\nabla_\Theta J(\Theta) = -\frac{1}{N}\sum_{n=1}^{N}\sum_{t=1}^{L}(R_L - b)\nabla_\Theta \log \pi_\Theta\left(a_t|\mathbf{s_t}\right) \tag{4}$$

On this basis, the attention mechanism is introduced into the strategy network, and the Encoder-Decoder framework is adopted. Use Bi-LSTM [4] as the encoder model, LSTM as the decoder model, and the output of the structured representation model as the input of the strategy network, because the core logic of the attention model is from focusing on the whole to focusing on the core. The purpose of this article is the same, so the combination of reinforcement learning and Soft Attention mechanism to make up for the shortcomings of the predecessors in the traditional attention model in the text classification process, relying heavily on the scoring function [1]. After introducing the attention mechanism, the corresponding actions in each state are output as shown in Fig. 1.

$$S_t = f(S_{t-1}, Y_{t-1}, C_t) \tag{5}$$

$$C_t = \sum_{j=1}^{T}\alpha_{tj}h_j \tag{6}$$

$$\alpha_{tj} = \frac{exp(e_{tj})}{\sum_{k=1}^{T} exp(e_{tk})} \tag{7}$$

$$e_{tj} = g(S_{t-1}, h_j) \tag{8}$$

Where h_j is the hidden vector of the input, f is the tanh activation function; C is the attention distribution; and α_{tj} is the attention obtained by each input.

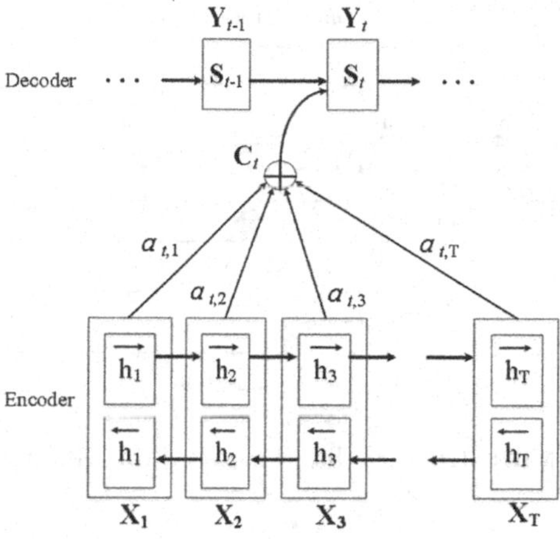

Fig. 1. Soft attention mechanism

After introducing the attention mechanism, the global observation can be better, so that the generated action sequence is optimized in two aspects of the strategy gradient and the attention mechanism, thereby improving the model effect.

3 Information Distilled of Korean (IDK)

3.1 The Main Idea of the Model

Our ultimate goal is to reconstruct more concise Korean sentences by finding important, task-related words, and at the same time get a structural representation for Korean text classification, and improve text classification through optimized structured representation. While the text classification has been improved, the structured representation has also been optimized, and the two promote each other. The model consists of three parts: strategy gradient network, structured representation model, classification network. The strategy network adopts a random strategy to sample the actions corresponding to each state, sampling until the end of the sentence, and generating a sequence of actions for the current sentence. Then the structured representation model converts the action sequence into a structured representation. Based on this idea, the IDK model is proposed. The classification network classifies based on the obtained structured representation and provides the reward function calculation for the strategy network. Since a complete structured representation can be given to calculate the reward of the current structured representation, this process can be solved by the Policy Gradient method. The specific model is shown in Fig. 2.

The model is interleaved by three parts. The state representation of the strategy network comes from the structured model. The structured model is generated by the action sequence of the strategy network and the input of the sentence. The classification network is classified and predicted by the resulting

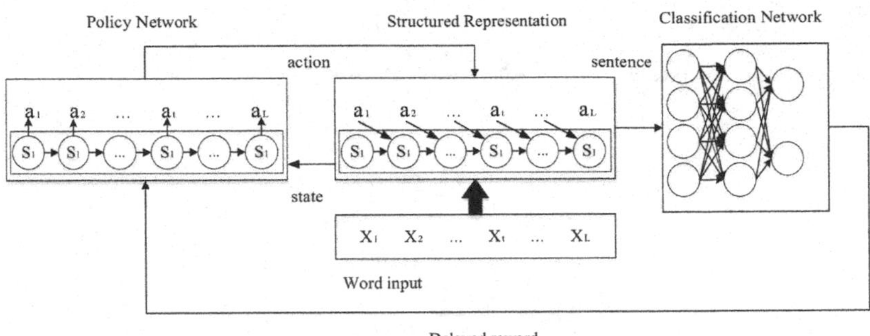

Fig. 2. Network model structure diagram based on reinforcement learning.

structured model. Strategy The network obtains the reward function value from the classification effect obtained by the classification network, thereby guiding the strategy to learn a better structured representation.

3.2 Model Specific Construction

Inspired by the ID-LSTM of *Tianyang et al.*, an attention mechanism was introduced into the original strategy network to double optimize the action sequence. The main idea of IDK proposed in our thoughts is to build a structured representation of a sentence by extracting important words and deleting irrelevant words in the sentence. In the well-known Chinese and English text processing tasks, there are many examples such as: "with", "and", "in", "of" and other stop words, such stop words rarely help complete text processing tasks, so it is necessary to refine important features in sentences. Different from the traditional method, this method does not create a stop word list, and deletes all stop words together. Because many stop words often constitute a special phrase structure, combing the logical relationship between the context and deleting it directly without filtering, it will cause the loss of language content and semantic information, so this method is chosen in our thoughts to purify the final representation form, thus Concentrate sentences to enhance the effectiveness of downstream classification tasks.

The IDK model converts the sequence of actions transferred from the strategy network into a structured representation of sentences, Given a sentence X shaped like $X = x_1 x_2 \cdots x_L$, After the sentence X is transferred to the strategy network, each action a_i corresponding to the word position x_i is selected from keeping the current word or deleting the current word, which satisfies the following rules:

$$\mathrm{S_t}, \mathrm{C_t}, = \begin{cases} \mathrm{S_{t-1}}, \mathrm{C_{t-1}}, & a_t = Delete \\ \Phi\left(\mathrm{S_{t-1}}, \mathrm{C_t}, \mathrm{Y_{t-1}}\right), & a_t = Retain \end{cases} \tag{9}$$

The Φ represents the function of the entire model (including gating unit and update function), S is the hidden state corresponding to the Decoder cell unit; Y is the output corresponding to the hidden state of the cell unit; C is the hidden state distribution of the Encoder cell unit; When deleting a word, the storage unit and hidden state attention distribution of the current position will be copied from the previous position.

For classification, the last hidden state of the IDK model is used as the input of the classification network, where $\mathbf{W_s} \in R^{d \times K}$, $\mathbf{b_s} \in R^{K}$ is the parameter of the classification network, d is the dimension of the hidden state, is the label of the category, K is the number of classification clusters, the classification network is based on the IDK model The obtained structured representation produces a probability distribution on the class label, as shown in Eq. 10.

$$P(y\,|X\,) = \mathrm{softmax}(\mathbf{W}_{\mathrm{s}}\mathrm{S}_{\mathrm{L}} + \mathrm{b}_{\mathrm{s}}) \tag{10}$$

To calculate reward, take the logarithm of the output probability calculated by the classification network in Eq. 10, as shown in Eq. 11, where c_g stands for classification label. In order to make the model use as few words as possible, the two items in the formula are controlled by calculating the ratio of the number of deleted words in the sentence to the length of the sentence. Maintain accuracy and balance the two effects of using a few words, Where L' represents the number of deleted words, and γ represents the hyperparameter between 0 and 1 that balances the two terms.

$$R_L - \log P\,(c_g|X) + \gamma L'/L \tag{11}$$

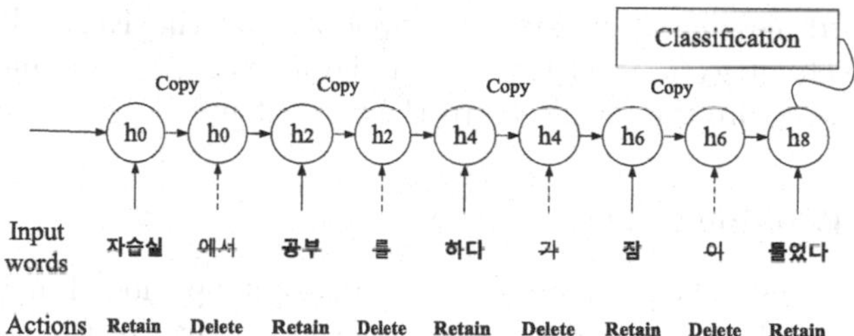

Fig. 3. ID-Korean model

As shown in Fig. 3, after word recognition is performed on Korean text, only important words are retained, and the ratio of the amount retained to the number deleted is controlled by the reward function.

4 Experimental Results and Analysis

4.1 Data Set Description

The data set used in the experiments in this article comes from the corpus constructed by the laboratory to undertake the "China-Korea Science and Technology Information Processing Comprehensive Platform" project. It is further organized into a corpus composed of abstracts of Korean scientific and technological literature. There are about 30,000 documents, divided into 13 categories such as animals, oceans, and aerospace. Each Category randomly selects documents according to a 7:3 ratio to form a training set and a test set [9,13]. The details of the data set are shown in Table 1.

Table 1. Data set introduction

Category	Number of entries	Category	Number of entries
Animal	4582	Botany	6172
Microorganism	5472	Biotechnology	1215
Biomedical science	2752	Climate	708
The marine environment	810	Geology	1735
Marine technology	819	Materials engineering	781
Measurement technology	1728	Aerospace	4436
Others	1478		

For Korean corpora, sentences are composed of phrases separated by spaces, and these phrases are usually followed by auxiliary words or endings. According to the grammatical characteristics of Korean, In the preprocessing process, the Hannanum word segmentation system developed by the Korea University of Science and Technology is used to cut out the auxiliary words and endings in the phrase, and restore the predicate to the word itself.

4.2 Model Training

When training a classification network, a cross-entropy loss function is used, in which the probability distribution of the ground truth of the corresponding sentence is coded by one-hot, as shown in Eq. 12.

$$\mathcal{L} = -\sum_{X \in \mathcal{D}} \sum_{y=1}^{K} \hat{p}(y, X) \log P(y|X) \tag{12}$$

GloVe training is used to initialize the word vector in the representation model [10], the dimension is set to 256 dim, and it is updated together with other parameters. When using gradient descent to update parameters, the speed of model learning depends on the learning rate and partial derivative value, To smooth the update of Policy Gradient, multiply the suppression factor γ by Eq. 4 and set it to 0.1, γ is set to 0.2 in the IDK, Eq. 11, In the training process, the Adam optimizer [7] is used to optimize the parameters, the learning rate is 0.0005, the Dropout tailoring is used before the classification network classification, the probability is 0.5, and the mini-batch is 5.

In the model training process, the classification accuracy rate using the IDK model changes with the number of iterations as shown in Fig. 4. The text classification accuracy rate is about 68% at the beginning of the training, and the accuracy rate increases as the number of iterations increases. When the number of iterations is between 400 and 600, the classification accuracy of the model rises fastest. After 800 iterations, the classification accuracy of multilingual text tends to be stable, indicating that the training of the neural network model has converged. At this time, the text classification of the IDK model The accuracy rate reached 83.23%.

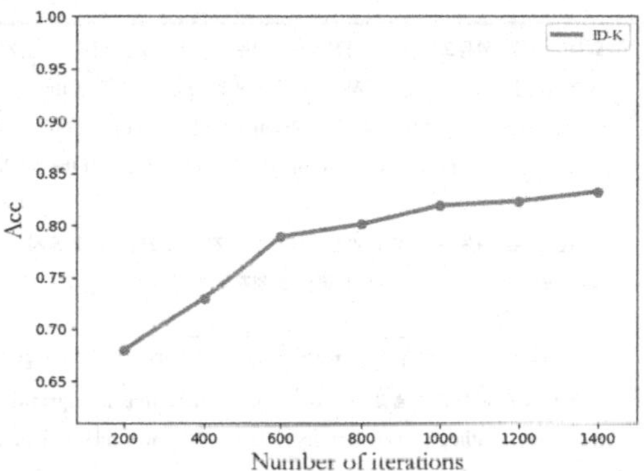

Fig. 4. Soft attention mechanism

4.3 Comparative Experiment

In the comparative experiment, a varicty of baselines were selected: basic neural network model CNN without specific structure; LSTM; Bi-LSTM; attention model Self-Attention [8]. The dimension of the word vector used by these baselines is the same as this article, and the effect is shown in Table 2.

Table 2. Accuracy under different classifiers

Models	ACC	Models	ACC
CNN	78.5	T-BLSTM-CNN	81.68
LSTM	74.6	Self-Attention	82.91
Bi-LSTM	78.14	**IDK**	**83.23**

As shown in Table 2, the classification effect shows that: in different models, our method performs well in classification. When comparing with previous methods, we combine reinforcement learning and attention model to use a self-discovery structure and Optimize the structured representation model for text classification, Different from the predecessors who only focused on the sequence model and its optimization, this paper designs the model from two aspects of reinforcement learning and attention. Its classification effect also proves the effectiveness and necessity of representation learning and text structuring.

4.4 Examples of Structured Presentation Results

Original	도시는 인류 활동의 영향을 가장 크게 받는 지구의 표면 이고 도시 시스템의 탄소 순환은 세계 및 지역의 탄소 순환에서 중요한 위치와 역할을 한다（Cities are the surface of the earth, which is most affected by human activity, and the carbon cycle of urban systems plays an important position and role in the carbon cycle of the world and regions.）
IDK	**도시는 인류** ~~활동의~~ **영향을** ~~가장 크게~~ **받는 지구의 표면 이고 도시 시스템의 탄소 순환은** ~~세계 및 지역의 탄소 순환에서~~ **중요한 위치와 역할을** ~~한다~~
Original	곤충 병원체에서 직접 샘플을 채취하고 검출 및 정량화하는 것은 곤충 유행병학 조사에서 병원체 풍도를 직접 반영할 수 있다（Taking, detecting, and quantifying samples directly from insect pathogens can directly reflect pathogen wind levels in insect epidemiological surveys.）
IDK	**곤충 병원체에서** ~~직접~~ **샘플을 채취하고 검출** ~~및~~ **정량화하는** ~~것은~~ **곤충 유행병학** ~~조사에서~~ **병원체 풍도를** ~~직접~~ **반영할** ~~수 있다~~

Fig. 5. Structural representation example

The specific structured example is shown in Fig. 5. In the IDK model, the strike through indicates the word to be deleted on the original text. The Korean text's mood words, auxiliary words, and some adjectives are deleted, and important part of the nouns are retained. The larger the model segmentation structure is, the closer it is to manual annotation, and the original text is more useful for downstream text classification tasks after being structured.

5 Conclusion

We combined reinforcement learning methods to learn Korean sentence representations by finding important words related to the task. In the framework of reinforcement learning and attention mechanism, this paper uses the IDK model, which is used to extract task-related words and express them in purified sentences. Among them, reinforcement learning uses the accuracy rate of text classification as a baseline to optimize the action sequence, and the action sequence can generate a text structure representation that is more suitable for classification. An attention mechanism is introduced in the process of action sequence generation to compensate for the variance of the reinforcement learning method. The disadvantage of being too large and difficult to fit, compared with the traditional attention model, not only has its advantages of taking into account the overall situation, but also adds a more ingenious way to improve the accuracy of downstream tasks, and the experimental results have performed well. Experiments show that our method can find important words related to tasks without explicit structure annotation. The model not only improves the effect of Korean text classification, but also works well in the task of processing Korean text important word recognition.

Acknowledgements. This work was supported by the National Language Commission Scientific Research Project (YB135-76); Yanbian University Foreign Language and Literature First-Class Subject Construction Project (18YLPY13).

References

1. Bahdanau, D., Cho, K., Bengio, Y.: Neural machine translation by jointly learning to align and translate. In: International Conference on Machine Learning. arXiv:1409.0473v7 (2015)
2. Bi, Y.: A research on Korean natural language processing. J. Chin. Inf. Process. **25**(06), 166–169+182 (2011)
3. Chung, J., Gulcehre, C., Cho, K., Bengio, Y.: Empirical evaluation of gated recurrent neural networks on sequence modeling. In: NIPS 2014 Workshop (2014)
4. Graves, A., Schmidhuber, J.: Frame wise phoneme classification with bidirectional LSTM and other neural network architectures. Neural Netw. **18**(5–6), 602–610 (2005)
5. Joulin, A., Grave, E., Bojanowski, P., Mikolov, T.: Bag of tricks for efficient text classification. In: The European Chapter of the ACL (EACL), pp. 427–431 (2017)
6. Keneshloo, Y., Ramakrishnan, N., Reddy, C.K.: Deep transfer reinforcement learning for text summarization. Society for Industrial and Applied Mathematics. arxiv:1810.06667v2 (2019)
7. Kingma, D.P., Ba, L.J.: A method for stochastic optimization. In: International Conference on Learning Representations (2015)
8. Lin, Z., et al.: A structured self-attentive sentence embedding. In: International Conference on Learning Representations (2017)
9. Tian, M., Zhao, Y., Cui, R.: Identifying word translations in scientific literature based on labeled bilingual topic model and co-occurrence features. In: Sun, M., Liu, T., Wang, X., Liu, Z., Liu, Y. (eds.) CCL/NLP-NABD - 2018. LNCS (LNAI), vol. 11221, pp. 76–87. Springer, Cham (2018). https://doi.org/10.1007/978-3-030-01716-3_7
10. Pennington, J., Socher, R., Manning, C.D.: Glove: global vectors for word representation. In: EMNLP 2014, pp. 1532–1543 (2014)
11. Sutton, R.S., McAllester, D.A., Singh, S.P., Mansour, Y.: Policy gradient methods for reinforcement learning with function approximation. In: NIPS 2000, pp. 1057–1063 (2000)
12. Zhang, T., Huang, M., Zhao, L.: Learning structured representation for text classification via reinforcement learning. In: The Association for the Advance of Artificial Intelligence (2018)

13. Meng, X., Cui, R.: Multilingual text classification method based on bidirectional long-short memory unit and convolutional neural network. Comput. Appl. Res. **132**(04), 1–6 (2019)
14. Kim, Y.: Convolutional neural networks for sentence classification. In: The 2014 Conference on Empirical Methods in Natural Language Processing, pp. 1746–1751 (2014)
15. Yang, Z., Yang, D., Dyer, C., He, X., Smola, A., Hovy, E.: Hierarchical attention networks for document classification. In: Annual Conference of the North American Chapter of the Association for Computational Linguistics, pp. 1480–1489 (2016)
16. Yogatama, D., Blunsom, P., Dyer, C., Grefenstette, E., Ling, W.: Learning to compose words into sentences with reinforcement learning. In: International Conference on Learning Representations (2017)
17. Zhu, X., Sobihani, P., Guo, H.: Long short-term memory over recursive structures. In: International Conference on Machine Learning, pp. 1604–1612 (2015)

Mongolian Questions Classification Based on Multi-Head Attention

Guangyi Wang[1], Feilong Bao[1,2], and Weihua Wang[1,2(✉)]

[1] College of Computer Science, Inner Mongolia University, Hohhot, China
wanggycs@163.com, {csfeilong,wangwh}@imu.edu.cn
[2] Inner Mongolian Key Laboratory of Mongolian Information Processing Technology, Hohhot, China

Abstract. Question classification is a crucial subtask in question answering system. Mongolian is a kind of few resource language. It lacks public labeled corpus. And the complex morphological structure of Mongolian vocabulary makes the data-sparse problem. This paper proposes a classification model, which combines the Bi-LSTM model with the Multi-Head Attention mechanism. The Multi-Head Attention mechanism extracts relevant information from different dimensions and representation subspace. According to the characteristics of Mongolian word-formation, this paper introduces Mongolian morphemes representation in the embedding layer. Morpheme vector focuses on the semantics of the Mongolian word. In this paper, character vector and morpheme vector are concatenated to get word vector, which sends to the Bi-LSTM getting context representation. Finally, the Multi-Head Attention obtains global information for classification. The model experimented on the Mongolian corpus. Experimental results show that our proposed model significantly outperforms baseline systems.

Keywords: Question classification · Mongolian · Morpheme · Multi-Head Attention mechanism

1 Introduction

When people read a specific sentence on a flyer or some magazine, they can understand the context or intent of the sentence. And they can also extract information from the sentence. How to make a computer think like a human. Natural Language Processing (NLP) and Natural Language Understanding (NLU) study how to make the computer understand the semantics of natural language. The computer uses natural language to communicate with people to realize human-machine interaction. Deep learning models have achieved state-of-the-art performance in various natural language processing tasks such as text summarization [13], question answering [3] and machine translation [6]. In recent years, question answering is a key technology in intelligent applications. It has aroused widespread concern. Pipeline the first task of question system is to classify the

M. Sun et al. (Eds.): CCL 2020, LNAI 12522, pp. 273–284, 2020.
https://doi.org/10.1007/978-3-030-63031-7_20

domain of the dialogue after the user enters the message (text or voice). Question classification divides questions into several semantic categories. The machine gets a predicted category of the dialogue and the system returns a concise and accurate answer. The understanding of questions provides constraints for improving the accuracy of question answering system. In [11], the authors have studied the influence of each part of the question answering system on the system performance. The question classification recognition has the greatest influence on the system performance. Therefore, to get a good question answering system, it is necessary to design a high accuracy model of question classification.

However, the research of the Mongolian questions classification is very fewer. The reason is that Mongolian corpus is scarce and there is no public Mongolian corpus. Data collected from internet are noisy and uncertain in terms of coding and spelling. The word-formation is different from Chinese and English. It consists of roots, stems and affixes. These problems result in unlimited vocabulary. The existing short text classification methods are not effective. How to classify the questions accurately is a complicated problem.

In this article, the training data were crawled from the Mongolian web sites. After cleaning the invalid data, we constructed a question classification data set. We propose a method of the Mongolian question classification, which combines the Bi-LSTM model with the Multi-Head Attention mechanism. As shown in Fig. 1, the model is named MA-B. To better learn semantic information from sentences, we introduce the morphemes representation. The character vector and the morpheme vector are concatenated to get word vector. It sends to Bi-LSTM getting context representation. The Multi-Head Attention mechanism extracts relevant information from different dimensions. In the classification layer, we use the softmax classifier to output the probability of each category.

The paper is organized as follows: Sect. 2 gives the related work. Section 3 presents the question classification method in detail. Section 4 shows the experiments and results. Section 5 summarizes the full text and give some future works.

2 Related Work

Question classification is a kind of short text classification [1]. There have been many studies on questions classification. Chinese and English, which are rich in resources, have achieved good results. The traditional method was based feature engineering such as bag of words (BOW) and n-gram. Both were combined with term frequency-inverse document frequency (TF-IDF) and other element features as text features. However, these methods ignore the context semantic information. There were some methods based machine learning, including Nearest Neighbors (NN) [19], Naive Bayes [9], and Support Vector Machine (SVM) [2]. In [17], the authors utilized the external knowledge base for text classification. In recent years, researchers have tried to extract semantic information from sentences via deep learning. The combination of TextCNN [5], TextRNN [7], LSTM [18], TextGCN [20], with word embedding has been widely used in text classification.

There are some researches on rare resource languages to classify questions. For example, Uyghur is also a few resource language and have complex word-formation. In [12], the authors proposed a method of Uyghur short text classification based reliable sub-word morphology. Mongolian language processing has been further developed, such as morphological segmentation [16], spelling correction [8], named entity recognition [15]. The Mongolian question classification needs to be solved urgently.

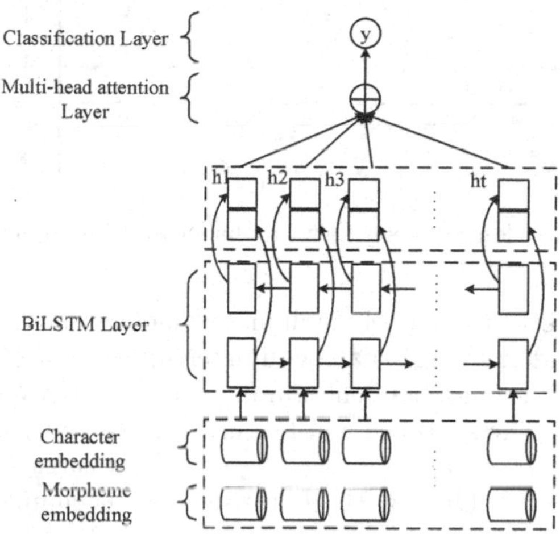

Fig. 1. The model architecture of MA-B.

3 Model Architecture

In this section, we will introduce this model from bottom to up. The Multi-Head Attention mechanism can fully capture the long-distance text features. But it is difficult to deal with the sequence information. The recurrent neural network can effectively obtain the context order information of sequences. It can effectively supplement the Multi-Head Attention mechanism. As depicted in Fig. 1, MA-B model is proposed by combining Bi-LSTM network with Multi-Head Attention mechanism.

3.1 Morpheme Vector

Mongolian is a kind of agglutinative language, which consists of roots, stems and suffixes. The Chinese words need to be segmented, which is called Chinese word segmentation [21]. There are natural spaces between words in Mongolian, but morphological segmentation is needed in Mongolian because the root and stem suffixes of Mongolian words are connected with many different endings.

The Mongolian word formation features result in unlimited vocabulary. This paper uses Latin to deal with Mongolian. The contrast between Latin characters and Mongolian letters is shown in Fig. 2.

Mongolian alphabet	Latin letter	Mongolian alphabet	Latin letter	Mongolian alphabet	Latin letter	Mongolian alphabet	Latin letter	Mongolian alphabet	Latin letter
ᠠ	a	ᠧ	E	ᠺ	k	ᠮ	m	ᠲ	t
ᠡ	e	ᠨ	n	ᠻ	K	ᠯ	l	ᠳ	d
ᠢ	i	ᠩ	N	ᠼ	C	ᡀ	L	ᠶ	y
ᠣ	q	ᠪ	b	ᠽ	Z	ᡁ	Z	ᠴ	c
ᠤ	v	ᠫ	p	ᠾ	H	ᡂ	Q	ᠵ	j
ᠥ	o	ᠸ	w	ᠿ	R	ᠰ	s	ᠷ	r
ᠦ	u	ᠹ	f	ᠭ	g	ᠱ	x	ᠬ	h

Fig. 2. Comparison between Latin alphabet and Mongolian alphabet.

In this paper, we introduce Mongolian morphemes representation. The suffix is segmented by identifying a narrow uninterrupted space (NNBS) (U+202F, Latin: "-") to make it an independent training unit. As shown in Fig. 3, after segmentation the suffix, the sentence will be turned into *"kqmpani -y'in havli -y'in homun ebedcileged genedte nasv barajai, tegun -u" ori ogcege -y'i hen egurgelehu boged bvcagahv yqsqtai"*. The length of this sentence is changed to 19 units.

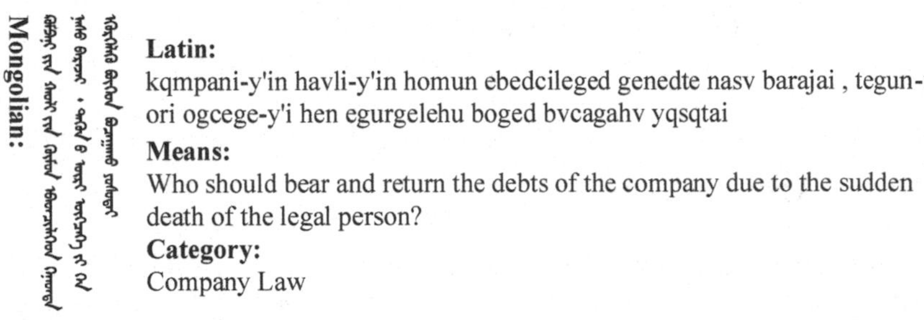

Fig. 3. Example of traditional Mongolian script, Latin transliterature, category tag and their meanings.

The Word2vec is a common tool for training word vectors. The Word2vec [10] contains CBOW (Continuous Bag of Word) and Skip-gram. This paper uses the Skip-gram model to train morpheme vectors. Given a sequence of morphemes $\mathbf{m} = m_1, ..., m_T \in M$. The output of the model is a probability distribution. The morpheme skip-gram model predict contextual morphemes when given current morpheme. The formula is as follows:

$$\frac{1}{T}\sum_{t=1}^{T}\sum_{-c\leq j\leq c, j\neq 0} \log p(m_{t+j} \mid m_t) \tag{1}$$

where c is the size of the context window for the current central morpheme m_t. The simplest formulation of the probability $p(m_{t+j}|m_t)$ is:

$$p(o \mid c) = \frac{\exp({u_o}^T v_c)}{\sum_{m=1}^{M} \exp({u_m}^T v_c)} p \tag{2}$$

where o is the ids of the output morpheme, c is the ids of the central morpheme, u is the output morpheme vector, v is the input morpheme vector, and M is the morphemes set.

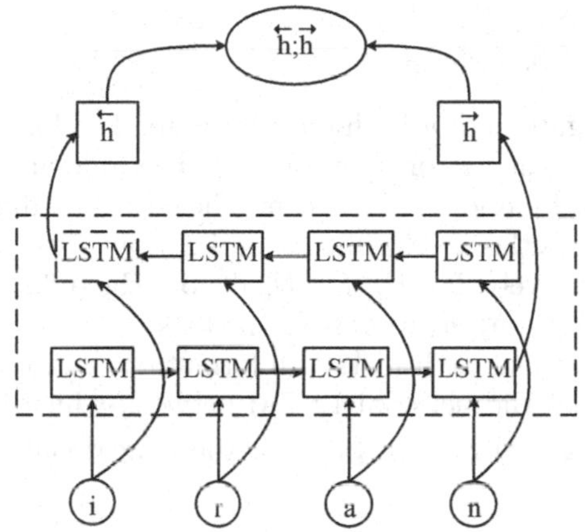

Fig. 4. The character embedding of Mongolian morpheme.

3.2 Character Vector

To better represent the semantic information in sentences, we use the Bi-LSTM model to learn the character embedding from training data. The character Bi-LSTM network consists of forward LSTM layer and backward LSTM layer. The forward layer can learn word prefix information. And the backward layer learns the morphological information. Both layers are connected to the same output layer. We get the character representation. As shown in Fig. 4 is the structure of Bi-LSTM character embedding network.

3.3 Bi-LSTM Layer

LSTM [4] network is a special type of recursive neural network, which can capture the context order information of the sequence and solve the problem of long dependency. LSTM is a variant of RNN. It introduces some gates to solve the gradient problem. LSTM calculates an output vector according to the current input and the output of the previous unit. The output vector is then used as input to the next unit.

LSTM is mainly composed of four parts: storage unit c_t, input gate i_t, output gate o_t, and forget gate f_t. Those gates control the proportion of history to omit or to store in the next time stamp. LSTM calculates the output vector based on the current input and the output of the previous unit, which is then used as the input of the next unit. The calculation formula is as follows:

$$\begin{aligned} f_t &= \sigma(W_{(f)}x_t + U_{(f)}h_{t-1} + b_{(f)}) \\ i_t &= \sigma(W_{(i)}x_t + U_{(i)}h_{t-1} + b_{(i)}) \\ o_t &= \sigma(W_{(o)}x_t + U_{(o)}h_{t-1} + b_{(o)}) \\ c_t &= \tilde{c} + i_t \odot \tanh(W_{(c)}x_t + U_{(c)}h_{t-1} + b_{(c)}) \\ \tilde{c} &= f_t \odot c_{t-1} \\ h_t &= o_t \odot \tanh(c_t) \end{aligned} \tag{3}$$

where i_t is the input gate and o_t is the output gate. The forget gate f_t is a reset memory unit. x_t the input vector. h_t represents the hidden unit vector. σ is the point product sigmoid function. $\odot$ represents the corresponding multiplication of elements. W_i, W_f, W_o is the weight matrix of the input gate, the forget gate, and the output gate respectively. U_f, U_i, U_c, U_o denote the different weight matrices for hidden h_t. And b_i, b_f, b_c, b_o represent the bias.

The LSTM can only encode historical information, but it is often not enough. The paper adopted the Bidrectional LSTM network which is composed of forward LSTM and backward LSTM. So, h is the concatenate of $\overleftarrow{h_t}$, $\overrightarrow{h_t}$ and h is shown as below.

$$h = \overleftarrow{h_t} + \overrightarrow{h_t} \tag{4}$$

where $\overrightarrow{h_t}$ is the forward output vector and $\overleftarrow{h_t}$ is backward output vector.

3.4 Multi-head Attention Layer

In recent years, *Transformer* model is very popular, which used in NLP tasks [14]. It uses the Multi-Head Attention mechanism. The Multi-Head Attention is the optimization of the traditional attention mechanism and it is used to fully capture the features of long distance and obtain the global information. It firstly projects the input into multiple feature spaces, then compute correlation score and utilize the scores to weight context representation, finally concatenates vectors weighted as output.

The input of Multi-Head Attention mechanism consists of Q (queries), K (keys) and D (dimension). The merging vector output from Bi-LSTM layer is the input of Q, K and V. Then Q, K, V are linearly transformed and finally input into scaled dot-product attention (SDPA). This process calculates one head at a time. As shown in Fig. 5, the model independently compute dot product attention for each part $head_i$. The details are described below.

$$SDPA(Q,K,V) = softmax\left(\frac{QK^T}{\sqrt{d_k}}\right)V \tag{5}$$

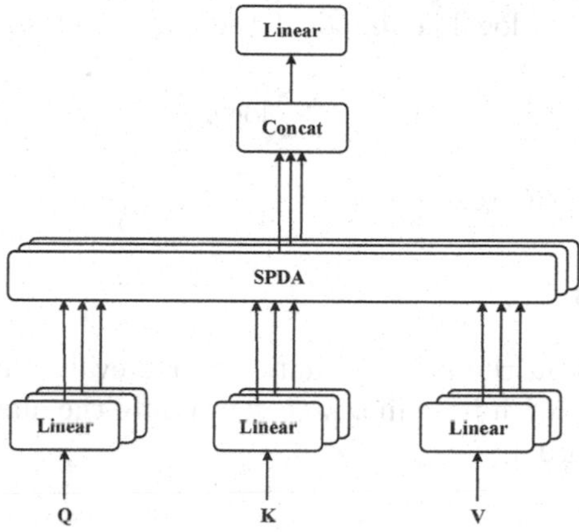

Fig. 5. The flowchart of scaled dot-product attention.

where *softmax* is a normalization function. The calculation formula is as follows:

$$softmax(g(Q,K)) = \frac{e^{g(Q,K)}}{\sum_i e^{g(Q,K_j)}} \tag{6}$$

where $g(Q,K)$ represents the similarity between Q and K. Similarity calculation is obtained by Q and K point product operation.

Then, all the scaled dot-product attention results of m times, are concatenated and the value obtained by a linear transformation is used as the result of the Multi-Head Attention model.

$$head_i = SDPA\left(QW_i^Q, QW_i^K, VW_i^V\right) \tag{7}$$

$$MA(Q,K,V) = Concat(head_1, \cdots, head_h) \tag{8}$$

where W_i^Q, W_i^K and W_i^V are projection matrices corresponding to Q, K and V respectively.

3.5 Classification Layer

Questions classification is a multi classification problem. The classification layer consists of two parts: a linear layer and a softmax layer. The text vector h can be used as features for questions classification.

$$y = softmax(W_h h + b_h) \tag{9}$$

We use the negative log likelihood of the correct classification as training loss.

$$L = -\sum_{t} \log y_{ti} \tag{10}$$

where i is the label of the text t.

4 Experiments

Our model is trained on the selected data set. By evaluating the classification results and comparing with baseline, we can evaluate the questions classification performance of the model.

4.1 Setting up

The training data mainly comes from China Mongolian News Network, People's Daily Online (Mongolian version), China Mongolian Broadcasting Network, China Judgements Online (Mongolian version) and other web sites. After removing duplicate data and cleaning invalid data, 115688 sentences were obtained by manual correction and annotation. The data of question classification is divided into eleven categories, as shown in Table 1. We divided the dataset into train, dev and test with the percent 80%, 10% and 10%, respectively.

Table 1. The data is divided into eleven categories.

Label	Categories	Number	Label	Categories	Number
0	Marriage and Family	10359	6	Property Disputes	9435
1	Labor Disputes	9621	7	Infringement	11258
2	Traffic Accident	11421	8	Company Law	9900
3	Credit and Debt	9401	9	Medical Disputes	8743
4	Criminal Defense	13020	10	Administrative Litigation	13872
5	Contract Disputes	8658			

4.2 Evaluation Metrics

Question classification is a multi classification task, so we use *precision, recall* and F_1 as the evaluation index. These metrics are calculated as:

$$\begin{aligned} P &= \frac{TP}{TP+FP} \\ R &= \frac{TP}{TP+FN} \\ F_1 &= \frac{2PR}{P+R} \end{aligned} \tag{11}$$

where TP is the number of correctly predicted question sentences. FP is the number of sentences that predicted as question sentences, but in actuality those are negative class. If the prediction is failed, and the positive class is predicted as a false negative (FN). F_1 is the harmonic mean of precision and recall.

4.3 Results

In this paper, TextCNN, Bi-LSTM and Attention-BiLSTM model are used as baselines. TextCNN [5] applies Convolutional Neural Networks (CNN) to text classification tasks. The key information in sentences is extracted by using multiple different size kernels. So it can better capture the local correlation. TextCNN is a commonly used baseline. Bi-LSTM and Attention-BiLSTM are commonly used models to extract text features. Attention is essentially an automatic weighted summation mechanism that makes the model more capable of handling long sequences.

The experiment is divided into two forms: 1) whether the combination of character vector and morpheme vector affects the performance of the model. 2) whether the introduction of Multi-Head Attention mechanism into the model affects the performance of the model. The experimental results are shown in Table 2.

Table 2. Comparison of experimental results.

Model	Character embedding	Morpheme embedding	P(%)	R(%)	F_1(%)
TextCNN	Yes	No	82.57	79.36	80.93
TextCNN	Yes	Yes	83.27	81.42	82.33
Bi-LSTM	Yes	No	83.22	83.95	83.58
Bi-LSTM	Yes	Yes	84.56	83.93	84.24
Att-BiLSTM	Yes	No	84.67	83.91	84.31
Att-BiLSTM	Yes	Yes	85.13	84.89	85.01
MA-B	Yes	No	86.58	86.01	86.29
MA-B	Yes	Yes	**86.71**	**86.51**	**86.61**

We compare the results from the table:

1) Introducing morpheme features in the embedding layer can improve performance. The F_1 value of MA-B model remains the highest among all models. About 1.6% improvement compared with the highest Att-BiLSTM model in the baseline model.
2) In the whole model, the introduction of Multi-Head Attention mechanism can effectively improve the model classification performance. Compared with Bi-LSTM model, our model is improved by about 2.2%. Compared with

Att-BiLSTM model, our model's classification ability is also significantly enhanced.

The reasons for the above results are as follows:

1) When judging the questions categories of sentences, we mainly consider the semantic information of sentences. In Mongolian word formation, morpheme vector can learn more syntactic and semantic information. Therefore, the introduction of morpheme features into the model will have a good performance.
2) Compared with the baseline model, the advantage of MA-B model is to use BiLSTM network to obtain the internal relationship between the front and back directions of sentences and get local information. The long-distance feature is fully captured by Multi-Head Attention mechanism, and relevant information is learned from different dimensions and representation subspaces.

5 Conclusion

In this paper, Bi-LSTM and Multi-Head Attention mechanism are used to model Mongolian corpus texts. By combining the ability of multi head attention to obtain global information with the ability of Bi-LSTM to obtain local sequence information, a better effect has been achieved. At the same time, in order to make the model better learn the text semantic information, Mongolian morphemes representation are further introduced.

However, there is a lot of room for improvement in the field of Mongolian questions classification. From the experiment, it can be seen that the introduction of pre training morphemes features has a good effect. In the future, feature engineering can be further reduced by using pretraining language models. At the same time, the research of Mongolian question intention recognition provides a good foundation for Mongolian question answering system in the future.

Acknowledgements. The project (Nos. 2018YFE0122900, CGZH2018125, 2019GG372, 2020GG0046) are supported by Inner Mongolia Science & Technology Plan; National Natural Science Foundation of China (Nos. 61773224); Natural Science Foundation of Inner Mongolia (Nos. 2018MS06006, 2020BS06001).

References

1. Alsmadi, I.M., Gan, K.H.: Review of short-text classification. Int. J. Web Inf. Syst. **15**(2), 155–182 (2019)
2. Cristianini, N., Shawe-Taylor, J.: An Introduction to Support Vector Machines and Other Kernel-Based Learning Methods. Cambridge University Press, Cambridge (2010)
3. He, X., Golub, D.: Character-level question answering with attention. In: Su, J., Carreras, X., Duh, K. (eds.) Proceedings of the 2016 Conference on Empirical Methods in Natural Language Processing, EMNLP 2016, Austin, Texas, USA, 1–4 November 2016, pp. 1598–1607. The Association for Computational Linguistics (2016)

4. Hochreiter, S., Schmidhuber, J.: Long short-term memory. Neural Comput. **9**(8), 1735–1780 (1997). https://doi.org/10.1162/neco.1997.9.8.1735
5. Kim, Y.: Convolutional neural networks for sentence classification. In: Moschitti, A., Pang, B., Daelemans, W. (eds.) Proceedings of the 2014 Conference on Empirical Methods in Natural Language Processing, EMNLP 2014, Doha, Qatar, 25–29 October 2014, A meeting of SIGDAT, a Special Interest Group of the ACL, pp. 1746–1751. ACL (2014)
6. Kudo, T.: Subword regularization: improving neural network translation models with multiple subword candidates. In: Gurevych, I., Miyao, Y. (eds.) Proceedings of the 56th Annual Meeting of the Association for Computational Linguistics, ACL 2018, Melbourne, Australia, 15–20 July 2018, vol. 1: Long Papers, pp. 66–75. Association for Computational Linguistics (2018)
7. Liu, P., Qiu, X., Huang, X.: Recurrent neural network for text classification with multi-task learning. In: Kambhampati, S. (ed.) Proceedings of the Twenty-Fifth International Joint Conference on Artificial Intelligence, IJCAI 2016, New York, NY, USA, 9–15 July 2016, pp. 2873–2879. IJCAI/AAAI Press (2016)
8. Lu, M., Bao, F., Gao, G., Wang, W., Zhang, H.: An automatic spelling correction method for classical Mongolian. In: Douligeris, C., Karagiannis, D., Apostolou, D. (eds.) KSEM 2019. LNCS (LNAI), vol. 11776, pp. 201–214. Springer, Cham (2019). https://doi.org/10.1007/978-3-030-29563-9_19
9. McCallum, A., Nigam, K., et al.: A comparison of event models for Naive Bayes text classification. In: AAAI-1998 Workshop on Learning for Text Categorization, vol. 752, pp. 41–48. Citeseer (1998)
10. Mikolov, T., Chen, K., Corrado, G., Dean, J.: Efficient estimation of word representations in vector space. In: Bengio, Y., LeCun, Y. (eds.) 1st International Conference on Learning Representations, ICLR 2013, Scottsdale, Arizona, USA, 2–4 May 2013, Workshop Track Proceedings (2013)
11. Moldovan, D.I., Pasca, M., Harabagiu, S.M., Surdeanu, M.: Performance issues and error analysis in an open-domain question answering system. ACM Trans. Inf. Syst. **21**(2), 133–154 (2003)
12. Parhat, S., Ablimit, M., Hamdulla, A.: Uyghur short-text classification based on reliable sub-word morphology. IJRIS **11**(3), 250–255 (2019)
13. Rush, A.M., Chopra, S., Weston, J.: A neural attention model for abstractive sentence summarization. In: Màrquez, L., Callison-Burch, C., Su, J., Pighin, D., Marton, Y. (eds.) Proceedings of the 2015 Conference on Empirical Methods in Natural Language Processing, EMNLP 2015, Lisbon, Portugal, 17–21 September 2015, pp. 379–389. The Association for Computational Linguistics (2015)
14. Vaswani, A., et al.: Attention is all you need. In: Guyon, I., et al. (eds.) Advances in Neural Information Processing Systems 30: Annual Conference on Neural Information Processing Systems 2017, Long Beach, CA, USA, 4–9 December 2017, pp. 5998–6008 (2017). http://papers.nips.cc/paper/7181-attention-is-all-you-need
15. Wang, W., Bao, F., Gao, G.: Learning morpheme representation for Mongolian named entity recognition. Neural Process. Lett. **50**(3), 2647–2664 (2019)
16. Wang, W., Fam, R., Bao, F., Lepage, Y., Gao, G.: Neural morphological segmentation model for Mongolian. In: 2019 International Joint Conference on Neural Networks (IJCNN), pp. 1–7. IEEE (2019)
17. Wang, X., Chen, R., Jia, Y., Zhou, B.: Short text classification using Wikipedia concept based document representation. In: 2013 International Conference on Information Technology and Applications, pp. 471–474. IEEE (2013)

18. Xiao, L., Wang, G., Zuo, Y.: Research on patent text classification based on word2vec and LSTM. In: 2018 11th International Symposium on Computational Intelligence and Design (ISCID), vol. 1, pp. 71–74. IEEE (2018)
19. Yang, Y., Liu, X.: A re-examination of text categorization methods. In: Proceedings of the 22nd Annual International ACM SIGIR Conference on Research and Development in Information Retrieval, pp. 42–49 (1999)
20. Yao, L., Mao, C., Luo, Y.: Graph convolutional networks for text classification. In: Proceedings of the AAAI Conference on Artificial Intelligence, vol. 33, pp. 7370–7377 (2019)
21. Zhou, J., Wang, J., Liu, G.: Multiple character embeddings for Chinese word segmentation. In: Alva-Manchego, F.E., Choi, E., Khashabi, D. (eds.) Proceedings of the 57th Conference of the Association for Computational Linguistics, ACL 2019, Florence, Italy, 28 July–2 August 2019, vol. 2: Student Research Workshop, pp. 210–216. Association for Computational Linguistics (2019)

Language Resource and Evaluation

The Annotation Scheme of English-Chinese Clause Alignment Corpus

Shili Ge[1], Xiaoping Lin[2], and Rou Song[1,3](✉)

[1] Laboratory of Language and Artificial Intelligence, Guangdong University of Foreign Studies, Guangzhou 510420, China
geshili@gdufs.edu.cn, songrou@126.com
[2] Center for Linguistics and Applied Linguistics, Guangdong University of Foreign Studies, Guangzhou 510420, China
lxpteresa@126.com
[3] College of Information Science, Beijing Language and Culture University, Beijing 100083, China

Abstract. A clause complex consists of clauses, which are connected by component sharing relations and logic-semantic relations. Hence, clause-complex level structural transformations in translation are concerned with the expression adjustment of these two types of relations. In this paper, a formal scheme for tagging structural transformations in English-Chinese translation is designed. The annotation scheme include 3 steps operated on two grammatical levels: parsing an English clause complex into constructs and assembling construct translations on the clause complex level; translating constructs independently on the clause level. The assembling step involves 2 operations: performing operation functions and inserting Chinese words. The corpus annotation shows that it is feasible to divide structural transformations in English-Chinese translation into 2 levels. The corpus, which unfolds formally the operations of clause-complex level structural transformations, would help to improve the end-to-end translation of complicated sentences.

Keywords: Clause-complex · Structural transformations · Corpus annotation · Machine translation

1 Introduction

The grammatical levels of a natural language include morpheme, word, group/phrase, clause, and clause complex. Units of a higher level are made up of units of a lower level. Therefore, the central task for machine translation is language transformations on each grammatical level between languages. So far, there have been many studies on group/phrasal- and clausal-level structures and structural transformations. However, clause-complex level (CC-level) structures and structural transformations are far less discussed.

Halliday [1] describes the structures of English clause complex based on the theory of Systemic-Functional Grammar. Wang [2] carries out an in-depth study on the structures of Chinese complex sentence in comparison with English. Luo [3] points out

M. Sun et al. (Eds.): CCL 2020, LNAI 12522, pp. 287–299, 2020.
https://doi.org/10.1007/978-3-030-63031-7_21

that clauses should be considered as the translation units in English-Chinese translation. These studies are enlightening, but they are limited to theoretical illustrations and discussions. Song and Ge [4] study clause complex for language engineering. They put forward and demonstrate the PTA (Parsing-Translating-Assembling) model for English-Chinese translation on the CC-level, which is only a tentative idea and has not been tested through corpus annotation. Ge and Song [5] clarify the concept of Component Sharing, define clause and clause complex based on this concept, and propose the design of the annotation scheme for English-Chinese Clause Alignment Corpus (ECCA Corpus). Yet, the details of the annotation scheme and specification of the ECCA Corpus still need further study and exploration, especially on the structural transformations between English and Chinese clause complexes and their annotation.

A clause complex consists of clauses, but many clauses are not connected linearly because there are shared components between them. In order to present the alignment of English and Chinese clauses, it is necessary to show how English and Chinese clauses correspond under various component sharing mechanisms. In ECCA Corpus, the correspondence relationship between English and Chinese clauses is shown through the annotation process of CC-level structural transformations, including construct analyzing, construct translating, and construct and component translations assembling. The work of this paper completes the annotation scheme, including defining the operation unit of CC-level structural transformations, i.e. constructs, specifying the content of each annotation step, formalizing assembling operations, and summarizing the operation functions used and the Chinese words inserted.

It is believed that ECCA Corpus is significant for theoretical linguistics and cognitive linguistics by providing samples for comparing CC-level structures and studying structural transformations between English and Chinese. Meanwhile, the corpus is believed to be significant in application. Although machine translation has been greatly improved with data-driven approaches, it still fails to produce satisfying results when it comes across long sentences with complicated structures. This corpus explores the feasibility of and practical ways for mechanical transformations on the CC-level. It is hoped that the knowledge of CC-level structural transformations may help to improve the performance of machine translation in dealing with complicated sentences.

The remainder of this paper is organized as follows: Sect. 2 introduces the objective of annotation, Sect. 3 introduces the annotation scheme, Sects. 4 and 5 present operation functions and inserted Chinese words applied in annotation; Sect. 6 provides relevant statistical results, and Sect. 7 concludes the paper.

2 Clause-Complex Level Structural Transformations

The ECCA Corpus is designed to annotate CC-level structural transformations between English and Chinese. In most linguistic theories, a clause complex is generally regarded as a group of clauses combined together based on logic-semantic relations. This being the case, CC-level structural transformations during translation should involve only reordering of clauses, which are usually organized in different logical ways between languages. However, there is another important transformation that should be noticed, i.e. the transformation of naming-telling structural relations.

Example 1: There are fewer than 100 potential customers for supercomputers priced between $15 million and $30 million -- presumably the Cray-3 price range.
Chinese Translation: 价格在１５００万美元至３０００万美元之间的超级计算机的潜在客户不到１００家，这个区间是克雷３号机大概的价格范围。
Machine Translation: 价格在１５００万美元到３０００万美元之间的超级计算机的潜在客户不到１００家——大概是Cray-3的价格区间。

In Example 1, the English clause complex contains a "modified component & modifying component" structure and a "described component & describing component" structure. As stated in Fang et al. [6], the modifying and describing components are tellings, while the modified and described ones are namings. The two namings are highlighted in grey. The modifying component, which closely follows behind its modified naming, "supercomputers", is marked with a single underline. The describing component, which closely follows its described naming, "between $15 million and $30 million", is marked with a wave underline. It can be seen that the described naming is embedded inside the previous modifying telling component. In the Chinese translation, the translation of the modifying telling component "priced between $15 million and $30 million" is reordered and placed before the translation of its modified naming, "supercomputers". Thus, the translation of the describing component, "– presumably the Cray-3 price range", could not share its described naming as it does in the English text. To deal with the problem, the described naming is reproduced in the Chinese translation as a generalized form "这个区间" and combined with the translation of its describing component into a new clause. However, the machine translation does not reproduce the described naming and thus fails to translate the "described component & describing component" structure correctly. This example shows that the adjustment of naming-telling relationship is no less important than logic-semantic relationship adjustment in CC-level structural transformations.

Previous corpus studies prove that naming-telling structures are prevalent in both Chinese and English clause complexes. Although the two languages share the same types of naming-telling structures, they have different distributions of the structure types [7]. As a result, naming-telling structure adjustment is often necessary in English-Chinese translation. Meanwhile, the two languages arrange clauses in different logical ways, which leads to the other kind of structural transformation.

To sum up, the annotations of CC-level structural transformations are to demonstrate the adjustment of naming-telling structures and logical expressions in English-Chinese translation.

3 Design of the Annotation Scheme

The CC-level structural transformations of Example 1 are illustrated in Fig. 1.

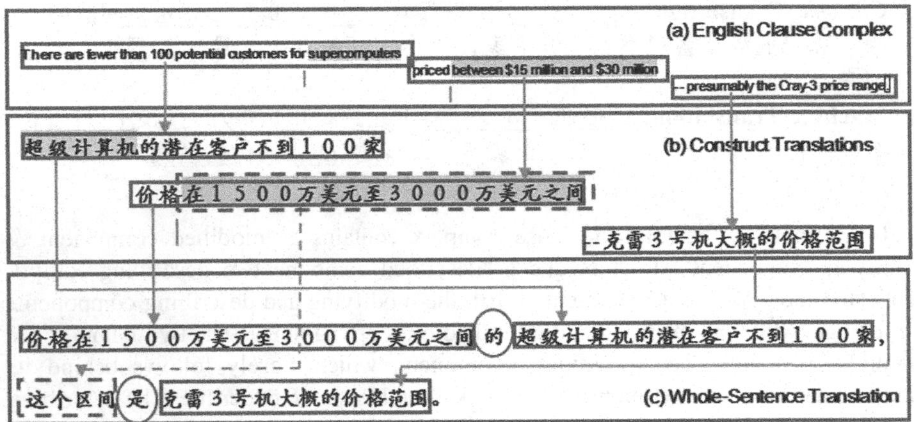

Fig. 1. CC-level structural transformations of Example 1

In Fig. 1-(a), the English clause complex is firstly segmented into three constructs based on naming-telling structural analysis. The grey parts are namings, whose left-boundaries are marked by the symbol "|" below the line. Tellings modifying or describing these namings take up new lines and are indented to the right after their namings. This way of demonstrating the naming-telling relationship is called Newline-Indent Schema.

Each line in Fig. 1-(a) is considered as one construct for making up the English clause complex, and they are translated independently in Fig. 1-(b). Each line of translations in Fig. 1-(b) is called a Construct Translation. Construct Translations are also displayed in the Newline-Indent Schema, with the translations of tellings indented to the right side after the translations of their namings. The arrows between Fig. 1-(a) and 1-(b) start from each English construct and point to their Chinese counterparts. Figure 1-(c) shows the whole-sentence translation. The solid line arrows between Fig. 1-(b) and 1-(c) start from each construct translation, and point to their new positions in the whole-sentence translation. The dash line arrow starts from the translation of a naming and points to its generalized form. The circles in Fig. 1-(c) mark the insertion of the particle "的" and the linking verb "是".

The graphic demonstration in Fig. 1 clearly displays how the English clause complex is transformed step by step into a Chinese one. However, the demonstration is quite complicated, hard to be annotated and not convenient for statistical analysis. Hence, a more formal annotation scheme for annotating structural transformations is designed.

The formal annotation scheme follows the 3 steps in the graphic demonstration: (1) segment English clause complexes into constructs and display them in Newline-Indent Schema; (2) translate independently each construct into Chinese; (3) rearrange construct translations for a whole-sentence translation.

The structural transformations are to be annotated at the end of each line of the whole-sentence translation. The parts that make up the whole-sentence translation are encoded as numbers, and the operations implemented on these parts are tagged as operation functions. In this way, structural transformations could be annotated formally. The following is a detailed illustration of the designs.

Whole-Sentence Translation of Example 1:
价格在１５００万美元至３０００万美元之间的超级计算机的潜在客户不到１００家，//2+的+1
这个区间是克雷 3 号机大概的价格范围。**//sum(2.2)+是+delt(3)**

As shown above, structural transformations are tagged after the symbol "//" at the end of each line. The numbers represent the parts making up the whole-sentence translation. For example, the number 2 of "2+的+1" represent the second line of construct translations, namely "价格在1500万美元至3000万美元之间". The number 2.2 of "**sum(2.2)+是+delt(3)**" represent the second section of the second line of construct translations, namely "在1500万美元至3000万美元之间". In the annotation scheme, the translations of namings are usually processed as a single unit. When the translation of a naming is positioned within a construct translation, the construct translation is segmented by the translation of this naming into several parts, which includes the naming translation, the parts before and/or after the naming translation. These segments are named as component translations. The component translations on the nth line are encoded from left to right as n.1, n.2, and n.3 etc. In this example, the second line of construct translation contains the translation of a naming at its end, and thus it is divided into two components. The component before the naming translation is encoded as 2.1, while the naming translation is encoded as 2.2. From this example, it can be seen that the parts making up a whole-sentence translation include construct translations and component translations. These two types of constituents in translations are the basic units to be dealt with by operation functions, and thus they are called operation units in this paper.

As for operation functions, they are used to mark the operations implemented on operation units. The symbol "+" means linking two operation units. The function "sum(2.2)" means turning the encoded component 2.2, namely "在1500万美元至3000万美元之间",, into a more generalized expression "这个区间". The function "delt(3)" means deleting the dash in the translation of the encoded construct 3, namely "--克雷3号机大概的价格范围". The designing of operation functions will be discussed in detail in Sect. 4.

Additionally, it is noted that the translation of every construct in the second step is independent of its context. Certainly, the disambiguation of a certain word still need reference to its context, but it is not allowed to add extra words, delete words or change the structures based on the context.

4 Operation Functions

There are two types of operations for CC-level structural transformations: (1) processing and assembling the operation units, and (2) inserting Chinese words. The first type of operation is annotated as operation functions, which will be discussed in this section. The second type of operation will be discussed in Sect. 5.

Operation functions are written in the format of FunctionName(x) or FunctionName(x,y), in which FunctionName specifies the operation to be implemented, while x and y specify the objects to be processed, which are all called operation units.

Twenty operation functions are designed, which involve 6 types of operations: link, reorder, add, delete, rewrite, and substitute. The 20 operation functions are listed in Table 1.

Table 1. Operation functions

Operation types	Operation functions
Link	concatenate(x,y) (i.e. x + y)
Reorder	demonstrated with the codes of operation units
Add	corcj(x), corcj2(x), prd(x)
Delete	ignore(x) (i.e. *x), delcj(x), delcj2(x), delpn(x), delt(x)
Rewrite	det(x), ndet(x), sum(x), pron(x), rel(x), paren(x), n2v(x)
Substitute	rpw(x,y), r2n(x,y), n2r(x,y)

Of all these functions, link and reorder are common operations in almost all processed whole-sentence translations. The usage of these two functions is shown above in Example 1. Other functions are divided into two types based on their adjustments to clause complex structures. Some of the two classes are discussed with examples in the following subsections. Due to limited space, the functions not discussed in this paper can be referred to Song et al. [8].

4.1 Operation Functions for Transforming Naming-Telling Structures

Due to different distributions of naming-telling structural types, it is often necessary to transform naming-telling structures during English-Chinese translation. Generally, there are 3 ways to rearrange English tellings in Chinese translations: (1) inserting the telling translation as a modifier on the left of its naming translation, (2) keeping the telling translation as a statement or a description on the right of its naming translation, (3) reproducing the naming and rendering it another way before linking it with the telling translation. Of these 3 ways, the previous two requires only the link and reorder operations. When it comes to the third way, extra processing is needed, namely to reproduce the naming and render it in certain forms. This is because in a clause complex, a naming, if referred to more than once, should take different forms for its respective occurrence. To be more specific, a naming usually appear at first in its full name or its indefinite form, and then appear in its definite form, as a pronoun, or as a more generalized form. The operation functions det(x), ndet(x), pron(x) and sum(x) are specially designed for rewriting a naming. Table 2 presents definitions of operation functions used to transform naming-telling structures.

Table 2. Operation functions for transforming naming-telling structures

Operation types	Operation functions	Definition
Rewrite	det(x)	change x into its definite form
Rewrite	ndet(x)	change x into its indefinite form
Rewrite	pron(x)	change x into a corresponding pronoun
Rewrite	sum(x)	change x into a more generalized term
Rewrite	rel(x)	concretize x based on the current context
Delete	ignore(x)	delete the relative pronoun/adverb in x
Substitute	rpw(x,y)	replace the relative pronoun/adverb in x with y

The usage of sum(x) has been illustrated in Example 1. The usage of ignore(x) and rpw(x,y) will be discussed in the following.

Since attributive clauses do not have clear semantic meanings by themselves, they need special treatment in annotation. In an attributive clause, the relative pronoun is only a formal substitute for the antecedent, and it is meaningless by itself. As a result, attributive clauses cannot be translated independent of context theoretically. To handle the problem, it is specified that relative pronouns in capitalized forms should be used to occupy the positions where the translations of antecedents should have been in construct translations.

In most cases, capitalized relative pronouns occupy the positions of a subject at the beginning of construct translations. Hence, the ignore(x) function is used to delete the capitalized relative pronouns before construct translations are linked with the translations of their namings.

Sometimes, capitalized relative pronouns occupy positions in the middle of construct translations. In this case, the function rpw(x,y) should be used to replace relative pronouns with the translations of their antecedents. Such substitutions are operable since capitalized relative pronouns are identifiable with their special forms. Example 2 shows the usage of this function.

Example 2: The Company has proposed an internal reorganization plan in Chapter 11 bankruptcy proceedings, under which it would remain an independent company.

(1) Newline-Indent Schema of English Clause Complex:
The … an internal reorganization plan in Chapter 11 bankruptcy proceedings,
| under which it would …company.

(2) Construct Translations:
该……提出了一个内部重组计划，
| 根据WHICH 它将仍为一个独立公司。

(3) Whole-Sentence Translation:
该公司已在第 1 1 章破产程序提出了一个内部重组计划，//1
根据该计划它将仍为一个独立公司。//rpw(2,sum(1.2))

The second line in Example 2-(1) is an attributive clause, with "an internal reorganization plan" as its antecedent. In this example, the antecedent is a naming while the attributive clause is its telling. In Example 2-(3), the result of "sum(1.2)" is a generalized term for "一个内部重组计划", namely "该计划"(this plan). The function "rpw(2,sum(1.2))" means replacing "WHICH" in the second line of construct translations with "该计划".

4.2 Operation Functions for Transforming Logical Expressions

English and Chinese clause complexes differ in logical expressions in the following 3 aspects: (1) clausal order, (2) the use of logical conjunctions, and (3) naming sharing of logically-related clauses. These differences may give rise to different translation problems, and thus different functions are designed to deal with them (Table 3).

Table 3. Operation functions for transforming logical expressions

Operation types	Operation function	Definition
Substitute	r2n(x,y)	replace the pronoun in x with the corresponding noun in y
Substitute	n2r(x,y)	replace the noun in x with the corresponding pronoun in y
Add	corcj(x)	add the matched conjunction for the first one in x
Add	corcj2(x)	add the matched conjunction for the second one in x
Delete	delcj(x)	delete the first conjunction in x
Delete	delcj2(x)	delete the second conjunction in x
Delete	delpn(x)	delete the relevant pronoun in x

Firstly, English and Chinese clause complexes have different clausal orders. The differences lie in two aspects: (1) In English, main clauses are usually placed before subordinate clauses, while it is the opposite in Chinese. (2) In English, quotation verbs are placed after or between quotations, while in Chinese, quotation verbs are usually placed before quotations. In the annotation scheme, the operation of reorder is demonstrated by the line numbers referring to clause translations. Sometimes, the reorder of clauses is accompanied with the necessity of changing referential order. The two functions r2n(x,y) and n2r(x,y) are specially designed for dealing with this situation.

Example 3: Yields may blip up again before they blip down because of recent rises in short-term interest rates.

(1) Newline-Indent Schema of English Clause Complex:
Yields may blip up again
before they blip down
because of recent rises in short-term interest.

(2) Construct Translations:
收益率可能会再次上升
它们在下降之前
因为最近短期利率上升。

(3) Whole-Sentence Translation:
因为最近短期利率上升，//3
收益率在下降之前，//r2n(2,1)
它们可能会再次上升。//n2r(1,2)

In Example 3, the English clausal orders should be adjusted in the Chinese translation. The rearrangement of clausal orders is displayed in Fig. 2.

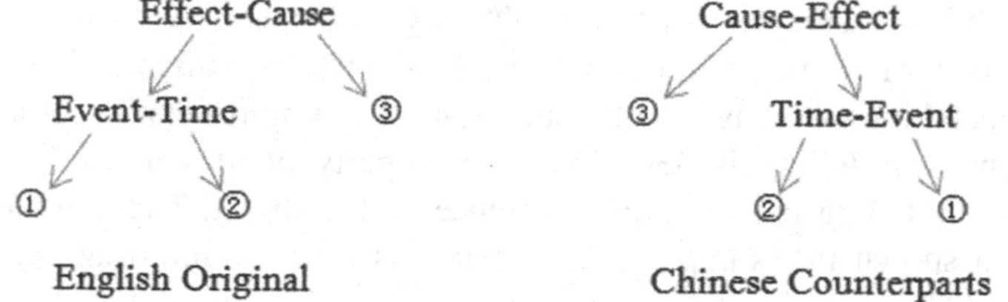

Fig. 2. Logical orders of clauses in Example 3 and its Chinese counterparts

The exchange of clausal orders is demonstrated with the exchange of orders of line numbers. As the first line shown in Example 3-(3), the number "3" at its end means that this line comes from the third line of construct translations. Meanwhile, the interclasual order between the first and second line of construct translations has also been changed in Example 3-(3). The first line of construct translations with the noun "收益率" is placed after the second line with the pronoun "它们". However, in general terms, the line with a pronoun is supposed to appear after the line with the noun it refers to. Hence, it is necessary to exchange the noun and pronoun concerned in the two lines. The function r2n(2,1) means replacing the pronoun "它们" in second line of construct translations with the corresponding noun "收益率" in the first line. The function n2r (1,2) means replacing the noun "收益率" in first line of construct translations with the corresponding noun "它们" in the second line.

However, the whole-sentence translation above is not optimal. A better whole-sentence translation is shown as the following.

因为最近短期利率上升，//3
所以收益率在下降之前可能会再次上升。//corcj(3)+r2n(2,1)+delpn(n2r(1,2))

In this new whole-sentence translation, line 2 and line 3 in the original whole-sentence translation are combined into one by deleting the pronoun "它们". The deletion of the pronoun is tagged as delpn(n2r(1,2)), which means deleting the pronoun in the result of n2r(1,2). With the operation of this function, the result of n2r(1,2), namely "它们可能会再次上升", is turned into "可能会再次上升". Meanwhile, the conjunction "所以" is added, matching that of the third line of construct translations. This addition of a conjunction is tagged as corcj(3).

Example 3 shows relevant functions for dealing with English-Chinese differences on clausal orders and on the use of logical conjunctions.

5 Inserted Chinese Words

The inserted Chinese words are function words such as linking verbs, particles, conjunctions and prepositions. The Chinese words can be classified into 2 types based on their functions: (1) words indicative of argumentative relations between operation units, and (2) words indicative of logical relations between operation units (Table 4).

Of words indicative of argumentative relations between operation units, the three most frequently used Chinese words are "的", "是" and "即". "的" is inserted between an attribute and its modified noun phrase. "是" and "即" are often inserted between a naming and its telling which is usually an appositive in English clause complexes. "是" is applied when the telling is describing the property of its naming, while "即" is applied when the telling has the same reference as its naming. The usage of "的" and "是" has been shown in Example 1. Example 4 is to show the usage of the logical conjunction "并".

Table 4. Inserted Chinese words

Words Indicative of Argumentative Relations Between Operation Units	
的	connect a modifier with its modified noun
是	connect a naming with its descriptive noun phrase
即	connect a naming with its parallel noun phrase
有	introduce a naming, functioning like "there be" in English
在	connect a naming with its telling as a location
将	inserted before a naming whose telling is about the processing or recognition of the naming
在…中	a frame in which a naming used as a location is embedded
Words Indicative of Logical Relations Between Operation Units	
使	logical component inserted before a telling which is the logical result of a behavior or an event
而	logical component inserted between two verb phrases which are in the cause-effect relations
就	logical component inserted before a telling which is the logical result of a behavior or an event
并	logical component suggesting a further movement

Example 4: Mrs. Yeargin concedes that she went over the questions in the earlier class, adding: "I wanted to help all" students.

(1) Newline-Indent Schema of English Clause Complex:

Mrs. Yeargin concedes
            that she went over the questions in the earlier class,
      adding:
            "I wanted to help all" students.

(2) Construct Translations:

叶尔金太太承认
            她在上一届的班里复习了那些问题，
      补充说：
            "我想帮助所有"学生。

(3) Whole-Sentence Translation:

叶尔金太太承认她在上一届的班里复习了那些问题，//1+2
      并补充说：//并+3
            "我想帮助所有"学生。//4

There are two clauses in this example. One clause is constituted by lines 1 and 2 in Example 4-(1), and the other is constituted by the naming in line 1, i.e. "Mrs. Yeargin", and the telling, lines 3 and 4. Semantically speaking, the second clause is the continuation of the first one, involving the action to be taken after that of the first clause. In the English clause complex, the logical relation is presented by using an infinite verb for the action in the second clause, namely "adding", to lower the grammatical hierarchical level of the clause. However, in Chinese, there is no such grammatical device as changing verb forms. Therefore, the logical conjunction "并" is added for connecting the two clauses logically.

6 Statistical Data

So far, we have annotated 2108 clause complexes on 136 documents from English Penn Treebank. Of the annotated clause complexes, 336 contain only one clause. Of the clause complexes containing more than one clause, 532 do not involved CC-level structural transformations. Therefore, only a total of 1240 clause complexes are annotated with relevant functions and Chinese words, accounting for 58.82% of the 2108 clause complexes.

The frequency of each operation function in ECCA Corpus is shown in Table 5. The number of each inserted Chinese word is also counted. The most frequently used words, "的", "是" and "即", appear for 486, 112 and 22 times, respectively. Other inserted Chinese words are used for less than 5 times.

Table 5. Frequency of operation functions in ECCA Corpus

Function	*x	pron(x)	sum(x)	det(x)	delt(x)	delpn(x)
Freq.	361	136	103	56	32	23
Function	rpw(x,y)	corcj(x)	r2n(x,y)	paren(x)	n2r(x,y)	delcj(x)
Freq.	20	20	16	14	11	9
Function	n2v(x)	prd(x)	rel(x)	ndet(x)	corcj2(x)	delcj2(x)
Freq.	8	6	6	4	1	1

7 Conclusions and Discussions

Component sharing relations and logic-semantic relations are organized differently in English and Chinese clause complexes. As a result, during English-Chinese translation, it is necessary to adjust the expressions of these two relations with some structural transformations on the clause complex level. This paper divides English-Chinese clause complex translation into two grammatical levels. On the clause complex level, an English clause complex is parsed into constructs, and the translations of these constructs are assembled into a whole-sentence translation. On the clause level, each construct is translated independently. The two-level translation mechanism, including operation functions and inserted Chinese words used in the assembling step, has been designed formally and proved feasible with corpus manual annotation.

By designing the two-level translation mechanism, this paper follows a common strategy for AI problem solving, namely to decompose a complicated task into sequential simple tasks. It is believed that this mechanism could reduce the demanded data scale and calculation complexity for machine-learning-based machine translation, since the task of translating a sentence is decomposed into simple tasks of translating and assembling shorter constructs. Meanwhile, although the mechanism cannot produce perfect results in some cases, it is an explainable translation process and thus is worth further exploring.

The work present in this paper is only initial. In the future, efforts will be made to enlarge the corpus size, improve the quality of annotated translations, provide multiple translation alternatives, design algorithms for realizing operation functions and discover linguistic knowledge based on the ECCA Corpus.

Acknowledgements. This research is supported by National Natural Science Foundation of China (61672175).

References

1. Halliday, M.A., Matthiessen, C.M.I.M.: An Introduction to Functional Grammar, 3rd edn. Edward Arnold, London (2004)
2. Wang, L.: The Complete Works of Wang Li Volume 8: Chinese Grammar Theory. Zhonghua Book Company, Beijing (2012)
3. Luo, X.: Unit of transfer in translation. Foreign Lang. Teach. Res. **4**, 32–37 (1992)
4. Song, R., Ge, S.: English-Chinese translation unit and translation model for discourse-based machine translation. J. Chin. Inf. Process. **29**(5), 125–136 (2015)
5. Ge, S., Song, R.: English-Chinese clause alignment corpus tagging system based on corpus annotation. J. Chin. Inf. Process. **34**(6), 27–35 (2020)
6. Fang, F., Ge, S., Song, R.: Error Analysis of English-Chinese Machine Translation. In: Sun, M., Huang, X., Lin, H., Liu, Z., Liu, Y. (eds.) CCL/NLP-NABD -2016. LNCS (LNAI), vol. 10035, pp. 35–49. Springer, Cham (2016). https://doi.org/10.1007/978-3-319-47674-2_4
7. Ge, S., Song, R.: The naming sharing structure and its cognitive meaning in Chinese and English. In: Xiong, D., Duh, K., Agirre, E., Aranberri, N., Wang, H. (eds.) Proceedings of the 2nd Workshop on Semantics-Driven Machine Translation (SedMT 2016), pp. 13–21. Association for Computational Linguistics (ACL), Stroudsburg (2016)
8. Song, R., Ge, S., Chen, X., Lin, X.: English-Chinese clause alignment corpus annotation guidelines. Technical Report of Collaborative Innovation Center for Language Research & Service of Guangdong University of Foreign Studies, Guangzhou (2020)

Categorizing Offensive Language in Social Networks: A Chinese Corpus, Systems and an Explanation Tool

Xiangru Tang[1,2,3], Xianjun Shen[1,2,3](✉), Yujie Wang[1,2,3], and Yujuan Yang[1,2,3]

[1] School of Computer, Central China Normal University, Wuhan, China
xjshen@mails.ccnu.edu

[2] National Language Resources Monitoring and Research Center for Network Media, Beijing, China

[3] Hubei Provincial Key Laboratory of Artificial Intelligence and Smart Learning, Wuhan, China

Abstract. Recently, more and more data have been generated in the online world, filled with offensive language such as threats, swear words or straightforward insults. It is disgraceful for a progressive society, and then the question arises on how language resources and technologies can cope with this challenge. However, previous work only analyzes the problem as a whole but fails to detect particular types of offensive content in a more fine-grained way, mainly because of the lack of annotated data. In this work, we present a densely annotated data-set COLA (**C**ategorizing **O**ffensive **LA**nguage), consists of fine-grained insulting language, anti-social language and illegal language. We study different strategies for automatically identifying offensive language on COLA data. Further, we design a capsule system with hierarchical attention to aggregate and fully utilize information, which obtains a state-of-the-art result. Results from experiments prove that our hierarchical attention capsule network (HACN) performs significantly better than existing methods in offensive classification with the precision of 94.37% and recall of 95.28%. We also explain what our model has learned with an explanation tool called Integrated Gradients. Meanwhile, our system's processing speed can handle each sentence in 10 ms, suggesting the potential for efficient deployment in real situations.

Keywords: Offensive language · Capsule network · Text classification

1 Introduction

In modern society, the occupation of offensive language on the online world, such as social media, is becoming a paramount concern. Offensive language differs considerably, ranging from pure abuse to more rigorous types of writing. Thus, offensive language is hard to be automatically identified. However, it's essential to track this; for example, the appearance of offensive language on social media

M. Sun et al. (Eds.): CCL 2020, LNAI 12522, pp. 300–315, 2020.
https://doi.org/10.1007/978-3-030-63031-7_22

is related to hate crimes in a real social situation [17]. Moreover, it can be pretty troublesome to distinguish fine-grained offensive language because few general definitions exist [4].

Recently, researchers have proposed some guidelines to identify the type and the attributes of offensive language [25]. However, the online world's offensive language is a general category containing specific examples of profanity or insult. In our work, "Offensive language" in the online world is defined in more detail and fine-grained. And to the best of our knowledge, though offensive language identification being a burgeoning field, there is no data-set yet for Chinese.

"Offensive" is pretty much something people identify as against morals, very inappropriate, or disrespectful. However, "offensive" is a broad general term and does not define the precise extent or the limits of its application. Thus, we classify the term "offensive" into three categories: "insulting," "antisocial" and "illegal" through stepwise refinement. "Insulting" is something rude, insensitive and/or offensive, directed at another person or group of people. This emphasizes that the content is a direct attack against specific others. "Antisocial" is harmful to organized society, or the language describes a behavior deviating from the social norm for long. "Illegal" language means it violates the language policy. Where the language policy refers to the government through legislation or policies to formally decide how languages are used. However, the language policies of each country are not completely consistent.

Thus, two questions arise a) how LRs can cope with the large numbers of offensive language in the online world, and b) can LT provide means to process and respond promptly to such language data streamed in a huge amount at high speed? Firstly, there is no existing data-set for the Chinese language to provide for correctives of hate speeches, cyberbullying, or fake news. Then, current methods can not produce highly precise results for detecting offensive content and behaviors. Also, They used inflexible proprietary APIs, which is hard to reproduce. On the other hand, there is a real need for methods to detect and deal with online words quickly because of the enormous amount of data created every day.

In this context, we present a sizeable Chinese classification corpus of offensive language called COLA. Then, we employ a deep dilated capsule network to extract hierarchical structure. We further design hierarchical attention to aggregate and fully utilize information within a hierarchical representation. Correctly, each sentence is embedded into capsules and incrementally distilled into task-relevant categories during the hierarchical attention process. What is more, we present an explanation tool, which proves that our work for the Chinese language seizes the pattern of offensive language in some points, and almost correctly identifies different varieties of offensive language, like hate speech and cyberbullying.

In summary, our work aims at answering the two questions a) how LRs can cope with a huge amount of offensive language in the online world, and b) can LT provide means to process such language data at high speed? The major contributions are highlighted as follows:

- We describe COLA, the first Chinese offensive language classification dataset. COLA is designed to study how language resources and technologies can cope with this offensive language challenge in the online world. It is now publicly available.
- We propose a hierarchical attention capsule network (HACN), where the hierarchical attention mechanism is introduced to model the hierarchical structure. It is inspired by capsule, with modifications to handle the words explicitly (Fig. 1).
- We show that our HACN model surpasses state-of-the-art methods for classification on COLA. Furthermore, our presented explanation tool clearly explains what our model has learned.

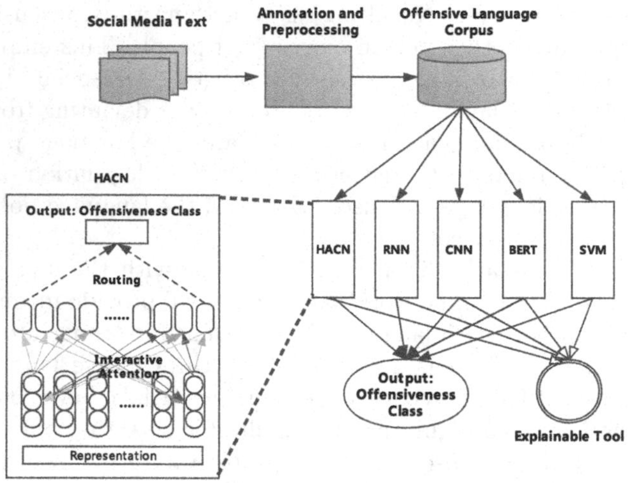

Fig. 1. Data processing pipeline system and the architecture of hierarchical attention capsule network (HACN).

2 Related Work

2.1 Corpus

Some previous works have discussed how to identify the offensive language, but in that literature, the offensive language is ranging from aggression to cyberbullying, toxic comments, and hate speech. In the following, we explain each of these open public challenges briefly.

SemEval-2019 Task: In Task 6 of SemEval 2019, they propose three separate sub-tasks. A sub-task is Offensive Language Selection, the other is Categorization of Offensive Language, and the last is Offensive Language Target Recognition. SemEval-2019 Task 6 is called OffensEval, and the collection methods of their data are explained in [26]. Additionally, it collected more than 14100 posts of sentences.

Aggression Identification (TRAC): The TRAC study [14] provided players with a data set containing a training set and a validation set. They are composed of 15,000 Facebook posts and comments annotated in English and Hindi. For the test set, two different sets are used, one from Facebook and the other from Twitter. It aims at distinguishing three types of data: non-aggressive, covert aggressive, and over-aggressive.

Hate Speech Recognition: In [3,5,15], they present a Abusive language selection task. Specially, [4] provided the hate speech recognition data set, which contains more than 24000 English tweets marked as non-offensive, hate speech, and profanity.

Offensive Language: The data-set provided by GermEval [23] focused on offensive language recognition in German tweets. The study showed a data set of more than 8,500 tagged tweets. This data set is used to perform binary classification task of distinguishing between offensive and non-offensive information. Besides, the second task divided offending tweets into three categories: profanity, insult, and abuse. While similar to our work, there are three important differences: (i) we have a third level in our hierarchy, (ii) we use different labels in the second level, and (iii) we focus on Chinese.

Toxic Comments (Kaggle): Kaggle holds a Toxic Comment Classification Challenge as an open dataset. The dataset in this competition was extracted from the comments of Wikipedia, and it was formed in six categories: toxicant, severe toxic, identity hate, threat, insult, obscene. Moreover, the data set is also employed outside of the competition [7], treated as an external training resource for the TRAC, as mentioned above [6].

However, each of these tasks tackles a particular challenge of detecting offensive language. Thus, we present a new dataset, hoping it could become a valuable resource for improving offensive language categorizing.

2.2 Classification

Traditional classification methods are designed with rich features and syntactic structures to achieve the classification task [12]. But, these feature-based methods are labor-intensive, and the performance depends largely on the quality of the features. Recently, deep learning methods are becoming popular for aspect-level sentiment classification. Recurrent Neural Networks (RNNs) are the most commonly used technique for this task [21]. The attention mechanism is further introduced to model the target-context association [22]. Recently, CNN-based models have shown the strengths inefficiency to tackle the aspect-level sentiment classification [11,16,24]. However, all the previous methods utilize static pooling operation or attention mechanism to locate the keywords, which fails to handle the overlapped features. We introduce vector-based feature representation and feature clustering to address this.

Capsule network was proposed to improve the representational limitations of CNN and RNN by extracting features in the form of vectors. The technique was firstly proposed in [10]. But is mainly devised for the image processing domain. Introducing capsules allows us to utilize a routing mechanism instead of pooling operation to generate high-level features, which is a more efficient way for features encoding. The routing module is able to cluster features in an iterative way, which achieved impressive performance recognizing highly overlapped digits. Several types of capsule networks have been proposed for natural language processing. [27] investigated capsule networks for text classification. They also found that capsule networks exhibit significant improvement when transferring single-label to multi-label text classification. However, interactive word-level attention is not considered in these typical capsule routing methods.

3 Data Collection

In this section, we describe the data set and how we annotate data set (Table 1).

3.1 Overview

Table 1. Statistics of the four classes in COLA data. Number of sentences in train set, test set, valid set.

Data	Train	Test	Valid	Total
Neutral	5357	1700	1546	8603
Insulting	5075	1660	1493	8228
Antisocia	841	303	218	1362
Illegal	327	96	91	514
Total	11600	3759	3348	18707

We create a large-scale data-set that annotates offensive texts in Chinese. The texts are crawled from Youtube and Weibo: 18.7k comments in total. Three annotators categorised these texts in four classes: neutral, insulting, antisocial, and illegal. We build a Chinese dataset from social media that people can communicate on Internet, such as Sina Weibo[1] and YouTube comments. Our released COLA contains user-generated comments from different social media platforms, and as we know, it is the first of its kind. And, the dataset is marked as capture different types of offensive language. We propose four automatic classification systems, each designed to work for the Chinese language.

[1] https://en.wikipedia.org/wiki/Sina_Weibo.

3.2 Data Acquisition

With more than 1000 comments and more than 10000 views as the thresholds, we selected 20 popular Chinese videos from YouTube. Furthermore, from the comments below the video are crawled through Google YouTube V3 API, which is offered by Google for researchers to collect comments. And a total of 20000 comments were received. We store the 20000 comments, and then we clean the data. We first convert the traditional Chinese character in the data set to simplified Chinese characters, and then filter out the useless data with messy codes and HTML tags.

There are some technologies we employ to crawl the data. Firstly, we retrieve 81718 Chinese sentences from Weibo and YouTube reviews in JSON format, and contain information such as timestamp, URL, text, user, re-tweets, replies, full name, id, and likes. Extensive processing is carried out to remove all the noisy sentences. We apply the following pre-processing steps: the documents are tokenized using NLTK, the URLs and mentioned users are removed, and all letters are converted to lower-case. As a result, a dataset of 18,707 offensive language sentences is created. Nevertheless, social media companies all have some methods to prevent crawlers. These methods can be divided into three categories: analyzing the headers of web page requests, monitoring the behavior of users visiting the website, and adjusting the directory and data loading methods. Corresponding to that, we adopt three approaches to crawl the data. For the first one, we could directly add HEADERS and REFERER to the code to bypass the check. And the same IP visits the same page multiple times in a short period, or the same account performs the same operation multiple times in a short time may cause the second one situation. For this situation, we can use the IP proxy to resolve it. We can use a browser to analyze the requests for the last situation. If we can obtain the AJAX request, then we can use the above two methods to resolve and obtain the corresponding data. However, if we cannot get AJAX requests, we can call the selenium + phantomjs framework and call its browser kernel to simulate human operations and JS scripts that trigger the page.

3.3 Annotation

We construct the data-set which comes from the hot issues of YouTube comments and Weibo. And web crawler gets our the data we needed. The data is annotated by three volunteers. After analyzing all the data, more than 18,707 sentences are selected. Then, we remove invalid tokens in the text, like HTML tags and emoticons, and treat the text as the preliminary data for hand-operated annotation. After that, the vocabulary was divided into three categories: insulting language, antisocial language, illegal language. Due to the special combination of sensitive words, the standard of language structure is pretty vague. We note that different people may have different understanding of the text during the process of annotation. It means the boundary of the same word may be different. Thus, three people are asked to annotate the same text to ensure accuracy. We should note that inter-annotator agreement and intra-annotator agreement have been considered for the coherence of annotations While annotating.

4 Proposed Methods

In this part, we describe the categorizing task to be performed, how we perform the task and the excellent methods are proposed for especially this task.

4.1 Task

The recognition and classification of offensive language in the online world can be realized as a multiple classification task. In this section, we describe several proposed neural networks in details. The aim of aspect-level sentiment classification is to predict the class y of a sentence. In our task, the $y \in \{Neutral, Insulting, Antisocial, Illegal\}$.

4.2 Baseline Systems

Several baseline models are evaluated in Table 2.

SVM: For training our SVM classifier, scikit-learn[2] machine learning in Python library is used for benchmarking. During our experiments, we carry out 10-fold cross validation. We select the Linear SVM formulation, known as C-SVC and the value of the C parameter is 1.0.

RNN: RNN is the high-efficiency method to solve classification problem in NLP tasks. In this paper, we adopt GRU, which has great superiority compared to LSTM and basic RNN. In the final Multi Layer Perceptron layer, 128 neurons are used for classification. And Sigmoid activation function is applied to the final layer.

CNN: We adopt word-level CNN model which has 1D convolution layer with 150 filters and kernel size 6, dropout 0.2, cross entropy loss function and four dense layers with ReLU, tanh, sigmoid and softmax activation respectively.

BERT: BERT has displayed its great advantage of text representation in many NLP tasks. We fine-tune the task-specific components, such as a softmax classifier with BERT or deem BERT model as a feature extractor. First of all, we pack the input features as $H_0 = e_1, \cdots, e_T$, where $e_t (t \in [1, T])$ is the group of the token embedding, position embedding and segment embedding corresponding to the input token x_t. Then the L transformer layers is introduced to refine the characteristics of the token layer by layer. Specifically, the representations $H^1 = h_1^l, \cdots, h_T^l$ at the l-th ($l \in [1, L]$) layer are calculated below:

$$H^l = \boldsymbol{Transformerl}(H^{l-1}) \tag{1}$$

We treat H^L as the contextualized representations of the input tokens.And use them to execute the downstream task's predictions.

[2] https://scikit-learn.org/stable/.

4.3 Challenge

However, there are still several limitations of the current approaches for offensive language categorizing. Firstly, little attention has been paid to the imbalance of different classes, which are essential and challenging because in categorizing tasks, it will be hard to capture the critical pattern of a specific class without sufficient data. Since the COLA data-set is unbalanced, the neural network may not have enough training examples of "illegal language" to learn. Consequently, it cannot catch the feature and structure of the "illegal language".

Second, existing research on offensive language detection cannot accurately detect offensive content because one sentence expresses multiple polarities, resulting in overlapped feature representation. The highly over-lapped features will confuse the classifier seriously, and the three types of specific offensive language do not have quite a considerable distinction. However, most existing methods only keep the most potent feature by max-pooling operation or utilize attention mechanisms to find the keywords, which fails to distinguish the over-lapped features.

Third, the dissemination and use of online platforms have grown significantly in every minute. Thus, the speed of the system is what we aspire to enhance, especially in this task. The systems are required to detect quickly and deal with this type of content in a short time. In this way, we should consider both accuracy and speed as the evaluation metric in a real application.

4.4 Hierarchical Attention Capsule Network

An original capsule network is a group of neurons obtained from the output of the convolutional operation performed on word representation h_n^a and h_n^c. So, the output of the capsule is a vector representing different properties of the same objective. The routing method [10] is employed in our model, and except for the high-dimensional output M, there is one more activation probability in our capsule.

We have already decided on the outputs of all the capsules C_L in the first capsule layer, and we now want to decide which capsules C_{L+1} to active in the layer above and how to assign each active low-level capsule to one active higher-level capsule. The vector-based features get clustered in the high-level capsules where the outputs of high-level capsules play the role of Gaussians, and the output vectors of low-level capsules play the role of the data points. To establish a semantic relationship model between aspect terms and context, we further devise an interactive attention-based routing mechanism.

Firstly, every primary capsule i is transformed by W_{ij} to cast a vote $V_{ij} = M_i W_{ij}$ for the output of high-level capsule j. Moreover, we can get the mean μ_j of the votes from the input capsules, and the variance σ_j about that mean for each dimension h:

$$\mu_j^h = \frac{\sum_i R_{ij} V_{ij}}{\sum_i R_{ij}} \tag{2}$$

$$\sigma_j^{h^2} = \frac{\sum_i R_{ij}(V_{ij}^h - \mu_j^h)^2}{\sum_i R_{ij}} \tag{3}$$

Then, we can calculate the activation probability of capsule j by:

$$c_j^h = \left(\beta_u + log(\sigma_j^h)\right) \sum_i R_{ij} \tag{4}$$

$$a_j = \boldsymbol{sigmoid}(\lambda(\beta_\alpha - \sum_h c_j^h)) \tag{5}$$

In there, μ_j^h is the $h^t h$ component of the capsule $j$âs vectorized output M_j, and β_uã β_α are trainable parameters.

Then, for the part of capsule routing procedure, we propose a hierarchical attention to capture the hierarchical representation. In particular, the scaled dot-product attention is used to map a set of key-value pairs and the query to a weight on the word-level token. The queries are the averaged representation of the word representation h_c are transformed to dimension d_k by trainable parameters:

$$\alpha_n = \frac{\boldsymbol{exp}\frac{k_n^c \times q_a}{\sqrt{d_k}}(q_a, k_n^c)}{\sum_{n=1}^{N} \boldsymbol{exp}\frac{k_n^c \times q_a}{\sqrt{d_k}}(q_a, k_n^c)} \tag{6}$$

We use a spread margin loss, L_k for each top-level capsule k to directly maximize the gap between the activation of the target class. Overall, our loss function L is the sum of the losses of all mentioned capsules:

$$L = \sum_{k \neq t} (\boldsymbol{max}(0, m - (a_t - a_k))) \tag{7}$$

5 Explanation Tool

The explanation for an algorithm is importance in some fields, such as clinical and financial decisions, because its results affect people directly. The European Union brings the *right to explanation* regulation [8] into force in 2018, which is a right to be given an explanation for an output of the algorithm. For example, a person who applies for a loan and is denied may ask for an explanation. However, machine learning algorithm is data driven, even the algorithm designers have no idea how it works, especially for the deep neural networks with thousands of parameters. Nowadays, a large amount of explanation methods [2,9,13,18,19] have been proposed to reveal the behavior of deep neural networks. Post hoc methods especially the attribution methods, is a big branch of explanation methods, which assign the output score to the contributions of input features.

$$\begin{aligned} \mathrm{IG}(x; F)_i = & \frac{x_i - x_i'}{m} \\ & \times \sum_{k=1}^{m} \frac{\partial F(x' + k/m \times (x_i - x_i'))}{\partial x_i}. \end{aligned} \tag{8}$$

We utilize *iNNvestigate* [1] a toolbox for explanation methods to evaluate on different tasks. Especially, we employ Integrated Gradients [20] as a tool, which is similarly to GradInput and computes the average value while the input varies along a linear path from a baseline x' to x. It solve the problem of Sensitivity property violation. The baseline x' is defined by the user and often chosen to be zero. m is the number of steps in the Riemman approximation of the integral. Our explanation tool is released for other researchers to use.

Fig. 2. Classification results with explanation tool.

6 Experiments and Results

For the training details, the method of cross-entropy loss was used in CNN, RNN, and BERT. And we used Adam as the optimizer, and the learning rate

Table 2. Experimental Results and comparisons of our capsule networks and baselines.

Model	Classes	Precision	Recall	F_1
SVM	Neutral	0.7895	0.8211	0.8050
	Insulting	0.8041	0.8444	0.8238
	Antisocial	0.7917	0.2000	0.3193
	Illegal	0.2188	0.1923	0.2047
	Weighted	0.7786	0.7966	0.7824
CNN	Neutral	0.9289	0.9524	**0.9405**
	Insulting	0.9559	0.9657	0.9607
	Antisocial	0.8056	0.9062	**0.8529**
	Illegal	0.4286	0.0380	0.0698
	Macro	0.7797	0.7156	0.7060
	Weighted	0.9270	0.9369	0.9281
	Micro	0.9369	0.9369	0.9369
RNN	Neutral	0.9171	0.9371	0.9270
	Insulting	0.9608	0.9446	0.9526
	Antisocial	0.7120	0.9271	0.8054
	Illegal	0.3415	0.1772	0.2333
	Macro	0.7328	0.7456	0.7296
	Weighted	0.9192	0.9233	0.9202
	Micro	0.9233	0.9233	0.9233
BERT	Neutral	0.9200	0.9459	0.9328
	Insulting	0.9666	0.9771	**0.9718**
	Antisocial	0.7593	0.8542	0.8039
	Illegal	0.7902	0.5861	**0.6730**
	Macro	0.8590	0.8408	0.8454
	Weighted	0.9260	0.9284	0.9259
	Micro	0.9284	0.9284	0.9259
Capsule	Weighted	0.9334	0.9419	0.9376
HACN	Weighted	0.9437	0.9528	**0.9486**

is set as 0.001. And also, we added dropout and early stop trick. The dropout trick randomly abandons a certain proportion of nodes in the training process to prevent the occurrence of over-fitting. Finally, dropout is Adopted as 0.5. The effects of the model are evaluated by early stop technology on the validation set after each iteration. When the validation set's evaluation result no longer improves in N consecutive rounds, the iterative process is truncated, and the process of training is suspended. The number of N has impacts on the time consumed and Whether the model is converge or not. We implemented our models with PyTorch, moreover, we finetuned the pre-trained based model, BERT, on two NVIDIA RTX 2080Ti GPUs. We also performed an ablation analysis of our HACN model. The weighted F1 drops 1.1% when we remove the hierarchical attention module.

We carry out several experiments during the evaluation phase, and the best experiment is taken into account for the evaluation phase. The systems are evaluated with the official competition metric, Precision, Recall, and Macro Averaged F1 score. What is more, this task is multiple classifications, so macro average, weighted average, micro average are also employed (Fig. 2).

In the filtering of insulting language in Chinese, the best performing model achieves a macro averaged F1-score of 94.05%. In the situation of antisocial language, the best performing system for the Chinese achieves a macro averaged F1-score of 97.18%. Finally, in the detection of illegal language, the state of the art NLP model for the Chinese is up to a macro averaged F1-score of 67.30%. However, we adopted a weighted average as a uniform evaluating method due to the unbalanced classes. Our hierarchical attention capsule network (HACN) is the best-performed model with an excellent result of 94.86%.

7 Analysis

In our experiment, we can find that RNN has an acceptable performance, however, there are some obvious shortcomings. When handling unbalanced data, such as a high recall rate, fail to classify certain classes because the class lacks visible character.

It is also necessary to remark that the experiments are performed with the default parameters. Thus there is an additional field for improvement with some finetuning, which we plan to consider for future research. Moreover, we note that the label distribution is extremely imbalanced because there might be a bias introduced by the algorithms.

Extending from the experiment above, we present a comparison experiment in Figure, where we record the valid accuracy over time and spot trends with different systems. Figure 3 illustrates that our proposed HACN model can quickly converge to its stable equilibrium values. In the meantime, the starting point shows that HACN can get a promising result in a short time (after the first epoch).

Figure 4 shows times spent in the testing step for 3,759 test sentences when using different systems. The curves generated with these results suggested that our proposed HACN model can achieve the classification at the quickest speed. Considering the deployment in a sound system, we only compared the testing step. Nevertheless, we believe we still have substantial advantages in training because there is no need to pretrain the model on a large number of data compared with BERT.

Furthermore, We show a deep analysis of the mis-classified cases in the evaluation process of our experiments. Thus, we do a manual analysis for those misclassified samples. This analysis aims at getting a deep comprehension of the

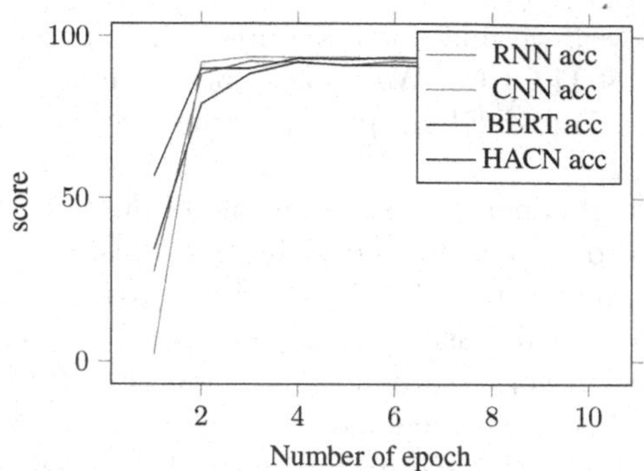

Fig. 3. Valid accuracy performance of each Epoch

areas our classifiers are lacking in. Our model fails to classify some metaphorical offensive words. It is usual for human to use euphemisms to tone down swear words in some situations. For some other cases, our model classifies some profanity text as offensive, which is actually not offensive. The classifier could miss these word variants, especially when the word variant is the only offensive word in the given sentence. In another situation, the word "sucks" is the only word that is often used offensively. However, the given tweet is not offensive because the author only describes their mood instead of insulting someone else. These misclassifications seem to indicate that the classifier reacts to trigger words with negative connotations but may not be capable of interpreting the words concerning the broader context.

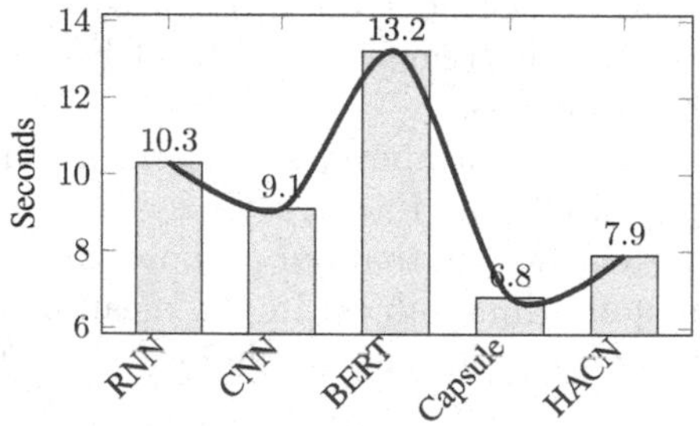

Fig. 4. Time costing in testing

8 Conclusion and Future Work

This work presents a Chinese corpus of offensive language crawled from microblog entries and video comments and manually categorized into 4 categories, and several models, including an allegedly novel architecture: Hierarchical

Attention Capsule Network, for classification tested on the corpus. We describe the data-set (with simple baselines) and then talk about the modeling with both standard and non-standard methods and tools for explanations. For the dataset, we present an annotated corpus of offensive language in the online world, consisting of sentences and the corresponding annotations. The corpus consists of 18707 sentences annotated with four classes, including neutral language, insulting language, antisocial language, and illegal language. We also present several systems used for the classification of offensive language. The baselines are SVM, RNN, CNN, and BERT. What's more, we present a novel capsule network (HACN) with hierarchical attention to model the semantic structure. The best F1 score of 94.86% is achieved when using HACN. Finally, we propose an explanation tool to illustrate what our systems have learned.

Identifying offensive language in the online world is also interdisciplinary, as it overlaps with psychology, sociology, and economics, while also raising legal and ethical questions, so we expect it to attract a broader audience. Thus, in the future, we would like to bring the ideas and research achievements of other related fields to deliver and share technology and solutions for offensive language from online user-generated content.

Acknowledgements. This research is supported by the National Language Commission Key Research Project (ZDI135-61), the National Natural Science Foundation of China (No.61532008 and 61872157), and the National Science Foundation of China (61572223).

References

1. Alber, M., et al.: Innvestigate neural networks! arXiv preprint arXiv:1808.04260 (2018)
2. Ancona, M., Ceolini, E., Öztireli, C., Gross, M.: A unified view of gradient based attribution methods for deep neural networks. In: NIPS 2017-Workshop on Interpreting, Explaining and Visualizing Deep Learning. ETH Zurich (2017)
3. Burnap, P., Williams, M.L.: Cyber hate speech on Twitter: an application of machine classification and statistical modeling for policy and decision making. Policy Internet **7**(2), 223–242 (2015)
4. Davidson, T., Warmsley, D., Macy, M., Weber, I.: Automated hate speech detection and the problem of offensive language. In: Eleventh International AAAI Conference on Web and Social Media (2017)
5. Djuric, N., Zhou, J., Morris, R., Grbovic, M., Radosavljevic, V., Bhamidipati, N.: Hate speech detection with comment embeddings. In: Proceedings of the 24th International Conference on World Wide Web, pp. 29–30. ACM (2015)
6. Fortuna, P., Ferreira, J., Pires, L., Routar, G., Nunes, S.: Merging datasets for aggressive text identification. In: Proceedings of the First Workshop on Trolling, Aggression and Cyberbullying (TRAC-2018), pp. 128–139 (2018)
7. Georgakopoulos, S.V., Tasoulis, S.K., Vrahatis, A.G., Plagianakos, V.P.: Convolutional neural networks for toxic comment classification. In: Proceedings of the 10th Hellenic Conference on Artificial Intelligence, p. 35. ACM (2018)
8. Goodman, B., Flaxman, S.: European union regulations on algorithmic decision-making and right to explanation. AI Mag. **38**(3), 50–57 (2017)

9. Guidotti, R., Monreale, A., Ruggieri, S., Turini, F., Giannotti, F., Pedreschi, D.: A survey of methods for explaining black box models. ACM computing surveys (CSUR) **51**(5), 93 (2018)
10. Hinton, G.E., Krizhevsky, A., Wang, S.D.: Transforming auto-encoders. In: Honkela, T., Duch, W., Girolami, M., Kaski, S. (eds.) ICANN 2011. LNCS, vol. 6791, pp. 44–51. Springer, Heidelberg (2011). https://doi.org/10.1007/978-3-642-21735-7_6
11. Huang, B., Carley, K.M.: Parameterized convolutional neural networks for aspect level sentiment classification. arXiv preprint arXiv:1909.06276 (2019)
12. Jiang, L., Yu, M., Zhou, M., Liu, X., Zhao, T.: Target-dependent Twitter sentiment classification. In: Proceedings of the 49th Annual Meeting of the Association for Computational Linguistics: Human Language Technologies-Volume 1, pp. 151–160. Association for Computational Linguistics (2011)
13. Koh, P.W., Liang, P.: Understanding black-box predictions via influence functions. In: Proceedings of the 34th International Conference on Machine Learning-Volume 70, pp. 1885–1894. JMLR. org (2017)
14. Kumar, R., Ojha, A.K., Malmasi, S., Zampieri, M.: Benchmarking aggression identification in social media. In: Proceedings of the First Workshop on Trolling, Aggression and Cyberbullying (TRAC-2018), pp. 1–11 (2018)
15. Kwok, I., Wang, Y.: Locate the hate: detecting tweets against blacks. In: Twenty-Seventh AAAI Conference on Artificial Intelligence (2013)
16. Li, X., Bing, L., Lam, W., Shi, B.: Transformation networks for target-oriented sentiment classification. arXiv preprint arXiv:1805.01086 (2018)
17. Müller, K., Schwarz, C.: Fanning the flames of hate: social media and hate crime. Available at SSRN 3082972 (2018)
18. Olah, C., Mordvintsev, A., Schubert, L.: Feature visualization. Distill **2**(11), e7 (2017)
19. Ribeiro, M.T., Singh, S., Guestrin, C.: Why should i trust you?: Explaining the predictions of any classifier. In: Proceedings of the 22nd ACM SIGKDD International Conference on Knowledge Discovery and Data Mining, pp. 1135–1144. ACM (2016)
20. Sundararajan, M., Taly, A., Yan, Q.: Axiomatic attribution for deep networks. In: Proceedings of the 34th International Conference on Machine Learning-Volume 70, pp. 3319–3328. JMLR. org (2017)
21. Tang, D., Qin, B., Feng, X., Liu, T.: Effective LSTMs for target-dependent sentiment classification. arXiv preprint arXiv:1512.01100 (2015)
22. Wang, Y., Huang, M., Zhao, L., et al.: Attention-based LSTM for aspect-level sentiment classification. In: Proceedings of the 2016 Conference on Empirical Methods in Natural Language Processing, pp. 606–615 (2016)
23. Wiegand, M., Siegel, M., Ruppenhofer, J.: Overview of the germeval 2018 shared task on the identification of offensive language (2018)
24. Xue, W., Li, T.: Aspect based sentiment analysis with gated convolutional networks. arXiv preprint arXiv:1805.07043 (2018)

25. Zampieri, M., Malmasi, S., Nakov, P., Rosenthal, S., Farra, N., Kumar, R.: Predicting the type and target of offensive posts in social media. arXiv preprint arXiv:1902.09666 (2019)
26. Zampieri, M., Malmasi, S., Nakov, P., Rosenthal, S., Farra, N., Kumar, R.: Semeval-2019 task 6: identifying and categorizing offensive language in social media (offenseval). arXiv preprint arXiv:1903.08983 (2019)
27. Zhao, W., Ye, J., Yang, M., Lei, Z., Zhang, S., Zhao, Z.: Investigating capsule networks with dynamic routing for text classification. arXiv preprint arXiv:1804.00538 (2018)

LiveQA: A Question Answering Dataset Over Sports Live

Qianying Liu[1,2], Sicong Jiang[1(✉)], Yizhong Wang[1,3], and Sujian Li[1]

[1] Key Laboratory of Computational Linguistics, MOE, Peking University, Beijing, China
512580728@qq.com, lisujian@pku.edu.cn

[2] Graduate School of Informatics, Kyoto University, Kyoto, Japan
ying@nlp.ist.i.kyoto-u.ac.jp

[3] University of Washington, Seattle, USA
yizhongw@cs.washington.edu

Abstract. In this paper, we introduce LiveQA, a new question answering dataset constructed from play-by-play live broadcast. It contains 117k multiple-choice questions written by human commentators for over 1,670 NBA games, which are collected from the Chinese Hupu (https://nba.hupu.com/games.) website. Derived from the characteristics of sports games, LiveQA can potentially test the reasoning ability across timeline-based live broadcasts, which is challenging compared to the existing datasets. In LiveQA, the questions require understanding the timeline, tracking events or doing mathematical computations. Our preliminary experiments show that the dataset introduces a challenging problem for question answering models, and a strong baseline model only achieves the accuracy of 53.1% and cannot beat the dominant option rule. We release the code and data of this paper for future research. (code: https://github.com/PKU-TANGENT/GAReader-LiveQA), (data: https://github.com/PKU-TANGENT/LiveQA).

Keywords: Question answering · Machine comprehension · Live text

1 Introduction

The research of question answering (QA), where a system needs to understand a piece of reading material and answer corresponding questions, has drawn considerable attention in recent years. While various QA datasets have been constructed to study how a QA system can understand a specific passage, the common sense knowledge and so on [4,10,12,13], most questions in these datasets could be given their answers by extracting from a few relevant sentences so that the model only needs to find a small set of supporting evidences, whose temporal ordering does not effect the final answer. In other words, these questions

Q. Liu—Equal contribution.

M. Sun et al. (Eds.): CCL 2020, LNAI 12522, pp. 316–328, 2020.
https://doi.org/10.1007/978-3-030-63031-7_23

1: 第一节两队得分之和能否达到 51 分或更多？
Will the sum of the scores of the two teams in the first quarter reach 51 points or more?

2: 本场勒布朗-詹姆斯能否得到 35 分或更多？
Will Lebron James get 35 points or more in this match?

3: 暂停回来，开拓者首次进攻会不会得分？
Will the Portland Trailblazers score in the first attack after the timeout ends?

4: 本节比赛双方还能否再次命中三分球？
Will another three-pointer take place in the rest of the quarter?

Fig. 1. Question examples from the LiveQA dataset.

are raised only considering a fixed document. However, in the real-life question answering, a question could have its **timelines**. To infer the answer, a good model needs to understand series of timeline information. For example, the question "how many points did Lebron James have?" would have different answers based on the time when the question was asked during a basketball game, and the answer would continuously change during the game. The other question "Which team would first earn 10 points?" would require a system to track down information of scoring points along the timeline until one team achieves 10 points.

According to the analysis above, we consider the timeline-based question answering problem as a gap which has not been covered by existing datasets. Thus, in this work we hope to construct a dataset where passages and questions both have timelines and question respondents are required to judge what information should be gathered for the questions involved in a timeline. Such a timeline inference-involved QA dataset introduces a new research line of reading comprehension, that evaluates the ability of understanding temporal information of a QA model.

Additionally, the real-world questions are often involved in some math calculation, such as addition, subtraction and counting. To answer the questions correctly, one not only needs to locate some specific sentences, but also do calculation or comparison on the extracted evidence. For example, "How many points did the winner team win?" needs one system to perform subtraction on the final score to get the correct answer.

To these ends, we construct a QA dataset *LiveQA* based on a Hupu-live-broadcasting-dataset, which is a set of Chinese live-broadcasting passages of NBA. Hupu is a sports news website that has live-broadcasting for basketball games. In the Hupu-live-broadcasting, the host of one sport game describes the details of the game vividly with emotion and different sentence structures, and presents many game-related quizzes during the game. We collect the description texts and their quizzes into LiveQA. Answering the quizzes requires one model to correctly understand the timeline information of the context: some quizzes ask about information of one-whole quarter of the game or which player reaches a certain score earlier. Thus, the model needs to fully understand the temporal information of the live-broadcasting and then performs inference based on the temporal information. Figure 1 shows four question examples in the LiveQA dataset. Answering the first two questions requires an addition math operation, and the 3rd and 4th questions need comparison operation. Meanwhile, we can see that all these questions are time-dependent and require temporal inference.

In summarize, the main characteristics of our LiveQA dataset include the following two aspects. Firstly, the questions are time-awared. The model needs temporal inference to obtain the final answer. Secondly, in our dataset, reading comprehension is not limited to extracting a few specific text spans from the document, but is involved with math calculation. These characteristics make LiveQA challenging for previous QA systems to answer its questions. In this paper, we present an analysis of the resulting dataset to show how these characteristics appear in the data. We also show how questions are involved with temporal inference, and these questions also require mathematical inference. To demonstrate how these characteristics affect the performance of the QA model, we design a pipeline method, which first tries to find supporting sentences and then uses a strong baseline multi-hop inference model named Gated-Attention Reader, to judge the baseline performance on LiveQA. Our experimental results show that such strong baseline model only slightly exceeds random choice, which achieve 53.1% and cannot beat the dominant option rule. The analysis and experimental results show how this dataset can effectively examine how a QA system can perform multi-hop temporal and mathematical inference, which is not covered by previous studies.

The following of this paper is organized as follows: In Sect. 2, we give a brief introduction of current QA research lines and research on live text processing. In Sect. 3, we describe how we constructed the dataset. In Sect. 4, we give statistics of the dataset and analyse the timelineness and mathematical inference in the data. In Sect. 5, we give evaluation results of baseline models and error analysis.

2 Related Works

In this section, we mainly introduce the various QA datasets which can be categorized as datasets with extractive answers, datasets with descriptive answers and datasets with multiple-choice questions.

2.1 Datasets with Extractive Answers

A number of QA datasets consist of numerous documents or passages which have considerable length. Each passage is equipped with several questions, answers of which are segments of the passage. The goal of a reading comprehension model is to find the correct text span. In other words, it may offer a begin position and an end position in the passage instead of generating the words itself. Such corpora are regarded as datasets with extractive answers.

The most famous dataset of this kind is Stanford Question Answering Dataset (SQuAD) [13]. SQuAD v1.0 consists of 107,785 question-answer pairs compiled by crowdworkers from 536 Wikipedia articles, and is much larger than previous manually labeled datasets. Over 50,000 unanswerable questions are added in SQuAD v2.0 [12]. It is more challenging for existing models because they have to make more unreliable guesses. As performances on SQuAD have become a common way to evaluate models, some experts regard SQuAD as the ImageNet [2] dataset in the NLP field.

Another frequently used dataset with extractive answers is CNN/Daily Mail dataset [5], which was released by Google DeepMind and University of Oxford in 2015. One shining point of it is that each entity is anonymised by using an abstract entity marker to prevent models from using word-level information or n-gram models to find the answer rather than comprehending the passage.

Text	Question	Choices	Answer
......			
哨响！克莱-汤普森逼得太紧了！吃到一次犯规！			
勇士要个长暂停！！！！！			
	停回来，勇士第一轮进攻能否得分？（罚球也算，直到球权转换）	能/不能	不能
稍等！！！			
第一节还有7分29秒！			
	本节勇士队最后一分是否由伊戈达拉获得?	是/不是	不是
好的！！！比赛继续！！！			
哈登走上罚球线！！！			
两罚都有！！14-9			
利文斯顿弧顶控球！！！			
......			

Fig. 2. A partial example of LiveQA timeline.

CBT [6], NewsQA [15], TriviaQA [8] and many other datasets can also be categorized into this class. They constitute a high proportion of MRC datasets, and can test the abilities of extractive models in various ways. Thus, we aim to construct a novel dataset, on which extractive models are likely to make mistakes in looking for the location of an answer, that the dataset can open a new research line for question answering by testifying the ability of models to understand timelineness.

2.2 Datasets with Descriptive Answers

Instead of selecting a span from the passage, datasets with descriptive answers require a reading comprehension model to generate whole and stand-alone sentences. These corpora are more closer to reality, because most questions in the real world cannot be solved simply by presenting a span or an entity. This kind of dataset is getting popular nowadays, and may be the trend of the development of MRC datasets.

MS MARCO (Microsoft MAchine Reading Comprehension) [11] is a dataset released by Microsoft in 2016. This dataset aims to address the questions and documents in the real world, as its questions are sampled from Bing's search query logs and its passages are extracted from web documents retrieved by Bing. The questions in MS MARCO are about ten times as many as SQuAD, and each question is equipped with a human generated answer. The dataset also includes unanswerable questions. All of the above characteristics make MS MARCO worthy of trying.

NarrariveQA [9] is another dataset with descriptive answers released by DeepMind and University of Oxford in 2017. The dataset consists of stories, books and movie scripts, with human written questions and answers based solely on human-generated abstractive summaries. Answering such questions requires readers to integrate information which may distribute across several statements throughout the document, and generate a cogent answer on the basis of this integrated information. In other words, they test that the reader comprehends language, not just that it can pattern match. We judge it a referential advantage of a dataset, so LiveQA requires the ability of tracking events as well as we show in Fig. 2, which will be detailedly introduced in following sections.

2.3 Datasets with Multiple-choice Questions

Datasets with descriptive answers have various advantages, but they are relatively difficult to evaluate the system performance precisely and objectively. Thus, corpuses with more gradable QA-pairs are also needed, which leads to the development of datasets with multiple-choice questions. Through diversified types of questions, these datasets can examine almost every ability of a reading comprehension model mentioned above and are easier to get a conclusive score. Many datasets of this kind have been released in recent years, and they have covered multiple domains. For example, RACE [10] and CLOTH [17] are collected from English exams, MCTest [14] is sampled from friction stories, and ARC [1] is extracted from science-related sentences. However, there is still not a reliable dataset which is built on sports events for MRC. Thus, our LiveQA dataset has the potential for filling several gaps in the field of MRC.

2.4 Live Text Processing

Previously, various studies have been conducted on automatically generate sports news from live text commentary scripts, which has been seen as a summarization task. Zhang et al. [20] proposed an investigation on summarization of sports news from live text commentary scripts, where they treat this task as a special kind of document summarization based on sentence extraction in a supervised learning to rank framework. Yao et al. [18] further verify the feasibility of a more challenging setting to generate news report on the fly by treating live text input as a stream for sentence selection. Wan et al. [16] studied dealing with the summarization task in Chinese. All these studies focuses on using the live text commentary scripts as the input of summarization and selecting sentences to form the summary. So far, we are the first to point out the importance of timelineness and mathematical reasoning in understanding live text commentary scripts.

3 LiveQA: Dataset Construction

In this section we introduce how to construct LiveQA from the raw Hupu text and present the corpus statistics. The whole process of building LiveQA mainly includes crawling the raw data and acquiring the game texts with corresponding quizzes.

3.1 Data Crawling

In Hupu, each game has a unique ID which is connected with its url. We collected the IDs from the Hupu's live schedule pages. Their formats are *games/year-month-date*. There are links to all the NBA games so that their IDs can be saved. After we saved the IDs into a file, we used the web debugging tool Fiddler to get a sample of the url of a game, and then changed the IDs in the url to make access to all the games. We are authorized by the legal department of hupu website to construct the dataset for only academical purpose.

3.2 Data Processing

Most previous datasets usually do not care for the storing positions of the passages and their questions. But in our dataset, the quizzes and the contexts shouldn't be separated because the time (the position) one quiz occurs is quite important for the final answer. If we separate the quizzes and their contexts, most quizzes may have different answers and cannot be answered even by human. Here we use some rules to clean the dataset. The lines starting with '@' are always interactions between the host and some active readers, which are irrelevant to the game. During the half-time break, the host will give out some "gift" questions to please the readers waiting for the second-half. Some of the questions appear like normal quizzes, but they need information outside the game to answer them, thus we exclude them from the data (i.e.. which team won more matches in the history?). Usually they have a prefix – 中场福利 in common. Besides, we exclude the descriptions of pictures from our data.

3.3 Data Structure

Here we give an explanation for the structure for each independent data sample.

For each live-stream of one match, the timeline data is sorted in time order, where the questions are inserted into the corresponding timeline position so that the timeline features of the questions could be inferred. As we show in Fig. 2, the plain content text and the question text share the same timeline, but question records have choices and answers along with the text. For each record in the timeline, it either contains a piece of live-stream text or a question bonded with the corresponding choices and the correct answer. Each question has two answer choices.

4 Dataset Statistics

Table 1. The details of statistics of the dataset.

Element	Count
Document	1,670
Sentences in total	1786616
Sentences in average	1069.83
Quizzes in total	117050
Quizzes in average	70.09

We show the statistics of the dataset in Table 1. The LiveQA dataset contains 1,670 documents, each of which has 70.09 quizzes and 1069.83 sentences on average. Next we analyze the questions from two different views. First, we simply classify the questions according to the positions of their answers. In general, some of the questions can be solved by extracting information from neighboring sentences, which involves a time period of the origin game. Such questions occupy 68.6% of all the questions. Some questions can be replied only by summarizing all the information after the game ends and occupy about 30.6%. Still, there exists a small percentage (0.8%) of questions which are impossible to be answered from the passage. Table 2 lists some examples for each type of questions.

Because most of the questions are associated with some numerical data in the game, we also classify the questions according to how the numerical data is performed. Four types of operations are commonly used including: *Comparison*, *Calculation*, *Inference* and *Tracking*. Then the questions are correspondingly classified are introduced in the following subsections. We also give some examples in Table 3.

4.1 Comparison

To answer the comparison questions, we usually need to find the comparative figures for the corresponding objects. For example, the commentator asks which of the two players will score more or which team will win. The second row in Table 3 belong to the *Comparison* questions. The easiest way to solve this kind of questions is to find the two figures appearing in the text and comparing them. It is likely to acquire such figures after the game ends, and the specific figures usually appear together in a summary of the game in the end. Thus, matching techniques are still necessary to the final answer.

Table 2. Questions statistics and examples sorted by the location of evidence.

Question type	Proportion	Example	Translation
Answered after the game ends	30.6%	本场森林狼能否赢快船4分或更多?	Will the Timberwolves beat the Clippers by more than 4 points?
		本场比赛谁会赢?	Which team will win?
Answered through the context	68.6%	第二节谁先命中三分球?	Which team will make a three-pointer first in the second quarter?
		首节最后一分会不会由罚球获得?	Will the last point in the first quarter scored through a free-throw?
Impossible to answer	0.8%	第二节比赛开始1分30秒时间内会不会有三分球命中?	Will a three-pointer be made in the first 90s of the second quarter?
		本场比赛会不会在北京时间10时58分之前结束?	Will the game end before 10:58 a.m.?

Table 3. Questions statistics and examples sorted by the inference process.

Question type	Proportion	Example	Translation
Comparison	16.6%	勒布朗-詹姆斯本场能否得到26分或更多?	Will Lebron James get 26 points or more in this game?
		本场谁的得分会更高?	Who will get higher score in this game?
Calculation	25.4%	本场凯尔特人能否赢猛龙3分或更多?	Will the Celtics beat the Raptors by more than 3 points?
		本场两队总得分能否达到207分或更多?	Will the total score of the two teams reach 207 points or more?
Inference	28.5%	暂停回来，雷霆队首次进攻能否得分?	After the timeout, will the Thunder score in their first round of attack?
		第二节比赛雷霆队最后一分会不会由威斯布鲁克得到?	Will the last point of the Thunder in the second quarter be got by Westbrook?
Tracking	29.5%	太阳队能否在本场命中8个或更多三分球?	Will the Suns make 8 three-pointers or more in this game?
		凯文-乐福首节犯规数会不会达到2次?	Will Kevin Love commit 2 fouls or more in the first quarter?

4.2 Calculation

The *Calculation* questions require extracting two or three figures and calculating their sum or difference. They differ from the Comparison questions in two ways – the figures are more scattered and a calculation step is needed. This means that a respondent has to look for more information efficiently. After the figures are obtained, if a respondent misjudges the type or the direction of the calculation, he will still probably get a wrong answer. Similar to the *Comparison* questions, the *Calculation* questions are mainly dependent on the correct sentences where the figures are located. These two kinds of questions are relatively easy compared to those ones which are not based on certain sentences. The second row in Table 3 give two example questions.

4.3 Inference

The third and fourth type of questions require the ability of summarizing and tracking information. A question of the third type needs a respondent to infer some figures through the text. For example, a question may be "After this time-out, will the Cavaliers score in the first round of attack?". The commentator obviously will not say that "The Cavaliers scored 2 points." or "The Cavaliers didn't score." A respondent may get the answer as "JR Smith makes a 2-point shot." Another example is "Will the last point of this quarter be scored through a free throw?" The information comes from the text of "Anthony Davis makes his second free throw ... The match ends!". It is impossible to get a reasonable answer by matching.

4.4 Tracking

The Tracking questions require more scattered information. A respondent should collect and accumulate specific information from a part of the passage, as the question is based on events happening repeatedly in a quarter or half of the game. For example, some questions ask about how many free-throws a player A will make in a quarter. As this figure does not appear in the passage, a respondent needs to count how many times the event 'A makes a free-throw' occurs. In other words, it is necessary to track events relevant to the player 'A' and 'free-throw'. When the player(A) is replaced with one team name, the new question is even more difficult because the information about each player belonging to the team should be tracked. Therefore, information tracking leads this kind of questions to be the most challenging ones in the dataset.

5 Baseline Models and Results

5.1 Models

To evaluate the QA performance on the LiveQA dataset, we implement 3 baseline models. The first is based on random selection, where the system randomly chooses a choice as the answer. The second is to choose the dominant option of each question. More concretely, 80.0% of questions are in format of 'yes' and 'no', where 57.8% has the answer 'no'. For the other multiple choice questions, 50.6% of them take the second option as the right answer. Thus, for 'yes/no' questions, we choose 'no', otherwise we choose the second option.

We also build a neural-network style baseline for our dataset to evaluate how state-of-the-art QA systems perform on the LiveQA dataset. Due to the uniqueness of our dataset, most of existing machine comprehension models are not suitable to it. For example, the QANet [19] model, which used to be a state-of-art model of SQuAD [13], is unavailable because it predicts the probability distribution of an answer's starting position and ending position in the context. But in LiveQA, a number of right answers do not directly appear in the context (e.g. an answer in format of 'can' or 'cannot'). Up to now, none of machine reading comprehension models has been designed for a dataset with consideration of timeline and mathematical computations. That means that the existing ones will not be likely to perform well on our dataset. The closest work to ours is multi-hop question answering, and thus we use a novel model Gated-Attention Reader [3] to experiment on LiveQA.

Gated-Attention Reader (GA) is an attention mechanism which uses multiplicative interactions between the query embedding and intermediate states of a recurrent neural network reader. GA enables a model to scan one document and the questions iteratively for multiple passes, and thus the multi-hop structure can target on most relevant parts of the document. It used to be the state-of-art model of several datasets, such as CNN/Daily Mail dataset [5] and CBT dataset [7].

The full context, which is usually composed of more than 1,000 sentences on average, is too heavy for GA as input. To apply GA to our dataset, we propose a pipeline method to first extract a set of candidate evidence sentences from the full content, and then apply the GA model on this set of sentences to predict the final answer. We employ TF-IDF style matching score to extract 50 most relevant sentences as the supporting evidence. To improve the accuracy of selecting the evidence candidates, if the question clearly requires some information after the game ends, we use the ending part of the content as the input.

Specifically, taken the embedding representation of a token, the Bi-directional Gated Recurrent Units (BiGRU) process the sequence in both forward and backward directions to produce two sequences of token-level representations, which are concatenated at the output as the final representation of the token. To perform multi-hop inference, the GA model reads the document and the query over k horizontal layers, where layer k receives the contextual embeddings $X_{(k-1)}$ of the document from the previous layer. At each layer, the document representation $D^{(k)}$ is computed by taking the full output of a document BiGRU where the previous layer embedding $X_{(k-1)}$ is the input. At the same time, a layer-specific query representation $Q^{(k)}$ is computed as the full output of a separate query BiGRU taking the query embedding Y as the input. The Gated-Attention is applied to $D^{(k)}$ and $Q^{(k)}$ to compute the contextual embedding $X^{(k)}$.

$$X^{(k)} = GAttn(BiGRU(X^{(k-1)}), BiGRU(Y)) \tag{1}$$

After obtaining the query-awared document representation, we perform answer prediction by matching the similarity of answer and content. We use bidirectional Gated Recurrent Units to encode the candidate answers into vectors $A(i)$, and then we compute matching score between summarized document and candidates using a bilinear attention. Finally we calculate the probability distribution of the options with softmax. The operations are similar to those in RACE [10].

$$s = softmax([Blin(A^i, D^{(k)});]_n^{i=1}) \tag{2}$$

5.2 Model Evaluation

Table 4. The results of different baseline models on the test set. Random denotes randomly selecting an answer. Dominate denotes selecting the dominate option. GA denotes the gated-attention reader.

Model	Acc
Random	50.0%
Dominant	**56.4%**
GA	53.1%

For the three baseline models, performance is reported with the accuracy on the test set in Table 4. The random selection method (Random) scores 50.0%, while the dominant option method (Dominate) reaches a score of 56.4%, which shows that our dataset does not have a certain pattern for the answers. Meanwhile, GA, which is a strong baseline for previous question answering problems, failed to perform better than the dominant option method and only achieves a score of 53.1%. Such results show that our dataset is challenging and needs further investigation for model design. In future work, how to incorporate temporal information and mathematical calculation into a QA model is the focus.

5.3 Case Study

In this subsection, we further analyze the prediction ability of the GA model. Table 5 shows some prediction cases in experimental results. From the first two questions, we can see that the model gives the correct answers when judging the result of a specific event. But for the other three questions which involve multiple events, the model fails to answer them correctly. A possible explanation is that, although GA is designed for multi-hop inference, it lacks ability in both information tracking and math calculation, which makes it difficult for the model to track down some complicated events.

We can see, for reading comprehension models that extract answers based on the similarity between the answer and the content, they would fail on LiveQA due to the fact that they cannot track down temporal information nor perform mathematical calculation. To outperform existing models on LiveQA, the system should consider focusing on tracking information of a certain event through the timeline. It should also have the ability to perform mathematical inference between different contents.

Table 5. Cases in the experimental results

Question	Translation	Correct answer	Answer given by the model
跳球之争！本场比赛哪支球队获得第一轮进攻球权?	Jump ball fight! Which team will win the chance of the first round of offence?	勇士(The Warriors)	勇士(The Warriors)
湖人全场总得分是奇数还是偶数?	Will the total score of the Lakers at the end of the game be odd or even?	奇数(odd)	奇数(odd)
尼克杨第二节能否命中3分球?	Can Nick Young make a three pointer in the second quarter?	能(Yes)	不能(No)
第三节结束，76人能否领先湖人4分或更多?	At the end of the third quarter, Will the 76ers lead the Lakers by 4 points of more?	不能(No)	能(Yes)
谁先获得30分?	Who will score his 30th point earlier?	24分的哈登(James Harden who has got 24 points)	25分的托马斯(Isaiah Thomas who has got 25 points)

6 Conclusion

In this paper, we present LiveQA, a question answering dataset constructed from play-by-play live broadcast. LiveQA can evaluate a machine reading comprehension model in its ability to understand the timeline, track events and do mathematical calculation. It consists of 117k questions, which are time-dependent and need math inference. Due to the novel characteristics, it is hard for existing QA models to perform well on LiveQA. We expect our dataset will stimulate the development of more advanced machine comprehension models.

Acknowledgement. We thank the anonymous reviewers for their helpful comments on this paper. This work was partially supported by National Natural Science Foundation Project of China (61876009), National Key Research and Development Project (2019YFB1704002), and National Social Science Foundation Project of China (18ZDA295). The corresponding author of this paper is Sujian Li.

References

1. Clark, P., et al.: Think you have solved question answering? try arc, the ai2 reasoning challenge. arXiv preprint arXiv:1803.05457 (2018)
2. Deng, J., Dong, W., Socher, R., Li, L.J., Li, K., Li, F.F.: Imagenet: a large-scale hierarchical image database. In: IEEE Conference on Computer Vision & Pattern Recognition (2009)
3. Dhingra, B., Liu, H., Yang, Z., Cohen, W.W., Salakhutdinov, R.: Gated-attention readers for text comprehension. arXiv preprint arXiv:1606.01549 (2016)

4. Dunn, M., Sagun, L., Higgins, M., Guney, V.U., Cirik, V., Cho, K.: Searchqa: A new q&a dataset augmented with context from a search engine. arXiv preprint arXiv:1704.05179 (2017)
5. Hermann, K.M., et al.: Teaching machines to read and comprehend. In: Advances in Neural Information Processing Systems, pp. 1693–1701 (2015)
6. Hill, F., Bordes, A., Chopra, S., Weston, J.: The goldilocks principle: reading children's books with explicit memory representations. Computer Science (2015)
7. Hill, F., Bordes, A., Chopra, S., Weston, J.: The goldilocks principle: reading children's books with explicit memory representations. arXiv preprint arXiv:1511.02301 (2015)
8. Joshi, M., Choi, E., Weld, D.S., Zettlemoyer, L.: Triviaqa: A large scale distantly supervised challenge dataset for reading comprehension (2017)
9. Kočiský, T., Schwarz, J., Blunsom, P., Dyer, C., Grefenstette, E.: The narrativeqa reading comprehension challenge (2017)
10. Lai, G., Xie, Q., Liu, H., Yang, Y., Hovy, E.: Race: Large-scale reading comprehension dataset from examinations. arXiv preprint arXiv:1704.04683 (2017)
11. Nguyen, T., Rosenberg, M., Xia, S., Gao, J., Li, D.: Ms marco: a human generated machine reading comprehension dataset (2016)
12. Rajpurkar, P., Jia, R., Liang, P.: Know what you don't know: Unanswerable questions for squad. arXiv preprint arXiv:1806.03822 (2018)
13. Rajpurkar, P., Zhang, J., Lopyrev, K., Liang, P.: Squad: 100,000+ questions for machine comprehension of text. arXiv preprint arXiv:1606.05250 (2016)
14. Richardson, M., Burges, C.J., Renshaw, E.: Mctest: a challenge dataset for the open-domain machine comprehension of text. In: Proceedings of the 2013 Conference on Empirical Methods in Natural Language Processing, pp. 193–203 (2013)
15. Trischler, A., Tong, W., Yuan, X., Harris, J., Suleman, K.: Newsqa: a machine comprehension dataset (2016)
16. Wan, X., Zhang, J., Yao, J., Wang, T.: Overview of the NLPCC-ICCPOL 2016 shared task: sports news generation from live webcast scripts. In: Lin, C.-Y., Xue, N., Zhao, D., Huang, X., Feng, Y. (eds.) ICCPOL/NLPCC -2016. LNCS (LNAI), vol. 10102, pp. 870–875. Springer, Cham (2016). https://doi.org/10.1007/978-3-319-50496-4_80
17. Xie, Q., Lai, G., Dai, Z., Hovy, E.: Large-scale cloze test dataset created by teachers. arXiv preprint arXiv:1711.03225 (2017)
18. Yao, J.G., Zhang, J., Wan, X., Xiao, J.: Content selection for real-time sports news construction from commentary texts. In: Proceedings of the 10th International Conference on Natural Language Generation, pp. 31–40 (2017)
19. Yu, A.W., Dohan, D., Luong, M.T., Zhao, R., Chen, K., Norouzi, M., Le, Q.V.: Qanet: Combining local convolution with global self-attention for reading comprehension. arXiv preprint arXiv:1804.09541 (2018)
20. Zhang, J., Yao, J.G., Wan, X.: Towards constructing sports news from live text commentary. In: Proceedings of the 54th Annual Meeting of the Association for Computational Linguistics (Volume 1: Long Papers), pp. 1361–1371 (2016)

Chinese and English Elementary Discourse Units Recognition Based on Bi-LSTM-CRF Model

Yancui Li[1,2](✉), Chunxiao Lai[1], Jike Feng[1], and Hongyu Feng[1]

[1] College of Information Engineering, Henan Institute of Science and Technology, Xinxiang, Henan, China
li.yancui@qq.com

[2] Key Laboratory of Advanced Theory and Application in Statistics and Data Science (East China Normal University), Ministry of Education, Shanghai, China

Abstract. Elementary Discourse Unit (EDU) recognition is the basic task of discourse analysis, and the Chinese and English discourse alignment corpus is helpful to the studies of EDU recognition. This paper first builds Chinese-English parallel discourse corpus, in which EDUs are annotated and aligned. Then, we present the framework of Bi-LSTM-CRF EDUs recognition model using word embedding, POS and syntactic features, which can combine the advantage of CRF and Bi-LSTM models. The results show that F1 is about 2% higher than the traditional method. Compared with CRF and Bi-LSTM, the Bi-LSTM CRF model can combine the advantages of them and obtains satisfactory results for Chinese and English EDUs recognition. The experiment of feature contribution shows that using all features together can get best result, the syntactic feature outperforms than other features.

Keywords: Elementary discourse units · Bi-LSTM-CRF · Chinese-English · Alignment corpus

1 Introduction

Discourse analysis is helpful for the performance of machine translation, question answering, summarization and other application. EDU recognition is a basic work in discourse analysis task. Only by recognition EDU, can we make further discourse analysis or other works. At present, the existing Chinese-English parallel corpus only align paragraphs, sentences and other linguistic units, but do not annotate bilingual EDUs alignment, which due to EDU recognition is mainly carried out on monolingual. However, EDUs recognition on Chinese and English is vital to bilingual analysis, machine translation et al. For Example1 is a bilingual sentence of Chinese and English, Chinese EDUs are numbered sequentially by e1, e2 and e3, and English EDUs are marked by e1′, e2′ and e3′. Obviously, e1 and e1′, e2 and e2′, e3 and e3′ are alignment pair.

Supported by organization National Natural Science Foundation of China (61502149).

M. Sun et al. (Eds.): CCL 2020, LNAI 12522, pp. 329–343, 2020.
https://doi.org/10.1007/978-3-030-63031-7_24

Example 1 A) [京杭运河古来繁华，] ^{e1}[两岸商贾云集，]^{e2} [贸易发达。]^{e3}

B) [The Beijing - Hangzhou Grand Canal has been prosperous since ancient times,] $^{e1'}$ [with both sides of the bank swarming with merchants] $^{e2'}$ [and well - developed trade.]$^{e3'}$

The main work of this paper is recognition the EDUs of Chinese and English as much as possible. The following is the contribution of this paper:

We annotate Chinese and English discourse alignment corpus, which is first corpus contain EDUs alignment information as far as we know;

We get satisfactory results without any handcraft feature by using Bi-LSTM-CRF Model;

We conduct to find out the contribution of various model and features.

This paper combines existing research and Chinese-English discourse alignment corpus to identify and analyze Chinese-English EDUs. Section 2 builds Chinese-English EDUs alignment corpus; Section 3 describes the Bi-LSTM-CRF model and the framework this paper used; Section 4 reports and analyzes the experimental results; Section 5 overviews the related work; Finally, Sect. 6 summarizes this paper and points out the future research direction.

2 Chinese-English Alignment Corpus

2.1 Chinese-English EDUs Alignment Methods

In order to represent the discourse, the first task is to define the EDUs. Inspired by the work of Li et al. (2014) and Feng (2013), we give the definition of Chinese EDUs. Firstly, a clause should contain more than one predicate, expressing not less than one proposition. Secondly, one EDU should have propositional function to another EDU. Finally, a clause should be segmented by some punctuation. As for English EDU, it is the corresponding content of Chinese EDU.

When annotate, we dividing Chinese sentence into parts, and adapting the alignment strategy of the source language is preferred. That is to say, it is segmented according to the established Chinese EDUs, and then align in English. Therefore, EDUs in Chinese and English sentences is correspondence.

Such as Example 1, recognition and alignment are achieved under the guidance of this principle. Since Chinese EDUs are preferential when making alignment rules, some sentences with widely ranges may appear in English translation. These EDUs are not adjacent in English sentences, it will affect the alignment of Chinese and English EDUs. EDUs cannot be completely corresponding in this case, and the solution is to align the main parts.

2.2 Chinese-English Alignment Corpus

According to the alignment annotation principle mentioned in the Sect. 2.1. We annotate alignment corpus of Chinese and English. Corpus select from Xinhua daily, and we have marked 100 Chinese-English translation documents. The Chinese-English parallel corpus is marked with Chinese as the main language, supplemented the parallel EDUs by English.

Due to the marked Chinese-English alignment corpus has many contents, and experiments are mainly for EDUs, this paper mainly introduces the annotation principle of EDU in corpus. After practical operation and analysis, the following three points are obtained:

1) The meaning of English and Chinese sentences. According to the logical semantic relations, the corresponding relations of the adjacent EDUs in the alignment corpus can be found respectively, and the relationship is used to divide and align English-Chinese corpus.
2) Structure. Combined with the structure of Chinese language and English language, the order of subject-verb-object in English-Chinese is consistent, and the translation of some noun clauses and adverbial clauses are also consistent, so it is possible to find out the corresponding words in English-Chinese so as to find the corresponding sentence components in English-Chinese for division.
3) Following the punctuation clues. In the translated English corpus, the punctuation in English is mostly consistent with that in Chinese. And according to the distribution of punctuation, the meaning of the text and the translated English EDUs can be more clearly inferred.

There are 100 documents, 513 paragraphs, 899 Chinese sentences, 1 281 English sentences and 2 153 Chinese-English EDU pairs which have been effectively marked. The Chinese EDU average length is 11 words, while the English EDU average length is 20 words. In the paper, the preset program is used to automatically find the parent node information of English EDUs, and the search is carried out in the automatic syntactic analysis tree of Stanford. The method of search is to look up the words from the beginning and the end of the English clauses successively until a common parent node is found in the syntactic tree. By the way of making statistics on the information of parent node which can be found, it is not difficult to find that the main syntactic structure which can make Chinese EDUs corresponding to English clauses are S, VP, NP, PP etc. The syntactic structure and occurrence frequency corresponding to English EDUs are shown in Fig. 1. From Fig. 1 we can see most of EDUs syntactic tag are S and VP, which is consistent with the definition of our EDU.

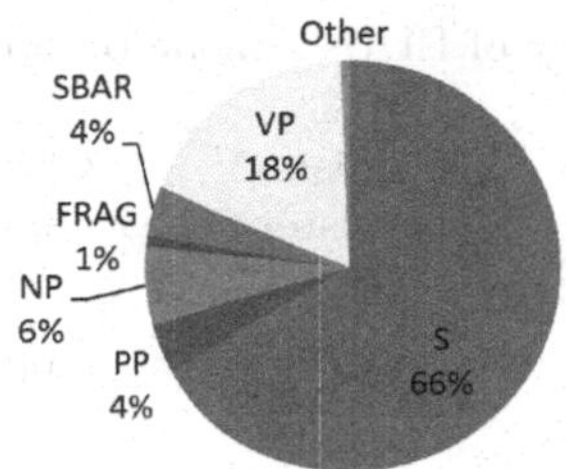

Fig. 1. Syntactic structure distribution of the EDUs

2.3 Tagging Strategies and Consistency

Two senior students of Chinese department carried out annotation training under the guidance of the project supervisor. 20 parallel paragraphs were randomly selected from the Xinhua daily to mark training corpus. We developed a platform for EDU annotation. The annotation training is mainly composed of three stages: 1) The tutor demonstrates the annotation of 10 documents, and explains the main annotation strategies, the annotation method and the operation of the annotation platform; 2) Two students mark the remaining 20 documents respectively; 3) Two students respectively proofread the 60 documents marked by themselves with the tutor, and the proofreading was completed in three times, mainly discussing the existing problems and the strategies and methods of correction and annotation. On this basis, the two students annotated the whole corpus together.

In the annotation, we employ left to right segment and alignment method. Consistency is a major criterion of annotation quality. The alignment EDUs annotation evaluation should take into account the recognition consistency and alignment consistency. So, the consistency of Chinese annotation, English annotation, and Chinese-English alignment annotation of the two annotators are considered:

Chinese consistency: consistency of two annotators on the same Chinese text.

English consistency: consistency of two annotators on the same English text.

Chinese-English alignment consistency: consistency of Chinese annotation on the same text by two annotators and the consistency of corresponding English alignment annotation.

We use Method1 and Method2 to compute the consistency.

Method1: computes the consistency of all possible annotations. There are punctuation marks at the recognition positions of Chinese EDUs, and punctuation marks that may be used as recognition marks. The recognition of EDUs in English is not based on punctuation, any space can be calculated as the recognition mark.

Method2: calculating the consistency of intersection ($A \cap B$) in all ($A \cup B$). Sentence Position = "X1...X2 — Y1...Y2", calculate the case that A and B mark the same position of recognition. Compared with method 1, this method is more accurate and can unify the evaluation criteria of Chinese and English EDUs recognition.

Table 1. The consistency of EDUs annotation for Chinese and English

	Chinese consistency	English consistency	Chinese-English alignment consistency
Method1	0.972	0.992	–
Method2	0.968	0.930	0.909

As shown in Table 1, recognition alignment shows good consistency, with Chinese alignment up to 0.972/0.968, English alignment up to 0.992/0.930. Even under the strictest circumstances of Method2, Chinese-English alignment up to the consistency rate of 0.909.

It is worth noting under the Method1, English consistency is better than Chinese, with 0.992 > 0.972. Under the Method2, Chinese better than English, this is because the consistency in the calculation, Chinese punctuation only for limited computation, but the English is for any Spaces.

However, the reality is that Chinese alignment is better than English with the same alignment evaluation criteria. This can be shown under Method2 (0.968 > 0.930), because Chinese recognition is marked by punctuation, which is relatively easy. However, English recognition is not marked by punctuation, and it is easy to recognition incorrectly. Therefore, Method2 can more accurately reflect the difference in bilingual alignment effect compared with Method1.

3 The Model of EDU Recognition Based on Bi-LSTM-CRF

In this section, we introduced the Bi-LSTM-CRF model we used, which is the combination of CRF and Bi-LSTM and have been used in several NLP task.

3.1 CRF

CRF is extension of both Hidden Markov Models and Maximum Entropy Model (Lafferty et al. 2001). It often solves some NLP problems, such as word recognition and image recognition. EDUs recognition is a sequence labeling problem. One solution is that it can assign each word in the sentence with label Y (word is EDU boundary) or N (word is not EDU boundary). CRF is a sequence labelling model with flexible feature space. Therefore, with given feature set and labeled training data, the CRF model solve EDUs recognition task. The model is defined as Eq. (1):

$$p(Y|X) = \frac{1}{Z(X)} exp(\sum_{k} \lambda_k f_k) \tag{1}$$

In Eq. (1), Z(X) is a probability normalization factor conditioned on X. λ_k is the corresponding weight of the feature set. f_k is the input sequence sentences, and Y is the output label of Y or N.

3.2 Bi-LSTM

RNN is a model suitable for sequence data, which uses previous and current state to determine the final output. However, in practical applications, RNN has only short-term memory because the gradient vanishing and exploding problem. Hochreiter and Schmidhuber (1997) propose LSTM network, a variant of RNN to solve this problem.

Figure 2 illustrates a single LSTM memory cell. We can see that it contains input, forget and output gate. The gates determine the current information, in a certain proportion or discarded, transferred to the next moment. Through the gate, LSTM can remove or add information to the cellular state. Therefore, they can solution the data long range dependencies problem.

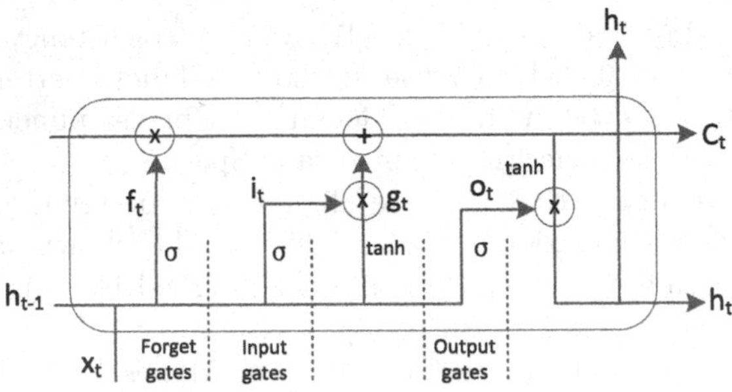

Fig. 2. LSTM memory cell

LSTM memory cell is implemented as the Eq. (2):

$$\begin{aligned}
&i_t=(W_{ii}x_t+b_{ii}+W_{hi}h_{(t-1)}+b_{hi})\\
&f_t=(W_{if}x_t+b_{if}+W_{hf}h_{(t-1)}+b_{hf})\\
&g_t=\tanh(W_{ig}x_t+b_{ig}+W_{hg}h_{(t-1)}+b_{hg})\\
&o_t=(W_{io}x_t+b_{io}+W_{ho}h_{(t-1)}+b_{ho})\\
&c_t=f_t c_{(t-1)}+i_t g_t\\
&h_t=o_t \tanh(c_t)
\end{aligned} \tag{2}$$

As shown in Eq. (2), the logistic sigmoid function is denoting as σ. it is the input gate. ct is cell vectors. i_t decides the information will be stored in c_t. ft is forgot gate, and it decides the information can through from the previous cell. o_t is output gate, it decides the information output to the current hidden state h_t. W is the weight matrix, and b is bias vectors of each gate. They are learned during training. $\oplus$ denotes the vector concatenation.

For sequence tagging task, Graves and Jürgen (2005) utilize a bidirectional LSTM (Bi-LSTM) network. Bi-LSTM is extended on the basis of LSTM, and it contains two difference direction layers. The sequence $\overline{h} = (\overline{h_1 h_2}......\overline{h_n})$ of the Bi-LSTM layer is obtained past and future input features by the forward and backward LSTM. The LSTM allows more context dependent information than LSTM.

3.3 Bi-LSTM-CRF

We describe our Bi-LSTM-CRF models in details. Figure 3 shows the Bi-LSTM-CRF framework. As we can see from Fig. 3, there are input layer, embedding layer, Bi-LSTM layer, CRF layer and output layer. First, words in sentences and their features are vectorized. Secondly, the Bi-LSTM model is fed with feature vectors to learn contextual features from the forward and backward directions. Then, Bi-LSTM output result is input to CRF layer. Finally, the CRF layer predicts the globally optimal clause sequence. In addition, to reduce the influence of overfitting, we add a dropout layer at ends of the Bi-LSTM model.

Bi-LSTM-CRF expands the CRF layer on the basis of Bi-LSTM. The performance of CRF model in sequence annotation tasks has been verified. In this model, the Bi-LSTM through Bi-LSTM layer makes full use of past and future

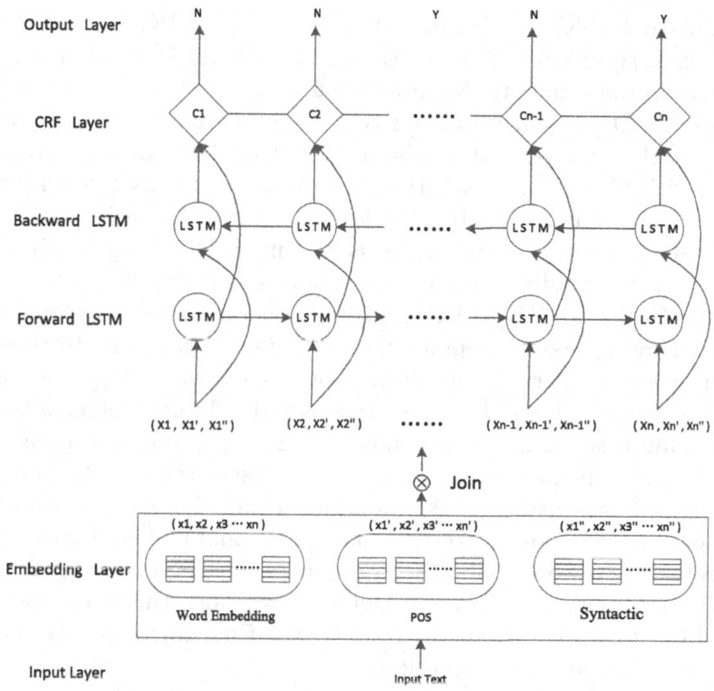

Fig. 3. The framework of Bi-LSTM-CRF Model

information, and CRF layer make use of tag in-formation. So, this model can predict the current tag by incorporate the advantage of Bi-LSTM and CRF.

4 Experiment Results

4.1 Experiment Setting

The input layers of our models are the input text. We give the vector representations of words, Part of speech (POS) and Syntactic. Word embeddings are pretrained using skip-n-gram, a variation of word2vec (Mikolov et al. 2013) that sensitive to the order of word. These embeddings are adjusted during training. We find improvements using pretrained word embeddings. For English, the embedding dimension we used is 200. For Chinese, we use pre-trained vector files from People's Daily News, the embedding dimension is 300 (Li et al. 2018). We use dropout training to avoid the model depending on one representation too strongly, and find it is import to result.

POS is the process of marking a word as nouns, verbs, adjectives, adverbs, etc. POS is used in many NLP task and proved very useful. Syntactic is the component that takes input sentence and give the grammatical tree structure of sentence, which is widely used to understanding written language, discourse parsing et al. For example, syntactic can output the phrases tag of words.

We use Stanford coreNLP (Manning et al. 2014) to get POS and syntactic feature, it can give the POS and syntactic tag of each word. In this paper, we use parent phrase tag as syntactic feature simplify.

The task of EDUs recognition is giving a tag to every word in a sentence. A single EDU could span several words in a sentence. Sentences can represent in the Y (Yes) or N (No) format, where each word is labeled as Y label if the word is the end of EDU, and as N label if it is the beginning or inside of EDU word.

We exploit standard training methods for our model. Using AdaGrad (Duchi 2011) as stochastic gradient decent. Calculate derivatives from standard back propagation (Goller and Kuchler 1996). In order to prevent over fitting, we regularize our model using dropout method (Srivastava et al. 2014), and fixed rate 0.5 for dropout layer. We obtain improvements after using dropout.

We set the initial AdaGrad learning rate as 0.01. The dimension of pre-trained word embedding is set as 200. The dimension of LSTM hidden state as 200. The W and b are randomly initialized with a uniform distribution in the range (−0.01, 0.01). We use publicly available 200 dimensional embeddings trained for English, there are total 40 000 words. We use 300 dimensional embeddings for Chinese, there are total 3 589 words. The Bi-LSTM units set 256, epoch set 200.

In our experiment, for Chinese EDUs recognition, there are total 12 581 words, 32 POS tags and 29 syntactic tags. For English, there are total 4 106 words, 47 POS tags and 29 syntactic tags.

4.2 Experiment Results

In this section, EDU recognition is carried out in our Chinese-English alignment corpus. There are total of 100 documents, 513 paragraphs and 2 153 EDUs were involved. The recognition of English word is 42 122 in total, among which there are 2 153 positive labels. The ratio of positive and negative examples is 19.6:1 for English. Overall, the average length of the English EDUs is about 20 words, while the Chinese EDUs is 11 words. The experiment splits instances into 10 parts, and use 8 parts for training, 1 part for verification and 1 part for testing. The features we used are word embedding, POS tag and syntactic tag. The recognition results of Chinese words boundary are indicated in Table 2.

Table 2. Chinese EDU words boundary recognition results

Model	P	R	F1
Li's Maxent	87.4	93.6	90.4
CRF	86.7	96	91.1
Bi-LSTM	**95.4**	89.8	92.5
Bi-LSTM-CRF	92.3	**94.4**	**93.4**

In Table 2, the best results are highlighted bold for each metric. From Table 2, we can see that by combining Bi-LSTM, pretrained embedding, and CRF on the top of the framework, our Bi-LSTM-CRF model outperforms best of all. We obtain the satisfactory results with the F1 93.4% and R 94.4% by using Bi-LSTM-CRF model.

Table 3 shows the English EDUs recognition result. For the purpose of comparison, we list Li's (Li et al. 2013) Maxent model results together with ours CRF and Bi-LSTM, especially our Bi-LSTM-CRF model results for comparison. The best F1 is 94.4% using Bi-LSTM-CRF model.

Table 3. English EDU words boundary recognition results

Model	P	R	F1
Li's Maxent	86.5	78.7	82.4
CRF	87.4	91	89.1
Bi-LSTM	94.0	91.9	92.9
Bi-LSTM-CRF	**95.5**	**93.4**	**94.4**

Figure 4 comprise the result of F1 between Chinese and English for different models. We can see the best model is Bi-LSTM-CRF model, by joint decoding label sequence can benefit the final performance of neural network models, followed by Bi-LSTM and CRF. The reason is that EDUs recognition is sequence tag task, Bi-LSTM and CRF classifier perform better than traditional Maxent classifier.

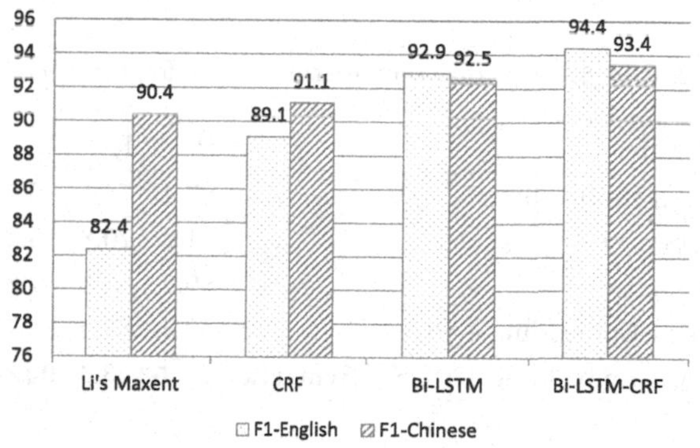

Fig. 4. Comparison of F1 between Chinese and English for different models

Figure 4 shows that English EDUs recognition result is higher than Chinese using Bi-LSTM or Bi-LSTM-CRF, the reason is that the pretrained embedding of Chinese words are more than English, with Chinese 35 598 where English

4 000, the two is 10 times difference. But for using Maxent or CRF model, Chinese EDUs identification F1 is higher than English.

4.3 The Contribution of Features

In order to investigate the contribution of the features, we give experiments specifically targeted at features for EDUs recognition. Table 4 shows the performance of P, R, F1 for Chinese separately using different feature, and Table 5 gives the results of English.

Table 4 and Table 5 show that syntactic feature outperform than other features, the F1 can reach 81.6% and 81.8% for Chinese and English. The reason is that both in Chinese and English, most EDU word syntactic labels contain IP and VP syntactic, while word with syntactic NP, PP and LCP are not EDU boundary. Syntactic information is highly related with EDUs recognition than other information. The combine of all features performance best both in Chinese and English, that means the more information used, the better the results.

Table 4. The different feature result for Chinese

Features	P	R	F1
Word Embedding	65.2	88.6	75.1
POS	70.1	80.2	74.8
Syntactic	81.1	82.1	81.6
Word Embedding + POS	76.7	90.7	83.1
Word Embedding + POS + Syntactic	92.3	**94.4**	**93.4**

Table 5. The different feature result for English

Features	P	R	F1
Word Embedding	87.4	74.5	80.4
POS	71.2	79.8	75.3
Syntactic	80.2	83.5	81.8
Word Embedding + POS	90.4	87.1	88.7
Word Embedding + POS + Syntactic	**95.5**	93.4	**94.4**

POS is the commonly used in NLP task, from the results, we find it is also useful for EDU recognition. As shown in Table 5, only using word embedding feature, we can get F1 80.4% for English. We also find that word embedding feature is useful than syntactic feature for English, mainly because Chinese word is sparing. And Chinese EDUs boundary usually have punctuation, which have IP tag, so syntactic feature is useful than word embedding feature for Chinese.

According to the results, we know that using word embedding, POS and syntactic feature together, we can get best result, it proves the effectiveness of our features.

4.4 Discussion

There are about 6% EDUs recognition error, and we discuss the reason as follows. There are two cases of errors: one is negative instances are recognized as positive instances. The other is positive instances are recognized as negative instances. From the recognition consistency compute method of Sect. 2.3, we notice the punctuation plays an important role in EDUs recognition, especially in Chinese. For example, if the front words of comma are the subject of the sentence, therefore the position of this comma is not EDU boundary. But when using our model, the syntactic of the words is IP, which may lead to mistake recognition.

In EDUs recognition, it is difficult to distinguish EDUs from complex sentence structure. For example, if you believe that "在…以后，终于(In...After that, finally)" is a connective that expresses the relation of succession. It can be considered as an EDU. However, traditional grammar generally analyzes it as an adverbial, a part of the syntactic structure. This is transition between textual structure and syntactic structure. We currently follow the traditional grammar, leaving the analysis of this situation to the syntactic structure.

For the automatic alignment of Chinese and English EDUs, we found that most of EDUs are sequence alignment, only about 4% of EDUs adjusted sequentially when from Chinese to English. So, for EDUs alignment, the main problem is EDUs recognition, which is influence on the result of automatic alignment EDUs. The difficulty of EDUs alignment is that EDUs does not correspond and adjust in order, which needs further research.

This paper only does Chinese and English EDUs recognition respectively, but does not do Chinese-English EDUs alignment. Once EDUs are identified, the next step is to align, and since EDUs are basically one-to-one, EDUs alignment can be turned into a machine translation or classification problem.

5 Related Work

Due to the emergence of discourse corpus, there have been a lot of researches on the recognition of English discourse. One of the corpora which are widely used is Rhetorical Structure Theory Discourse Treebank (RSTDT) building by Carlson et al. (2003), the other is Penn Discourse Treebank (PDTB) annotated by PDTB Research Group (2007). The RST represents a discourse as a tree, with phrases or clauses as EDU. PDTB adopts the predicate-argument view, with two spans as its arguments.

Due to the EDUs in RST consecutive annotation, the EDUs automatic identification on RSTDT is also called EDUs recognition, and now there is much research on it and the results are ideal, more representative research results include: Soricut and Marcus (2003) adopt statistics method for recognition, the F1 of EDUs recognition on the automatic syntax tree and standard syntax tree are 83.1% and 84.7%. Hernault et al. (2010) give a discourse recognition model based on sequential data annotation. They use lexical and syntactic features get the F1 94%, which is close to 98% of the F1 of manually. According to the above we can know that recognition accuracy of EDUs on RSTDT is relatively high, and there is little room for further improvement. For the un-sequential annotation of arguments on PDTB, not all the discourse is covered. So, some researchers propose to replace the whole argument with the argument center in the recognition of argument (Wellner and Pustejovsky 2007; Elwell and Baldridge 2008; Wellner 2009). And other researches put forward to the point of identifying sentences that contain arguments (Prasad et al. 2010), the recognition accuracy of Arg1 and Arg2 are 65% and 85% (Xu 2013). Braund et al. (2017) research whether syntax help discourse segmentation, the results show that dependency information is less useful than expected, but they provide a fully scalable, robust model that only relies on part-of-speech information, and show that it performs well across languages in the absence of any gold-standard annotation.

Deep learning method has made breakthroughs in many NLP tasks in recent years. Among them, Cyclic Neural Network (RNN) is a typical sequence marking model, and it is proposed by Goller and Kuchler (1996). However, RNN is limited by gradient disappearance and gradient explosion, Hochreiter and Schmidhuber (1997) come up with the variation of RNN which is named Long Short-Term Memory (LSTM). Because it only gets one-way contextual information, Graves and Schmidhuber (2005) raise the Bi-directional Long Short-Term Memory (Bi-LSTM), and applied it to speech identification. Bi-LSTM can effectively utilize past and future features in a specific time range. On the other hand, Conditional Random Field (CRF) algorithm which is put forward by Lafferty et al. (2001) has been widely applied in NLP recent years. In sequence marking tasks, CRF can take into account the anteroposterior dependence between adjacent labels of output. Considering the above reasons, there are some studies attempting to combine Bi-LSTM and CRF to build model for sequence data (Ji Me et al. 2018). Bi-LSTM and CRF hybrid model were first applied to the sequence labeling task of NLP by Huang et al. (2015), Ma and Hovy (2016) focus Bi-LSTM, CRF and CNN models and apply them to sequence marking tasks. Bi-LSTM-CRF model is applied in identifying biomedicine named entity (Greenberg et al. 2018), The effectiveness of the model in sequence marking tasks is gradually verified.

There are few discourse corpora in Chinese to mark EDU information (Zhang et al. 2014; Li et al. 2014). At present, the task of EDU recognition is few referred. Zhang et al. (2014) only identified the relation, but no relevant result about argument identification. Li et al. (2014) research on Chinese EDUs recognition based on comma, and Chinese EDUs recognition result can reach 90%. Ge Haizhu et al. (2019) proposes a Chinese EDU recognition approach based on theme-rheme theory, which can pay more attention on the internal structure of EDU, and the F1 score is 89.96%. However, limited by bilingual corpus, there is no EDUs recognition of both Chinese and English research.

6 Conclusion

The discourse alignment corpus of Chinese-English is annotated in this paper. The corpus has a complete EDU definition, annotation method, quality assurance and available scale. The corpus we annotated in this paper is the basic task of EDUs recognition.

Then we developed an EDUs recognition system using Bi-LSTM-CRF model. Our neural model achieved satisfactory results for Chinese and English EDU recognition. To our knowledge, we are among the first to develop an effective neural network-based approach to recognize EDUs for both Chinese and English. We input word embedding, POS and syntactic feature to our model in order to improve the result. By incorporating these features, our model can extract EDUs automatically and high quality. The F1 can reach 93.4% and 94.4% for Chinese and English separately, which is reaching the practical using.

This model can also be generalized to solve other problems. In the future, we will improve the effect of recognition Chinese and English EDUs, then try to automatic align them.

Acknowledgements. This paper is supported by the National Natural Science Foundation of China (61502149), by the Open Research Fund of Key Laboratory of Advanced Theory and Application in Statistics and Data Science (East China Normal University), Ministry of Education (KLATASDS1806), as well as the high-level talent research project of Henan Institute of Science and Technology (2017039).

References

Braud, C., Lacroix, O., Anders, S.: Does syntax help discourse segmentation? Not so much. In: Conference on Empirical Methods in Natural Language Processing, Copenhagen, Denmark, pp. 2432–2442. Association for Computational Linguistics (2017)

Carlson, L., Marcu, D., Okurowski, M.E.: Building a discourse-tagged corpus in the framework of rhetorical structure theory. In: van Kuppevelt, J., Smith, R.W. (eds.) Current and New Directions in Discourse and Dialogue. Text, Speech and Language Technology, vol. 22, pp. 85–112. Springer, Dordrecht (2003). https://doi.org/10.1007/978-94-010-0019-2_5

Duchi, J., Hazan, E., Singer, Y.: Adaptive subgradient methods for online learning and stochastic optimization. J. Mach. Learn. Res. **12**(7), 257–269 (2011)

Elwell, R., Baldridge, J.: Discourse connective argument identification with connective specific rankers. In: IEEE International Conference on Semantic Computing, pp. 198–205 (2008)

Feng, W.H.: Alignment and annotation of Chinese-English discourse structure parallel corpus. J. Chin. Inf. Process. **27**(6), 158–165 (2013)

Ge, H.Z., Kong, F., Zhou, G.D.: Chinese elementary discourse unit recognition based on theme-rheme theory. J. Chin. Inf. Process. **33**(8), 20–27 (2019)

Goller, C., Kuchler, A.: Learning task-dependent distributed representations by backpropagation through structure. In: IEEE International Conference on Neural Networks, Washington, DC, USA, pp. 347–352 (1996)

Graves, A., Schmidhuber, J.: Framewise phoneme classification with bidirectional LSTM and other neural network architectures. Neural Netw. **18**(5), 602–610 (2005)

Greenberg, N., Bansal, T., Verga, P., McCallum, A.: Marginal likelihood training of BiLSTM-CRF for biomedical named entity recognition from disjoint label sets. In: Proceedings of the Conference on Empirical Methods in Natural Language Processing, Brussels, Belgium, pp. 2824–2829. Association for Computational Linguistics (2018)

Hernault, H., Bollegala, D., Ishizuka, M.: A sequential model for discourse segmentation. In: Gelbukh, A. (ed.) CICLing 2010. LNCS, vol. 6008, pp. 315–326. Springer, Heidelberg (2010). https://doi.org/10.1007/978-3-642-12116-6_26

Hochreiter, S., Schmidhuber, J.: Long short-term memory. Neural Comput. **9**(8), 1735–1780 (1997)

Huang, Z., Xu, W., Yu, K.: Bidirectional LSTM-CRF models for sequence tagging. Computation and Language (2015)

Ji, M., Kuzman, G., David, W: State-of-the-art Chinese word recognition with Bi-LSTMs. In: Proceedings of the Conference on Empirical Methods in Natural Language Processing, Melbourne, Australia, pp. 4902–4908. Association for Computational Linguistics (2018)

Lafferty, J., Mccallum, A., Pereira, F.: Conditional random fields: probabilistic models for segmenting and labeling sequence data. In: Proceedings of the Eighteenth International Conference on Machine Learning, pp. 282–289. Morgan Kaufmann Publishers Inc (2001)

Li, S., Zhao, Z., Hu, R., et al.: Analogical reasoning on Chinese morphological and semantic relations. In: Proceedings of the 56th Annual Meeting of the Association for Computational Linguistics, Melbourne, Australia, pp. 138–143. Association for Computational Linguistics (2018)

Li, Y.C., Feng, W.H., Sun, J., et al.: Building Chinese discourse corpus with connective-driven dependency tree structure. In: Proceedings of Empirical Methods in Natural Language Processing, Doha, Qatar, pp. 2105–2114. Association for Computational Linguistics (2014)

Li, Y.C., Feng, W.H., Zhou, G.D., et al.: Research of Chinese clause identification based on comma. Acta Scientiarum Naturalium Universitatis Pekinensis **49**(1), 7–14 (2013)

Ma, X., Hovy, E.: End-to-end sequence labeling via Bi-directional LSTM-CNNs-CRF. In: Proceedings of the Meeting of the Association for Computational Linguistics, Berlin, Germany, pp. 1064–1074. Association for Computational Linguistics (2016)

Manning, C.D., Mihai, S., John, B., et al.: The stanford core NLP natural language processing toolkit. In: Proceedings of the 52nd Annual Meeting of the Association for Computational Linguistics, Baltimore, Maryland, pp. 55–60. Association for Computational Linguistics (2014)

Mikolov, T., Sutskever, I., Chen, K., et al.: Distributed representations of words and phrases and their compositionality. In: Advances in Neural Information Processing Systems 26, pp. 3111–3119 (2013)

PDTB Research Group. The Penn discourse Treebank 2.0 annotation manual. IRCS Technical Reports Series (2007)

Prasad, R., Joshi, A.K., Webber, B.L.: Exploiting scope for shallow discourse parsing. In: Proceedings of the Seventh International Conference on Language Resources and their Evaluation, Valletta, Malta, pp. 2076–2083 (2010)

Soricut, R., Marcus, D.: Sentence level discourse parsing using syntactic and lexical information. In: Proceedings of the 2003 Conference of the North American, pp. 149–156 (2003)

Wellner, B., Pustejovsky, J.: Automatically identifying the arguments of discourse connectives. In: EMNLP-CoNLL, Prague, Czech Republic, pp. 92–101. Association for Computational Linguistics (2007)

Srivastava, N., Hinton, G., Krizhevsky, A., et al.: Dropout: a simple way to prevent neural networks from overfitting. J. Mach. Learn. Res. **15**(1), 1929–1958 (2014)

Wellner, B.: Sequence models and ranking methods for discourse parsing. Faculty of the Graduate School of Arts and Sciences Brandeis University Computer Science James Pustejovsky, Brandeis University (2009)

Xu, F.: Research of key issues in english discourse structure analysis. Soochow University (2013)

Zhang, M.Y., Qin, B., Liu, T.: Chinese discourse relation semantic taxonomy and annotation. J. Chin. Inf. Process. **28**(2), 28–36 (2014)

Social Computing and Sentiment Analysis

Better Queries for Aspect-Category Sentiment Classification

Yuncong Li[1], Cunxiang Yin[1], Sheng-hua Zhong[2](✉), Huiqiang Zhong[1], Jinchang Luo[1], Siqi Xu[1], and Xiaohui Wu[1]

[1] Baidu Inc., Beijing, China
{liyuncong,yincunxiang,zhonghuiqiang,luojinchang,xusiqi01, wuxiaohui02}@baidu.com

[2] College of Computer Science and Software Engineering, Shenzhen University, Shenzhen, China
csshzhong@szu.edu.cn

Abstract. Aspect-category sentiment classification (ACSC) aims to identify the sentiment polarities towards the aspect categories mentioned in a sentence. Because a sentence often mentions more than one aspect category and expresses different sentiment polarities to them, finding aspect category-related information from the sentence is the key challenge to accurately recognize the sentiment polarity. Most previous models take both sentence and aspect category as input and query aspect category-related information based on the aspect category. However, these models represent the aspect category as a context-independent vector called aspect embedding, which may not be effective enough as a query. In this paper, we propose two contextualized aspect category representations, Contextualized Aspect Vector (CAV) and Contextualized Aspect Matrix (CAM). Specifically, we use the coarse aspect category-related information found by the aspect category detection task to generate CAV or CAM. Then the CAV or CAM as queries are used to search for fine-grained aspect category-related information like aspect embedding by aspect-category sentiment classification models. In experiments, we integrate the proposed CAV and CAM into several representative aspect embedding-based aspect-category sentiment classification models. Experimental results on the SemEval-2014 Restaurant Review dataset and the Multi-Aspect Multi-Sentiment dataset demonstrate the effectiveness of CAV and CAM.

Keywords: Aspect-category sentiment classification · Contextualized Aspect Vector · Contextualized Aspect Matrix

1 Introduction

Sentiment analysis [9,10] is an important task in Natural Language Processing (NLP). It deals with the computational treatment of opinion, sentiment, and subjectivity in text. Aspect-based sentiment analysis [13–15] is a branch of sentiment

Y. Li, C. Yin—Equal contribution.

M. Sun et al. (Eds.): CCL 2020, LNAI 12522, pp. 347–358, 2020.
https://doi.org/10.1007/978-3-030-63031-7_25

analysis and aspect-category sentiment analysis (ACSA) is a subtask of it. In ACSA, there are a predefined set of aspect categories, and a predefined set of sentiment polarities. Given a sentence, the task aims to predict the aspect categories mentioned in the sentence and the corresponding sentiments. Therefore, ACSA contains two subtasks: aspect category detection (ACD) that detects aspect categories in a sentence and aspect-category sentiment classification (ACSC) that categorizes the sentiment polarities with respect to the detected aspect categories. Figure 1 shows an example, "Staffs are not that friendly, but the taste covers all". ACD detects the sentence mentions two aspect categories: *service* and *food*, and ACSC predicts the sentiment polarities to them: negative and positive respectively. In this work, we focus on ACSC, while ACD as an auxiliary task is used to find coarse aspect category-related information for the ACSC task.

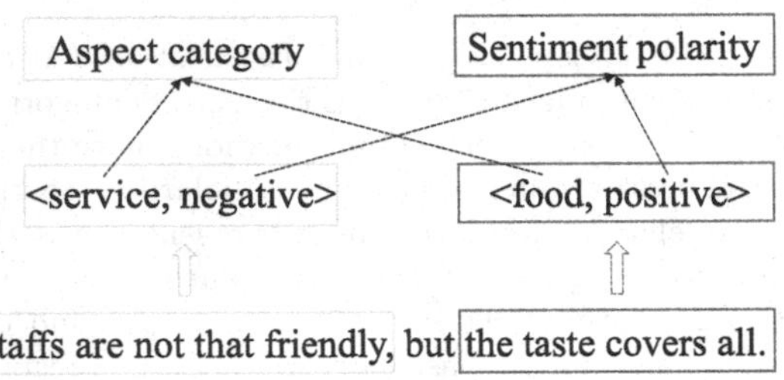

Fig. 1. An example of aspect-category sentiment analysis.

Because a sentence often mentions more than one aspect category and expresses different sentiment polarities to them, to accurately recognize the sentiment polarities, most previous models [1,3,4,6,8,16,20–24] take both sentence and aspect category as input and query aspect category-related information based on the aspect category, then generate aspect category-specific representations for aspect-category sentiment classification. However, these models represent the aspect category as a context-independent vector called aspect embedding (AE). These models can be called aspect embedding-based models. Since aspect embedding only contains the global information of aspect category and loses the context-dependent information, it is semantically far away from the words in the sentence, and may not be effective enough as a query to search for aspect category-related information for the ACSC task. These models may be improved by replacing the aspect embedding with context-dependent aspect category representations.

The HiErarchical ATtention (HEAT) network [1] used context-dependent aspect category representations to search for aspect category-related information for the ACSC task and obtained better performance. The context-dependent aspect category representations are generated by concatenating the aspect embedding and the aspect term representation in a sentence. An aspect term is a word or phrase that appears in the sentence explicitly indicating an aspect category. For the example in Fig. 1, the aspect terms are "Staffs" and "taste" indicating aspect category *service* and *food* respectively. However, the HEAT network requires aspect term annotation information that the data for ACSC usually does not have. Moreover, the HEAT network ignores the situation where the aspect category is mentioned implicitly in sentences without any aspect term, making aspect category representations degenerate to context-independent representations in this situation.

In this paper, we propose two novel contextualized aspect category representations, Contextualized Aspect Vector (CAV) and Contextualized Aspect Matrix (CAM). CAV or CAM contain context-dependent information even though there are no aspect terms in sentences, and aspect term annotation information is not required to generate them. Concretely, we use the coarse aspect category-related information found by the ACD task to generate CAV or CAM. Then CAV or CAM as queries are used to search for fine-grained aspect category-related information like aspect embedding by aspect-category sentiment classification models. Specifically, we first use an attention-based aspect category classifier to obtain the weights of the words in a sentence, which indicate the degree of correlation between the aspect categories and the words. Then, we get CAV by combining the weighted sum of the word representations with corresponding aspect embedding. That is to say, CAV contains two kinds of representations of an aspect category: context-independent representation and context-dependent representation, which capture global information and local information respectively. Since CAV may lose details of the words, we also propose an aspect category matrix representation, called Contextualized Aspect Matrix (CAM), which is a not-sum version of CAV.

In summary, the main contributions of our work can be summarized as follows:

- We propose two novel contextualized aspect category representations, Contextualized Aspect Vector (CAV) and Contextualized Aspect Matrix (CAM). They include the global information and local information about the aspect category and are better queries to search for aspect category-related information for aspect category sentiment classification (ACSC). To the best of our knowledge, it is the first time to represent aspect category as matrix.
- We experiment with several representative aspect embedding-based models by replacing the aspect embedding with CAV or CAM. Experimental results on the SemEval-2014 Restaurant Review dataset and the Multi-Aspect Multi-Sentiment dataset demonstrate the effectiveness of CAV and CAM.

2 Related Work

In this section, we first present a brief review about aspect-category sentiment classification. Then, we show the related study on context-aware aspect embedding that is a kind of context-dependent aspect category representation for targeted aspect based sentiment analysis (TABSA).

2.1 Aspect-Category Sentiment Classification

Many models [1,3,4,6,8,16,18–24] have been proposed for the aspect-category sentiment classification (ACSC) task. Wang et al. [21] proposed an attention-based LSTM network for aspect-level sentiment classification. Tay et al. [20] introduced a word-aspect fusion attention layer to attend based on associative relationships between sentence words and aspect categories. Xue et al. [24] proposed to extract sentiment features with convolutional neural networks and selectively output aspect category related features for classification with gating mechanisms. Xing et al. [23] proposed a novel variant of LSTM, which incorporates aspect information into LSTM cells in the context modeling stage. Liang et al. [8] proposed a novel Aspect-Guided Deep Transition model, which utilizes the given aspect category to guide the sentence encoding from scratch. Jiang et al. [4] proposed new capsule networks to model the complicated relationship between aspects and contexts. To force the orthogonality among aspect categories, Hu et al. [3] proposed constrained attention networks (CAN) for multi-aspect sentiment analysis. To avoid error propagation, some joint models [6,18,22] have been proposed, which perform aspect category detection (ACD) and aspect-category sentiment classification (ACSC) jointly. Li et al. [6] proposed an end-to-end machine learning architecture, in which the ACD task and the ACSC task are interleaved by a deep memory network. Wang et al. [22] proposed the aspect-level sentiment capsules model (AS-Capsules), which utilizes the correlation between aspect and sentiment through shared components including capsule embedding, shared encoders, and shared attentions. The capsule embedding is similar to the aspect embedding. All these models represented aspect category as context-independent representations, which may benefit from CAV or CAM.

Closely related to our method is the HiErarchical Attention (HEAT) network proposed by Cheng et al. [1], in which an aspect attention extracts the aspect term information, and then a context-dependent aspect category representation generated based on the aspect term information is used to guide the sentiment attention to better allocate aspect-specific sentiment words of the text. However, extracting aspect term information requires additional aspect term annotation information. In addition, HEAT ignores the situation where the aspect category is mentioned implicitly in texts. There are also some models that don't rely on aspect embedding. Schmitt et al. [18] also proposed a joint model, in which different aspect categories have different sentiment classifiers to generate aspect category-specific representations. Sun et al. [19] constructed an auxiliary sentence from the aspect and converted ABSA to a sentence-pair classification task.

2.2 Context-Aware Aspect Embedding

Context-aware aspect embedding is a kind of context-dependent aspect category representation [7]. Liang et al. [7] proposed an embedding refinement method to generate context-aware target embedding and aspect embedding for targeted aspect based sentiment analysis (TABSA) [17], which utilizes a sparse coefficient vector to adjust the embeddings of target and aspect from the context and yields the state-of-the-art performance in this task. However, their method relies on context-aware target embedding to generate aspect embedding, and can't be applied in the ACSC task directly.

3 Method

In this section, we describe our proposed two contextualized aspect category representations, Contextualized Aspect Vector (CAV) and Contextualized Aspect Matrix (CAM), in detail.

Motivated by the process that people search for information through search engines: before finding the result they want, they usually try different words and adjust their queries based on previous results, the process to generate CAV or CAM consists of two steps. In the first step, the ACD task as an auxiliary task is used to find coarse aspect category-related information. In the second step, the coarse aspect category-related information is used to optimize original query (e.g. aspect embedding). Specifically, an attention-based aspect category classifier generates the weights of the words in a sentence about all predefined categories. Then the weights are used to generate CAV and CAM. The framework of our proposed method is demonstrated in Fig. 2.

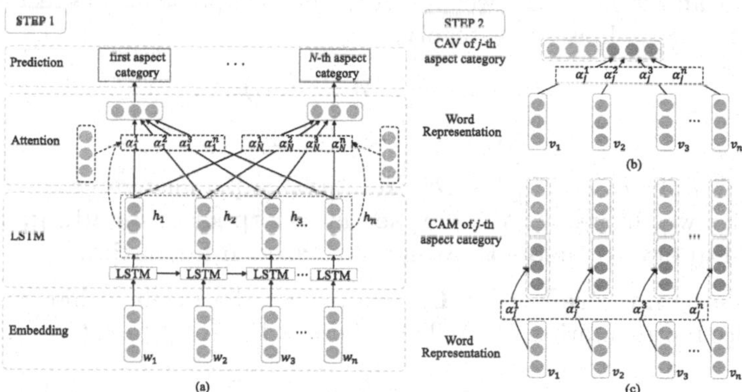

Fig. 2. (a) shows the attention-based aspect category classifier, which generates the weights of the words in a sentence about all predefined aspect categories. (b) and (c) show how to generate CAV and CAM based on the weights and the original representations of the words respectively.

3.1 Coarse Aspect Category-Related Information

In this step, the ACD task is used to find coarse aspect category-related information. It is a multi-label classification problem, and can be formulated as follows.

There are N predefined aspect categories $A = \{A_1, A_2, ..., A_N\}$ in the dataset. Given a sentence, denoted by $S = \{w_1, w_2, ..., w_n\}$, the task checks each aspect $A_j \in A$ to see whether the sentence S mentions it.

An attention-based aspect category classifier is used for this task, because it can offer the weights of the words in a sentence about all predefined categories indicating which word is related to which aspect category. The overall architecture of the model is illustrated in Fig. 2(a). The model contains four modules: embedding layer, LSTM layer, attention layer, and aspect category prediction layer. All aspect categories share the embedding layer and the LSTM layer, and different aspect categories have different attention layers and prediction layers.

Embedding Layer: The input of this layer is a sentence consisting of n words $\{w_1, w_2, ..., w_n\}$. With an embedding matrix U, the input sentence is converted to a sequence of vectors $X = \{x_1, x_2, ..., x_n\}$, where $U \in R^{d \times |V|}$, d is the dimension of the word embeddings, and $|V|$ is the vocabulary size.

LSTM Layer: The word embeddings of the sentence are then fed into a LSTM [2] layer, which outputs hidden states $H = \{h_1, h_2, ..., h_n\}$. At each time step i, the hidden state h_i is computed by:

$$h_i = LSTM(h_{i-1}, x_i) \tag{1}$$

The size of the hidden state is also set to be d.

Attention Layer: This layer takes the output of the LSTM layer as input, and produce an attention [25] weight vector for each predefined aspect category. Formally, for the j-th aspect category:

$$M_j = tanh(W_j H + b_j) \tag{2}$$

$$\alpha_j = softmax(u_j^T M_j) \tag{3}$$

where $W_j \in R^{d \times d}$,$b_j \in R^d$,$u_j \in R^d$ are learnable parameters, and $\alpha_j \in R^n$ is the attention weight vector. **We can see u_j as aspect embedding, which is the initial query for aspect category-related information.**

Aspect Category Prediction Layer: We use the weighted hidden state as the sentence representation for ACD prediction. For the j-th category:

$$r_j = H\alpha_j^T \tag{4}$$

$$\hat{y}_j = sigmoid(W_j r_j + b_j) \tag{5}$$

where $W_j \in R^{d \times 1}$ and $b_j \in R$.

Loss: As each prediction is a binary classification problem, the loss function for the N aspect categories of the sentence is defined by:

$$L(\theta) = -\sum_{j=1}^{N} y_j log\hat{y}_j + (1 - y_j)log(1 - y_j) + \lambda||\theta||_2^2 \tag{6}$$

where y_j is the correct label, λ is the L_2 regularization factor, N is the number of total aspect categories and θ contains all the parameters.

3.2 Context-Dependent Aspect Category Representations

In this step, the attention weight vectors offered by the ACD task is used to generate contextualized Aspect Vector (CAV) and Contextualized Aspect Matrix (CAM). **They are the results of optimizing the initial query based on context-dependent information**. Figure 2(b) and Fig. 2(c) show how to generate CAV and CAM respectively. Given a sentence representation $V = \{v_1, v_2, \ldots, v_n\}$ from an ACSC model and the attention weight vectors of all predefined aspect categories offered by the ACD task, CAV of the j-th aspect category is computed by:

$$v_{CAV_j} = [v_{CAVG_j}; v_{CAVL_j}] \tag{7}$$

$$v_{CAVL_j} = \sum_{i=1}^{n} v_i \alpha_j^i \tag{8}$$

where $v_i \in R^{d_l}$ and d_l is the dimension of the word representations, $v_{CAVG_j} \in R^{d_g}$ and $v_{CAVL_j} \in R^{d_l}$ are the global representation and the local representation respectively, d_g is the dimension of the global aspect category representation, v_{CAVG_j} is initialized randomly and learned during training ACSC models like aspect embedding, and α_j^i indicates the weight of the i-th word about the j-th aspect category. V can be the output of the embedding layer or the sentence encoder in ACSC models.

Because the aspect category representation vectors, such as aspect embedding, often are repeated as many times as there are words in the sentence and concatenated to the word representations of the sentence, we also propose the Contextualized Aspect Matrix (CAM), which can be directly concatenated to the word representations and retains more details of the words. For the j-th aspect category, M_{CAM_j} is computed by:

$$M_{CAM_j} = \{[v_{CAVG_j}; v_1\alpha_j^1], [v_{CAVG_j}; v_2\alpha_j^2], ..., [v_{CAVG_j}; v_n\alpha_j^n]\} \tag{9}$$

where v_{CAVG_j} is the same as it in CAV.

Then the CAV or CAM as queries are used to search for fine-grained aspect category-related information like aspect embedding by ACSC models. Figure 3 shows how to integrate CAV and CAM into AT-LSTM [21].

4 Experiments

4.1 Datasets

In order to evaluate the effectiveness of our methods, we conduct experiments on the SemEval-2014 Restaurant Review (Restaurant-2014) dataset [15] and the Multi-Aspect Multi-Sentiment for Aspect Category Sentiment Analysis (MAMS-ACSA) dataset [4]. The Restaurant-2014 is a widely used dataset. However, most sentences in Restaurant-2014 contain only one aspect category or multiple aspect

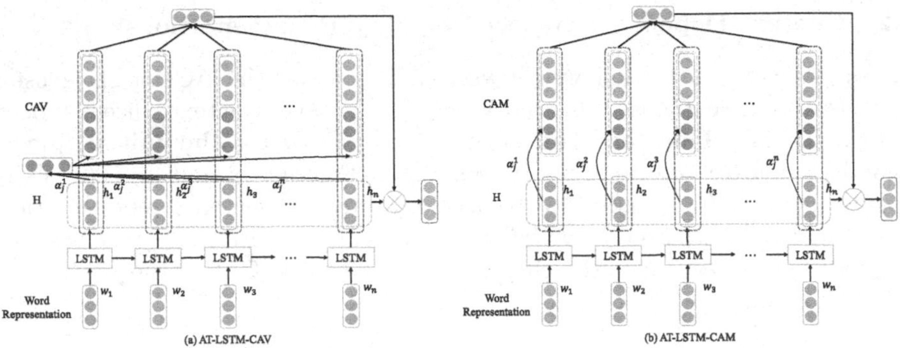

Fig. 3. AT-LSTM-CAV and AT-LSTM-CAM, which are obtained by replacing the aspect embedding in AT-LSTM [21] with CAV and CAM respectively.

categories with the same sentiment polarity, which makes ABSA task degenerate to sentence-level sentiment analysis. To mitigate the problem, Jiang et al. [4] released the MAMS-ACSA dataset, all sentences in which contain multiple aspects with different sentiment polarities. Since there is no official development set for the Restaurant-2014 dataset, we use the split offered by Xue et al. [24]. Statistics of these two datasets are given in Table 1.

Table 1. Statistics of the datasets.

Dataset		Positive	Negative	Neutral	Total
Restaurant-2014	Train	1855	733	430	3018
	Validation	324	106	70	500
	Test	657	222	94	973
MAMS-ACSA	Train	1929	2084	3077	7090
	Validation	241	259	388	888
	Test	245	263	393	901

4.2 Implementation Details

We implement our models in PyTorch [11]. For all models, including the aspect category classifier and the aspect-category sentiment classification models, we use the pre-trained 300d Glove embeddings [12] to initialize word embeddings, which is fixed in all models. We use Adam optimizer [5] with learning rate 0.001 to train all models. We set L_2 regularization factor $\lambda = 0.00001$. The batch sizes are set to 32 and 64 for the Restaurant-2014 dataset and the MAMS-ACSA dataset respectively. For CAV and CAM, d_g is equivalent to d_l. For the aspect category sentiment classification models, we replace the aspect embedding with the CAV or CAM, just adjust the parameters to make the dimensions matching, and use hyper-parameter settings described in original papers. The aspect category classifier and the aspect-category sentiment classification models are trained in a pipeline manner. That is to say, the aspect category classifier is first trained, then the aspect-category sentiment classification models are trained, where the attention weights offered by the aspect category classifier are used to generate CAV or CAM. We fine-tune the hyper-parameters for all baselines on the validation set. We run all models for 5 times and report the average results on the test datasets.

4.3 Comparison Methods

We select the following methods as baseline models:

AE-LSTM [21] first get the aspect-aware sentence embedding by concatenating the aspect embedding with each word embedding. Then the aspect-aware sentence embedding is fed into a LSTM layer. The final sentence representation is the last hidden state of the LSTM layer.

AT-LSTM [21] models the sentence via a LSTM model. Then it combines the hidden states from the LSTM with the aspect embedding to generate the attention vector. The final sentence representation is the weighted sum of the hidden states.

ATAE-LSTM [21] further extends AT-LSTM by taking the aspect-aware sentence embedding as input.

CapsNet [4] is a capsule network that can model the complicated relationship between aspect categories and contexts and obtains state-of-the-art performance on the MAMS-ACSA dataset. It also takes the aspect-aware sentence embedding as input.

Our methods:

***-CAV** replace the aspect embedding in the baseline models with CAV.

***-CAM** replace the aspect embedding in the baseline models with CAM

4.4 Results and Analysis

Experimental results are illustrated in Table 2. From Table 2 we draw the following conclusions. First, we observe that most models with CAV obtain better performance. Specifically, by replacing the aspect embedding with CAV, our proposed methods outperform their counterparts in 5 of 8 results. Compared original models, AT-LSTM-CAV and ATAE-LSTM-CAV improves the performance by 3.9% and 3.4% on the Restaurant-2014 dataset respectively. AE-LSTM-CAV, AT-LSTM-CAV and ATAE-LSTM-CAV improves the performance by 3.9%, 6.6% and 2.5% on the MAMS-ACSA dataset respectively. In addition, AT-LSTM-CAV obtains the best performance on Restaurant-2014. Second, most models with CAM also obtain better performance. Specifically, by replacing the aspect embedding with CAM, most of our proposed methods outperform their counterparts. AE-LSTM-CAM, AT-LSTM-CAM and ATAE-LSTM-CAM

Table 2. Results of the ACSC task in terms of accuracy (%). "*" refers to citing from Tay et al. [20]. "†" refers to citing from Jiang et al. [4]. Best scores are marked in bold.

Method	Restaurant-2014	MAMS-ACSA
AE-LSTM	76.876 (±2.037)	63.019 (±2.318)
AE-LSTM-CAV	76.711 (±0.963)	66.970 (±0.824)
AE-LSTM-CAM	80.493 (±1.422)	70.721 (±0.717)
AT-LSTM	77.9*	66.436†
AT-LSTM-CAV	**81.891 (±0.493)**	73.052 (±1.551)
AT-LSTM-CAM	80.740 (±0.681)	**75.539 (±0.657)**
ATAE-LSTM	77.8*	70.634†
ATAE-LSTM-CAV	81.172 (±0.398)	73.141 (±1.499)
ATAE-LSTM-CAM	81.829 (±0.784)	73.452 (±1.217)
CapsNet	81.110 (±0.492)	73.986†
CapsNet-CAV	77.246 (±0.696)	69.700 (±0.659)
CapsNeT-CAM	80.417 (±0.558)	75.117 (±0.203)

improves the performance by 3.6%, 2.8% and 4% on the Restaurant-2014 dataset, by 7.7%, 9.1% and 2.8% on the MAMS-ACSA dataset, respectively. AT-LSTM-CAM and CapsNeT-CAM surpass the state-of-the-art baseline mode CapsNeT (+1.6% and +1.1% respectively) on the MAMS-ACSA dataset. Third, CAM outperform CAV in 7 of 8 results. This is because CAM retains more details of the words. Finally, we observe that, in 4 of 6 results, CAV leads to performance drop when aspect category sentiment classification models use it to get aspect-aware sentence embedding by concatenating it with each word embedding. Specifically, compared to AE-LSTM, AT-LSTM-CAV and CapsNet, AE-LSTM-CAV, ATAE-LSTM-CAV and CapsNet-CAV reduce by 0.2%, 0.7% and 4.6% on the Rest14 dataset. Compared to CapsNet, CapsNet-CAV reduces by 4.2% on the MAMS-ACSA dataset. The possible reason is that, in this situation, every word representation contains all aspect category-related information of the sentence, which leads to the sentence encoder, such as LSTM [2], to concentrate on the aspect category-related information and discard the aspect category-related sentiment information. It suggests that CAV be best used in attention mechanisms.

4.5 Attention Visualizations

Figure 4 displays the performance of the attention to find aspect category-related words for the ACSC task. Sentence 1 shows that the attention can find the aspect terms for different aspect categories obviously. In sentence 2, while the aspect term for the aspect category *service* is "taste", the attention finds "friendly" that is more useful than "taste" for the ACSC task. The sentence 3 don't have any aspect term for the aspect category *price*, however, the attention also finds the useful word "cheap".

Sentence id	Aspect category	Attention weights
1	food	0.06 0.07 0.63 0.22 I go to Sushi Rose for fresh **sushi** and great **portions** all at a reasonable price
	price	1.00 I go to Sushi Rose for fresh sushi and great portions all at a reasonable price
2	food	0.04 0.93 Staffs are not that friendly, but the taste covers all.
	service	0.99 **Staffs** are not that friendly, but the taste covers all
3	price	0.99 I thought the food isn't cheap at all compared to Chinatown

Fig. 4. Visualization of attention weights of different aspect categories in the ACD task. The numbers on the top of words are the attention weights of the words. The weights greater than 0.01 are labeled. The bold words are the labeled aspect terms. The color depth expresses the important degree of the word.

5 Conclusion

In this paper, we propose two novel contextualized aspect category representations, Contextualized Aspect Vector (CAV) and Contextualized Aspect Matrix (CAM). They include both the global information and local information about the aspect category and are better queries to search for aspect category-related information for the ACSC task. Moreover, CAV or CAM contain context-dependent information even though there are no aspect terms in sentences, and aspect term annotation information is not required to generate them. We experiment with several representative aspect embedding-based models by replacing the aspect embedding with CAV or CAM. Experimental results on the SemEval-2014 Restaurant dataset and the Multi-Aspect Multi-Sentiment (MAMS) dataset show that the variants with CAV or CAM obtain better performance. In future works, we will explore the performance of CAV and CAM with knowledge from open knowledge graphs on the ACSC task.

References

1. Cheng, J., Zhao, S., Zhang, J., King, I., Zhang, X., Wang, H.: Aspect-level sentiment classification with heat (hierarchical attention) network. In: CIKM, pp. 97–106 (2017)
2. Hochreiter, S., Schmidhuber, J.: Long short-term memory. Neural Comput. **9**(8), 1735–1780 (1997)
3. Hu, M., et al.: Can: constrained attention networks for multi-aspect sentiment analysis. In: EMNLP-IJCNLP, pp. 4593–4602 (2019)
4. Jiang, Q., Chen, L., Xu, R., Ao, X., Yang, M.: A challenge dataset and effective models for aspect-based sentiment analysis. In: EMNLP-IJCNLP, pp. 6281–6286 (2019)
5. Kingma, D.P., Ba, J.: Adam: a method for stochastic optimization. arXiv preprint arXiv:1412.6980 (2014)
6. Li, C., Guo, X., Mei, Q.: Deep memory networks for attitude identification. In: WSDM, pp. 671–680 (2017)

7. Liang, B., Du, J., Xu, R., Li, B., Huang, H.: Context-aware embedding for targeted aspect-based sentiment analysis. In: Proceedings of the 57th Annual Meeting of the Association for Computational Linguistics, Florence, Italy, pp. 4678–4683. Association for Computational Linguistics, July 2019
8. Liang, Y., Meng, F., Zhang, J., Xu, J., Chen, Y., Zhou, J.: A novel aspect-guided deep transition model for aspect based sentiment analysis. In: EMNLP-IJCNLP, pp. 5572–5584 (2019)
9. Liu, B.: Sentiment analysis and opinion mining. Synth. Lect. Hum. Lang. Technol. **5**(1), 1–167 (2012)
10. Pang, B., Lee, L.: Opinion mining and sentiment analysis. Found. Trends® Inf. Retrieval **2**(1–2), 1–135 (2008)
11. Paszke, A., et al.: Automatic differentiation in pytorch (2017)
12. Pennington, J., Socher, R., Manning, C.D.: Glove: global vectors for word representation. In: EMNLP, pp. 1532–1543 (2014)
13. Pontiki, M., et al.: Semeval-2016 task 5: aspect based sentiment analysis. In: Proceedings of the 10th International Workshop on Semantic Evaluation (SemEval-2016), pp. 19–30 (2016)
14. Pontiki, M., Galanis, D., Papageorgiou, H., Manandhar, S., Androutsopoulos, I.: Semeval-2015 task 12: aspect based sentiment analysis. In: Proceedings of the 9th International Workshop on Semantic evaluation (SemEval 2015), pp. 486–495 (2015)
15. Pontiki, M., Galanis, D., Pavlopoulos, J., Papageorgiou, H., Androutsopoulos, I., Manandhar, S.: SemEval-2014 task 4: aspect based sentiment analysis. In: Proceedings of the 8th International Workshop on Semantic Evaluation (SemEval 2014), pp. 27–35 (2014)
16. Ruder, S., Ghaffari, P., Breslin, J.G.: A hierarchical model of reviews for aspect-based sentiment analysis. In: EMNLP, pp. 999–1005 (2016)
17. Saeidi, M., Bouchard, G., Liakata, M., Riedel, S.: SentiHood: targeted aspect based sentiment analysis dataset for urban neighbourhoods. In: Proceedings of COLING 2016, the 26th International Conference on Computational Linguistics: Technical Papers, pp. 1546–1556. The COLING 2016 Organizing Committee, Osaka, Japan, December 2016
18. Schmitt, M., Steinheber, S., Schreiber, K., Roth, B.: Joint aspect and polarity classification for aspect-based sentiment analysis with end-to-end neural networks. In: EMNLP, pp. 1109–1114 (2018)
19. Sun, C., Huang, L., Qiu, X.: Utilizing BERT for aspect-based sentiment analysis via constructing auxiliary sentence. NAACL **1**, 380–385 (2019)
20. Tay, Y., Tuan, L.A., Hui, S.C.: Learning to attend via word-aspect associative fusion for aspect-based sentiment analysis. In: AAAI (2018)
21. Wang, Y., Huang, M., Zhu, X., Zhao, L.: Attention-based LSTM for aspect-level sentiment classification. In: EMNLP, pp. 606–615 (2016)
22. Wang, Y., Sun, A., Huang, M., Zhu, X.: Aspect-level sentiment analysis using as-capsules. In: The World Wide Web Conference, pp. 2033–2044 (2019)
23. Xing, B., et al.: Earlier attention? aspect-aware LSTM for aspect sentiment analysis. arXiv preprint arXiv:1905.07719 (2019)
24. Xue, W., Li, T.: Aspect based sentiment analysis with gated convolutional networks. In: ACL (Volume 1: Long Papers), pp. 2514–2523 (2018)
25. Yang, Z., Yang, D., Dyer, C., He, X., Smola, A., Hovy, E.: Hierarchical attention networks for document classification. In: NAACL, pp. 1480–1489 (2016)

Multimodal Sentiment Analysis with Multi-perspective Fusion Network Focusing on Sense Attentive Language

Xia Li[1,2(✉)] and Minping Chen[2]

[1] Guangzhou Key Laboratory of Multilingual Intelligent Processing, Guangzhou, China

[2] School of Information Science and Technology, Guangdong University of Foreign Studies, Guangzhou, China

{xiali,minpingchen}@gdufs.edu.cn

Abstract. Multimodal sentiment analysis aims to learn a joint representation of multiple features. As demonstrated by previous studies, it is shown that the language modality may contain more semantic information than that of other modalities. Based on this observation, we propose a Multi-perspective Fusion Network(MPFN) focusing on Sense Attentive Language for multimodal sentiment analysis. Different from previous studies, we use the language modality as the main part of the final joint representation, and propose a multi-stage and uni-stage fusion strategy to get the fusion representation of the multiple modalities to assist the final language-dominated multimodal representation. In our model, a Sense-Level Attention Network is proposed to dynamically learn the word representation which is guided by the fusion of the multiple modalities. As in turn, the learned language representation can also help the multi-stage and uni-stage fusion of the different modalities. In this way, the model can jointly learn a well integrated final representation focusing on the language and the interactions between the multiple modalities both on multi-stage and uni-stage. Several experiments are carried on the CMU-MOSI, the CMU-MOSEI and the YouTube public datasets. The experiments show that our model performs better or competitive results compared with the baseline models.

Keywords: Multimodal sentiment analysis · Multimodal fusion · Sense Attentive Language

1 Introduction

Multimodal sentiment analysis is a task of predicting sentiment of a video, an image or a text based on multiple modal features. With the increase of short videos on the internet, such as Douyin, YouTube, etc., multimoal sentiment analysis can be used to analyze the opinions of the public based on the speaker's language, facial gestures and acoustic behaviors.

M. Sun et al. (Eds.): CCL 2020, LNAI 12522, pp. 359–373, 2020.
https://doi.org/10.1007/978-3-030-63031-7_26

Based on the successes in video, image, audio and language processing, multimodal sentiment analysis has been studied extensively and produced impressive results in recent years [6–8,20,25]. The core of the multimodal sentiment analysis is to capture a better fusion of different modalities. Different methods are proposed to fuse the multimodal features and help to capture the interactions of the modalities. Tensor Fusion Network [22] is proposed to obtain raw unimodal representations, bimodal interactions and tri-modal interactions in the form of 2D-tensor and 3D-tensor simultaneously. Low-rank Fusion Network [7] is then proposed to alleviate the drawback of the large amount of parameters by low-rank factor. Although the above methods achieved good results, they treat all modalities equally and fuse the modalities in the same contribution. We find that language modality always contain more semantic information for sentiment analysis, that's why most of ablation experiments of previous studies [8,15,22] show that when using features from only one modality, the model using language features performs much better than using vision features or acoustic features.

In this paper, we take the assumption that the language modality contains more information than that of the vision and acoustic modalities. We regard language as the major modality and hope to use other modalities to assist the language modality to produce better performance for multimodal sentiment analysis. To this end, we propose a multi-perspective fusion network for multimodal sentiment analysis focusing on sense attentive language. Our model focuses on two aspects: (1) getting rich semantic language representation through the fusion of the sense level attention of language guided by other modalities. (2) learning comprehensive multimodal fusion from multiple perspectives, as well as keeping the enhanced language representation.

In order to get rich semantic information of the language modality, we incorporate a sense-level attention network into the model to obtain a more elaborate representation of the language. Generally speaking, there are many words which have more than one sense and their different senses may lead to different sentiment of a text in different context. Previous studies try to distinguish the ambiguities of a word from the text modality [21,27] using HowNet [4] and LIWC [12], while we hope the sense of a word can be distinguished not only by the context of the text but also by fusion of other modalities(video and acoustic). As an example shown in Fig. 1, we hope to predict the sentiment of the language "It would make sense". As can be seen, the word "sense" in the language modality has a higher attention weight which could be guided by the "smile face of the vision modality" and "high sound audio modality", and also by the "common sense" of the word "sense", which expresses more positive sentiment.

For the effectiveness of modal fusion, the key problem is to model the gap between different modalities and to learn a better multimodal fusion. In this paper, we propose a multi-stage and uni-stage strategy to fuse the multiple modalities in order to capture the interactions between multi-stage sharing information and global information integrated from uni-stage fusion. For multi-stage fusion, we use CNN with different window sizes to capture the multimodal fusion of consecutive temporals within different windows respectively. As for uni-stage

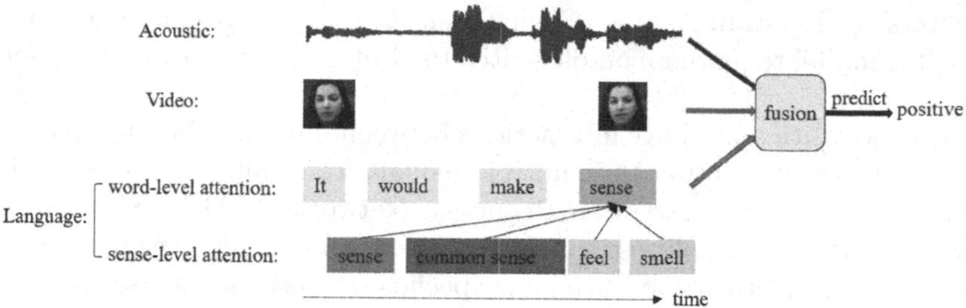

Fig. 1. The sense-level attention and word-level attention of the text "It would make sense" learned by our model. The first line is the acoustic modality, the second line is the video modality. The third line and the last line are the language modality, in which the third line is the original sentence, and the last line presents the senses of word "sense". Darker color means greater weight.

fusion, we first perform a projection operation on the concatenation of the LSTM outputs of three modalities, then attention mechanism is applied to learn the different contributions of the multimodal features at each temporal and produce a summary, which is regarded as the global multimodal fusion. The main contributions of our work are as follows:

1) To the best of our knowledge, this is the first time to use WordNet to reduce ambiguity for the task of multimodal sentiment analysis, which dynamically learn the different weights of the sense words to produce sense-attentive language presentation.
2) We propose to take language as the major modality and learn multimodal fusion from multi-stage and uni-stage perspective. Our final representation not only contains multimodal fusion, but also keeps the language representation, which is helpful in multimodal sentiment analysis.
3) Our model outperforms the baseline models on the CMU-MOSI, the CMU-MOSEI and the YouTube datasets and the ablation study shows the effectiveness of each components in our model.

2 Related Work

Compared with conventional text-based sentiment analysis, sentiment analysis with multiple modalities achieves significant improvements [1]. One of the most challenging task in multimodal sentiment analysis is to learn a joint representation of multiple modalities.

Earlier work uses fusion approaches such as concatenation of multi-modality features [5,11], while recent studies propose more sophisticated fusion approaches. Poria et al. [15] propose a LSTM-based model to capture contextual information. Zadeh et al. [22] propose a Tensor Fusion Network to explicitly aggregate unimodal, bimodal and trimodal interactions. Liu et al. [7] propose a Low-rank Fusion Network to alleviate the drawback of the large amount of parameters by low-rank factor. Chen et al. [2] propose a Gated Multimodal Embedding model to learn an on-off switch to filter noisy or contradictory modalities.

As the modalities can have interactions between different timestamps, several models are proposed to fuse the multiple modals from different views. Zadeh et al. [25] propose a Multi-attention Recurrent Network (MARN) to capture the interaction between modalities at different timestamps. Zadeh et al. [23] propose a Memory Fusion Network to learn view-specfic interactions and use an attention mechanism called the Delta-memory Attention Network (DMAN) to identify the cross-view interactions. Liang et al. [6] propose a Recurrent Multistage Fusion Network (RMFN) to model cross-modal interactions using multi-stage fusion approach, in which each stage of fusion focuses on a different subset of multimodal signals, learning increasingly discriminative multimodal representations.

Recently, Pham et al. [14] propose to learn joint representations based on translations between modalities. They use a cycle consistency loss to ensure that the joint representations retain maximal information from all modalities. Instead of directly fusing features at holistic level, Mai et al. [8] propose a strategy named 'divide, conquer and combine' for multimodal fusion. Their model performs fusion hierarchically to consider both local and global interactions. Wang et al. [20] propose a Recurrent Attended Variation Embedding Network (RAVEN) to model expressive nonverbal representations by analyzing the negrained visual and acoustic patterns. Tsai et al. [19] introduce a model that factorizes representations into two sets of independent factors: multimodal discriminative and modality-specic generative factors to optimize for a joint generative-discriminative objective across multimodal data and labels.

Although previous studies propose many effective approaches, most of them treat all modalities equally during the learning of multimodal fusion, which are different from our approach. In our model, we propose a sense-level attention network to learn different word representation under different senses. With the sense-attentive word representation, we can learn enhanced language representation. In addition, we try to learn sufficient multimodal fusion through multi-stage fusion and uni-stage fusion, as well as keeping the language representation to form our final representation.

3 Our Model

Our model consists of three components: sense attentive language representation which is regarded as the main representation of the multimodal fusion; multi-stage multimodal fusion which is designed to capture the interactions between the sharing information on the multi-stage; uni-stage multimodal fusion which is used to capture the global fusion information. The whole architecture of our model is shown in Fig. 2. In the following sections, we will introduce the sense-level attention network in Sect. 3.1, and describe the multi-stage multimodal fusion and the uni-stage multimodal fusion strategy in Sect. 3.2. Section 3.3 describes the final representation and model training.

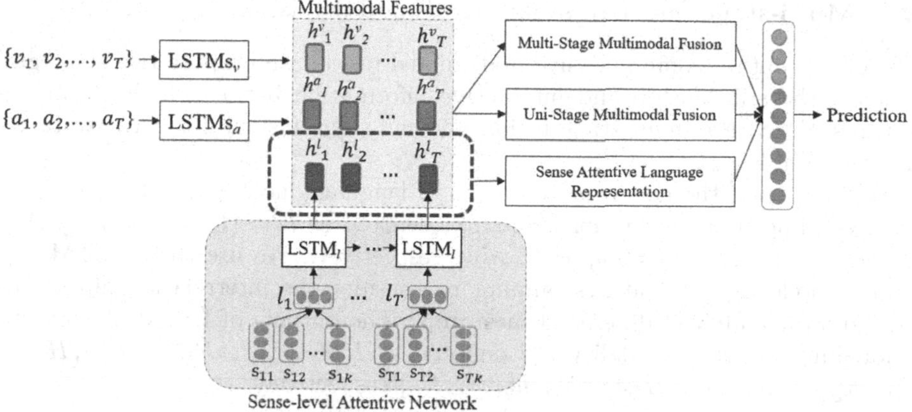

Fig. 2. The whole architecture of our model. The sense-level attention is used to learn the different importance of the sense words of each word in the language modality and produce a sense-attentive representation of language. LSTM layers are then used to model the features from language, vision and acoustic modalities. Three blocks are used to learn multi-stage multimodal fusion, uni-stage multimodal fusion and language representation respectively, which are concatenated to form the final representation.

3.1 Sense-Level Attention Network

As language has rich semantic information, a word may has different senses in different contexts, which may make the sentiment of a sentence totally different. However, the word's embedding representation is unique in the pretrained embeddings. In order to let the model to better distinguish different meanings of a same word, similar to the work of [21,27], we use WordNet to get k number of different senses of a word into the model. If a word don't have any sense in WordNet, we input k number of original words into the model. If there are more than k number of senses for the word, we take the first k number of senses in order and pad the sense sequence with the original word if the number of senses of the word is less than k. We denote the sense sequence of the i-th word in the sentence as $S_i = \{s_{i1}, s_{i2}, \ldots, s_{ik}\}$. The word senses and the original word are converted into embeddings to be input into the model. Then attention mechanism is used to learn the importance weight of different senses of a word and the weighted sum of the embeddings of different senses forms the new representation l_i of the word, as shown in Eqs. (1–3), where W_i and u_i are the trainable weights, b_i is the bias.

$$o_{ij} = relu\left(W_i s_{ij} + b_i\right) \tag{1}$$

$$\alpha_{ij} = softmax\left(u_i o_{ij}\right) \tag{2}$$

$$l_i = \sum_{j=1}^{k} a_{ij} s_{ij} \tag{3}$$

3.2 Multi-stage and Uni-stage Multimodal Fusion

In order to obtain comprehensive multimodal fusion, we propose two strategies to learn the relationship and interactive information between multiple modal features, which are multi-stage fusion and uni-stage fusion. The two strategies are shown in Fig. 3.

After getting the new representation of language modality and the original features of acoustic and vision modality, denoted as $L = \{l_1, l_2, \ldots, l_T\}, A = \{a_1, a_2, .., a_T\}$ and $V = \{v_1, v_2, .., v_T\}$ respectively. We use three LSTM layers for modeling the features, aiming to consider the interrelationship of the individual modality in different timestamps. The outputs of LSTM of acoustic, vision and language modality are denoted as $H_A = \{h_1^a, h_2^a, \ldots, h_T^a\}, H_V = \{h_1^v, h_2^v, \ldots, h_T^v\}$ and $H_L = \{h_1^l, h_2^l, \ldots, h_T^l\}$ respectively.

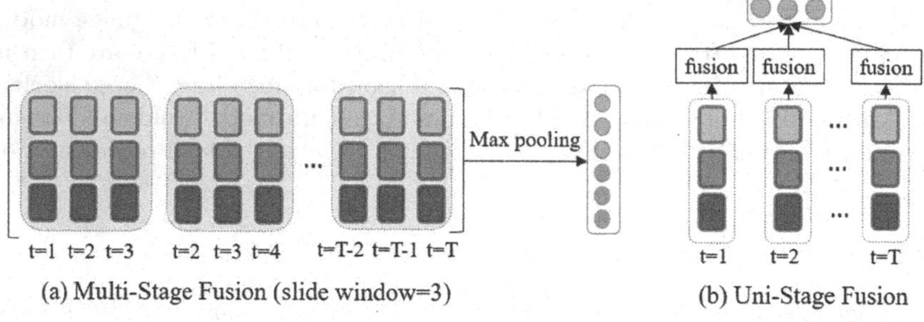

Fig. 3. The strategies of multimodal fusion proposed in our model. The multi-stage fusion aims to capture the interactions of the shared multimodal information in different timestamps. The uni-stage fusion aims to capture the global interactions of multimodal features fused within the same timestamp.

Multi-stage Multimodal Fusion. First we concatenate h_i^a, h_i^v and h_i^l, then we use different CNN layers with different window sizes to learn the multi-stage shared fusion. For CNN with window size 1, we aim to model the relationship between the three modalities timestamp by timestamp, through which we can get the fusion about the word, facial expression and speech tone of the speaker at the same timestamp. For CNN with window size bigger than 1, we aim to model the relationship between the three modalities within several timestamps. We perform maxpooling operation on top of the CNNs respectively and concatenate the results, getting the multi-stage shared multimodal fusion $h_{multi-stage}$. The convolution operation of CNN is shown in Eq. (5–6), where W_z and b_z are trainable weights and bias respectively, w is the window size, f is activation function which is *relu* in our implementation and [ ] denotes for concatenation.

$$h_i = \left[h_i^a, h_i^v, h_i^l\right] \tag{4}$$

$$z_i = f\left(W_z\left[h_i : h_{i+w-1}\right] + b_z\right) \tag{5}$$

$$Z_w = maxpooling\left(\left[z_1, z_2, ..., z_T\right]\right) \tag{6}$$

As stated above, we use different CNN layers with different window sizes following maxpooling operation, getting Z_w representation ($w = 1, 2, ...$), finally Z_w are concatenated to form the multi-stage fusion $h_{multi-stage}$.

Uni-stage Multimodal Fusion. The uni-stage fusion is applied to learn the different contributions of the multimodal feature at each temporal and produce a summary, which is regarded as the global multimodal fusion. We use another block to learn uni-stage multimodal fusion. Specifically, as shown in Eq. (7), we use a non-linear projection layer to project features of three modalities into the same space.

$$h_i^{'} = f\left(W_f\left[h_i^a, h_i^v, h_i^l\right] + b_f\right) \tag{7}$$

where W_f is the trainable weights, b_i is the bias, f is *relu* activation function and [ ] denotes for concatenation. Then we perform attention operation on the projected results $h_i^{'}$ to get a summary about of which stages the multimodal features are most important for sentiment analysis, as shown in Eqs. (8–10).

$$o_i = tanh\left(W_a h_i^{'} + b_i\right) \tag{8}$$

$$\alpha_i = softmax\left(u_a o_i\right) \tag{9}$$

$$h_{uni-stage} = \sum_{i=1}^{T} a_i h_i^{'} \tag{10}$$

where α_i is the attention weight of timestamp i. We use the attention weights to perform weighted sum on $h_i^{'}$, getting the uni-stage multimodal fusion $h_{uni-stage}$.

3.3 Final Representation and Model Training

As mentioned before, we believe that language modality contains richer information than other modalities, thus we perform attention operation on H_L to get the final language representation h_l. At last we concatenate h_l, the multi-stage multimodal fusion $h_{multi-stage}$ and uni-stage multimodal fusion $h_{uni-stage}$ to form the final representation h_{final}. The final representation is input to a fully-connected layer and a prediction layer to get the output, as shown in Eqs. (11–12):

$$h_{final}^{'} = relu\left(W_1 h_{final} + b_1\right) \tag{11}$$

$$y = f(W_2 h_{final}^{'} + b_2) \tag{12}$$

where W_1 and W_2 are trainable weights, b_1 and b_2 are biases. f is $softmax$ function for classification task. For regression task, we don't need activation function. y is the prediction.

4 Experiments

4.1 Dataset

We conduct several experiments on the CMU-MOSI [26] dataset, the CMU-MOSEI [24] dataset and the YouTube [10] dataset. The CMU-MOSI dataset contains 93 videos from the social media website, each of which comes from a different speaker who is expressing his or her opinions towards a movie. The videos in CMU-MOSI dataset are split into 2199 video clips, and each clip has a sentiment label $y \in [-3, 3]$, which represents strongly positive (labeled as +3), positive (+2), weakly positive (+1), neutral (0), weakly negative (−1), negative (−2), strongly negative (−3) respectively. The CMU-MOSEI dataset is a made up of 23,043 movie review video clips taken from YouTube. Following [8], we consider positive, negative and neutral sentiments in the paper. The YouTube dataset is collected from YouTube which contains 269 video clips. The statistical information of the three datasets is shown in Table 1.

Table 1. The statistical information of the experimental dataset.

Dataset	CMU-MOSI	CMU-MOSEI	YouTube
#Train	1284	15920	173
#Valid	229	2291	36
#Test	686	4832	60

4.2 Evaluation Metrix

Following previous work, we use different evaluation metrix on different datasets. For CMU-MOSI, we conduct experiments on binary classification task, multi-class classification task and regression task. For binary classification, we report accuracy and F1 score, whereas for multi-class classification we only report accuracy. For regression task, we report Mean Absolute Error (MAE) and Pearson's Correlation (Corr). For all the metrics, higher values denote better performance, except MAE where lower values denote better performance. For CMU-MOSEI and YouTube datasets, we conduct 3 classification task and report accuracy and F1 score.

4.3 Experimental Details

For all datasets, 300-dimensional GloVe embeddings [13] are used to represent the language features; Facet[1] library is used to extract a set of visual features

[1] https://imotions.com/biosensor/fea-facial-expression-analysis/.

and COVAREP [3] is used to extract acoustic features. We use WordNet to get 4 sense words for each word. Note that we add a constraint that the sense words should contain the original word. Besides, in WordNet, sense may contains more than one word, if this happen we use the average embedding of the words in the sense as the representation of the sense. The sizes of hidden states of LSTMs encoding language features, vision features and acoustic features are 100, 10 and 30 respectively. We use CNNs with window size 1 and 3 respectively to learn the multi-stage multimodal fusion and the filter number of CNN is set to 50. The batch size is set to 32, 16 and 16 for CMU-MOSI, CMU-MOSEI, YouTube datasets respectively, and the initial learning rate is set to 0.0008, 0.0003 and 0.0001 for the three datasets respectively. For CMU-MOSI dataset, we use L1 loss as training loss, for other two datasets, we use cross entropy loss as training loss. We report the experimental results predicted by the model which performs best on the validation set.

4.4 Baseline Models

We use several models as our baselines to compare with our model. Firstly, we use THMM [10] and MV-HCRF [18] as the traditional baseline models. THMM [10] concatenates language, acoustic and vision features and then uses HMM for classification. MV-HCRF [18] is an extension of the HCRF for Multi-view data, explicitly capturing view-shared and view specific sub-structures. Secondly, we use MV-LSTM [17], BC-LSTM [15], CAT-LSTM [16], GME-LSTM [2], TFN [22], CHFusion [9], LMF [7], MFN citech26ZadehLMPCM18, RMFN [6] and MARN [25] as the early neural network based compared models. Lastly, we use several previous state of the art models as our baseline models. MCTN [14] learns joint representations of multi-modalities by cyclic translations between modalities. HFFN [8] proposes a hierarchical feature fusion network, named 'divide, conquer and combine' to explore both local and global interactions in multiple stages. MFM [19] is proposed to optimize for a joint generative-discriminative objective across multimodal data and labels.

4.5 Experimental Results

Experimental Results on the CMU-MOSI Dataset. The results of our model and baseline models on the CMU-MOSI dataset is shown in Table 2. As is shown, the neural network based models outperform traditional machine learning models with a large margin. Among all models, our model achieves the second best performance on accuracy and F1 score of binary classification and accuracy of 7 classification, and our model achieves the best performance on MAE and Pearson's correlation of regression task compared with the baseline models. Specifically, our model achieves competitive results compared with HFFN on binary classification task, and outperforms MCTN, which is the best model on MAE among the baseline models by 4.5% on MAE. For Pearson's correlation (Corr), our model outperforms RMFN which achieves the best performance on Corr among the baselines by 3.9%. As for seven classification task, we achieve the second best performance. The overall experimental results on the CMU-MOSI dataset show the effectiveness of our model.

Table 2. Experimental results of different models on the CMU-MOSI dataset.

Model	Binary		Regression		7-class
	Acc	F1	MAE	Corr	Acc
THMM [10]	50.7	45.4	–	–	17.8
MV-HCRF [18]	65.6	65.7	–	–	24.6
MV-LSTM [17]	73.9	74.0	1.019	0.601	33.2
BC-LSTM [15]	73.9	73.9	1.079	0.581	28.7
GME-LSTM [2]	76.5	73.4	0.955	–	–
TFN [22]	74.6	74.5	1.040	0.587	28.7
LMF [7]	76.4	75.7	0.912	0.668	32.8
RMFN[6]	78.4	78.0	0.922	0.681	**38.3**
MARN [25]	77.1	77.0	0.968	0.625	34.7
MFN [23]	77.4	77.3	0.965	0.632	34.1
MFM [19]	78.1	78.1	0.951	0.662	36.2
MCTN [14]	79.3	79.1	0.909	0.676	–
HFFN [8]	**80.2**	**80.3**	–	–	–
MPFN(Ours)	80.0	80.0	**0.864**	**0.720**	37.0

Experimental Results on the YouTube Dataset. Table 3 shows the experimental results of our model and the baseline models on the YouTube dataset. The YouTube is a very small dataset, as shown in Table 1, not all neural network based models outperform traditional machine learning models both on accuracy and F1 score. However, compared with the baseline models, our model achieves the best performance on both accuracy and F1 score, which outperforms the previous state-of-the-art model MFM by 1.7% on accuracy and 3.5% on F1 score. Although the YouTube dataset is very small, our model can achieve the best performance among the baseline models.

Experimental Results on the CMU-MOSEI Dataset. For the CMU-MOSEI dataset, we conduct experiments on 3 classification tasks. We present the experimental results of different models in Table 4. As we can see, our model achieves the best performance on both accuracy and F1 score, which outperforms HFFN by 0.93% on accuracy and 0.6% on F1 score, and outperforms BC-LSTM by 0.53% on accuracy and 0.63% on F1 score. Note that the CMU-MOSEI is the largest dataset in this paper. In addition, we can see that although CAT-LSTM and LMF achieve relative good performance on accuracy, their performance on F1 score is much worse than that on accuracy. Our model can achieve both good performance on accuracy and F1 score. Experimental results on the CMU-MOSEI dataset and the YouTube dataset show that our model can adapt to both small data and large data.

Table 3. Experimental results of different models on the YouTube dataset

Model	Acc	F1
THMM [10]	42.4	27.9
MV-HCRF [18]	44.1	44.0
MV-LSTM [17]	45.8	43.3
BC-LSTM [15]	45.0	45.1
TFN[22]	45.0	41.0
MARN [25]	48.3	44.9
MFN [23]	51.7	51,6
MCTN [14]	51.7	52.4
MFM [19]	53.3	52.4
MPFN(Ours)	**55.0**	**55.9**

Table 4. Experimental results of different models on the CMU-MOSEI dataset.

Model	Acc	F1
BC-LSTM [15]	60.77	59.04
TFN [22]	59.40	57.33
CAT-LSTM [16]	60.72	58.83
CHFusion [9]	58.45	56.90
LMF [7]	60.27	53.87
HFFN [8]	60.37	59.07
MPFN(Ours)	**61.30**	**59.67**

4.6 Ablation Studies

In order to investigate the impact of various components in our model, we conduct several ablation experiments on the CMU-MOSI dataset, which are shown in Table 5. In the experiment, we remove one kind of component of our full model each time. Specifically, we remove the sense-level attention (denoted as MPFN-no-sense-att), the multi-stage multimodal fusion (denoted as MPFN-no-multi-stage-fusion), the uni-stage multimodal fusion (denoted as MPFN-no-uni-stage-fusion) and final language representation (denoted as MPFN-no-language-final) respectively.

As shown in Table 5, once we remove any component of our model, the performance will decline. For example, if we remove the sense-level attention and use the original word embedding as word representation, the performance of our model will drop by 1.0% on accuracy, 1.4% on F1 score of binary classification task, 3.8% on MAE, 2.5% on Corr, and 2.9% on accuracy of 7 classification task on the CMU-MOSI dataset. This observation suggests that using WordNet and sense-level attention to dynamically learn the word representation is effective.

In terms of multimodal fusion, we can see that if we remove the multi-stage fusion block or the uni-stage fusion block, the performance of our model will also drop, which indicates that both multi-stage fusion and uni-stage fusion are important for multimodal sentiment analysis. Furthermore, it seems that the multi-stage multimodal fusion plays a more important role than uni-stage multimodal fusion on the CMU-MOSI dataset.

Last but not least, we remove the final language representation which is concatenated with the multimodal fusion representation to see whether this operation is useful. The experimental results prove our early assumption. As we mentioned, ablation studies of previous researches show that if only using features of one modality as input, the model which use language modality features as input performs best. If only using multimodal fusion representation to form the final representation, some intra-modality information of language will be lost during fusion process. Concatenating the final language representation with the multimodal fusion representation to form the final representation can address this problem.

Table 5. Ablation studies on the CMU-MOSI dataset.

Model	Binary		Regression		7-class
	Acc	F1	MAE	Corr	Acc
MPFN-no-sense-att	79.0	78.6	0.902	0.695	34.1
MPFN-no-multi-stage-fusion	79.0	79.0	0.882	0.698	36.9
MPFN-no-uni-stage-fusion	79.3	79.3	0.888	0.711	33.5
MPFN-no-language-final	79.3	79.3	0.899	0.714	34.4
MPFN(Ours)	**80.0**	**80.0**	**0.864**	**0.720**	**37.0**

4.7 Discussion

In order to investigate how each modality effects the performance of our model, we conduct several experiments to compare the performance of our model using unimodal, bimodal and multimodal features, as shown in Table 6.

For unimodal features, we can see that our model only using sense attentive language representation outperforms the model that only using audio features or video features with significant margin, which is consistent with our early assumption that language modality is dominant. For bimodal features, we can infer that when integrating language modality with acoustic modality or vision modality, the performance of the model outperforms that of only using language representation, which indicates that acoustic and vision modalities play auxiliary roles and the multi-perspective multimodal fusion can improve the performance of the model. However, when using audio features and video features as input, the performance of the model is still much worse than that of only using language

Table 6. The performance of our model using unimodal, bimodal and multimodal features.

Modality	Source	Binary		Regression		7-class
		Acc	F1	MAE	Corr	Acc
Unimodal	Audio	57.1	56.2	1.396	0.196	16.0
	Video	57.3	57.3	1.431	0.137	16.2
	Sense attentive language	79.0	79.1	0.922	0.689	34.0
Bimodal	Sense attentive language + Audio	79.7	79.6	0.881	0.701	34.7
	Sense attentive language + Video	79.6	79.6	0.915	0.714	32.9
	Audio + Video	59.0	59.0	1.391	0.176	19.7
Multimodal	**Sense attentive language + Audio + Video**	**80.0**	**80.0**	**0.864**	**0.720**	**37.0**

modality, which again proves that language modality is the most important modality in this task.

When cooperating three modalities, our full model MPFN achieves the best performance among the different combinations, which demonstrates the effectiveness of multi-perspective multimodal fusion proposed in this paper.

5 Conclusion

In this paper, we propose a novel multi-perspective fusion network focusing on sense attentive language for multimodal sentiment analysis. Evaluations show that using our proposed multi-stage and uni-stage fusion strategies and using sense attentive language representation can improve performance on multimodal sentiment analysis for the CMU-MOSI, CMU-MOSEI and YouTube data. Our model also achieves a new state-of-the-art in the YouTube and CMU-MOSEI dataset on accuracy and F1 measure metrics compared with the baseline models. The experimental results using different modal combinations also show that the proposed sense attentive language modal achieves the most significant performance improvement on the CMU-MOSI dataset, especially on the 7-classification results, indicating that the sense attentive language modal plays an important role in multimodal sentiment analysis task. Like most of other models, our approach also focuses on the multimodal data with the same length of stamp. In the future, we will investigate a novel fusion of multimodal data with different length of stamp.

Acknowledgments. This work is supported by National Natural Science Foundation of China (No. 61976062) and the Science and Technology Program of Guangzhou (No. 201904010303).

References

1. Baltrusaitis, T., Ahuja, C., Morency, L.: Multimodal machine learning: a survey and taxonomy. IEEE Trans. Pattern Anal. Mach. Intell. **41**(2), 423–443 (2019)

2. Chen, M., Wang, S., Liang, P.P., Baltrusaitis, T., Zadeh, A., Morency, L.: Multimodal sentiment analysis with word-level fusion and reinforcement learning. In: Proceedings of the 19th ACM International Conference on Multimodal Interaction, ICMI 2017, pp. 163–171 (2017)
3. Degottex, G., Kane, J., Drugman, T., Raitio, T., Scherer, S.: COVAREP - a collaborative voice analysis repository for speech technologies. In: IEEE International Conference on Acoustics, Speech and Signal Processing, ICASSP 2014, pp. 960–964 (2014)
4. Dong, Z.: Knowledge description: what, how and who. In: Proceedings of International Symposium on Electronic Dictionary, vol. 18 (1988)
5. Lazaridou, A., Pham, N.T., Baroni, M.: Combining language and vision with a multimodal skip-gram model. In: The 2015 Conference of the North American Chapter of the Association for Computational Linguistics: Human Language Technologies, NAACL-HLT 2015, pp. 153–163 (2015)
6. Liang, P.P., Liu, Z., Zadeh, A., Morency, L.: Multimodal language analysis with recurrent multistage fusion. In: Proceedings of the 2018 Conference on Empirical Methods in Natural Language Processing, EMNLP 2018, pp. 150–161 (2018)
7. Liu, Z., Shen, Y., Lakshminarasimhan, V.B., Liang, P.P., Zadeh, A., Morency, L.: Efficient low-rank multimodal fusion with modality-specific factors. In: Proceedings of the 56th Annual Meeting of the Association for Computational Linguistics, ACL 2018, pp. 2247–2256 (2018)
8. Mai, S., Hu, H., Xing, S.: Divide, conquer and combine: hierarchical feature fusion network with local and global perspectives for multimodal affective computing. In: Proceedings of the 57th Conference of the Association for Computational Linguistics, ACL 2019, pp. 481–492 (2019)
9. Majumder, N., Hazarika, D., Gelbukh, A.F., Cambria, E., Poria, S.: Multimodal sentiment analysis using hierarchical fusion with context modeling. Knowl. Based Syst. **161**, 124–133 (2018)
10. Morency, L., Mihalcea, R., Doshi, P.: Towards multimodal sentiment analysis: harvesting opinions from the web. In: Proceedings of the 13th International Conference on Multimodal Interfaces, ICMI 2011, pp. 169–176 (2011)
11. Ngiam, J., Khosla, A., Kim, M., Nam, J., Lee, H., Ng, A.Y.: Multimodal deep learning. In: Proceedings of the 28th International Conference on Machine Learning, ICML 2011, pp. 689–696 (2011)
12. Pennebaker, J.W., Booth, R.J., Francis, M.E.: Linguistic inquiry and word count: Liwc [computer software]. Austin, TX: liwc. net 135 (2007)
13. Pennington, J., Socher, R., Manning, C.D.: Glove: global vectors for word representation. In: Proceedings of the 2014 Conference on Empirical Methods in Natural Language Processing, EMNLP 2014, pp. 1532–1543 (2014)
14. Pham, H., Liang, P.P., Manzini, T., Morency, L., Póczos, B.: Found in translation: learning robust joint representations by cyclic translations between modalities. In: The Thirty-Third AAAI Conference on Artificial Intelligence, AAAI 2019, pp. 6892–6899 (2019)
15. Poria, S., Cambria, E., Hazarika, D., Majumder, N., Zadeh, A., Morency, L.: Context-dependent sentiment analysis in user-generated videos. In: Proceedings of the 55th Annual Meeting of the Association for Computational Linguistics, ACL 2017, pp. 873–883 (2017)
16. Poria, S., Cambria, E., Hazarika, D., Majumder, N., Zadeh, A., Morency, L.: Multilevel multiple attentions for contextual multimodal sentiment analysis. In: 2017 IEEE International Conference on Data Mining, ICDM 2017, pp. 1033–1038 (2017)

17. Rajagopalan, S.S., Morency, L.-P., Baltrušaitis, T., Goecke, R.: Extending long short-term memory for multi-view structured learning. In: Leibe, B., Matas, J., Sebe, N., Welling, M. (eds.) ECCV 2016. LNCS, vol. 9911, pp. 338–353. Springer, Cham (2016). https://doi.org/10.1007/978-3-319-46478-7_21
18. Song, Y., Morency, L., Davis, R.: Multi-view latent variable discriminative models for action recognition. In: 2012 IEEE Conference on Computer Vision and Pattern Recognition, pp. 2120–2127 (2012)
19. Tsai, Y.H., Liang, P.P., Zadeh, A., Morency, L., Salakhutdinov, R.: Learning factorized multimodal representations. In: 7th International Conference on Learning Representations, ICLR 2019 (2019)
20. Wang, Y., Shen, Y., Liu, Z., Liang, P.P., Zadeh, A., Morency, L.: Words can shift: dynamically adjusting word representations using nonverbal behaviors. In: The Thirty-Third AAAI Conference on Artificial Intelligence, AAAI 2019, pp. 7216–7223 (2019)
21. Xie, R., Yuan, X., Liu, Z., Sun, M.: Lexical sememe prediction via word embeddings and matrix factorization. In: Proceedings of the Twenty-Sixth International Joint Conference on Artificial Intelligence, IJCAI 2017, pp. 4200–4206 (2017)
22. Zadeh, A., Chen, M., Poria, S., Cambria, E., Morency, L.: Tensor fusion network for multimodal sentiment analysis. In: Proceedings of the 2017 Conference on Empirical Methods in Natural Language Processing, EMNLP 2017, pp. 1103–1114 (2017)
23. Zadeh, A., Liang, P.P., Mazumder, N., Poria, S., Cambria, E., Morency, L.: Memory fusion network for multi-view sequential learning. In: Proceedings of the Thirty-Second AAAI Conference on Artificial Intelligence, AAAI 2018, pp. 5634–5641 (2018)
24. Zadeh, A., Liang, P.P., Poria, S., Cambria, E., Morency, L.: Multimodal language analysis in the wild: CMU-MOSEI dataset and interpretable dynamic fusion graph. In: Proceedings of the 56th Annual Meeting of the Association for Computational Linguistics, ACL 2018, pp. 2236–2246 (2018)
25. Zadeh, A., Liang, P.P., Poria, S., Vij, P., Cambria, E., Morency, L.: Multi-attention recurrent network for human communication comprehension. In: Proceedings of the Thirty-Second AAAI Conference on Artificial Intelligence, AAAI 2018, pp. 5642–5649 (2018)
26. Zadeh, A., Zellers, R., Pincus, E., Morency, L.: MOSI: multimodal corpus of sentiment intensity and subjectivity analysis in online opinion videos. Computing Research Repository arXiv:1606.06259 (2016)
27. Zeng, X., Yang, C., Tu, C., Liu, Z., Sun, M.: Chinese LIWC lexicon expansion via hierarchical classification of word embeddings with sememe attention. In: Proceedings of the Thirty-Second AAAI Conference on Artificial Intelligence, AAAI 2018, pp. 5650–5657 (2018)

CAN-GRU: A Hierarchical Model for Emotion Recognition in Dialogue

Ting Jiang(✉), Bing Xu(✉), Tiejun Zhao(✉), and Sheng Li(✉)

Laboratory of Machine Intelligence and Translation School of Computer Science and Technology, Harbin Institute of Technology Harbin, Harbin, China
jiangting_hit@163.com, {hitxb,tjzhao,lisheng}@hit.edu.cn

Abstract. Emotion recognition in dialogue systems has gained attention in the field of natural language processing recent years, because it can be applied in opinion mining from public conversational data on social media. In this paper, we propose a hierarchical model to recognize emotions in the dialogue. In the first layer, in order to extract textual features of utterances, we propose a convolutional self-attention network(CAN). Convolution is used to capture n-gram information and attention mechanism is used to obtain the relevant semantic information among words in the utterance. In the second layer, a GRU-based network helps to capture contextual information in the conversation. Furthermore, we discuss the effects of unidirectional and bidirectional networks. We conduct experiments on Friends dataset and EmotionPush dataset. The results show that our proposed model(CAN-GRU) and its variants achieve better performance than baselines.

Keywords: CAN-GRU · Attention mechanism · Dialogue emotion recognition.

1 Introduction

As an important component of human intelligence, emotional intelligence is defined as the ability to perceive, integrate, understand and regulate emotions [15]. Emotion is the essential difference between human and machine, so emotion understanding is an important research direction of artificial intelligence. As the most common way for people to communicate in daily life, dialogue contains a wealth of emotions. Recognising the emotions in the conversation is of great significance in intelligent customer service, medical systems, education systems and other aspects.

According to [17], textual features usually contain more emotional information than video or audio features, so we focus on the emotion analysis of dialogue text and aims to recognize the emotion of each utterance in dialogues.

There are some challenges in this task. First, the length of an utterance may be too long, making it difficult to capture contextual information. Furthermore, a dialogue usually contains lots of utterances, therefore, it's hard to grasp

M. Sun et al. (Eds.): CCL 2020, LNAI 12522, pp. 374–387, 2020.
https://doi.org/10.1007/978-3-030-63031-7_27

long-term contextual relations between utterances. Second, the same word may express different emotions in different contexts. For example, in Table 1, while in different dialogues, the word 'Yeah' can express three different emotions, that is , joy, neutral and suprise. To tackle these challenges, we propose a hierarchical model based on convolutional attention network and gated recurrent unit (CAN-GRU). Existing works pay little attention to the extraction of semantic information within an utterance. In this work, we focus on this problem. Our proposed model can extract n-gram information by CNNs and use self-attention to capture contextual information within an utterance in the first layer. Moreover, we utilize a GRU-based network to model the sequence of utterances in the second layer, which can fully combine the context when analyzing utterance emotion and solve the problem of long-term dependence between texts at the same time.

Table 1. The word 'Yeah' expresses different emotions in the different contexts.

Speaker	Utterance	Emotion
Phoebe	Can I tell you a little secret?	neutral
Rachel	**Yeah!**	**joy**
Wayne	Hey Joey, I want to talk to you	neutral
Joey	**Yeah?**	**neutral**
Gary	Hey Chandler, what are you doing here?	suprise
Chandler	Gary, I'm here to report a crime	neutral
Gary	**Yeah?**	**suprise**
Chandler	It is a crime that you and I don't spend more time together	neutral

2 Related Work

Text emotion recognition is one of the most hot topic in natural language processing. Recent years,a lot of classical neural networks are used to tackle this problem. Such as Long Short-Term Memory Network [8], Gated Recurrent Unit Network [2] and textual Convolutional Neural Network [11]. However, these models don't perform well when the texts are too long, because it's hard to capture the long-range contextual information. Later, attention mechanism [1] is proposed to solve this problem. Recently, self-attention [20] is widely used since it can solve the long-term dependence problem of text effectively.

Recent years, more and more researchers focus on emotion recognition in conversation. This task aims to recognize the emotion of each utterance in dialogues. bcLSTM [18] extracts textual features by CNN and model the sequence of utterances by LSTM. Considering inter-speaker dependency relations, conversational memory network(CMN)[7] has been proposed to model the speaker-based emotion using memory network and summarize task-specific details by attention mechanisms. ICON [6] improves the CMN, it hierarchically models the self-speaker emotion and inter-speaker emotion into global memories. DialogueRNN [14] uses emotion GRU and global GRU to model inter-party relation, and uses party GRU to model relation between two sequential states of the same party. DialogueGCN [5] improves DialogueRNN by graph convolutional network, and it can hold richer context relevant to emotion. However, these models may be too complex for small textual dialogue datasets.

In this paper, we study on the EmotionX Challenge [9], Dialogue Emotion Recognition Challenge, which aims to recognize the emotion of each utterance in dialogues. According to the overview of this task, the best team [10] proposes a CNN-DCNN auto encoder based model, which includes a convolutional encoder and a deconvolutional decoder. The second place team [13] mainly uses BiLSTM with a self-attentive architecture on the top for the classiffication. The third place team [19] proposes a hierarchical network based on attention models and conditional random fields(CRF). For a meaningful comparison, we use the same dataset and metric as the challenge in our study.

3 Method

3.1 Task Definition

Given a dialogue $dia = \{u_1, u_2, ..., u_N\}$, where N is the number of utterances in the dialogue, $u_i = \{w_1, w_2, ..., w_L\}$ represents the $ith(1 \leq i \leq N)$ utterance in the dialogue that consists of L words, our goal is to analyze the emotion of each utterance in the dialogue.

To solve this task, we propose a hierarchical model CAN-GRU and extend three variants, CAN-GRUA, CAN-biGRU and CAN-biGRUA(illustrated in Fig. 1).

3.2 Text Feature Extraction

In this section, we discuss the first layer of the model. Like [18], we use convolutional neural network to extract the features of the utterance. Inspired by [4], in order to capture the contextual information of long text effectively, we use convolutional self-attention network(CAN) instead of traditional CNN network.

For query embedding Q, key embedding K and value embedding V involved in attention operation, they may need different effective features. And different effective information can be extracted in different convolution operations. So we obtain the Q, K and V embeddings by convolving the input word embeddings, instead of using the input word embeddings as the Q, K and V embeddings directly:

$$Q = f(conv(E, W_q) + b_q) \tag{1}$$

$$K = f(conv(E, W_k) + b_k) \tag{2}$$

$$V = f(conv(E, W_v) + b_v) \tag{3}$$

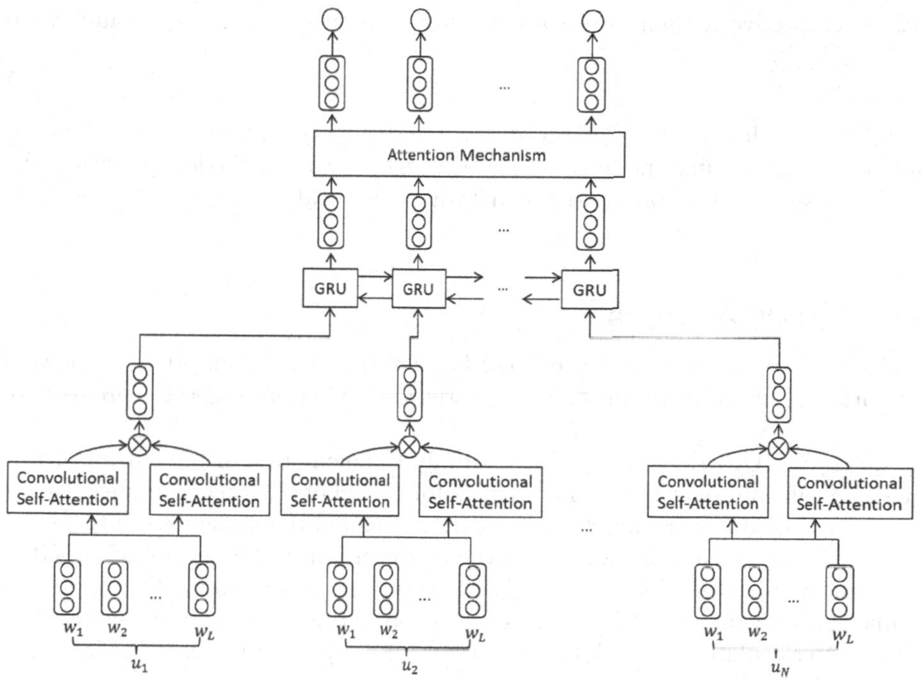

Fig. 1. The architecture of our proposed CAN-biGRUA. In the first layer, convolutional neural network and self-attention mechanism are used to extract text features. In the second layer, biGRU with an attentive architecture on the top is used to model the sequence of utterances in the dialogue.

In the equations above, E is the input word embeddings, $\{E, Q, K, V\} \in \mathbb{R}^{l \times d}$, where l means the length of the sentence and d means the embedding dimension. $\{W_q, W_k, W_v\} \in \mathbb{R}^{w \times n \times d}$, where w is the window size of filters and n is the feature maps of filters. $\{b_q, b_k, b_v\} \in \mathbb{R}^d$. $conv(E, W)$ means convolution operation between E and W. And f is the activation function.

After getting Q, K, V embeddings, we calculate semantic relations among words within the utterance by the scaled dot product attention operation. More specificly, Q and K operate to get the weight matrix. Then we scale this weight matix by $\sqrt{d}$. After that, softmax operation is conducted to obtain the standardized weight matrix, which is used to express the degree of attention between words in the sentence, then the normalized weight matrix is mutiplied with V to get the result $Z \in \mathbb{R}^{l \times d}$ of attention operation:

$$Z = softmax(\frac{QK^T}{\sqrt{d}})V \quad (4)$$

As mentioned in [4], attention mechanisms cannot capture complex interactions, because it is designed for creating weighted averages. So we do the Eqs. (1)–(3) twice to create Q_a, K_a, V_a and Q_b, K_b, V_b respectively, and get Z_a, Z_b by operate

Eq. (4) respectively, then perform elementwise multiplication on Z_a and Z_b to get $U \in \mathbb{R}^{l \times d}$:

$$U = Z_a \otimes Z_b \tag{5}$$

Finally, for each $U_i (1 \leq i \leq N)$ in $dia = \{U_1, U_2, ..., U_N\}$, we get the individual embedding e_{u_i} by max-pooling on the contextual word embeddings within the U_i. In this way, we obtain a set of utterance embeddings $\{e_{u_1}, e_{u_2}, ..., e_{u_N}\}$ in one dialogue.

3.3 Dialogue Modeling

In this section, we discuss the sencond layer of the model. Considering different networks, we propose the hierarchical model CAN-GRU and its three progressive variants.

CAN-GRU: In real life, when we analyze the emotion of the current utterance, we can only refer to the historical information of the past utterances in the conversation. So in our model, we use GRU to model the sequence of utterances in the dialogue, because it can memory and transmit historical information. GRU [2] is an improved model for the original recurrent neural networks and it performs well with simple calculation. At timestep t, it use reset gate R_t and update gate Z_t to calculate current hidden state S_t with input utterance embedding e_{u_t} and hidden state s_{t-1} at the previous time step.

$$R_t = \sigma(e_{u_t} W_{ur} + S_{t-1} W_{sr} + b_r) \tag{6}$$

$$Z_t = \sigma(e_{u_t} W_{uz} + S_{t-1} W_{sz} + b_z) \tag{7}$$

$$H_t = tanh(e_{u_t} W_{uh} + (R_t \otimes S_{t-1}) W_{sh} + b_h) \tag{8}$$

$$S_t = Z_t \otimes S_{t-1} + (1 - Z_t) \otimes H_t \tag{9}$$

where W, b are trainable parameters and $\otimes$ means elementwise mutilication.

CAN-GRUA: However, it is difficult to grasp long-term dependence between sentences when there are too many sentences in a conversation. That is, it is hard for the current utterance to capture the historical information contained in the distant utterance. To solve this problem, we connect an attention layer upon the GRU to obtain the influence degree of historical information on the emotion of the current utterance. If the weight calculated by attention mechanism tends to be large, it indicates that the preceding utterance have an important influence on the current utterance, so this preceding utterance should be given more attention.

$$\tilde{S}_t = \sum_{i=1}^{t-1} S_i \alpha_i \tag{10}$$

$$\alpha_i = \frac{exp(S_t S_i)}{\sum_{i=1}^{t-1} exp(S_t S_i)} \tag{11}$$

Here, $S_t \in \mathbb{R}^m$ is the current hidden state, where m is the dimension of the hidden state. $S_i \in \mathbb{R}^m$ is the preceding hidden state at time step i, $\tilde{S_t} \in \mathbb{R}^m$ is the attention result at time step t.

CAN-biGRU: In fact, when analyzing the emotion of the utterance, we can not only use the historical information before the utterance, but also the future information after the current utterance. This is because emotional tone is usually maintained and does not shift frequently within a conversation in a short time. If we only pay attention to the historical information, it may be difficult to analyze the emotion of the current utterance, while the future information can be helpful in the analysis. Therefore, using both historical and future information can help to capture a richer context. Bidirectional GRU(biGRU) is used to model the sequence of utterances abstracting contextual features forward and backward, which can provides context for emotion classification more effectively.

CAN-biGRUA: As mentioned before, biGRU also suffers from the difficulty of obtaining semantic connections between long sequences. So we connect a self-attention layer on the top of the hidden states of biGRU to take full advantage of global contextual information.

$$\tilde{S_t} = \sum_{i=1}^{N} S_i \alpha_i \tag{12}$$

$$\alpha_i = \frac{exp(S_t S_i)}{\sum_{i=1}^{N} exp(S_t S_i)} \tag{13}$$

Here, $S_t \in \mathbb{R}^m$ is the current hidden state, $S_i \in \mathbb{R}^m$ is the hidden state at time step i, N is the number of sentences in the dialogue, $\tilde{S_t} \in \mathbb{R}^m$ is the attention result at time step t.

3.4 Emotion Classification

As mentioned above, we get final representations of utterances $\{\tilde{S_1}, \tilde{S_2}, ..., \tilde{S_t}, ..., \tilde{S_N}\}$. Then we utilize a fully-connected layer and a softmax layer to get the emotion class of each utterance in a dialogue.

$$f_t = tanh(W_f \tilde{S_t} + b_f) \tag{14}$$

$$o_t = softmax(W_o \tilde{f_t} + b_o) \tag{15}$$

$$\hat{y_t} = \underset{i}{argmax}(o_t[i]), i \in [1, c] \tag{16}$$

where $W_f \in \mathbb{R}^{m \times m}$, $b_f \in \mathbb{R}^m$. $W_o \in \mathbb{R}^{m \times c}$, c is the number of emotion class, $b_o \in \mathbb{R}^c$, $o_t \in \mathbb{R}^c$, $\hat{y_t}$ is the predicted class for utterance u_t.

3.5 Training

Like [10], in order to solve the problem of emotion class imbalance, we use a weigted cross entropy loss as a minimization target to optimize the parameters in the model. We give higher weight to the loss of minority class data sample in the dataset.

$$Loss = \frac{1}{K}\sum_{k=1}^{K} weight_k loss_k \tag{17}$$

$$loss_k = -[y_k log(p_k) + (1 - y_k) log(1 - p_k)] \tag{18}$$

$$\frac{1}{weight_k} = \frac{count_i}{\sum_{i=1}^{c} count_i} \tag{19}$$

where K is the total number of samples, y_k is the ground-truth, p_k is the probability calculated in softmax layer, $count_i$ is the total number of samples in the same class as sample k.

4 Experiments

4.1 Datasets

We conduct experiments on two datasets provided by the EmotionX Challenge [9].

Friends[1]: The conversations in this dataset are from the Friends TV show transcripts. The dataset contains eight emotion categories: joy, anger, sadness, surprise, fear, disgust, neutral, and non-neutral.

EmotionPush(see footnote 1): The conversations in this dataset are from the facebook messenger logs after processing the private information. Emotion categories are the same as Friends dataset.

In the challenge [9], each dataset is divided into the training set with 720 dilogues, the validation set with 80 dialogues and the test set with 200 dialogues. Since there are few utterances for some emotions, the challenge only evaluate the performence of recognition for four emotions: joy, anger, sadness and neutral. Table 2 shows the distributions of train, validation, test samples and the distributions of the emotions for both datasets respectively.

4.2 Evaluation Metric

We use the unweighted accuracy(UWA) as the evaluation metric instead of the weighted accuracy(WA), the same as the challenge. This is because WA is easily

[1] http://doraemon.iis.sinica.edu.tw/emotionlines.

Table 2. Statistics of the datasets.

Dataset	Dialogue(Utterance)			Emotion				
	Train	Validation	Test	Anger	Joy	Sadness	Neutral	Others
Friend	720(10561)	80(1178)	200(2764)	759	1710	498	6530	5006
EmotionPush	720 (10,733)	80(1202)	200(2807)	140	2100	514	9855	2133

influced by the large proportion of neutral emotion and UWA can help to make a meaningful comparision.

$$UWA = \frac{1}{c}\sum_{i=1}^{c} a_i, WA = \sum_{i=1}^{c} weight_i a_i \tag{20}$$

where a_i is the accuracy of class i and $weight_i$ is the percentage of the class i.

4.3 Experimental Setting

We use 300-dimensional pre-trained GloVe[2] [16] word-embeddings which is trained from web data. We use three distinct convolution filters of sizes 3, 4, and 5 respectively, each having 100 feature maps. The dimension of the hidden states of the GRU is set to 300. We use adam [12] optimizer and set the initial learning rate as 1.0×10^{-4}. The learning rate is halved every 20 epochs during training. Dropout probability is set to 0.3.

4.4 Baselines

In experiments, we compare our proposed model with the following models.

CNN-DCNN: The winner of EmotionX Challenge [10]. The model contains a convolutional encoder and a deconvolutional decoder. The linguistic features enhance the latent feature of the model.

SA-LSTM: The second place of the challenge [13]. A self-attentive biLSTM network can provide information between utterances and the word dependency in each utterance.

HAN: The third place of the challenge [19]. LSTM with attention mechanism gets the sentence embedding. Another LSTM and CRF layer model the context dependency between sentence embeddings of the dialogue.

scGRU: We implement the basic model proposed by [18], but with a few changes. The same as [18], CNN is used to extract text features, but we use a contextual GRU network instead of a contextual LSTM network to model the sequences.

bcGRU: We implement the variant model proposed by [18], CNN is also used to obtain utterance features, but the biLSTM network used in the author's work is replaced by biGRU network.

[2] http://nlp.stanford.edu/projects/glove/.

4.5 Main Results

Table 3. Experimental results on Friend dataset and EmotionPush dataset.

Model	Friend					EmotionPush				
	Anger	Joy	Sadness	Neutral	UWA	Anger	Joy	Sadness	Neutral	UWA
CNN-DCNN	55.3	71.1	55.3	68.3	62.5	45.9	**76.0**	51.7	76.3	62.5
SA-BiLSTM	49.1	68.8	30.6	**90.1**	59.6	24.3	70.5	31.0	**94.2**	55.0
HAN	39.8	57.6	50.6	73.5	55.4	21.6	63.1	54.0	88.2	56.7
scGRU	51.6	68.7	44.8	72.6	59.4	49.8	68.2	57.1	75.4	62.6
bcGRU	54.1	69.8	43.5	73.4	60.2	50.1	71.4	61.6	71.8	63.7
CAN-GRU	56.2	67.0	**55.9**	71.4	62.6	52.4	70.6	59.8	74.5	64.3
CAN-biGRU	54.8	68.1	52.9	76.3	63.0	**55.7**	71.8	60.1	74.9	65.6
CAN-GRUA	**57.6**	70.2	53.7	76.2	64.4	53.2	72.1	61.5	78.3	66.3
CAN-biGRUA	56.4	**72.6**	54.4	77.8	**65.3**	54.3	73.8	**62.9**	77.4	**67.1**

Table 3 presents the performance of baselines and CAN-GRU along with its variants.

Baselines: Our implemented bcGRU model performes better than scGRU on both datasets. On the Emotionpush dataset, bcGRU's performance has surpassed CNN-DCNN, and it is the best model in baselines. On the Friend dataset, CNN-DCNN remains the best baseline.

CAN-GRU: In the first layer, it uses the convolutional self-attention mechanism to extract utterance features, and in the second layer, GRU is used to model the sequence of utterances. Compared with scGRU, it attains 3.2% and 1.7% improvement on the Friend dataset and EmotionPush dataset.

CAN-biGRU: Compared with CAN-GRU, it uses biGRU at the second layer and get improvements on the two datasets. CAN-biGRU achieves 2.8% and 1.9% improvements over bcGRU on the Friend dataset and the Emotionpush dataset respectively. Both the improvements of CAN-GRU and CAN-biGRU over baselines illustrate that the convolutional self-attention mechanism can capture contextual information in long text effectively.

CAN-GRUA: Compared with CAN-GRU, an attention mechanism is connected upon the GRU layer, which can help the model better capture the historical information of utterance and give high weight to important historical information. It gets 1.8% and 2.0% improvements over CAN-GRU on the two datasets.

CAN-biGRUA: At the top of biGRU, a self-attention mechanism is added to help calculate the importance of contextual information by using historical and future information when analyzing the current utterance emotion. This model achieves the best results, it improves 2.8% and 4.6% over baseline on the two datasets respectively.

Table 4. Experimental results for BERT and CAN-biGRU(*).

Model	Friend					EmotionPush				
	Anger	Joy	Sadness	Neutral	UWA	Anger	Joy	Sadness	Neutral	UWA
BERT	78.1	86.5	74.3	**90.3**	82.3	79.4	**89.7**	85.3	92.4	86.7
CAN-biGRU(*)	**81.2**	**87.4**	**78.7**	89.1	**84.1**	**82.8**	88.3	**87.6**	**94.1**	**88.2**

In addition, we use the pretrained model BERT [3] to get the word embeddings and input the pre-trained word embeddings into our CAN-biGRU, the experimental results are shown as the CAN-biGRU(*) in Table 4. As we can see, while BERT achieves a high degree of accuracy, our model can be further improved on the basis of BERT. CAN-biGRU(*) gets 1.8% and 1.5% improvements over BERT on the Friend dataset and the Emotionpush dataset respectively.

4.6 Case Study

Table 5. Some case comparisons of emotion recognition results by bcGRU and CAN-biGRUA

Speaker	Utterance	True label	bcGRU	CAN-biGRU
Phoebe	Oh, it's bad. It's really bad ... Which I do	sadness	**sadness**	**sadness**
Chandler	How's your room Rach?	neutral	**neutral**	**neutral**
Rachel	Everything's ruined ... blue sweater	sadness	neutral	**sadness**
Joey	Hey-hey-hey!	joy	**joy**	**joy**
Chandler	What are you doing?	neutral	**neutral**	**neutral**
Phoebe	We're just celebrating that Joey ... back	joy	neutral	**joy**
Phoebe	I'm sorry ... Check this out	neutral	sad	sad
Monica	No, Phoebe ... you play it at the wedding	neutral	**neutral**	**neutral**

In Table 5, we compare the emotion recognition results of bcGRU and CAN-biGRU. In the first case, 'bad' expresses strong emotion and both two model can recognize the sad emotion successfully. While there is no explicit emotion word in the third utterance, but the word 'ruined' delivers bad information, our CAN-biGRU can extract semantic information among words by CAN and gives the right prediction. In the second case, the word 'celebrating' in the third utterance express the joy emotion implicitly. Our model obtains the contextual information through the CAN, and makes the correct prediction. However, in the third case, both two model make false predictions for the utterance said by Phoebe, since the word 'sorry' expresses strong sad emotion. This shows CAN is still limited in such complicated semantic environment.

As shown in Table 6, we analyse some cases of the results of emotion recognition by our CAN-biGRUA. In the first two cases, our model can successfully

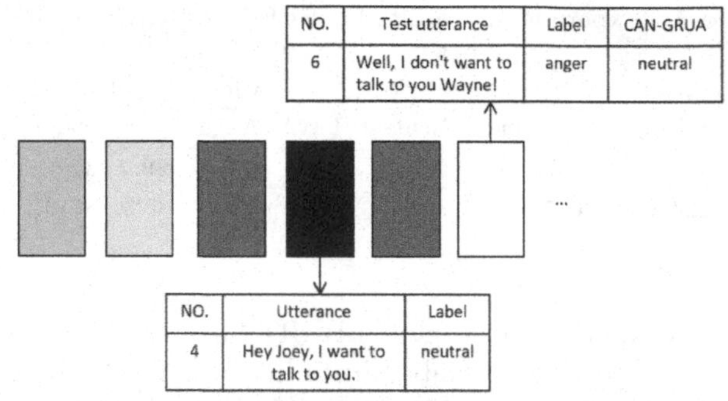

NO.
Test utterance
Label
CAN-GRUA
6
Well, I don't want to talk to you Wayne!
anger
neutral
...
NO.
Utterance
Label
4
Hey Joey, I want to talk to you.
neutral

(a) Recognition results and attention results of CAN-GRUA

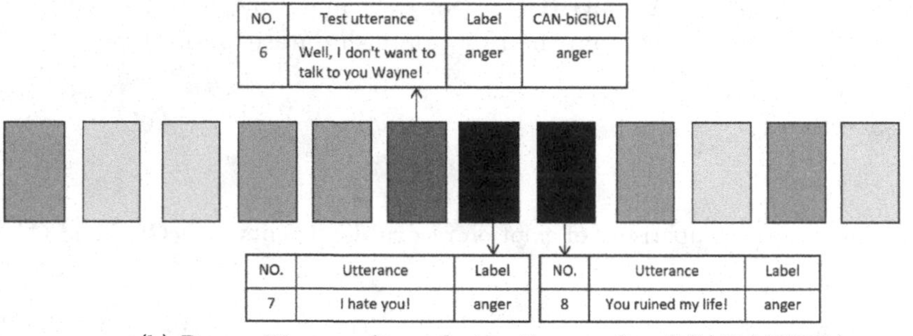

(b) Recognition results and attention results of CAN-biGRUA

Fig. 2. Comparison of recognition results and attention results between CAN-GRUA and CAN-biGRUA. Deeper color means higher attention.

recognise the emotion category of utterances. In the conversation of Monica and Joey, 'Yeah' expresses neutral emotion, while in the conversation of Chloe and Ross, 'Yeah' with '!' expresses stronger emotion and our model analyses its joy emotion effectively. However, in the third case, our model makes wrong classification for the first, third and fourth utterances. This dialogue contains three different emotions and emotion shifts frequently. The failure of our model indicates that although considering the context, model's ability to understand emotions in the complicated situation is limited and still needs improvement.

In Fig. 2, we compare the recognition results and attention results of CAN-GRUA and CAN-biGRUA for the sixth utterance in the dialogue. As we can see, CAN-GRUA only uses historical information and focuses on the fourth utterance, and it takes neutral emotion as a result. While CAN-biGRUA takes both historical and future information into account, and it mainly pays attention to the seventh and the eighth utterances which contain strongly anger emotion, so the model finnaly classifies the test utterance as anger emotion. This case shows that considering both historical information and future information can help model make better classifications.

Table 6. Some cases of emotion recognition results by CAN-biGRUA

Speaker	Utterance	True label	Predicted label
Monica	Hey, Joey, could you pass the cheese?	**neutral**	**neutral**
Joey	Yeah	**neutral**	**neutral**
Chloe	That's so great for you guys!	**joy**	**joy**
Ross	Yeah!	**joy**	**joy**
Chloe	Good luck, with your girlfriend	**neutral**	**neutral**
Monica	Ross, we can handle this	neutral	joy
Ross	Well,... be hurt over something that is so silly	**sadness**	**sadness**
Ross	I mean, enough of the silliness!	anger	sadness
Chandler	Well, why don't you tell her to stop being silly!	anger	sadness

5 Conclusion

In the paper, we propose a hierarchical model(CAN-GRU) to tackle emotion recognition in dialogues. Unlike existing works, we focus on semantic information extraction within utterance in the dialogue. N-gram features and relevant semantic information among words in the utterance are learned by the convolutional self-attention network in the first layer and the sequence of utterances is modeled by the GRU-based network in the second layer. We improve CAN-GRU to three variants, CAN-biGRU, CAN-GRUA and CAN-biGRUA. Experimental results show that attention mechanism can help to grasp long-term dependency in the contexts effectively. CAN-biGRUA achieves better results than CAN-GRUA demonstrates that it is necessary to consider both past and future information of the utterance. In the future, we will try to explore deeper semantic information in the context and focus more on emotion shift to solve the problem of poor performance of the model in complex situations.

Acknowledgments. The work of this paper is funded by the project of National key research and development program of China (No. 2018YFC0830700).

References

1. Bahdanau, D., Cho, K., Bengio, Y.: Neural machine translation by jointly learning to align and translate. In: 3rd International Conference on Learning Representations (2015)
2. Cho, K., et al.: Learning phrase representations using RNN encoder-decoder for statistical machine translation. In: Proceedings of the 2014 Conference on Empirical Methods in Natural Language Processing, pp. 1724–1734. ACL (2014)
3. Devlin, J., Chang, M., Lee, K., Toutanova, K.: BERT: pre-training of deep bidirectional transformers for language understanding. In: Proceedings of the 2019 Conference of the North American Chapter of the Association for Computational Linguistics: Human Language Technologies, NAACL-HLT 2019, Minneapolis, MN, USA, 2–7 June 2019, vol. 1 (Long and Short Papers) (2019)

4. Gao, S., Ramanathan, A., Tourassi, G.D.: Hierarchical convolutional attention networks for text classification. In: Proceedings of the Third Workshop on Representation Learning for NLP, Rep4NLP@ACL, pp. 11–23. Association for Computational Linguistics (2018)
5. Ghosal, D., Majumder, N., Poria, S., Chhaya, N., Gelbukh, A.F.: Dialoguegcn: a graph convolutional neural network for emotion recognition in conversation. In: Proceedings of the 2019 Conference on Empirical Methods in Natural Language Processing and the 9th International Joint Conference on Natural Language Processing, EMNLP-IJCNLP 2019, Hong Kong, China, 3–7 November 2019, pp. 154–164. Association for Computational Linguistics (2019)
6. Hazarika, D., Poria, S., Mihalcea, R., Cambria, E., Zimmermann, R.: ICON: interactive conversational memory network for multimodal emotion detection. In: Proceedings of the 2018 Conference on Empirical Methods in Natural Language Processing, pp. 2594–2604. Association for Computational Linguistics (2018)
7. Hazarika, D., Poria, S., Zadeh, A., Cambria, E., Morency, L., Zimmermann, R.: Conversational memory network for emotion recognition in dyadic dialogue videos. In: Proceedings of the 2018 Conference of the North American Chapter of the Association for Computational Linguistics: Human Language Technologies, vol. 1 (Long Papers), pp. 2122–2132. Association for Computational Linguistics (2018)
8. Hochreiter, S., Schmidhuber, J.: Long short-term memory. Neural Comput. **9**(8), 1735–1780 (1997)
9. Hsu, C., Ku, L.: Socialnlp 2018 emotionx challenge overview: Recognizing emotions in dialogues. In: Proceedings of the Sixth International Workshop on Natural Language Processing for Social Media, SocialNLP@ACL, pp. 27–31. Association for Computational Linguistics (2018)
10. Khosla, S.: EmotionX-AR: CNN-DCNN autoencoder based emotion classifier. In: Ku, L., Li, C. (eds.) Proceedings of the Sixth International Workshop on Natural Language Processing for Social Media, SocialNLP@ACL, pp. 37–44. Association for Computational Linguistics (2018)
11. Kim, Y.: Convolutional neural networks for sentence classification. In: Proceedings of the 2014 Conference on Empirical Methods in Natural Language Processing, pp. 1746–1751. ACL (2014)
12. Kingma, D.P., Ba, J.: Adam: a method for stochastic optimization. In: 3rd International Conference on Learning Representations (2015)
13. Luo, L., Yang, H., Chin, F.Y.L.: Emotionx-DLC: Self-attentive BiLSTM for detecting sequential emotions in dialogues. In: Proceedings of the Sixth International Workshop on Natural Language Processing for Social Media, SocialNLP@ACL, pp. 32–36. Association for Computational Linguistics (2018)
14. Majumder, N., Poria, S., Hazarika, D., Mihalcea, R., Gelbukh, A.F., Cambria, E.: Dialoguernn: An attentive RNN for emotion detection in conversations. In: The Thirty-Third AAAI Conference on Artificial Intelligence, AAAI 2019, the Thirty-First Innovative Applications of Artificial Intelligence Conference, IAAI 2019, the Ninth AAAI Symposium on Educational Advances in Artificial Intelligence, EAAI 2019, Honolulu, Hawaii, USA, 27 January – 1 February 2019, pp. 6818–6825. AAAI Press (2019)
15. Mayer, J.D., Roberts, R.D., Barsade, S.G.: Human abilities: emotional intelligence. Ann. Rev. Psychol. **59**, 507–536 (2008)
16. Pennington, J., Socher, R., Manning, C.D.: Glove: global vectors for word representation. In: Proceedings of the 2014 Conference on Empirical Methods in Natural Language Processing, pp. 1532–1543. ACL (2014)

17. Poria, S., Cambria, E., Gelbukh, A.F.: Deep convolutional neural network textual features and multiple kernel learning for utterance-level multimodal sentiment analysis. In: Proceedings of the 2015 Conference on Empirical Methods in Natural Language Processing, pp. 2539–2544. ACL (2015)
18. Poria, S., Cambria, E., Hazarika, D., Majumder, N., Zadeh, A., Morency, L.: Context-dependent sentiment analysis in user-generated videos. In: Proceedings of the 55th Annual Meeting of the Association for Computational Linguistics, vol. 1: Long Papers, pp. 873–883. Association for Computational Linguistics (2017)
19. Saxena, R., Bhat, S., Pedanekar, N.: EmotionX-Area66: predicting emotions in dialogues using hierarchical attention network with sequence labeling. In: Proceedings of the Sixth International Workshop on Natural Language Processing for Social Media, SocialNLP@ACL, pp. 50–55. Association for Computational Linguistics (2018)
20. Vaswani, A., et al.: Attention is all you need. In: Advances in Neural Information Processing Systems 30: Annual Conference on Neural Information Processing Systems, pp. 5998–6008 (2017)

A Joint Model for Aspect-Category Sentiment Analysis with Shared Sentiment Prediction Layer

Yuncong Li(✉), Zhe Yang(✉), Cunxiang Yin, Xu Pan, Lunan Cui, Qiang Huang, and Ting Wei

Baidu Inc., Beijing, China
{liyuncong,yangzhe08,yincunxiang,panxu,cuilunan, huangqiang03,weiting}@baidu.com

Abstract. Aspect-category sentiment analysis (ACSA) aims to predict the aspect categories mentioned in texts and their corresponding sentiment polarities. Some joint models have been proposed to address this task. Given a text, these joint models detect all the aspect categories mentioned in the text and predict the sentiment polarities toward them at once. Although these joint models obtain promising performances, they train separate parameters for each aspect category and therefore suffer from data deficiency of some aspect categories. To solve this problem, we propose a novel joint model which contains a shared sentiment prediction layer. The shared sentiment prediction layer transfers sentiment knowledge between aspect categories and alleviates the problem caused by data deficiency. Experiments conducted on SemEval-2016 Datasets demonstrate the effectiveness of our model.

Keywords: Aspect-category sentiment analysis · Aspect-Based Sentiment Analysis · Data deficiency.

1 Introduction

Aspect-category sentiment analysis (ACSA) is a subtask of aspect-based sentiment analysis (ABSA) [13–15]. ACSA aims to identify all the aspect categories mentioned in texts and their corresponding sentiment polarities. An aspect category (or simply aspect) is an entity E and attribute A pair, denoted by E#A. For example, in the text "The place is small and cramped but the food is fantastic.", the aspect categories mentioned in the text are *AMBIENCE#GENERAL* and *FOOD#QUALITY*, and their sentiment polarities are negative and negative, respectively.

Many methods have been proposed to address the ACSA task. However, most existing methods [1,11,16,19,21,24,26] divide the ACSA task into two subtasks: aspect category detection (ACD) which detects aspect categories in a text

Y. Li and Z. Yang—Equal contribution.

M. Sun et al. (Eds.): CCL 2020, LNAI 12522, pp. 388–400, 2020.
https://doi.org/10.1007/978-3-030-63031-7_28

and sentiment classification (SC) which categorizes the sentiment polarities with respect to the detected aspect categories, and perform these two tasks separately. Such two-stage approaches lead to error propagation, that is, errors caused by aspect category detection would affect sentiment classification. To avoid error propagation, previous studies [4,17,22] have proposed some joint models, which jointly model the detection of aspect categories and the classification of their polarities. Further more, given a text, these joint models detect all the aspect categories mentioned in the text and predict the sentiment polarities toward them at once. Although these joint models obtain promising performances, they train separate parameters for each aspect category and therefore suffer from data deficiency of some aspect categories. For example, the english laptops domain dataset from SemEval-2016 task 5: Aspect-based Sentiment Analysis [13] has a quarter of the aspect categories whose sample size are less than or equal to 2 (see Table 2). Previous joint models will under-fit on the aspect categories with deficient samples.

To solve the problem caused by data deficiency mentioned above, we propose a novel joint model, which contains a shared sentiment prediction layer. Our model is based on the observation that the sentiment expressions and their polarities of different aspect categories are transferable. For instance, in Table 1, the three aspect categories *LAPTOP#QUALITY*, *LAPTOP#PRICE*, and *LAPTOP#OPERATION_PERFORMANCE* have the same sentiment word "surprised" and the consistent polarity. The shared sentiment prediction layer transfers sentiment knowledge between aspect categories and alleviates the problem caused by data deficiency.

In summary, the main contributions of our work are as follows:

- We propose a novel joint model for the aspect category sentiment analysis (ACSA) task, which contains a shared sentiment prediction layer. The shared sentiment prediction layer transfers sentiment knowledge between aspect categories and alleviates the problem caused by data deficiency.
- Experiments conducted on SemEval-2016 Datasets demonstrate the effectiveness of our model.

Table 1. Different aspect categories have the same sentiment word which have the same sentiment polarity.

Aspect category	Text	Polarity
LAPTOP#QUALITY	...I was **surprised** at the overall quality and the price...	positive
LAPTOP#PRICE		positive
LAPTOP#OPERATION_PERFORMANCE	...I was **surprised** with the performance and quality of this HP Laptop...	positive
LAPTOP#QUALITY		positive

2 Related Work

Existing methods for Aspect-Category Sentiment Analysis (ACSA) can be divided into two categories: two-stage methods and joint models.

Two-stage methods perform the ACD task and the SC task separately. Zhou et al. [26] and Movahedi et al. [11] perform the ACD task. Zhou et al. [26] propose a semi-supervised word embedding algorithm to obtain word embeddings on a large set of reviews, which are then used to generate deeper and hybrid features to predict the aspect category. Movahedi et al. [11] utilize topic attention to attend to different aspects of a given text. Many methods [1,5,8,9,16,18,19,21,23,24,27] have been proposed for the SC task. Wang et al. [21] first propose aspect embedding (AE) and use an Attention-based Long Short-Term Memory Network (AT-LSTM) to generate aspect-specific text representations for sentiment classification based on aspect embedding. Ruder et al. [16] propose a hierarchical bidirectional LSTM (H-LSTM) to modeling the interdependencies of sentences in a review. Tay et al. [19] propose a method named Aspect Fusion LSTM (AF-LSTM) to model word-aspect relationships. Xue and Li [24] propose a model, namely Gated Convolutional network with Aspect Embedding (GCAE), which incorporates aspect information into the neural model by gating mechanisms. Jiang et al. [5] proposed new capsule networks to model the complicated relationship between aspects and contexts. All the two-stage methods have the problem of error propagation.

Joint models jointly model the detection of aspect categories and the classification of their polarities. Only a few joint models [4,17,22] have been proposed for ACSA. Schmitt et al. [17] propose two joint models: End-to-end LSTM and End-to-end CNN, which produce all the aspect categories and their corresponding sentiment polarities at once. Hu et al. [4] propose constrained attention networks (CAN), which extends AT-LSTM to multi-task settings and introduces orthogonal and sparse regularizations to constrain the attention weight allocation. As a result, the CAN achieves better sentiment classification performance. However, to train the CAN, we need to annotate the multi-aspect sentences with overlapping or nonoverlapping. Wang et al. [22] propose the aspect-level sentiment capsules model (AS-Capsules), which utilizes the correlation between aspect category and sentiment through shared components including capsule embedding, shared encoders, shared attentions and a shared recurrent neural network. These joint models train separate parameters for each aspect category, which results in that these models under-fit on the aspect categories with deficient samples.

3 Proposed Model

We first formulate the problem. There are N predefined aspect categories $A = \{A_1, A_2, \cdots, A_N\}$ and M predefined sentiment polarities $P = \{P_1, P_2, \cdots, P_M\}$ in the dataset. Given a sentence or a review, denoted by $S = \{w_1, w_2, \cdots, w_n\}$, the task aims to predict the aspect categories and the corresponding sentiment polarities, i.e., aspect-sentiment pairs$\{< A_j, P_k >\}$, expressed in the text. The overall model architecture is illustrated in Fig. 1, which contains six modules: embedding layer, Bi-LSTM layer, aspect attention layer, sentiment attention layer, aspect category prediction layer, and shared sentiment prediction layer. Then, we display the details of each module and introduce the training objective function.

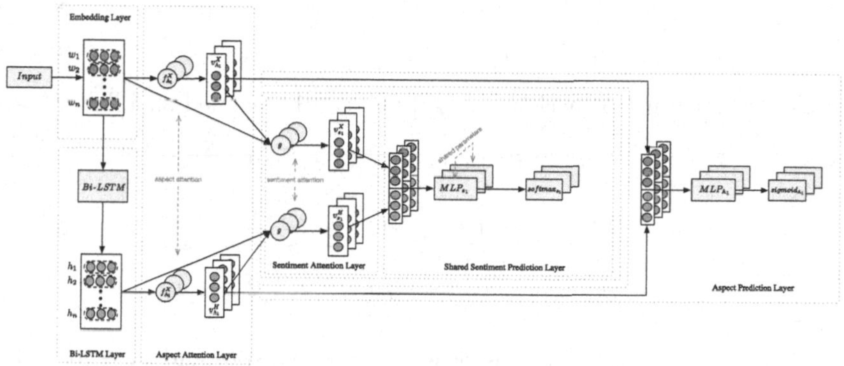

Fig. 1. Overall architecture of the proposed method.

3.1 Embedding Layer

The input to our model is a text consisting of n words $\{w_1, w_2, ..., w_n\}$. With a embedding matrix U, the input text is converted to a sequence of vectors $X = \{x_1, x_2, ..., x_n\}$. Where $U \in R^{d_w \times |V|}$, d_w is the dimension of the word embeddings, and $|V|$is the vocabulary size.

3.2 Bidirectional LSTM Layer

The word embeddings of the text are then fed into a Bidirectional LSTM [2] network (Bi-LSTM) with two LSTM [3] networks. We can obtain two hidden representations, and then concatenate the forward hidden state and backward hidden state of each word. Formally, given the sequence of vectors $X = \{x_1, x_2, \cdots, x_n\}$, Bi-LSTM outputs hidden states $H = \{h_1, h_2, \cdots, h_n\}$. At each time step $i = 1, 2, \cdots, n$, the hidden state h_i of the Bi-LSTM is computed by:

$$\overrightarrow{h_i} = \overrightarrow{LSTM}(\overrightarrow{h}_{i-1}, x_i) \tag{1}$$

$$\overleftarrow{h_i} = \overleftarrow{LSTM}(\overleftarrow{h}_{i+1}, x_i) \tag{2}$$

$$h_i = [\overrightarrow{h_i}, \overleftarrow{h_i}] \tag{3}$$

where $\overrightarrow{h_i} \in R^{d_s}$,$\overleftarrow{h_i} \in R^{d_s}$, $h_i \in R^{2d_s}$, and d_s denotes the size of the hidden state of LSTM.

3.3 Aspect Attention Layer

This layer applies an attention mechanism on the outputs of both the embedding layer and the Bi-LSTM layer and generates aspect-specific representations for the ACD task. Different aspect categories have different attention parameters. The process can be formulated as follows:

$$v_{A_j}^X = f_{A_j}^X(X), j = 1, \cdots, N \tag{4}$$

$$v_{A_j}^H = f_{A_j}^H(H), j = 1, \cdots, N \tag{5}$$

where f(·) is an attention mechanism [25] and can be defined as follows:

$$f(V) = v = \Sigma_{i=1}^n \alpha_i v_i \tag{6}$$

$$u_i = tanh(W_a v_i + b_a) \quad for \quad i = 1, 2, \cdots, n \tag{7}$$

$$\alpha_i = \frac{exp(u_i^T u_w)}{(\Sigma_{j=1}^n exp(u_j^T u_w)} \quad for \quad i = 1, 2, \cdots, n \tag{8}$$

where $V = \{v_1, \cdots, v_i, \cdots, v_n\}$ is a sequence of vectors and $v_i \in R^d$. $W_a \in R^{m \times d}$,$b_a \in R^m$, and $u_w \in R^m$ are the parameters of the attention mechanism. m is the dimensionality of the attention context vector, and d is the dimensionality of the input vector. Note that the vector v generated by $f(V)$ is a weighted sum of vectors in V and is in the same semantic space with them.

3.4 Aspect Category Prediction Layer

Aspect category prediction layer takes as input the concatenation of the aspect-specific representations at the embedding layer and the Bi-LSTM layer for the ACD task and predicts whether the text mentions the aspect categories. Formally, for the j-th aspect category:

$$v_{A_j} = [v_{A_j}^X, v_{A_j}^H] \tag{9}$$

$$\widehat{y}_{A_j} = \sigma(\widehat{W}_{A_j} ReLU(W_{A_j} v_{A_j} + b_{A_j}) + \widehat{b}_{A_j}) \tag{10}$$

$$\sigma(x) = \frac{1}{(1 + e^{-x})} \tag{11}$$

where W_{A_j},b_{A_j}, $\widehat{W}_{A_j}$, and $\widehat{b}_{A_j}$ are the parameters of the j-th aspect category. If $\widehat{y}_{A_j}$ is greater than the specified threshold τ, we judge that the j-th aspect category is mentioned by the text.

3.5 Sentiment Attention Layer

This layer generates aspect-specific text representations for the SC task based on aspect-specific text representations for the ACD task. For the j-th aspect category, its aspect-specific text representations for the SC task can be computed as follows:

$$v_{s_j}^X = g(X, v_{A_j}^X) \tag{12}$$

$$v_{s_j}^H = g(H, v_{A_j}^H) \tag{13}$$

where X and H are the outputs of the embedding layer and the Bi-LSTM layer, respectively. $v_{A_j}^X$ and $v_{A_j}^H$ are the aspect-specific text representations of the j-th aspect category for the ACD task at the outputs of the embedding layer and the Bi-LSTM layer respectively. g(·) is an attention mechanism [20] and can be defined as follows:

$$v_s = g(X, v_q) \tag{14}$$

$$\beta_i = \frac{exp(x_i^T v_q)}{\Sigma_{j=1}^n x_j^T v_q} \quad for \quad i = 1, 2, \cdots, n \tag{15}$$

$$v_s = \Sigma_{i=1}^n \beta_i x_i \tag{16}$$

where X is a sequence vectors $\{x_1, x_2, \cdots, x_n\}$, and v_q is the query vector of the attention. We use the dot product to compute attention weights because it does not import extra aspect-specific parameters. Since the query vector and the key vector of the attention are in the same semantic space in our model, the dot product is reasonable.

3.6 Shared Sentiment Prediction Layer

The aspect-specific text representation of the j-th aspect for the SC task is generated by concatenating the aspect-specific text representations of the j-th aspect category for the SC task at the outputs of the embedding layer and the Bi-LSTM layer. The representation are then fed to a fully connected layer with the ReLU activation function and then the output of the fully connected layer is fed to another fully connected layer with the softmax activation function to generate sentiment probability distribution. Formally, for the j-th aspect category:

$$v_{s_j} = [v_{s_j}^X, v_{s_j}^H] \tag{17}$$

$$\widehat{y}_{s_j} = softmax(\widehat{W}_s ReLU(W_s v_{s_j} + b_s) + \widehat{b}_s) \tag{18}$$

$$softmax(x)_i = \frac{exp(x_i)}{\Sigma_{k=1}^M exp(x_k)} \tag{19}$$

where W_s , b_s , $\widehat{W}_s$, and $\widehat{b}_s$ are the shared parameters of all aspect categories.

3.7 Loss

For the aspect category detection task, as each prediction is a binary classification problem, the loss function is defined by:

$$L_A(\theta) = -\Sigma_{j=1}^{N} y_{A_j} log\widehat{y}_{A_j} + (1 - y_{A_j})log(1 - \widehat{y}_{A_j}) \tag{20}$$

For the sentiment classification task, the loss function is defined by:

$$L_s(\theta) = -\Sigma_{j=1}^{N}\Sigma_{k=1}^{M} y_{s_{jk}} log(\widehat{y}_{s_{jk}}) \tag{21}$$

if the j-th aspect category is not mentioned in the text, $y_{s_{jk}} = 0$ for $k = 1, 2, \cdots, M$.

We jointly train our model for the two tasks. The parameters in our model are then trained by minimizing the combined loss function:

$$L_s(\theta) = L_A(\theta) + \eta L_s(\theta) + \lambda||\theta||_2^2 \tag{22}$$

where η is the weight of sentiment classification loss, λ is the L2 regularization factor and θ contains all the parameters except for Bi-LSTM layer's parameters. Furthermore, to avoid over-fitting, we adopt the dropout strategy to enhance our model.

4 Experiments

4.1 Datasets

We conduct experiments on four public datasets from SemEval-2016 task 5: Aspect-based Sentiment Analysis [13]:

CH-CAME-SB1 is a Chinese sentence-level dataset about digital cameras domain.

CH-PHNS-SB1 is a Chinese sentence-level dataset about mobile phones domain.

EN-REST-SB2 is an English review-level dataset about restaurants domain.

EN-LAPT-SB2 is an English review-level dataset about laptops domain.

We randomly split the original training set into training, validation sets in the ratio 9:1. We use quartiles to measure the distribution of the sample size of aspects in these datasets. Detailed statistics are summarized in Table 2. Particularly, for the three datasets, CH-CAME-SB1, CH-PHNS-SB1, and EN-LAPT-SB2, the sample size of 50% of the aspects are no more than 7.

4.2 Evaluation Metrics

We use micro-averaged F1-scores as the evaluation metric for both the ACSA and the ACD:

$$F_1 = \frac{2 * P * R}{P + R} \tag{23}$$

Table 2. Statistics of the datasets. #aspect and #polarity represent the number of predefined aspects and sentiment polarities, respectively. #train, #dev, and #test represent the sample size of training sets, validation sets, and test sets, respectively. #min indicates the minimum value of the aspect sample size. #Q1, #Q2, #Q3 are the first quartile, the second quartile, and the third quartile of the aspect sample size, respectively.

Dataset	#aspect	#polarity	#train	#val	#test	#min	#Q1	#Q2	#Q3
CH-CAME-SB1	75	2	1090	169	481	1.0	1.0	6.0	13.0
CH-PHNS-SB1	81	2	1152	181	529	1.0	1.8	4.5	23.3
EN-LAPT-SB2	88	4	355	40	80	1.0	2.0	7.0	20.0
EN-REST-SB2	12	4	301	34	90	20.0	36.5	68.5	177.0

where precision and recall are defined as:

$$P = \frac{|S \cap G|}{|S|} \tag{24}$$

$$R = \frac{|S \cap G|}{|G|} \tag{25}$$

Here S is the set of aspect-sentiment pairs or aspect category annotations (in ACSA and ACD, respectively) that a model returns for all the test texts, and G is the set of the gold (correct) aspect-sentiment pairs or aspect category annotations. To evaluate the SC task, we use the gold aspect category annotations to select sentiment polarities model predicts and calculated the accuracy.

4.3 Comparison Methods

We select the following methods for comparison.

End-to-end LSTM [17] performs the ACSA task, which jointly models the detection of aspects and the classification of their polarities in an end-to-end trainable neural network.

End-to-end CNN [17] is an CNN version of End-to-end LSTM, which replaces the Bi-LSTM in End-to-end LSTM with a convolutional neural network (CNN) described in [6].

AS-Capsules [22] utilizes the correlation between aspect category and sentiment through shared components including capsule embedding, shared encoders, shared attentions and a shared recurrent neural network.

SemEval-2016 Best is the best model for each subtask of SemEval-2016 task 5: Aspect based Sentiment Analysis [13].

Our Model – w/o Share was added to show the effectiveness of the shared sentiment prediction layer, which trains a separate sentiment prediction layer for each aspect.

4.4 Implementation Details

We implement all models in Keras. We set $\lambda = 0.01$ and gradient clipping norm to 5. Adam [7] optimizer is applied to minimize the loss. We apply a dropout of $p = 0.5$ after the embedding layer and the Bi-LSTM layer. Hidden layer size for Bi-LSTM is 100. We use 300-dimensional word embeddings. We use GloVe [12] embeddings which are pre-trained on an unlabeled corpus whose size is about 840 billion for English and Skip-Gram [10] embeddings which are pre-trained on the Baidu Encyclopedia dataset for Chinese. If an aspect is not mentioned, its corresponding sentiment label is set to a zero vector. We set threshold $\tau = 0.25$ for aspect category detection. While batch size is 32 on CAME-SB1 and CH-PHNS-SB1, batch size is 10 on EN-LAPT-SB2 and EN-REST-SB2. The sentiment classification loss weight is 1 on CH-CAME-SB1, CH-PHNS-SB1, and EN-LAPT-SB2, and is 0.6 on EN-REST-SB2. To reduce the randomness of results, we train each model three times and report their averaged scores.

Table 3. Results of the ACSA task in terms of micro-averaged F1-scores(%).

Models	CH-CAME-SB1	CH-PHNS-SB1	EN-LAPT-SB2	EN-REST-SB2
End-to-end cnn	34.96	19.87	36.52	66.03
End-to-end lstm	41.52	26.30	37.94	63.75
AS-Capsules	38.85	27.05	33.47	63.99
Our Model	**42.01**	**28.98**	**50.05**	68.24
– w/o Share	36.23	22.72	49.43	**68.28**

Table 4. Results of the SC task in terms of accuracy(%).

Models	CH-CAME-SB1	CH-PHNS-SB1	EN-LAPT-SB2	EN-REST-SB2
SemEval-2016 Best	80.45	73.34	75.05	81.93
End-to-end cnn	70.55	64.15	69.42	80.78
End-to-end lstm	75.12	67.36	72.00	80.03
AS-Capsules	76.96	71.56	69.05	76.794
Our Model	**82.54**	**76.50**	**75.91**	82.43
– w/o Share	69.09	59.86	70.53	**83.33**

4.5 Results

Table 3, Table 4, and Table 5 show our experimental results on the ACSA, SC, and ACD tasks, respectively. The best results are marked in bold.

Table 5. Results of the ACD task in terms of micro-averaged F1-scores(%).

Models	CH-CAME-SB1	CH-PHNS-SB1	EN-LAPT-SB2	EN-REST-SB2
SemEval-2016 Best	36.3	22.5	60.4	**83.9**
End-to-end cnn	47.83	26.64	42.25	76.20
End-to-end lstm	**52.98**	33.81	43.81	76.24
AS-Capsules	48.72	33.66	40.99	78.29
Our Model	51.81	**36.67**	**62.91**	81.65
– w/o Share	52.16	36.54	62.72	81.45

Table 3 shows the experimental results on the ACSA task, which show the overall performance of our joint model. We observe that our proposed joint model outperforms the baseline models on all datasets, which demonstrates the effectiveness of our model.

The experimental results on the SC task are in Table 4. First, we observe that our model surpasses all baseline models on all datasets, which indicates the effectiveness of our model predicting the sentiment polarities toward given aspect categories. Second, our model outperforms its variant (– w/o Share) by 13.45%, 16.64% and 5.38% on CH-CAME-SB1, CH-PHNS-SB1, and EN-LAPT-SB2 datasets, respectively. The reason is that the three datasets have many aspect categories which only have a few instances and benefit from parameter sharing. This shows that our shared parameter prediction layer can alleviate the problem caused by data deficiency. Meanwhile, our model obtains worse perfromance than its variant (– w/o Share) on the EN-REST-SB2 dataset. The possible reason is that the sample size of the aspect categories in the EN-REST-SB2 dataset is enough to train independent sentiment prediction parameters, and parameter sharing brings some noise between aspect categories.

Table 5 shows the results on the ACD task. Although we did not specifically optimize our model for the ACD task, our model still achieves competitive performance. Specifically, our model outperforms all baselines on the CH-PHNS-SB1 and EN-LAPT-SB2 datasets.

4.6 Case Studies

To have an intuitive understanding of our proposed shared sentiment prediction layer for the SC task, we use the EN-LAPT-SB2 dataset to illustrate the impact of knowledge transferring. The selected aspect category from the dataset is *OS#MISCELLANEOUS*. There are only two samples with both negative polarities in the training set, while there are two samples in the test set, whose polarities are negative and positive, respectively. Table 6 shows that our model can correctly predict the polarity of the sample with positive sentiment in the test set. After removing the shared sentiment prediction layer, – w/o Share fails to predict the polarity of the sample with positive sentiment, which confirms the importance of the shared sentiment prediction layer.

Table 6. Impact of the shared sentiment prediction layer on the sentiment prediction of the aspect category *OS#MISCELLANEOUS.*

	Text	Label	Our model	-w/o Share
Train set	...The only objection I have is that after you buy it the windows 7 system is a starter and charges for the upgrade...	negtive		
	...The flaws are, this computer is not for computer gamers because of the OS X...	negtive		
Test set	...The OS is easy, and offers all kinds of surprises	positive	positive	negtive
	...The free upgrade to Mountain Lion FAILED	negtive	negtive	negtive

5 Conclusion

In this work, we propose a novel joint model which contains a shared sentiment prediction layer. The shared sentiment prediction layer transfers sentiment knowledge between aspect categories and alleviates the problem caused by data deficiency. Experiments conducted on four datasets from SemEval-2016 task 5 demonstrate the effectiveness of our model. Furture work could consider introducing extra component that prevents the shared sentiment prediction layer from transfering aspect-specific sentiment knowledge.

References

1. Cheng, J., Zhao, S., Zhang, J., King, I., Zhang, X., Wang, H.: Aspect-level sentiment classification with heat (hierarchical attention) network. In: Proceedings of the 2017 ACM on Conference on Information and Knowledge Management, pp. 97–106 (2017). https://doi.org/10.1145/3132847.3133037
2. Graves, A., Mohamed, A.R., Hinton, G.: Speech recognition with deep recurrent neural networks. In: 2013 IEEE International Conference on Acoustics, Speech and Signal Processing, pp. 6645–6649. IEEE (2013). https://doi.org/10.1109/ICASSP.2013.6638947
3. Hochreiter, S., Schmidhuber, J.: Long short-term memory. Neural Comput. **9**(8), 1735–1780 (1997). https://doi.org/10.1162/neco.1997.9.8.1735
4. Hu, M.,et al.: Can: constrained attention networks for multi-aspect sentiment analysis. In: Proceedings of the 2019 Conference on Empirical Methods in Natural Language Processing and the 9th International Joint Conference on Natural Language Processing (EMNLP-IJCNLP), pp. 4593–4602 (2019). https://doi.org/10.18653/v1/D19-1467

5. Jiang, Q., Chen, L., Xu, R., Ao, X., Yang, M.: A challenge dataset and effective models for aspect-based sentiment analysis. In: Proceedings of the 2019 Conference on Empirical Methods in Natural Language Processing and the 9th International Joint Conference on Natural Language Processing (EMNLP-IJCNLP), pp. 6281–6286 (2019). https://doi.org/10.18653/v1/D19-1654
6. Kim, Y.: Convolutional neural networks for sentence classification. In: Proceedings of the 2014 Conference on Empirical Methods in Natural Language Processing (EMNLP), pp. 1746–1751. Association for Computational Linguistics, Doha, Qatar October 2014. https://doi.org/10.3115/v1/D14-1181, https://www.aclweb.org/anthology/D14-1181
7. Kingma, D.P., Ba, J.: Adam: a method for stochastic optimization. arXiv preprint arXiv:1412.6980 (2014)
8. Lei, Z., Yang, Y., Yang, M., Zhao, W., Guo, J., Liu, Y.: A human-like semantic cognition network for aspect-level sentiment classification. Proceedings of the AAAI Conference on Artificial Intelligence, vol. 33, pp. 6650–6657 (2019). https://doi.org/10.1609/aaai.v33i01.33016650
9. Liang, Y., Meng, F., Zhang, J., Xu, J., Chen, Y., Zhou, J.: A novel aspect-guided deep transition model for aspect based sentiment analysis. In: Proceedings of the 2019 Conference on Empirical Methods in Natural Language Processing and the 9th International Joint Conference on Natural Language Processing (EMNLP-IJCNLP), pp. 5572–5584 (2019). https://doi.org/10.18653/v1/D19-1559
10. Mikolov, T., Sutskever, I., Chen, K., Corrado, G.S., Dean, J.: Distributed representations of words and phrases and their compositionality. In: Advances in Neural Information Processing Systems, pp. 3111–3119 (2013)
11. Movahedi, S., Ghadery, E., Faili, H., Shakery, A.: Aspect category detection via topic-attention network. arXiv preprint arXiv:1901.01183 (2019)
12. Pennington, J., Socher, R., Manning, C.D.: Glove: global vectors for word representation. In: Proceedings of the 2014 conference on empirical methods in natural language processing (EMNLP), pp. 1532–1543 (2014). https://doi.org/10.3115/v1/D14-1162
13. Pontiki, M., et al.: Semeval-2016 task 5: aspect based sentiment analysis. In: Proceedings of the 10th International Workshop on Semantic Evaluation (SemEval-2016), pp. 19–30 (2016). https://doi.org/10.18653/v1/S16-1002
14. Pontiki, M., Galanis, D., Papageorgiou, H., Manandhar, S., Androutsopoulos, I.: Semeval-2015 task 12: aspect based sentiment analysis. In: Proceedings of the 9th International Workshop on Semantic Evaluation (SemEval 2015), pp. 486–495 (2015). https://doi.org/10.18653/v1/S15-2082
15. Pontiki, M., Galanis, D., Pavlopoulos, J., Papageorgiou, H., Androutsopoulos, I., Manandhar, S.: SemEval-2014 task 4: aspect based sentiment analysis. In: Proceedings of the 8th International Workshop on Semantic Evaluation (SemEval 2014), pp. 27–35. Association for Computational Linguistics, Dublin, Ireland August 2014. https://doi.org/10.3115/v1/S14-2004, https://www.aclweb.org/anthology/S14-2004
16. Ruder, S., Ghaffari, P., Breslin, J.G.: A hierarchical model of reviews for aspect-based sentiment analysis. In: Proceedings of the 2016 Conference on Empirical Methods in Natural Language Processing, pp. 999–1005 (2016). https://doi.org/10.18653/v1/D16-1103
17. Schmitt, M., Steinheber, S., Schreiber, K., Roth, B.: Joint aspect and polarity classification for aspect-based sentiment analysis with end-to-end neural networks. In: Proceedings of the 2018 Conference on Empirical Methods in Natural Language Processing, pp. 1109–1114 (2018). https://doi.org/10.18653/v1/D18-1139

18. Sun, C., Huang, L., Qiu, X.: Utilizing BERT for aspect-based sentiment analysis via constructing auxiliary sentence. In: Proceedings of the 2019 Conference of the North American Chapter of the Association for Computational Linguistics: Human Language Technologies, vol. 1 (Long and Short Papers), pp. 380–385 (2019). https://doi.org/10.18653/v1/N19-1035
19. Tay, Y., Tuan, L.A., Hui, S.C.: Learning to attend via word-aspect associative fusion for aspect-based sentiment analysis. In: Thirty-Second AAAI Conference on Artificial Intelligence (2018)
20. Vaswani, A., et al.: Attention is all you need. In: Advances in Neural Information Processing Systems, pp. 5998–6008 (2017)
21. Wang, Y., Huang, M., Zhu, X., Zhao, L.: Attention-based LSTM for aspect-level sentiment classification. In: Proceedings of the 2016 Conference on Empirical Methods in Natural Language Processing, pp. 606–615 (2016). https://doi.org/10.18653/v1/D16-1058
22. Wang, Y., Sun, A., Huang, M., Zhu, X.: Aspect-level sentiment analysis using as-capsules. In: The World Wide Web Conference, pp. 2033–2044 (2019). https://doi.org/10.1145/3308558.3313750
23. Xing, B., et al.: Earlier attention? aspect-aware LSTM for aspect sentiment analysis. arXiv preprint arXiv:1905.07719 (2019)
24. Xue, W., Li, T.: Aspect based sentiment analysis with gated convolutional networks. In: Proceedings of the 56th Annual Meeting of the Association for Computational Linguistics (vol. 1: Long Papers), pp. 2514–2523 (2018). https://doi.org/10.18653/v1/P18-1234
25. Yang, Z., Yang, D., Dyer, C., He, X., Smola, A., Hovy, E.: Hierarchical attention networks for document classification. In: Proceedings of the 2016 Conference of the North American Chapter of the Association for Computational Linguistics: Human Language Technologies, pp. 1480–1489 (2016). https://doi.org/10.18653/v1/N16-1174
26. Zhou, X., Wan, X., Xiao, J.: Representation learning for aspect category detection in online reviews. In: Twenty-Ninth AAAI Conference on Artificial Intelligence (2015)
27. Zhu, P., Chen, Z., Zheng, H., Qian, T.: Aspect aware learning for aspect category sentiment analysis. ACM Trans. Knowl. Discov. Data (TKDD) **13**(6), 1–21 (2019)

NLP Applications

Compress Polyphone Pronunciation Prediction Model with Shared Labels

Pengfei Chen(✉), Lina Wang, Hui Di, Kazushige Ouchi, and Lvhong Wang

Research and Development Center, Toshiba, China
{chenpengfei,wanglina,dihui}@toshiba.com.cn,
kazushige.ouchi@toshiba.co.jp, wlv1990@gmail.com

Abstract. It is well known that deep learning model has huge parameters and is computationally expensive, especially for embedded and mobile devices. Polyphone pronunciations selection is a basic function for Chinese Text-to-Speech (TTS) application. Recurrent neural network (RNN) is a good sequence labeling solution for polyphone pronunciation selection. However, huge parameters and computation make compression needed to alleviate its disadvantages. Meanwhile, Large-scale-labels classification leads to more complicated network and heavy computation cost. In contrast to existing quantization with low precision data format and projection layer, we propose a novel method based on shared labels, which focuses on compressing the fully-connected layer before Softmax for models with a huge number of labels in TTS polyphone selection. The basic idea is to compress large number of target labels into a few label clusters, which will share the parameters of fully-connected layer. Furthermore, we combine it with other methods to further compress the polyphone pronunciation selection model. The experimental result shows that for Bi-LSTM (Bidirectional Long Short Term Memory) based polyphone selection, shared labels model decreases about 52% of original model size and accelerates prediction by 44% almost without performance loss. It is worth mentioning that the proposed method can be applied for other tasks to compress model and accelerate calculation.

Keywords: Bi-LSTM · Polyphone pronunciation prediction · Model compression · Shared labels

1 Introduction

Polyphone pronunciation prediction is a basic module of G2P (Grapheme-to-Phoneme) in Chinese Text-to-Speech (TTS) system, which provides the right pronunciation for Chinese character. The algorithms of polyphone pronunciation prediction include dictionary-based algorithm, statistical machine learning-based algorithm like Conditional Random Field (CRF) in (Lafferty et al.) [4], and deep learning-based algorithm like RNN. Dictionary-based method may fail for polyphone words problem, such as "朝阳" can be read as "chao2yang2" and

M. Sun et al. (Eds.): CCL 2020, LNAI 12522, pp. 403–414, 2020.
https://doi.org/10.1007/978-3-030-63031-7_29

"zhao1yang2", and Out of Vocabulary (OOV) problem. CRF and RNN perform well for polyphone pronunciation selection with context features. However, CRF needs manually designed context features. Neural network always has more parameters, larger model size and more expensive computation cost. In the application for embedded device, small model size and quick computation are necessary. So compression and acceleration of neural network is a hot research field in recent years.

There are several methods to compress deep learning model: low precision of data format, quantification, pruning, low rank factorization, and knowledge distillation. Low precision of data format replaces double or float with 16-bit float, which can reduce model size by quarter or half, but it cannot reduce running time when corresponding computations are not supported by existing instruction sets. Quantification and pruning methods pack some weights into one and prune the weights close to zero. It can reduce the model size but cannot accelerate the computation. Low rank factorization changes the network structure by factorizing large matrix into small matrixes. It can reduce the model size and accelerate the computation.

Our interest focuses on how to compress model that contains huge number of labels, and each input has a fixed label set. For example, each polyphone Chinese character has several (less than 10) fixed pronunciations. If we take all pronunciations of polyphone characters as target labels, there are too many parameters in fully-connected layer before Softmax. Therefore, the computation of fully-connected layer and Softmax will be costly. One idea is to use 1 character-1 model method, which can reduce the network complexity and computation cost, but it needs larger memory because of too many polyphone characters. This inspired us how we can share parameters in a single model which has comparable size to one-character model.

We propose a method to share parameters by means of different characters sharing pronunciation labels. We randomly assign labels for each character's pronunciation under conditions of avoiding label conflicts and assuring that the target label set is small. This yields smaller model size and less computation cost. Then we train our Bi-LSTM model with newly tagged corpus, and compare it with other compression methods in model size, memory usage and decoding time. The experimental results show our method can compress the model to half size and accelerate computation speed while maintaining a comparable performance compared with original model, overcoming the problem of too many labels.

2 Related Work

LSTM (Hochreiter and Schmidhuber) [7] is excellent in sequence labeling by learning contextual information. Bi-LSTM can use the future and history information to improve performance. LSTM encodes the embedding of input sequentially into a vector, which will keep the history information. Bi-LSTM will concatenate the vectors of forward LSTM and backward LSTM, which can utilize future and history information. Followed by fully-connected layer and Softmax,

Bi-LSTM model can predict the label with the highest probability. Lample et al. [3] presented a neural architectures based on Bi-LSTM and CRF to predict name entity. Cai et al. [12] described a system composed of Bi-LSTM acting as an encoder and a prediction network for Chinese polyphone pronunciation prediction. The output size equals the number of all possible pronunciations of polyphone character.

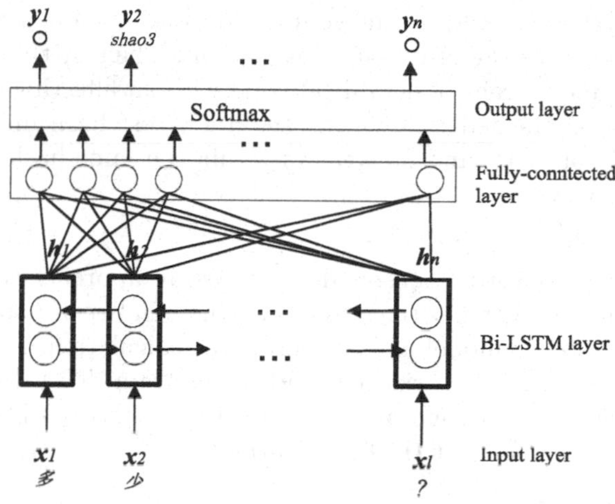

Fig. 1. Polyphone pronunciation sequence labeling by Bi-LSTM

There is a long history of model compression in Nature Language Processing (NLP) tasks. Many compression approaches have been proposed, including low precision data format and quantization, network pruning and parameter sharing, tensor decomposition, knowledge distillation. We briefly review the most popular methods in this section.

Low Precision Data Format and Quantization. Low precision data format can reduce the model size exponentially. Quantization compresses value to a less bits data to reduce the number of bytes of the weight parameter. For example, we can compress a float value to an 8-bit integer, one of 256 equally-sized intervals within the range. 16-bit, 8-bit, and even 1-bit quantization were proposed to compress network. Gupta et al. [8] used 16-bit wide fixed-point number representation to train neural network, without degradation in the accuracy of classification tasks. Courbariaux and Bengio [5] proposed a binarized Neural Networks (BNNs) with binary weights and activations, which can drastically reduce memory size, without any loss in classification accuracy.

Network Pruning and Parameter Sharing. Pruning can remove the parameters below threshold from network. Parameter sharing groups weights into hash brackets for sharing. Network pruning and parameter sharing not only can reduce the structure complexity of model, but also can improve generalization of network.

Grachev et al. [1] introduced pruning to network compression for language modeling, and compared performance and model size with baseline model. Han et al. [9] trained a network to learn which connections are important firstly. Then they pruned the unimportant connections and retrained the network to fine tune the weights of the remaining connections. Their method improved the energy efficiency and storage of neural networks without affecting accuracy. Chen et al. [10] presented HashedNets which used a low-cost hash function to group weights into hash buckets, and all weights within the same hash bucket share a single parameter value.

Tensor Decomposition. Tensor decomposition approaches can factorize weight matrix into smaller matrixes to reduce the number of parameters, which can compress the size of model and accelerate the calculation. Grachev et al. [1] compared low-rank (LR) factorization and tensor train (TT) decomposition in LSTM compression in language modeling. In their result, LR LSTM 650-650 is the most useful model for practical application.

Knowledge Distillation. Hinton et al. [2] first proposed knowledge distillation to transfer knowledge from teacher model to student model. This method can be used to improve the compressed model (student model) by exploiting original model (teacher model).

3 Shared Labels Model

3.1 Framework

Next we will introduce how to implement our novel proposal. For polyphone Chinese characters, there are as many as 1304 pronunciations in our corpus. 1305 labels, including "O" for non-polyphone characters, are used in our original model, which leads to the facts that the fully-connected layer before Softmax contains a large proportion of the weights and Softmax computation is costly. Considering the number of any polyphone pronunciations is less than 10, only 10 shared labels are used in our novel proposal to reduce the memory usage and computational cost.

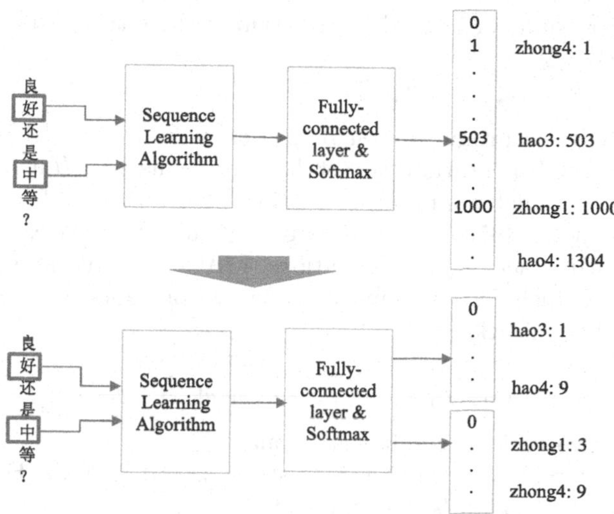

Fig. 2. Illustration of shared labels model

In our proposal, we use a character-based model for polyphone pronunciation prediction. We take a sentence as the input, such as "良好还是中等?". Each character's embedding comes from pre-trained embedding model. The inputs are fed into sequence learning algorithm to learn semantics and features, and the output label is predicted with 1 fully-connected layer and Softmax. The sequence learning method includes but not limited to Bi-LSTM. For the output layer, the original 1305 labels are mapped to 10 digital shared labels. For example, labels "hao3" and "hao4" for "好" are mapped to label-1 and label-9 respectively. The prediction is based on the 10 shared labels.

The digital labels for different polyphone characters can be the same but their real pronunciations may be different. For example, in the above sentence, "好 hao4" and "中 zhong4" share the digital label "9". We can translate the digital shared labels into real pronunciations with a dictionary based on the method of the following section.

3.2 Theoretical Analysis

Theoretically, the framework can reduce parameter number and accelerate computation speed. We use Bi-LSTM as sequence learning algorithm. Given sequence length L, character embedding size V, LSTM hidden size H, target label number N, the parameter number of Bi-LSTM layer is

$$N_1 = 2 * 4 * (V * H + H * H + H) \tag{1}$$

The parameter number of Fully-connected layer is

$$N_2 = 2 * H * N \tag{2}$$

The parameter number of model is the sum of these two parts:

$$N_{para} = N_1 + N_2 \quad (3)$$

For our polyphone pronunciation prediction task, the model is always not complex. In our implementation, we set L as 100, V as 100, H as 200 and N as 1305. The total parameter number of baseline model is 1,003,600, and that of shared labels model is 485,600, which reduces about 52% of size.

We take multiply-accumulate operations (MACCs) as measure of computations. One MACC includes one multiplication and one addition.

For vector multiplication:

$$y = w_0 * x_0 + w_1 * x_1 + \cdots + w_{n-1} * x_{n-1} \quad (4)$$

w and x are two vectors, result y is a scalar.

A dot-product between two vectors of size n uses n MACCs. For a sequence of length L, the total MACCs of our model is

$$N_{MACCs} = L * [2 * 4 * (V + H) * H + 2 * H * N] \quad (5)$$

According the Eq. (5), the total MACCs of baseline model is 100.2 million. In our shared labels method, the total MACCs is 48.4 million, reducing about 52% compared with the baseline model.

If we take Floating Point Operations (FLOPs) as measure, there will be more reduction in computation because of less operations in Softmax.

3.3 Modules of Polyphone Pronunciation Selection with Shared Labels

In the training phase, we need to convert pronunciation to digital label. As mentioned above, we map 1305 labels to 10 shared labels. The mapping relations are saved in two dictionaries consisting of character-pronunciation-label_ID and character-label_ID-pronunciation. Details will be described in the next section. Then we handle the corpus with shared labels. We train Bi-LSTM model with pre-processed corpus.

In inference phase, we get the digital label from the prediction of model. Then we replace the digital label with real pronunciation by looking up dictionary.

Assign Pronunciations to Label Clusters. The relation between polyphone characters and their pronunciations is N:N. So it is important to map 1305 pronunciations to 10 shared labels. We take ID 0 for non-polyphone characters, whose pronunciation can be determined by looking up dictionary. Meanwhile, keeping balanced data number in each shared label will benefit to training speed and performance of the model. We assume that the label IDs subject to random distribution.

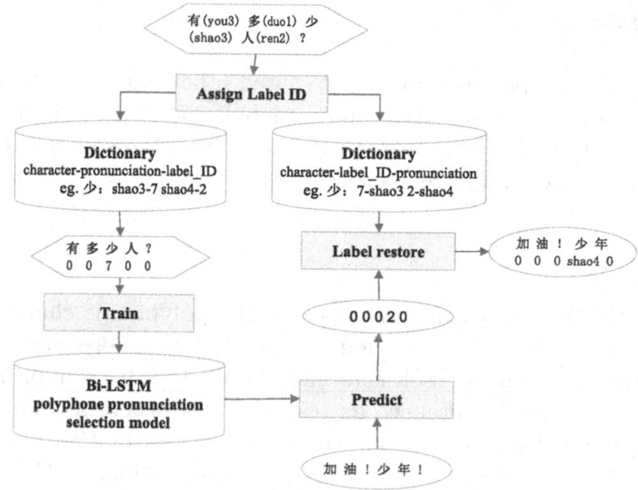

Fig. 3. Modules of polyphone pronunciation selection with shared labels model

The algorithm is as follow.

Algorithm 1. Assign Label ID for pronunciation

```
1: for char in poly_chars do
2:    for pron in char's prons do
3:       rand_id = randint(1,9)
4:       for char in homophone chars with this pron do
5:          for c_pron in char's prons do
6:             if c_pron.label_id == rand_id then
7:                goto 3
8:             else
9:                continue
10:            end if
11:            return rand_id
12:         end for
13:      end for
14:   end for
15: end for
```

Firstly, traverse each character's pronunciations (line 1,2), and assign a label randomly (line 3). If the label is the same with that of other pronunciations of current character and related homophone characters (line 4–6), reassign it randomly (line 7). Repeat this process until all pronunciations are assigned to a certain label. Because of randomness of label assignment, the labels distribution keeps balance.

4 Experiments

In this section, we compare our proposal shared labels model with standard Bi-LSTM model. Besides, we also test shared labels model combining with other methods, such as low precision float, projection layer. We compare their performance, memory usage, and speed respectively.

4.1 Data

Cai et al. [12] did their experiments with a public polyphonic character dataset, but it was unavailable when we tried to use it. So we use our own data as train and test sets. We have 188k sentences labeled with their pronunciations. We randomly select 1000 sentences as test set, and others as train set. There are 1127 polyphone characters in our corpus consisting of 1305 pronunciations (labels). Other Chinese characters are non-polyphone, which are labeled as "O", and their pronunciations are got by looking up dictionary.

4.2 Experimental Settings

Our experiments are done in tensorflow-GPU[1] version (train) and tensorflow-CPU version (test). Our CPU is Intel Xeon E5, and it does not support AVX512.

Baseline Bi-LSTM Model. The structure of the network is the same as in Sect. 3.2. In the training phase, we set the batch size to 16, learning rate to 0.1, and the dropout rate to 0.2. We adopt gradient descent optimization to learn the parameters.

Shared Labels Model. We have the same setting with baseline model, except the output size is 10.

Bi-LSTM with Projection Layer. For further compression, we add a projection layer between input layer and Bi-LSTM layer. Projection layer can factorize the big matrix into small matrixes, which can save memory and reduce the number of parameters.

[1] https://github.com/tensorflow/tensorflow.

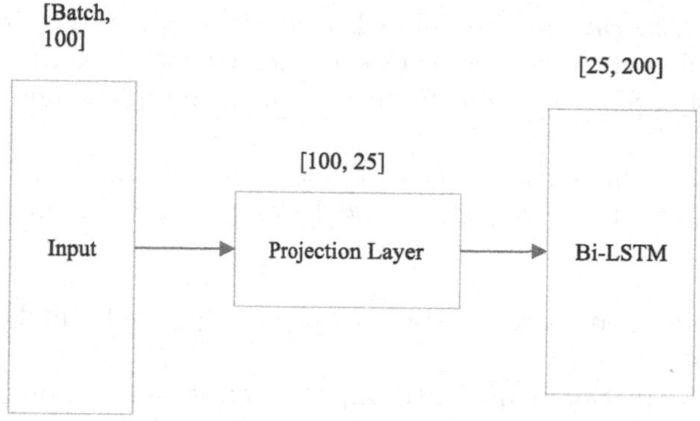

Fig. 4. Bi-LSTM with projection layer

Knowledge Distillation. Transfer learning is a promising method to improve the performance of compressed and simplified networks. We take 16-bit model with projection layer as student model and baseline as teacher model. We adopt the fusion of soft target and hard target as learning object.

Shared Labels in CRF. CRF is a statistical-based machine learning algorithm, which is popular used in sequence labeling problems. We use it to implement the polyphone selection with shared labels to check if it is workable.

4.3 Experiment Results

Model Size and Performance. We compare the models by F1-score and file size.

Table 1. F1-score and model size for different models

Model	F1-score	Model size(Kb)
Baseline (Bi-LSTM)	96.86	3925
+ Shared	96.78	1897
+ 16bit	96.85	1963
+ 16bit + Shared	96.75	948
+ 16bit + PL	94.40	1733
+ 16bit + PL + KD	94.96	1733
+ 16bit + Shared + PL	94.55	719

Note: Shared means shared labels model, PL means Bi-LSTM model with projection layer, KD means knowledge distillation.

From Table 1, we can see shared labels model is compressed to 48% of baseline model with no obvious performance loss. Combined with 16-bit float and shared label, the size of baseline model is further compressed to 24%, and F1-score only drops 0.11 point.

Compared with projection and knowledge distillation, shared model shows good result in model size reduction and keeps high performance at the same time.

a) By adding projection layer, 16-bit model is compressed a little, but the performance dropped a lot.
b) Knowledge distillation is useful to improve performance of projection model.
c) Compared with knowledge distillation, shared labels model has a much smaller size and comparable performance.
d) In general, shared labels outperforms projection in both model size and performance.

Memory Usage. We test the memory usage with open source tool Valgrind[2]. From Fig. 5, we can see that the memory usage of shared label model drops a lot compared with that of baseline model.

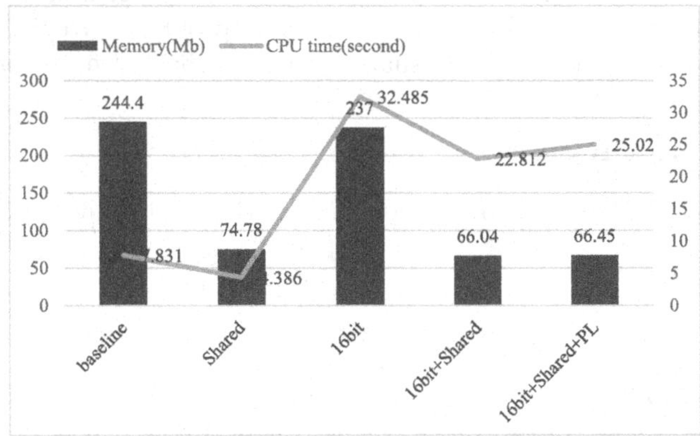

Fig. 5. Memory usage and CPU time of different models

Decoding Time. It takes less time for shared labels model compared with its counterpart model, with accelerating by 44% in Shared vs. baseline and 30% in 16bit+Shared vs. 16bit model. But for 16-bit model, the decoding is much slower than baseline, because our experiment machine does not support 16-bit float instructions.

[2] https://sourceware.org/git/?p=valgrind.git;a=summary.

Result on Shared Labels in CRF. For CRF with 1305 labels, the feature number is about 900 million which makes it unable to be loaded into memory as much as 256 GB. So the training cannot be continued. If we use 10 shared labels for CRF, the feature number is about 6,900,000, with an exponential reduction compared with previous model. The training speed is fast and its F1-score is as high as 0.9765 (Table 2).

Table 2. Result on CRF for polyphone pronunciation selection

Model	F1-score	Model size (Kb)
CRF	NA (unable to train)	
Shared-CRF	97.65	50292

5 Conclusion

We propose a novel shared labels method to compress polyphone pronunciation selection model. It decreases size of models consisting of huge number of labels by mapping labels to small shared labels. Our proposed method reduces the model size and memory usage remarkably, and accelerate decoding speed without performance loss compared with other methods.

In the future, we will verify the compressed model on embedded device and investigate other tasks which can apply this method.

References

1. Grachev, A.M., Ignatov, D.I., Savchenko, A.V.: Neural networks compression for language modeling. In: Pattern Recognition and Machine Intelligence, vol. 10597 (2017)
2. Hinton, G., Dean, J., Vinyals, O.: Distilling the knowledge in a neural network. In: NIPS Deep Learning and Representation Learning Workshop (2015)
3. Lample, G., Ballesteros, M., Subramanian, S., Kawakami, K., Dyer, C.: Neural architectures for named entity recognition. In: Proceedings of NAACL-HLT, pp. 260–270 (2016)
4. John, L., Andrew, M., Pereira, F.C.N.: Conditional random fields: probabilistic models for segmenting and labeling sequence data. In Proceedings ICML (2001)
5. Courbariaux, M., Bengio, Y.: Binarynet: training deep neural networks with weights and activations constrained to +1 or -1. CoRR, vol. abs/1602.02830 (2016)
6. Abadi, M., et al.: TensorFlow: large-scale machine learning on heterogeneous systems. Software available from tensorflow.org (2015)
7. Hochreiter, S., Schmidhuber, J.: Long short-term memory. Neural Comput. **9**(8), 1735–1780 (1997)

8. Gupta, S., Agrawal, A., Gopalakrishnan, K., Narayanan, P.: Deep learning with limited numerical precision. In: Proceedings of the 32Nd International Conference on International Conference on Machine Learning - vol. 37, ser. ICML 2015, pp. 1737–1746 (2015)
9. Han, S., Pool, J., Tran, J., Dally, W.J.: Learning both weights and connections for efficient neural networks. In: Proceedings of the 28th International Conference on Neural Information Processing Systems, ser. NIPS 2015 (2015)
10. Chen, W., Wilson, J., Tyree, S., Weinberger, K.Q., Chen, Y.: Compressing neural networks with the hashing trick. In: JMLR Workshop and Conference Proceedings (2015)
11. Yu, C., Wang, D., Zhou, P., Zhang, T.: A survey of model compression and acceleration for deep neural networks. In: IEEE Signal Processing Magazine, Special Issue on Deep Learning for Image Understanding (2019)
12. Cai, Z., Yang, Y., Zhang, C., Qin, X., Li, M.: Polyphone disambiguation for mandarin Chinese using conditional neural network with multi-level embedding features. INTERSPEECH (2019)

Multi-task Legal Judgement Prediction Combining a Subtask of the Seriousness of Charges

Zhuopeng Xu[2], Xia Li[1,2](✉), Yinlin Li[2], Zihan Wang[2], Yujie Fanxu[2], and Xiaoyan Lai[2]

[1] Guangzhou Key Laboratory of Multilingual Intelligent Processing, Guangzhou, China
xiali@gdufs.edu.cn

[2] School of Information Science and Technology, Guangdong University of Foreign Studies, Guangzhou, China
zhuopengxu@126.com, ksyzformy@126.com, zihanwang0703@126.com, yujiefanxu@126.com, avaxiaoyan@126.com

Abstract. Legal Judgement Prediction has attracted more and more attention in recent years. One of the challenges is how to design a model with better interpretable prediction results. Previous studies have proposed different interpretable models based on the generation of court views and the extraction of charge keywords. Different from previous work, we propose a multi-task legal judgement prediction model which combines a subtask of the seriousness of charges. By introducing this subtask, our model can capture the attention weights of different terms of penalty corresponding to charges and give more attention to the correct terms of penalty in the fact descriptions. Meanwhile, our model also incorporates the position of defendant making it capable of giving attention to the contextual information of the defendant. We carry several experiments on the public CAIL2018 dataset. Experimental results show that our model achieves better or comparable performance on three subtasks compared with the baseline models. Moreover, we also analyze the interpretable contribution of our model.

Keywords: Legal judgement prediction · Interpretable prediction · Subtask of the seriousness of charges · Multi-task learning

1 Introduction

Legal Judgement Prediction (LJP) aims to predict charge, law article and terms of penalty automatically based on the fact descriptions of the criminal cases. It can be used to help the court's judgement and provide legal guidance and assistance to the public.

In recent years, different methods have been proposed to improve the performance of legal judgement prediction task. Some previous studies need to design

M. Sun et al. (Eds.): CCL 2020, LNAI 12522, pp. 415–429, 2020.
https://doi.org/10.1007/978-3-030-63031-7_30

features manually [3,6–8] and some of neural network based models extract features automatically and achieve significant improvements [9,17,18]. However, there are still some challenging problems, including the improvement of the performance and the enhancement of the interpretability of the terms of penalty prediction.

For the improvement of the performance in terms of penalty prediction, previous studies use multi-task and joint learning to obtain the sharing information among different subtasks. Zhong et al. [18] propose a Directed Acyclic Graph structure with topological relations to capture the information attribution among three subtasks, which effectively improve the problem of insufficient fine-grained in LJP. For the enhancement of the interpretability, different solutions are proposed to the problem. Ye et al. [17] propose a Seq2Seq model to formulate LJP as a natural language generation problem. Their model take fact descriptions and charge labels as input and outputs the court's view. The outputs are used as an auxiliary information for practical judgement. Liu et al. [9] propose a multi-task learning model to incorporate charge keywords extracted by TF-IDF and TextRank. Their model has a good interpretability by introduced the keyword information.

Although different methods are proposed for the above two problems, we argue that some of the knowledge are known in LJP task and can be incorporated into the model for improving the performance and the interpretability of the prediction results.

The first one is the seriousness of charges. Actual judgement procedure tells us that the final decision of the terms of penalty is largely determined by the seriousness of the case, which depends on the case fact descriptions and the terms of penalty definition described in the article corresponding to the charge of the case. Inspired by the actual judgement procedure, we propose to design a subtask of the seriousness of charges which is determined by the charge and the terms of penalty for LJP. According to the scope of legal terms of penalty, we can easily divide a fact description into two categories: serious and less serious. Detailed descriptions and examples are given in Sect. 3.3. The new subtask is used to obtain attentions of different terms of penalty according to serious and less serious predicted by the subtask and let the model pay more attention to those important terms of penalty for the corresponding fact descriptions. As an example with predicted charge as "murder" which is shown in Fig. 1, we can see that our model captures more attention on "Death or life Imprisonment" with predicted serious label and on "7–10 years" with predicted less serious label, which is useful for the model selecting the right terms of penalty.

The second one is the defendant information which is known in the case fact descriptions. Previous studies focuses on the fact descriptions only (e.g., just using text words), ignoring the importance of the context information of the defendant. To this end, we propose to incorporate the position of the defendant into the model. By introducing the defendant position-aware embedding for the fact descriptions, we can capture more context information of the defendant which is helpful for the prediction of subtasks. The main contributions of our work are as follows:

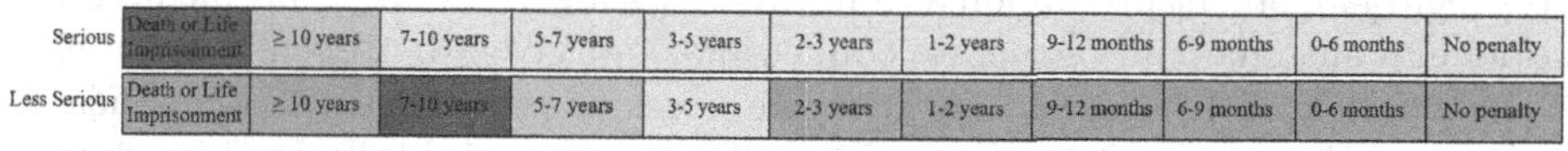

Serious	Death or Life Imprisonment	≥ 10 years	7-10 years	5-7 years	3-5 years	2-3 years	1-2 years	9-12 months	6-9 months	0-6 months	No penalty
Less Serious	Death or Life Imprisonment	≥ 10 years	7-10 years	5-7 years	3-5 years	2-3 years	1-2 years	9-12 months	6-9 months	0-6 months	No penalty

Fig. 1. Attentions of different terms of penalty for charge of "murder" generated by the proposed subtask. As can be seen, term of death or life imprisonment and terms of 7–10 years are paid more attention for serious murder and less serious murder respectively.

1) We propose a multi-task legal judgement prediction model combining a subtask of the seriousness of charges. By introducing this subtask, our model improves the performance and the interpretability of the terms of penalty prediction in LJP.
2) Based on the importance of defendant in the fact descriptions, we propose to incorporate the position information of the defendant into the model, making it capable of giving attention to the relevant context information of the defendant.
3) We carry several experiments on the CAIL2018 dataset. We will show that our proposed model achieves a better or comparable performance in all subtasks than the baseline models. We also give a discussion of our model's interpretability in terms of penalty prediction.

2 Related Work

Legal judgement prediction task usually includes three subtasks: charges prediction, law articles recommendation and terms of penalty prediction. We will review the work of legal judgement prediction from single-task based models and multi-task based models.

2.1 Single-Task Based Legal Judgement Prediction Models

In the models of single-task based legal judgement prediction, the core perspective is to use different encoding method to represent the fact descriptions more correctly. Luo et al. [10] propose an attention-based neural network with two hierarchical encoding structures to jointly model the fact descriptions and the top k relevant law articles. Their model achieves good performance for those simple cases, which indicates that the hierarchical encoding structure and introducing of law articles effectively improve the result of charge prediction. Hu et al. [2] propose an attribute-attentive charge prediction model. They incorporate the fact descriptions attributed by attention mechanism with the original text.

Their model performs well in few-shot charges and confusing charge pairs. Ye et al. [17] propose a label-conditioned Seq2Seq model with attention mechanism. The model take the fact descriptions and charge labels as input and formulates legal judgement prediction as a natural language generation problem. Their model can automatically generate court views and give a better interpretability of the prediction. In order to improve the terms of penalty prediction, Chen et al. [1] regard term prediction as a kind of regression problem. By introducing charge labels and using a structure of Deep Gating Network (DGN), their model achieves good results for the terms of penalty prediction.

2.2 Multi-task Based Legal Judgement Prediction Models

Most of above models are proposed for single task such as charge prediction or terms of penalty prediction. However, judge's actual judgement procedure tells us that different subtasks are often related with each other, like charge is related with law and charge is also related to the terms of penalty. To this end, different multitask based learning models are proposed to obtain the relationship information of different subtasks. Zhong et al. [18] propose a topological multitask learning framework for three subtasks of law articles, charges, and the terms of penalty. They formalized the dependencies among these subtasks as a Directed Acyclic Graph for neural network learning. Their model improves the problem of insufficient fine-grained of legal judgement prediction task. Yang et al. [15] propose a multi-perspective bi-feedback network with the word collocation attention mechanism. Liu et al. [9] propose a multi-task learning framework for legal judgement prediction. They use charge keywords extracted by TF-IDF and Text Rank as auxiliary information and use a hierarchical structure to decode the fact descriptions. Their model shows good interpretability because of the introduced charge keywords. Wang et al. [12] propose a hybrid attention model which combines the improved hierarchical attention network (iHAN) and the deep pyramid convolutional neural network (DPCNN) by ResNet. Their model achieves a good performance for the subtask of the terms of penalty. Xu et al. [14] take advantages of a novel graph neural network to distinguish confusing law articles and improve the capacity of the encoding of the fact descriptions. Zhong et al. [19] propose a model based on reinforcement learning, which can visualize the prediction process and give interpretable judgements by giving a process of QA judgement. Their model greatly improves the interpretability of legal judgement prediction task.

This paper focuses on multi-task legal judgement prediction. Different from previous studies, our work focuses on the scope of legal terms of penalty of the different seriousness in the law. We introduce the seriousness of charges as a subtask into the model. By introducing this subtask, it is expected that the prediction of the terms of penalty can obtain improvements not only on the performance but also on the interpretability of the prediction. In addition, in order to make a better judgement to the defendant, our model also combines the defendant's position information in the model.

3 Proposed Model

3.1 Architecture of Our Model

In this paper, we propose a multi-task legal judgement prediction model combining a subtask of the seriousness of charges, which consists of two parts. The first part is encoding layer, in which a defendant position-aware context information is incorporated into the fact descriptions representation. The second part is decoding layer, in which a subtask of the seriousness of charges is introduced to help obtain the attention of different terms of penalty corresponding to the predicted charge. Our model is shown in Fig. 2. In the following sections, we will introduce embedding of the defendant's position information in Sect. 3.2 and describe our design of subtask of the seriousness of charges in Sect. 3.3. Section 3.4 will describe model training and prediction.

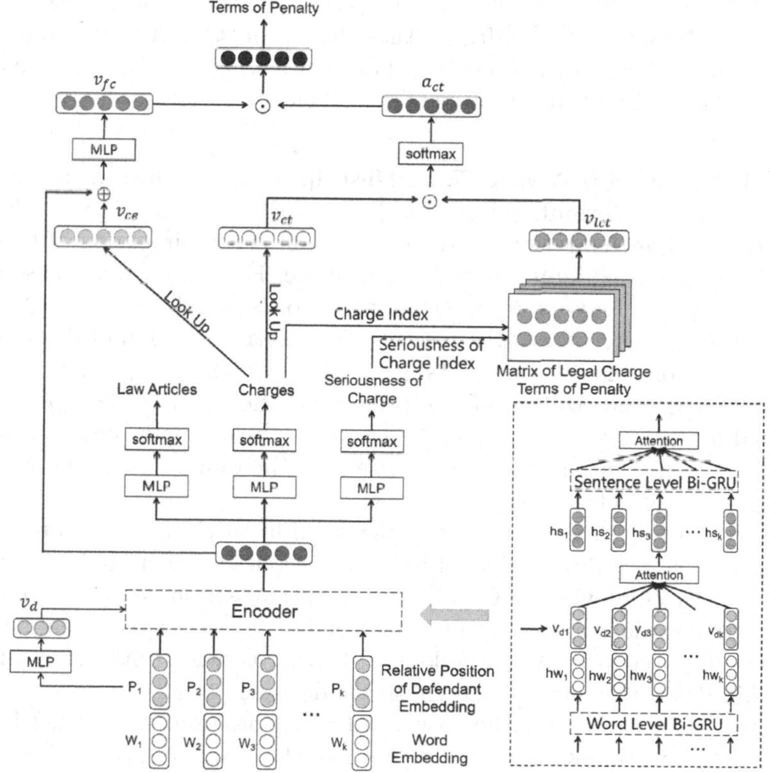

Fig. 2. The whole architecture of our model.

He Mou	hold	chopper	run after and cut	the hurt	Zheng MouMou	the defendant	He MouMou	in charge of	drive	pick up
贺某	持	砍刀	追砍	被害人	郑某某	被告人	贺某某	负责	驾车	接应
7	6	5	4	3	2	1	0	1	2	3

Fig. 3. An example of relative position of each word to the defendant in a sentence.

3.2 Embedding of the Defendant's Position Information

Design of the Defendant's Position Information. In order to obtain the context information of defendant in the fact descriptions, we use the relative position of each word to the defendant as an indicator to represent the context information of the defendant in a sentence. For example, as show in Fig. 3, the defendant is "贺某某 (He MouMou)" whose position is set to 0, the word "驾车 (drive)" whose position is 2 and the word "追砍 (run after and cut)" whose position is 4. We can see that the action "驾车 (drive)" is more relative to the defendant than that of "追砍 (run after and cut)". By incorporating the position, the model can learn to focus more on the action "驾车 (drive)" than the action "追砍 (run after and cut)". This kind of defendant position-aware fact descriptions representation has a better expression of the context information of defendant.

Defendant's Position-Aware Fact Descriptions Encoding. For a given fact descriptions, we formulate it as $d = \{s_1, ..., s_n\}$, in which $s_i \in \mathbb{R}^{L_w \times m}$ is the representation of vectorization of i-th sentence, m is the dimension of the word vector, L_w is the maximum length of a sentence. For sentence s_i, it is formulated as $s_i = \{w_{i1}, \ldots, w_{ik}\}$ represented by k words, in which $w_{ik} \in \mathbb{R}^m$ is the representation of word embedding vector. The relative position of defendant in document d is formulated as $p = \{sp_1, \ldots, sp_n\}$, in which $sp_i \in \mathbb{R}^{L_s \times n}$ is the representation of vectorization of i-th relative position of defendant of sentence formulated as $sp_i = \{wp_{i1}, \ldots, wp_{ik}\}$, $wp_{ik} \in \mathbb{R}^n$ is the representation of vectorization of k-th position in i-th sentence, n is the dimension of the vector of relative position of defendant.

As shown in Fig. 2, we employ a structure of hierarchical attention network [16] to encode the fact descriptions. Firstly, we encode each word in each sentence on word level by employing Bi-GRU network with attention. We then obtain the hidden representation of each sentence. Secondly, we encode each sentence of a document on sentence level by employing Bi-GRU network with attention, and then obtain the hidden representation of the document.

For word level encoding, the new word representation is obtained by concatenating the word embedding vector and the relative position of defendant vector. We formulate sentence consist of the new word representation as $s = \{x_1, \ldots, x_k\}$, in which x_k is obtained by w_k and wp_k formulated as $x_k = [w_k; wp_k]$. The w_k and wp_k represent the representation of k-th word in sentence s and the representation of the relative position of defendant of k-th word respectively. Then, we input the representation of sentence s into a word level Bi-GRU network, and then obtain the hidden output of sentence s

formulated as $hw = \{hw_1, hw_2, \ldots, hw_k\}$. At t time stamp, we concatenate the hidden output of the forward and backward GRU unit formulated as $hw_t = [\overrightarrow{h_t}, \overleftarrow{h_t}]$.

Defendant's Position-Aware Attention Enhancing. We combine the relative position of defendant vector into word level attention so that the hidden output of each GRU unit in sentence s can better capture the information of position of defendant. Firstly, we employ a multilayer perceptron to obtain the vector v_{dj} which represent the information of position of defendant in each unit in sentence s. Then, we concatenate the hw and v_{dj}. Employing a one-layer MLP, we obtain the new hidden output u_h. Finally, we obtain the hidden representation H_s of sentence s after obtaining the attention aw_t of the new hidden output u_h via softmax function. The W_w and b_w are the parameter of hidden layer projection, u_w is word level context vector. The calculation formula is shown in Eqs. (1)∼(4).

$$v_{dj} = MLP(sp_j) \tag{1}$$
$$u_h = tanh(W_w[hw, v_{dj}] + b_w) \tag{2}$$
$$aw = softmax(u_h^T u_w) \tag{3}$$
$$H_s = \sum_t aw_t hw_t \tag{4}$$

For sentence level encoding, we input each representation H_s of sentences into a sentence level Bi-GRU network with attention, and then obtain the final hidden representation v_f of fact descriptions.

3.3 Design of the Subtask of the Seriousness of Charges

Based on the definition of terms of penalty, we divide each charge into two categories: serious and less serious. Then we annotate each charge with two legal terms of penalty vectors, which have the same dimension with the prediction of terms of penalty subtask. We also annotate all the samples with the seriousness of chargs, then we can carry a new subtask of the seriousness of charges in the model.

Tagging Rules. First of all, we manually annotate the legal terms of penalty vectors of the two categories with serious and less serious. The tagging rules are as follows: when the legal terms of penalty is less serious, according to the actual terms of penalty described in law articles, we set the vector of less serious category of the corresponding charge. If there is no distinction between the seriousness of the charge of legal terms of penalty, the vectors of the corresponding serious and less serious legal terms of penalty are set as the same. When a charge includes several seriousness such as less serious, serious, very serious, etc., we combine the serious and more serious parts as the serious category.

Then, we annotate each sample with the label of seriousness. Given a sample, we can determine its corresponding range of legal terms of penalty based on the charge label, if the corresponding range is serious, the seriousness label of the sample is annotated 'serious'; if the corresponding range is less serious, then the seriousness label is annotated 'less serious'. A special case is that if the terms of penalty label is not within the scope of the serious and less serious, we will still annotate it as 'serious'.

Example Demonstration. In order to better illustrate our annotation rules, we give an example of tagging for the legal terms of penalty vector tagging of a specific charge. As shown in Fig. 4, take "故意伤害罪 (intentional assault)" as an example, according to the definition of the corresponding law article 234, we firstly divide the legal terms of penalty into the following three categories: 'less serious': fixed-term imprisonment of not more than three years, criminal detention or public surveillance; 'serious': fixed-term imprisonment of not less than three years but not more than 10 years; 'very serious': fixed-term imprisonment of not less than 10 years, life imprisonment or death. Based on our classification of seriousness, we combine the corresponding legal terms of penalty range of "serious" and "very serious", and the final "serious" legal terms of penalty text is: "fixed-term imprisonment of not less than 3 years, life imprisonment or death"; the "less serious" legal terms of penalty text is "fixed-term imprisonment of not more than three years". Then according to the 11 categories of the subtask of terms of penalty prediction, the corresponding legal terms of penalty range vectors are generated, the serious category vector of the legal terms of penalty of 'intentional assault' is [1, 1, 1, 1, 1, 0, 0, 0, 0, 0, 0], and the less serious category vector is [0, 0, 0, 0, 0, 1, 1, 1, 1, 1, 1].

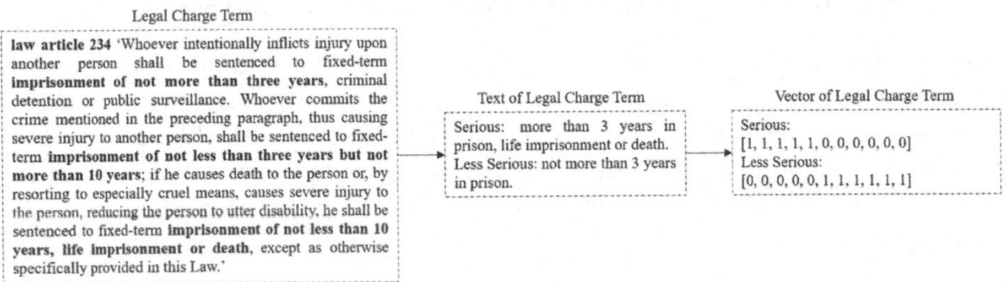

Fig. 4. An example of legal charge terms of penalty.

3.4 Model Training and Prediction

As shown in Sect. 3.2, after getting the final hidden representation of the fact descriptions v_f, we employ three different multilayer perceptrons to obtain the decoding vector of law articles, charges, seriousness of charge respectively. Then, as shown in Eq. (5), we input them into softmax function to get prediction results $\hat{y}_1$, $\hat{y}_2$ and $\hat{y}_3$ of three subtasks. As shown in Eqs. (6) and (7), the index vector of the corresponding charges and seriousness of charge was obtained by the prediction results of charges and seriousness of charge respectively.

$$\hat{y}_k = softmax(MLP_k(v_f)), k = 1, 2, 3 \tag{5}$$

$$i_{charge} = argmax(\hat{y}_2) \tag{6}$$

$$i_{seriousness} = argmax(\hat{y}_3) \tag{7}$$

According to the index of the prediction results of charges and seriousness of charge, we obtain legal charge term vector v_{lct}. Similar to word embedding, we obtain charge term vector v_{ct} and charge embedding vector v_{ce} in weight matrices W_{ct} and W_{ce} respectively which have different dimensions and perform joint learning in the model. Then, as shown in Eq. (8), we calculate charge term attention weight a_{ct} via charge term vector v_{ct} and legal charge term vector v_{lct}.

$$a_{ct} = softmax(v_{ct} \odot v_{lct}) \tag{8}$$

After concatenating the final hidden representation v_f and charge embedding vector v_{ce}, we input it into a multilayer perceptron. Then, we obtain the vector v_{fc} which is the fusion of fact descriptions and charges, as shown in Eq. (9).

$$v_{fc} = MLP([v_f, v_{ce}]) \tag{9}$$

Finally, as shown in Eq. (10), we do a hadamard product of v_{fc} and a_{ct} and obtain the final decoding vector v_t of terms of penalty. Then, as shown in Eq. (11), we input the vector v_t into softmax function to get prediction results $\hat{y}_4$.

$$v_t = v_{fc} \odot a_{ct} \tag{10}$$

$$\hat{y}_4 = softmax(v_t) \tag{11}$$

In the training process, we use cross-entropy loss function as the loss function of our model. After calculating each cross-entropy loss for each subtask, we sum each loss of different subtasks as the total loss. As shown in Eq. (12), i represents the i-th subtask, Y_i represents the total number of classes of the i-th subtask, and j represents the j-th class.

$$loss_{total} = -\sum_{i=1}^{4}\sum_{j=1}^{Y_i} y_{i,j} log(\hat{y}_{i,j}) \tag{12}$$

4 Experiments

4.1 Dataset

We use the CAIL2018[1] [13] dataset to be evaluated in this paper. Similar to the work of Zhong et al. [18], we do some relevant preprocess on the datasets. Firstly,

[1] https://github.com/china-ai-law-challenge/CAIL2018.

we filter out the crime data that contained multiple charges and multiple relevant law articles. Secondly, we remove the crime data with charges appeared less than 100 times in the datasets. Finally, similar to the work of TOPJUDGE [18], we divide the terms of penalty into 11 non-overlapping intervals. The detailed information of the CAIL2018 are shown in Table 1.

Table 1. Statistical information of the CAIL2018 dataset.

Dataset	Amount	Subtasks	Amount
Training set	101513	Charges	119
Testing set	26731	Law articles	103
Validation set	10818	Terms of penalty	11

4.2 Compared Models

In order to compare on three subtasks, we built a multi-task implementation on those not designed for multi-task baseline models. We use Bi-LSTM, TextCNN [4] and Hierarchical Attention Networks (HAN) [16] as three different structures to encode the fact descriptions. For HAN structure, we employ a word level of Bi-GRU network with attention and a sentence level of Bi-GRU network with attention to encode the fact descriptions. We employ three different multilayer perceptrons for multitask prediction for these three baselines. We use TOPJUDGE [18] and Few-Shot [2] as our another compared models based on their multi-task joint learning and additional auxiliary information design. For the Few-Shot model, we also employ three different multilayer perceptrons for multitask prediction.

4.3 Experimental Setting

In our experiment, we use THULAC [5] for word segmentation. We use skip-gram [11] for pre-training of all fact descriptions and get a pre-trained 200-dimensional matrix of word vectors. For the position of defendant, we embed each position into a 100-dimensional vector and perform joint training in the model. For the CNN-based and Bi-LSTM-based models in the baselines, we set the maximum document length to 512 words. For the HAN-based models, the maximum sentence length is set to 100 words, the maximum document length is set to 15 sentences. The unit dimension of hidden layer is set to 256, and the output dimension of each level vector is set to 256. In our model, we embed the maximum sentence length of the relative position of defendant with a blank vector. For training, we use the Adam optimizer to control stochastic gradient descent. The learning rate of the optimizer set to 0.001, the batch size set to 128, and the epoch set to 16. We select the model that performed best on the validation set, and report the results on the testing set.

4.4 Experimental Results

Similar to previous work, we use accuracy (Acc.), macro-precision (MP), macro-recall (MR) and macro-F1 (F1) as metrics in this paper, the final experimental results are shown in Table 2. As LJP task is a multi-label classification task, and there is an extremely unbalanced phenomenon among various categories in the CAIL2018 dataset, we mainly focus on the comparison of the results of macro-F1.

Table 2. Experimental results of our model and baselines.

Tasks	Law articles				Charges				Terms of penalty			
Metrics	Acc.	MP	MR	F1	Acc.	MP	MR	F1	Acc.	MP	MR	F1
Bi-LSTM	79.33	76.45	77.11	75.20	81.69	81.26	**81.77**	80.38	39.66	31.99	29.34	28.26
Text CNN	76.77	74.21	73.13	71.43	82.38	81.20	78.16	78.32	37.85	32.49	27.78	27.79
HAN	81.08	76.85	77.48	76.05	81.97	80.89	81.90	80.37	41.07	31.25	30.71	28.40
TOPJUDGE	**82.11**	76.14	75.82	75.01	82.40	79.48	79.21	78.29	40.04	32.74	30.45	29.59
Few-Shot	79.59	75.62	74.97	73.97	83.33	82.22	80.42	80.56	40.33	30.88	**33.38**	**30.65**
Our model	81.04	**78.43**	**77.27**	**76.49**	**84.47**	**82.42**	81.46	**81.14**	**41.96**	**34.89**	31.11	30.45

Firstly, we compare our model with Bi-LSTM, TextCNN and HAN models. As shown in Table 2, we can see our model achieves the best macro-F1 value in all three subtasks. And it shows that our model performs great results especially in the subtask of terms of penalty. Our model is 30.45% which is 2.19% higher than Bi-LSTM, 2.66% higher than TextCNN, 2.05% higher than HAN. The results prove the effectiveness of the subtask of seriousness of charge introduced in our model.

Secondly, we compare our model with TOPJUDGE model which is also a multi-task LJP model. As shown in Table 2, our model also achieves better performance in all three subtasks. Our model increases by 1.48% on law articles prediction subtask, 2.85% on charges prediction subtask and 0.86% on terms of penalty prediction subtask. This result shows that our model is ascending to a certain extent on three subtasks compared with the TOPJUDGE model.

Finally, we compare our model with the Few-Shot model which also uses an auxiliary information to help improve the performance of a subtask. We can see that our model increases by 2.52% on law articles prediction, 0.58% on charges prediction, and decreases by 0.2% on terms of penalty prediction which is comparable with the Few-Shot model. The results indicate that the overall performance of our model can be improved on the basis of improving term prediction results.

4.5 Ablation Studies

In order to analyze the influence of each part of our model, several ablation experiments are conducted in this paper. We remove four parts from our model to see the influences: 1) We remove the word level attention calculated by the position of defendant which is named as w/o drp_att. 2) We remove the whole part of using position of defendant, named as w/o drp_pos+drp_att, which means that the model only judges with the fact descriptions. 3) We remove the subtask of the seriousness of charges which is named as w/o seriousness to see the influence of the subtask to the whole model. 4) We remove the whole part of relative position of defendant and the subtask of the seriousness of charges, which means that the fact descriptions is only encoded by hierarchical attention networks and predicted by multitask learning, and the model is named as w/o drp_both+seriousness. The results of different parts of ablation studies are shown in Table 3.

Table 3. Results of ablation experiments.

Tasks	Law articles		Charges		Terms of penalty		Seriousness of charge	
Metrics	Acc.	F1	Acc.	F1	Acc.	F1	Acc.	F1
Our Model	81.04	76.49	84.47	81.14	41.96	30.45	87.18	80.09
w/o drp_att	81.71	76.3	83.49	81.31	41.63	29.95	86.88	80.05
w/o drp_pos+drp_att	81.68	75.53	83.28	81.02	41.66	29.84	86.87	79.57
w/o seriousness	81.28	76.59	83.87	81.51	41.41	28.41	/	/
w/o drp_both+seriousness	81.08	76.05	81.97	80.37	41.07	28.40	/	/

As shown in Table 3, when we remove the subtask of seriousness of charge (w/o seriousness), the macro-F1 of the subtask of terms of penalty prediction is reduced by 2.04%, which shows that the introducing of the subtask of seriousness of charge can significantly improve the result of the terms of penalty prediction. In addition, according to Table 3, we can see that the decoder with the introducing of the subtask of seriousness of charge is the most effective part in all additional components.

We also can see that after embedding the position information of defendant, the prediction results of charges and law articles can be improved. Moreover, compared with embedding the position information of defendant, using position information of defendant to improve word level attention can further improve the performance of the model in three subtasks. When we remove all the position information of defendant, the macro-F1 of the subtask of law articles prediction will decrease by 0.96%. This result shows that the position information of defendant can mainly improve the result of law articles prediction. In the end, when we combine the position information of defendant and the subtask of the seriousness of charges into a model, the performances of all three subtasks are improved.

4.6 Interpretability Analysis

In order to analyze the interpretability of our model, we choose a representative case to illustrate how the design of the subtask of the seriousness of charges can be improved in interpretability of the prediction of terms of penalty.

As shown in Fig. 5, given the fact descriptions, previous method will predict and give the terms of penalty directly without any auxiliary information. While in our model, firstly, we will preliminarily predict the prediction results of charges, law articles and the seriousness of charge. With the predicted charge and the seriousness of the charge, our model can determine the range of legal charge terms of penalty, this is important and useful for the judge and the public to get the auxiliary information of the terms of the penalty. Finally, the model outputs the prediction result of the terms of penalty. Compared with previous direct prediction process of terms of penalty, the prediction process of our model has a better interpretability of the prediction.

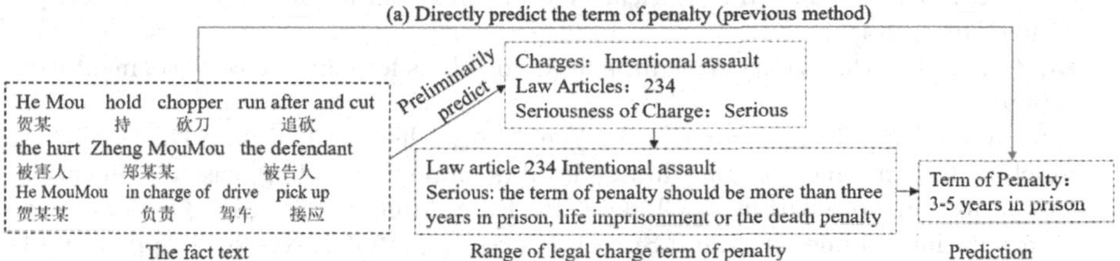

Fig. 5. Terms of penalty prediction process of our model compared with previous method.

5 Conclusion

In this paper, we propose to design and combine a subtask of the seriousness of charges for multi-task legal judgement prediction. Evaluations demonstrate the effectiveness of our model on charge prediction, law article recommendation and the terms of penalty prediction, indicating that the introduced subtask of the seriousness of charges and the sufficient encoding of the fact descriptions for the defendant are useful. Our model also shows the good interpretability on the task of terms of penalty prediction. In the future, we will explore a better method to incorporate the contextual information of defendant and investigate the usefulness of different subtasks for multi-task legal judgement prediction.

Acknowledgments. This work is supported by National Natural Science Foundation of China (No. 61976062), the Science and Technology Program of Guangzhou (No. 201904010303) and Special Funds for the Cultivation of Guangdong College Students' Scientific and Technological Innovation (No. pdjh2020a0197).

References

1. Chen, H., Cai, D., Dai, W., Dai, Z., Ding, Y.: Charge-based prison term prediction with deep gating network. In: Proceedings of the 2019 Conference on Empirical Methods in Natural Language Processing and the 9th International Joint Conference on Natural Language Processing, pp. 6361–6366 (2019)
2. Hu, Z., Li, X., Tu, C., Liu, Z., Sun, M.: Few-shot charge prediction with discriminative legal attributes. In: Proceedings of the 27th International Conference on Computational Linguistics, pp. 487–498 (2018)
3. Katz, D.M., Bommarito Ii, M.J., Blackman, J.: Predicting the behavior of the supreme court of the united states: A general approach. Plos One **12**(4) (2014)
4. Kim, Y.: Convolutional neural networks for sentence classification. In: Proceedings of the 2014 Conference on Empirical Methods in Natural Language Processing, pp. 1746–1751 (2014)
5. Li, Z., Sun, M.: Punctuation as implicit annotations for Chinese word segmentation (2009)
6. Lin, W.C., Kuo, T.T., Chang, T.J., Yen, C.A., Chen, C.J., Lin, S.D.: Exploiting machine learning models for Chinese legal documents labeling, case classification, and sentencing prediction in Chinese. In: Proceedings of the 24th Conference on Computational Linguistics and Speech Processing (ROCLING 2012), pp. 140–141 (2012)
7. Liu, C.L., Hsieh, C.D.: Exploring phrase-based classification of judicial documents for criminal charges in Chinese. In: Foundations of Intelligent Systems, pp. 681–690 (2006)
8. Liu, Y., Chen, Y., Ho, W.: Predicting associated statutes for legal problems. Inf. Process. Manag. **51**(1), 194–211 (2015)
9. Liu, Z., Zhang, M., Zhen, R., Gong, Z., Yu, N., Fu, G.: Multi-task learning model for legal judgment predictions with charge keywords. J. Tsinghua Univ. (Sci. Technol.) **59**(7), 497 (2019)
10. Luo, B., Feng, Y., Xu, J., Zhang, X., Zhao, D.: Learning to predict charges for criminal cases with legal basis. In: Proceedings of the 2017 Conference on Empirical Methods in Natural Language Processing, pp. 2727–2736 (2017)
11. Mikolov, T., Chen, K., Corrado, G., Dean, J.: Efficient estimation of word representations in vector space. Computer Science (2013)
12. Wang, W., Chen, Y., Cai, H., Zeng, Y., Yang, H.: Judicial document intellectual processing using hybrid deep neural networks. J. Tsinghua Univ. (Sci. Technol.) **59**(7), 505 (2019)
13. Xiao, C., et al.: CAIL2018: a large-scale legal dataset for judgment prediction. CoRR abs/1807.02478 (2018), http://arxiv.org/abs/1807.02478
14. Xu, N., Wang, P., Chen, L., Pan, L., Wang, X., Zhao, J.: Distinguish confusing law articles for legal judgment prediction. CoRR abs/2004.02557 (2020). https://arxiv.org/abs/2004.02557
15. Yang, W., Jia, W., Zhou, X., Luo, Y.: Legal judgment prediction via multi-perspective bi-feedback network. In: Proceedings of the Twenty-Eighth International Joint Conference on Artificial Intelligence, pp. 4085–4091 (2019)
16. Yang, Z., Yang, D., Dyer, C., He, X., Smola, A., Hovy, E.: Hierarchical attention networks for document classification. In: Conference of the North American Chapter of the Association for Computational Linguistics: Human Language Technologies, pp. 1480–1489 (2016)

17. Ye, H., Jiang, X., Luo, Z., Chao, W.: Interpretable charge predictions for criminal cases: Learning to generate court views from fact descriptions. In: Proceedings of the 2018 Conference of the North American Chapter of the Association for Computational Linguistics: Human Language Technologies, pp. 1854–1864 (2018)
18. Zhong, H., Guo, Z., Tu, C., Xiao, C., Liu, Z., Sun, M.: Legal judgment prediction via topological learning. In: Proceedings of the 2018 Conference on Empirical Methods in Natural Language Processing, pp. 3540–3549 (2018)
19. Zhong, H., Wang, Y., Tu, C., Zhang, T., Liu, Z., Sun, M.: Iteratively questioning and answering for interpretable legal judgment prediction. In: Proceedings of the AAAI Conference on Artificial Intelligence, vol. 34, pp. 1250–1257 (2020)

Clickbait Detection with Style-Aware Title Modeling and Co-attention

Chuhan Wu[1(✉)], Fangzhao Wu[2], Tao Qi[1], and Yongfeng Huang[1]

[1] Department of Electronic Engineering & BNRist, Tsinghua University, Beijing 100084, China
wuchuhan15@gmail.com, taoqi.qt@gmail.com, yfhuang@tsinghua.edu.cn
[2] Microsoft Research Asia, Beijing 100080, China
wufangzhao@gmail.com

Abstract. Clickbait is a form of web content designed to attract attention and entice users to click on specific hyperlinks. The detection of clickbaits is an important task for online platforms to improve the quality of web content and the satisfaction of users. Clickbait detection is typically formed as a binary classification task based on the title and body of a webpage, and existing methods are mainly based on the content of title and the relevance between title and body. However, these methods ignore the stylistic patterns of titles, which can provide important clues on identifying clickbaits. In addition, they do not consider the interactions between the contexts within title and body, which are very important for measuring their relevance for clickbait detection. In this paper, we propose a clickbait detection approach with style-aware title modeling and co-attention. Specifically, we use Transformers to learn content representations of title and body, and respectively compute two content-based clickbait scores for title and body based on their representations. In addition, we propose to use a character-level Transformer to learn a style-aware title representation by capturing the stylistic patterns of title, and we compute a title stylistic score based on this representation. Besides, we propose to use a co-attention network to model the relatedness between the contexts within title and body, and further enhance their representations by encoding the interaction information. We compute a title-body matching score based on the representations of title and body enhanced by their interactions. The final clickbait score is predicted by a weighted summation of the aforementioned four kinds of scores. Extensive experiments on two benchmark datasets show that our approach can effectively improve the performance of clickbait detection and consistently outperform many baseline methods.

Keywords: Clickbait detection · Style-aware title modeling · Co-attention

1 Introduction

Clickbait is a type of web content that is designed to attract users' attention and further entice them to click hyperlinks to enter specific webpages, such as news

M. Sun et al. (Eds.): CCL 2020, LNAI 12522, pp. 430–443, 2020.
https://doi.org/10.1007/978-3-030-63031-7_31

articles, advertisements and videos [6]. Several illustrative examples of clickbaits are shown in Fig. 1. We can see that the title of the first clickbait is written in a sensationalized way by using words with strong emotions like "MUST", and the title of the second clickbait is misleading because it does not match the content of the body. Clickbaits are commonly used by online publishers, because clickbaits can draw more attention to the online websites where they are displayed and improve the revenue by attracting more clicks on advertisements [9]. However, clickbaits are deceptive to users because the main content of clickbaits is often uninformative, misleading, or even irrelevant to the title, which is extremely harmful for the reading satisfaction of users [7]. Thus, clickbait detection is an important task for online platforms to improve the quality of their web content and maintain their brand reputation by improving user experience [3].

Many methods formulate clickbait detection as a binary detection task, and they mainly focus on modeling the content of online articles and the relevance between title and body [9,16,27]. For example, Zhou et al. [27] proposed to use a combination of bi-GRU network and attention network to learn representations of tweets posted by users for clickbait detection. Dong et al. [9] proposed a similarity-aware clickbait detection model, which learns title and body representations via an attentive bi-GRU network, and measures the global and local similarities between these representations for clickbait prediction. However, in these methods the stylistic patterns of titles (e.g., capitalization) are not taken into consideration, which are useful clues for identifying clickbaits [3]. In addition, they cannot model the interactions between the contexts in the title and body, which are important for measuring the title-body relevance for clickbait detection.

Title	7 Things You MUST Know About Exercise and Weight Loss	Covid-19 news in your area	You Won't Believe How Many Beloved Mom-and-Pop Restaurants are Closing
Body	The biggest challenge for an obese person is losing a few extra pounds. Well, people sometimes let themselves eat what they like...	Download our app today and get what you want! Consider joining this community as a helpful resource...	The pandemic has caused a lot of businesses to fold, especially independent restaurants, cafes, and coffee shops.

Fig. 1. Several illustrative examples of clickbaits.

Our work is motivated by the following observations. First, the content of webpage title and body is important for clickbait detection. For example, in the title of the third webpage in Fig. 1, the contexts like "You Won't Believe" are important indications of clickbaits because they express strong emotions. In addition, the body of this webpage is short and uninformative, which also implies that this webpage is a clickbait. Second, the stylistic patterns of title like the usage of numeric and capitalized characters can also provide useful clues for identifying clickbaits. For example, the title of the first webpage in Fig. 1 starts with a number "7" and it uses an all-capital word "MUST" to attract attention, both of which are commonly used by clickbaits. Therefore, modeling the stylistic patterns of title can help detect clickbaits more accurately. Third, there is inherent relatedness between the contexts within the title and body of the same webpage. For example, the words "Weight Loss" in the title of the first webpage in Fig. 1 have close relatedness with the words "losing" and "pounds" in the body. Modeling these interactions are helpful for measuring the relevance between title and body more accurately.

In this paper, we propose a clickbait detection approach with style-aware title modeling and co-attention (SATC), which can consider the interactions between contexts within title and body as well as the stylistic patterns of title. We first use Transformers to learn representations of title and body based on their content, and then compute a title content score and a body content score based on the representations of title and body, respectively. In addition, we propose to use a character-level Transformer to learn a style-aware title representation by capturing the stylistic patterns in the title, and we further compute a title stylistic score based on this representation. Besides, we propose to use a co-attention network to model the interactions between the contexts within title and body, and further enhance their representations by encoding their interaction information. We compute a title-body matching score based on the relevance between the interaction-enhanced representations of title and body. The final unified clickbait score is a weighted summation of the four kinds of scores, which jointly considers the content of title and body, the stylistic information of title, and the relevance between title and body. Extensive experiments on two benchmark datasets show that our approach can effectively enhance the performance of clickbait detection by incorporating the stylistic patterns of title and the title-body interactions.

2 Related Work

Automatic detection of clickbaits is important for online platforms to purify their web content and improve user experience. Traditional clickbait detection methods usually rely on handcrafted features to build representations of webpages [3–7,11,14,19]. For example, Chen et al. [7] proposed to represent news articles with semantic features (e.g.., unresolved pronouns, affective words, suspenseful language and overuse numerals), syntax features (e.g., forward reference and reverse narrative) and image features (e.g., image placement and emotional content). In addition, they incorporate users' behaviors on news, like reading time, sharing and commenting, to enhance news representation. They use various classification models like Naive Bayes and SVM to identify clickbaits based on the news and user behavior features. Biyani et al. [3] proposed to represent webpages using content features like n-gram features extracted from title and body, sentiment polarity features, part-of-speech features and numerals features. They also incorporate the similarities between the TF-IDF features of title and the first 5 sentences in the body. Besides, they consider the informality of title, the use of forward reference, and the URL of webpage as complementary information. They used Gradient Boosted Decision Trees (GBDT) to classify webpages based on their features. Potthast et al. [19] proposed to detect clickbaits on

Twitter. They used features like bag-of-words, image tags, and dictionary matchings to represent tweets, and used bag-of-words, readability and length features to represent the linked webpage. They also incorporated several metadata features like the gender of user. They compared several machine learning models including logistic regression, naive Bayes, and random forests for clickbait classification. However, these methods need heavy feature engineering, which depends on a large amount of domain knowledge. In addition, handcrafted features are usually not optimal in representing the textual content of webpages since they cannot effectively model the contexts of words.

In recent years, several approaches explore to use deep learning techniques for clickbait detection [1,2,8–10,16,23,26,27]. For example, Agrawal et al. [1] proposed a neural clickbait detection approach, which uses convolutional neural network (CNN) with max pooling techniques to learn representations of titles. Zhou et al. [27] proposed to use a bi-GRU network to learn contextual word representations, and use an attention network to select important words for learning informative tweet representations for clickbait detection. Kumar et al. [16] proposed to learn title representations with an attentive bi-GRU network, and used two Siamese networks to respectively measure the relevance between the title and body and the relevance between the associated image and body. They combined the title representation and the relevance vectors for final prediction. Dong et al. [9] proposed a similarity-aware clickbait detection model. They used a combination of bi-GRU network and attention network to learn title and body representations, and computed a similarity vector based on the global and local vector similarities between the representations of titles and bodies. They combined the title and body representations with the similarity vector for clickbait prediction. However, these methods do not consider the stylistic patterns of titles when learning their representations, which are important cues for clickbait detection. In addition, they do not consider the interactions between the contexts in the title and body, which are usually important for evaluating their relevance. Different from existing methods, our approach incorporates a character-level Transformer to capture the stylistic patterns of title, which can help recognize clickbaits more accurately. In addition, it can model the interactions between title and body via co-attention to enhance their representations.

3 Methodology

In this section, we introduce our proposed clickbait detection approach with style-aware title modeling and co-attention (SATC). The framework of our proposed *SATC* approach is illustrated in Fig. 2. It consists of four core modules, i.e., *content modeling*, *style modeling*, *interaction modeling* and *clickbait prediction*. The details of each module are introduced as follows.

3.1 Content Modeling

The *content modeling* module is used to learn the representations of title and body from their content. We respectively denote the sequences of words in title

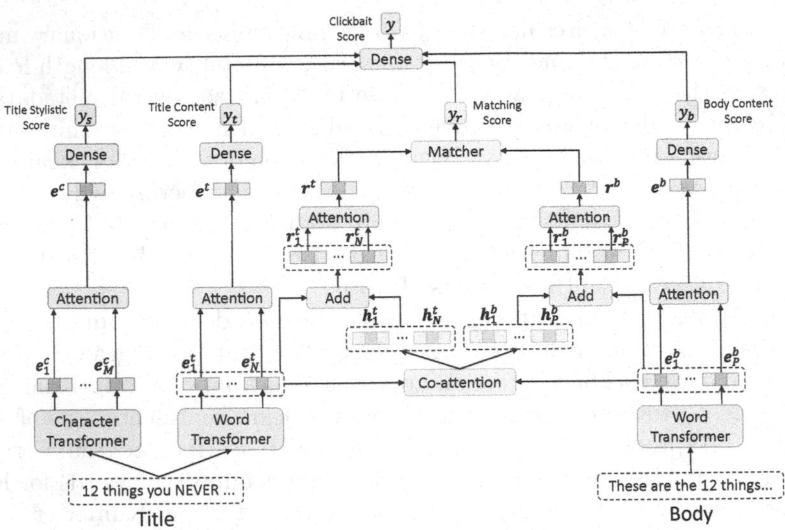

Fig. 2. The architecture of our *SATC* approach for clickbait detection.

and body as $[w_1^t, w_2^t, ..., w_N^t]$ and $[w_1^b, w_2^b, ..., w_P^b]$, where N and P respectively stand for the number of words in the title and body. In this module, we first use a word embedding layer to convert both word sequences into sequences of semantic vectors, which are denoted as $[\mathbf{w}_1^t, \mathbf{w}_2^t, ..., \mathbf{w}_N^t]$ and $[\mathbf{w}_1^b, \mathbf{w}_2^b, ..., \mathbf{w}_P^b]$. Usually the contexts of words in title and body are important for modeling their content. For example, in the title of the first webpage in Fig. 1, the contexts of the word "Loss" such as "Weight" and "Exercise" are useful clues for understanding that this word is about fitness rather than financial loss. Transformer [24] is an effective neural architecture for context modeling. Thus, we apply two independent Transformers to learn hidden representations of words in title and body by modeling their contexts. We denote the hidden representation sequences of words in title and body as $\mathbf{E}^t = [\mathbf{e}_1^t, \mathbf{e}_2^t, ..., \mathbf{e}_N^t]$ and $\mathbf{E}^b = [\mathbf{e}_1^b, \mathbf{e}_2^b, ..., \mathbf{e}_P^b]$, respectively. Different words in a title or body may have different importance for modeling the content. For instance, the word "MUST" in Fig. 1 is more important than the word "About" in learning title representation for clickbait detection. Thus, we apply attention mechanisms [25] to select words in the title and body to form unified representations for them (denoted as $\mathbf{e}^t$ and $\mathbf{e}^b$), which are respectively formulated as $\mathbf{e}^t = Attention([\mathbf{e}_1^t, \mathbf{e}_2^t, ..., \mathbf{e}_N^t])$ and $\mathbf{e}^b = Attention([\mathbf{e}_1^b, \mathbf{e}_2^b, ..., \mathbf{e}_P^b])$.

3.2 Style Modeling

The *style modeling* module is used to capture the stylistic patterns in the title to better identify clickbaits. Usually, there are some common patterns on the style of clickbait titles. For example, many clickbaits use all-capital words (e.g., "MUST", "NOT" and "THIS"), exclamation marks, and numeric characters to attract users' attention. Thus, it is very important to grasp these stylistic patterns in clickbait detection. To capture these patterns, we propose to use a character-level Transformer to learn style-aware title representations from its original characters. We denote the character sequence (including whitespace) of the title as $[c_1, c_2, ..., c_M]$, where M is the number of characters. We first convert these characters into their embeddings (denoted as $[\mathbf{c}_1, \mathbf{c}_2, ..., \mathbf{c}_M]$) via a character embedding layer, and then use a character Transformer to learn the hidden representations of these characters, which are denoted as $[\mathbf{e}_1^c, \mathbf{e}_2^c, ..., \mathbf{e}_M^c]$. Usually different characters may have different importance in style modeling. For example, in Fig. 1 the character "7" is more important than the character "a" in the word "and". Thus, we use a character-level attention network for character selection to build the style-aware title representation $\mathbf{e}^c$, which is formulated as $\mathbf{e}^c = Attention([\mathbf{e}_1^c, \mathbf{e}_2^c, ..., \mathbf{e}_M^c])$.

3.3 Interaction Modeling

The *interaction modeling* module is used to capture the interactions between title and body. For most webpages, the contexts in their titles usually have relatedness with the contexts in their bodies to a certain extent. For instance, the words "Restaurants" in the title of the third webpage in Fig. 1 have close relatedness with the words "businesses", "restaurants" and "cafes" in the body. These interactions are important cues for modeling the relevance between title and body, which is critical for clickbait detection. Thus, we propose to use a multi-head co-attention network to capture the interactions between title and body. More specifically, we first use the title word representation $\mathbf{E}^t$ as the query, and use the body word representation $\mathbf{E}^b$ as the key and value to compute a hidden representation $\mathbf{H}^t = [\mathbf{h}_1^t, \mathbf{h}_2^t, ..., \mathbf{h}_N^t]$, which summarizes the contexts within body and their interactions with the title. This process is formulated as $\mathbf{H}^t = MultiHead(\mathbf{E}^t, \mathbf{E}^b, \mathbf{E}^b)$. Next, we use the body word representation $\mathbf{E}^b$ as the query, and use the title word representation $\mathbf{E}^t$ as the key and value to compute an hidden representation $\mathbf{H}^b = [\mathbf{h}_1^b, \mathbf{h}_2^b, ..., \mathbf{h}_P^b]$ that conveys the contexts in title and their interactions with each word in body, which is formulated as $\mathbf{H}^b = MultiHead(\mathbf{E}^b, \mathbf{E}^t, \mathbf{E}^t)$. Then, we use the title-body interactions to enhance their representations. We add the hidden representation $\mathbf{H}^t$ to the original word representation $\mathbf{E}^t$ to form a unified one $\mathbf{R}^t$, i.e., $\mathbf{R}^t = \mathbf{E}^t + \mathbf{H}^t$. The unified body word representation $\mathbf{R}^b$ is obtained by $\mathbf{R}^b = \mathbf{E}^b + \mathbf{H}^b$. Similar to the *content modeling* module, we also use attention networks to obtain the interaction-enhanced representations of title and body (denoted as $\mathbf{r}^t$ and $\mathbf{r}^b$), which are formulated as $\mathbf{r}^t = Attention([\mathbf{r}_1^t, \mathbf{r}_2^t, ..., \mathbf{r}_N^t])$, and $\mathbf{r}^b = Attention([\mathbf{r}_1^b, \mathbf{r}_2^b, ..., \mathbf{r}_P^b])$, where $\mathbf{r}_i^t$ and $\mathbf{r}_i^b$ stand for the i-th vector in $\mathbf{R}^t$ and $\mathbf{R}^b$, respectively.

3.4 Clickbait Prediction

The *clickbait prediction* module is used to compute a clicbait score based on the representations of title and body. We first use a dense layer to compute a title content score y_t based on the content representation $\mathbf{e}^t$ of the title, which is formulated as $y_t = \mathbf{w}_t^\top \mathbf{e}^t + b_t$, where $\mathbf{w}_t$ and b_t are the kernel and bias parameters. We compute a body content score y_b based on $\mathbf{e}^b$ in a similar way, which is formulated as $y_b = \mathbf{w}_b^\top \mathbf{e}^b + b_b$, where $\mathbf{w}_b$ and b_b are parameters. Next, we use a matcher to compute a title-body matching score, which indicates the relevance between title and body. It takes the interaction-enhanced representations of title and body ($\mathbf{r}^t$ and $\mathbf{r}^b$) as the input, and outputs the matching score y_r. Following [17], we use dot-product to implement the matcher, and the score y_r is computed as $y_r = \mathbf{r}^t \cdot \mathbf{r}^b$. Then, we use another dense layer to compute a title stylistic score based on the style-aware title representation $\mathbf{e}^c$, which is formulated as $y_s = \mathbf{w}_s^\top \mathbf{e}^s + b_s$, where $\mathbf{w}_s$ and b_s are parameters. The final clickbait score y is a weighted summation of the aforementioned four scores and we use the sigmoid function for normalization, which is formulated as $y = sigmoid(\alpha_s y_s + \alpha_t y_t + \alpha_r y_r + \alpha_b y_b)$, where α_s, α_t, α_r and α_b are trainable parameters. For model training, we use binary cross-entropy as the loss function. By comparing the predicted clickbait score with the gold label, we can obtain the loss on the training samples, and further compute the gradients for update.

4 Experiments

4.1 Dataset and Experimental Settings

Our experiments are conducted on two benchmark datasets for clickbait detection. The first one is *Clickbait Challenge*[1], which is a dataset released by Clickbait Challenge 2017. This dataset contains the tweet texts posted by users and the content of the corresponding article. Each pair of tweet and article is annotated by 5 judgers, where each judger gives a clickbait score from 0 (non-clickabit) to 1 (clickbait) to this pair. Following [9], we regard the pairs with the mean score over 0.5 as clickbaits. The training set contains 19,538 pairs, and the validation set contains 2,495 pairs. Since the labels of the test set are not released, we evaluate the model on the current validation set, and randomly sample 10% of pairs in the training set for validation. The second one is *FNC*[2], which is released by the Fake News Challenge in 2017. In this dataset, each pair of title and body is labeled as "agree", "disagree", "discuss" or "unrelated". Following [9], we regard the pairs with "unrelated" labels as clickbaits. This dataset contains 49,972 pairs of titles and bodies for training and 25,413 pairs for test. We also use 10% of training samples for validation.

In our experiments, we use the pre-trained 300-dimensional Glove embeddings [18]. The character embeddings are 50-dimensional. The Transformers have two self-attention layers. We apply dropout [22] to the word and character embeddings at a ratio of 20%. We use Adam [15] as the optimizer (lr=0.01). These hyperparameters are searched according to the performance on the validation sets. Each experiment is repeated 5 times, and the average results in terms of accuracy, precision, recall and Fscore are reported.

[1] https://www.clickbait-challenge.org/.
[2] http://www.fakenewschallenge.org/.

4.2 Performance Evaluation

We compare our *SATC* method with several baseline methods, including: (1) DSSM [13], deep structured semantic model, where title is regarded as the query and body is regarded as document. The texts of title and body are represented by N-gram featuress. (2) CLSM [21], a variant of DSSM that uses CNN to learn text representations; (3) CNN [1,26], which detects clickbaits solely based on titles. Text-CNN is used to learn title representations. (4) LSTM [12], using LSTM networks to learn title and body representations for clickbait detection. (5) GRU-Att [27], using a combination of bi-GRU network and attention network to learn title representations for clickbait detection. (6) SiameseNet [16], which uses *GRU-Att* to learn title representations and uses Siamese networks to capture the relevance between title and body. (7) LSDA [9], which uses *GRU-Att* to learn title and body representations, and measures their relevance using the global and local similarities between the representation vectors of title and body.

Table 1. Performance comparison of different methods on the two datasets. *Improvement is significant at the level of $p < 0.01$.

Method	Clickbait Challenge				FNC			
	Accuracy	Precision	Recall	Fscore	Accuracy	Precision	Recall	Fscore
DSSM	0.817	0.655	0.661	0.658	0.747	0.894	0.740	0.811
CLSM	0.833	0.683	0.643	0.662	0.756	0.959	0.762	0.853
CNN	0.844	0.654	0.653	0.653	0.789	0.852	0.845	0.857
LSTM	0.827	0.642	0.621	0.631	0.868	0.925	0.884	0.913
GRU-Att	0.856	0.719	0.650	0.683	0.879	0.924	0.897	0.919
Siamese Net	0.844	0.695	0.688	0.691	0.859	0.920	0.877	0.907
LSDA	0.860	0.697	0.699	0.710	0.894	0.933	0.912	0.928
SATC*	**0.889**	**0.745**	**0.722**	**0.733**	**0.907**	**0.959**	**0.917**	**0.938**

The results on the two datasets are summarized in Table 1.[3] According to the results, we have several main findings. First, the methods that use neural networks to learn text representations (e.g., *CNN*, *LSTM*, *GRU-Att* and *SATC*) outperform the *DSSM* method that uses handcrafted features for text representation. It shows that handcrafted features are usually not-optimal in representing the textual content of webpages for clickbait detection. Second, the methods based on attention mechanisms (e.g., *GRU-Att* and *LSDA*) usually outperform the methods without attention (e.g., *CNN* and *LSTM*). This is probably because attention mechanism can select important contexts within title and body to learn more informative representations for them, which is beneficial for clickbait detection. Third, our approach can consistently outperform the compared baseline methods. This is because our approach can capture the stylistic patterns in the title to learn style-aware title representations, and meanwhile can

[3] Most results of baselines are taken from [9], except the result of Siamese Net on the *Clickbait Challenge* dataset since it is quite unsatisfactory. We report the results using our implementation instead.

model the interactions between contexts in title and body to help measure their relevance more accurately. In addition, Transformers may also have a greater ability than CNN, LSTM and GRU in context modeling. Thus, our method can detect clickbaits more effectively than baseline methods.

4.3 Influence of Different Scores

In this section, we conduct several ablation studies to explore the influence of the four clickbait scores. We compare the performance of our *SATC* approach by removing one of these scores in clickbait prediction. The results on the *Clickbait Challenge* and *FNC* datasets are respectively shown in Figs. 3(a) and 3(b). From the results, we find that the title content score plays the most important role. This is intuitive because clickbaits mainly rely on the content of their titles to attract users' attention and clicks. Thus, modeling the title content is critical for clickbait detection. In addition, we find the body content score is also important. This is because the body of many clickbaits may be misleading or uninformative. Thus, modeling the content of body is important for clickbait detection. Besides, the matching score is also useful for clickbait prediction. This is probably because the titles of some clickbaits do not perfectly match their bodies. Thus, modeling the relevance of title and body is useful for accurate clickbait detection. Moreover, we find the title stylistic score is also helpful. This is mainly because the stylistic patterns of title are important clues for identifying clickbaits, but these clues may not be captured by the content modeling module. Thus, the title stylistic score can provide complementary information to help detect clickbaits. These results verify the effectiveness of the four different clickbait scores in our approach.

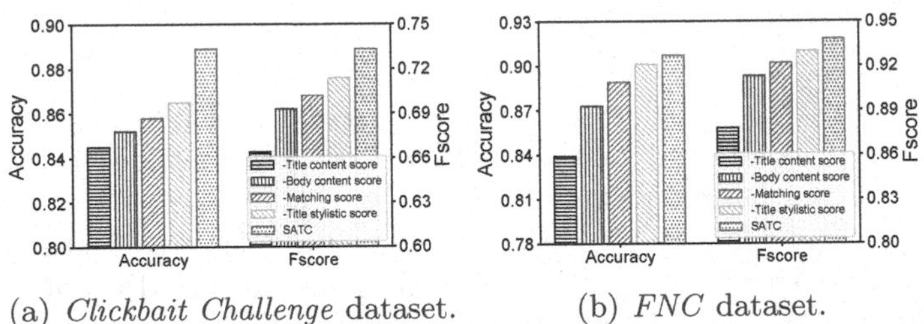

(a) *Clickbait Challenge* dataset. (b) *FNC* dataset.

Fig. 3. Influence of removing different scores in clickbait prediction.

4.4 Effectiveness of Attention Mechanism

In this section, we verify the effectiveness of the word-level attention, character-level attention and co-attention networks in our approach. More specifically, we compare the performance of our *SATC* approach and its variants without one kind of attention. The results on the *Clickbait Challenge* and *FNC* datasets are

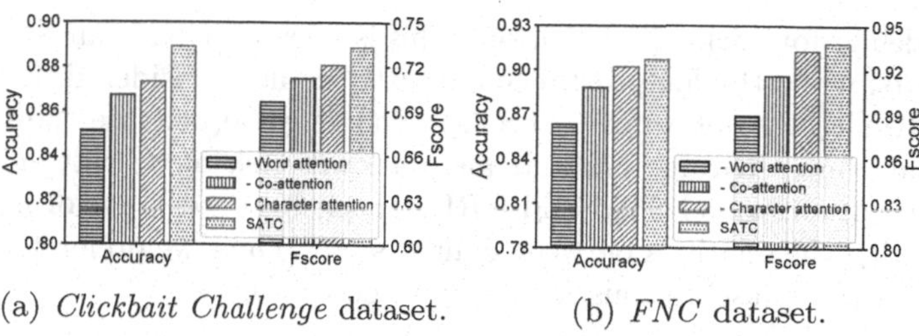

(a) *Clickbait Challenge* dataset. (b) *FNC* dataset.

Fig. 4. Effectiveness of different attention networks.

respectively shown in Figs. 4(a) and 4(b). We find that the word-level attention network is very helpful. This may be because different words are usually diverse in their informativeness and the work-level attention networks can attend to the important words in title and body, which can help learn more informative representations of them. In addition, the co-attention network can also effectively improve the model performance. This may be because the co-attention network can model the interactions of words in title and body and can further enhance the title and body representations by encoding interaction information, which is beneficial for evaluating the relevance between title and body. Besides, the character-level attention network can also improve the performance to some extent. This may be because different characters also have different importance in modeling the stylistic patterns of the title and the character-level attention network is able to select useful characters, which can help learn more informative style-aware title representations.

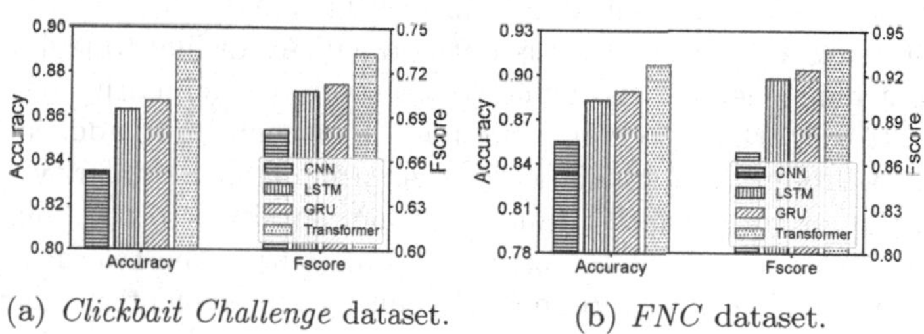

(a) *Clickbait Challenge* dataset. (b) *FNC* dataset.

Fig. 5. Effectiveness of Transformer in text modeling.

4.5 Effectiveness of Transformer

In this section, we verify the effectiveness of Transformers in text modeling in our approach. We compare the performance of *SATC* and its several variants using CNN, LSTM and GRU for text modeling, and the results are illustrated in Figs. 5(a) and 5(b). From the results, we find that using CNN is not optimal in text modeling for clickbait detection. This is because CNN can only capture local contexts, while the long-distance contexts are not considered. In addition, we find GRU slightly outperforms LSTM. This may be because the GRU networks contain fewer parameters and have a lower risk of overfitting. Besides, Transformer outperforms LSTM and GRU. This is because Transformer is effective in modeling the relations between contexts [24]. Thus, we prefer Transformer for learning text representations for clickbait detection.

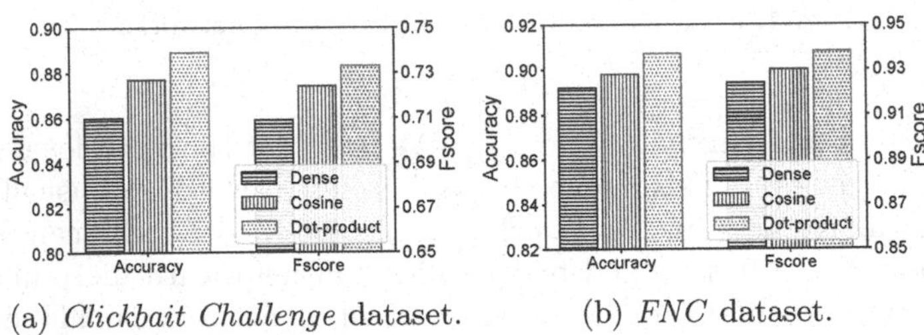

(a) *Clickbait Challenge* dataset. (b) *FNC* dataset.

Fig. 6. Influence of using different methods for computing matching scores.

4.6 Influence of Matching Methods

In this section, we explore the influence of using different methods to implement the matcher in our approach to compute the matching score. We compare the performance of *SATC* using dot-product, dense network and cosine similarity as the matcher. The results are illustrated in Figs. 6(a) and 6(b). From the results, we find that using a dense network is not optimal. According to [20], a possible reason is that dense network is difficult to measure the similarity between two vectors, and thereby the matching score may be inaccurate. In addition, we find that using dot-product is slightly better than using cosine similarity. This may be because the cosine similarity function is not sensitive to the length of the input vectors, which may not be optimal for measuring the relevance between the title and body. Thus, we choose dot-product to implement our matcher.

4.7 Case Study

In this section, we conduct several case studies to better understand the characteristics of our approach. The title, body, groundtruth and the predictions results of *GRU-Att*, *LSDA* and our *SATC* on several samples are shown in Table 2, and we have several findings. In Table 2, the first sample is a clickbait because its title does not match its body. However, since the *GRU-Att* method only considers the information of title, it fails to detect this clickbait. The other two methods that consider the relevance between title and body classify this sample correctly. Thus, it is important to model the title-body relevance for clickbait detection.

The title of the second sample in Table 2 contains a word with repeated characters to express strong emotion, which is an important indication of clickbaits. However, this word is out-of-vocabulary, making it difficult for the *GRU-Att* and *LSDA* methods to capture this clue. Thus, these methods fail to detect this clickbait. Different from them, our approach uses a character-level Transformer to capture the stylistic patterns in the title, and thereby can detect this clickbait at a high confidence. The third sample in Table 2 is not a clickbait because the title is formal and the title is relevant to the body. However, it is not easy to measure the relevance between the title and body of this sample without considering the interactions between their words, since the body does not frequently mention the words like "US" and "Watch" that appear in the title. Thus, the *LSDA* method, which does not consider the interactions between contexts, incorrectly classifies this sample as a clickbait. Since our approach uses a co-attention network to model title-body interactions, it classifies this sample correctly.

Table 2. The titles, bodies, labels and the predicted scores of different methods on several samples. 0 stands for non-clickbait and 1 stands for clickbait.

Title	Body	Label	**Prediction**		
			GRU-Att	LSDA	SATC
Report: NHL expansion to Las Vegas'a done deal'	Brain surgery recovery can be a gamble, but not everybody wakes up in the middle of the procedure...	1	0.07	0.88	0.95
The real-life Indiana Jane will make you soooooooooo jealous of her life	Meet the real-life Indiana Jane: American adventurer spends her life in dangerous jungles and uncharted wildernesses...	1	0.23	0.16	0.98
Apple Watch may be available outside US shortly after launch	Lately, Apple CEO has been making the rounds in Europe, stopping at various stores and chatting with employees. The last time we heard anything about his commentary on Apple Watch...	0	0.12	0.68	0.05

5 Conclusion

In this paper, we propose a clickbait detection approach with style-aware title modeling and co-attention, which can capture the stylistic patterns in the title and the interactions between the contexts in the title and body. We use Transformers to learn content representations of title and body, and respectively compute two content-based clickbait scores for them based on their representations. In addition, we propose to apply a character-level Transformer to capture the stylistic patterns of title for learning style-aware title representations, which are further used to compute a title stylistic score. Besides, we propose to use a co-attention network to model the relatedness between the contexts within title and body, and further combine their original representations with the interaction information to learn interaction-enhanced title and body representations, which are further used to compute a title-body matching score. The final clickbait score is predicted by a weighted summation of the four kinds of clickbait scores. Extensive experiments on two benchmark datasets show that our approach can effectively improve the performance of clickbait detection by using style-aware title modeling to capture stylistic information and co-attention networks to model title-body interactions.

Acknowledgments. Supported by the National Key Research and Development Program of China under Grant No. 2018YFC1604002, the National Natural Science Foundation of China under Grant Nos. U1936208, U1936216, U1836204 and U1705261.

References

1. Agrawal, A.: Clickbait detection using deep learning. In: 2016 2nd International Conference on Next Generation Computing Technologies (NGCT), pp. 268–272. IEEE (2016)
2. Anand, A., Chakraborty, T., Park, N.: We used neural networks to detect clickbaits: you won't believe what happened next! In: Jose, J.M., et al. (eds.) ECIR 2017. LNCS, vol. 10193, pp. 541–547. Springer, Cham (2017). https://doi.org/10.1007/978-3-319-56608-5_46
3. Biyani, P., Tsioutsiouliklis, K., Blackmer, J.: "8 amazing secrets for getting more clicks": Detecting clickbaits in news streams using article informality. In: AAAI (2016)
4. Bourgonje, P., Schneider, J.M., Rehm, G.: From clickbait to fake news detection: an approach based on detecting the stance of headlines to articles. In: Proceedings of the 2017 EMNLP Workshop: Natural Language Processing meets Journalism, pp. 84–89 (2017)
5. Cao, X., Le, T., et al.: Machine learning based detection of clickbait posts in social media. arXiv preprint arXiv:1710.01977 (2017)
6. Chakraborty, A., Paranjape, B., Kakarla, S., Ganguly, N.: Stop clickbait: detecting and preventing clickbaits in online news media. In: 2016 IEEE/ACM International Conference on Advances in Social Networks Analysis and Mining (ASONAM), pp. 9–16. IEEE (2016)
7. Chen, Y., Conroy, N.J., Rubin, V.L.: Misleading online content: recognizing clickbait as "false news". In: Proceedings of the 2015 ACM on Workshop on Multimodal Deception Detection, pp. 15–19 (2015)
8. Dimpas, P.K., Po, R.V., Sabellano, M.J.: Filipino and english clickbait detection using a long short term memory recurrent neural network. In: IALP, pp. 276–280. IEEE (2017)

9. Dong, M., Yao, L., Wang, X., Benatallah, B., Huang, C.: Similarity-aware deep attentive model for clickbait detection. In: Yang, Q., Zhou, Z.-H., Gong, Z., Zhang, M.-L., Huang, S.-J. (eds.) PAKDD 2019. LNCS (LNAI), vol. 11440, pp. 56–69. Springer, Cham (2019). https://doi.org/10.1007/978-3-030-16145-3_5
10. Fu, J., Liang, L., Zhou, X., Zheng, J.: A convolutional neural network for clickbait detection. In: 2017 4th International Conference on Information Science and Control Engineering (ICISCE), pp. 6–10. IEEE (2017)
11. Geçkil, A., Müngen, A.A., Gündogan, E., Kaya, M.: A clickbait detection method on news sites. In: 2018 IEEE/ACM International Conference on Advances in Social Networks Analysis and Mining (ASONAM), pp. 932–937. IEEE (2018)
12. Glenski, M., Ayton, E., Arendt, D., Volkova, S.: Fishing for clickbaits in social images and texts with linguistically-infused neural network models. arXiv preprint arXiv:1710.06390 (2017)
13. Huang, P.S., He, X., Gao, J., Deng, L., Acero, A., Heck, L.: Learning deep structured semantic models for web search using clickthrough data. In: CIKM, pp. 2333–2338 (2013)
14. Indurthi, V., Oota, S.R.: Clickbait detection using word embeddings. arXiv preprint arXiv:1710.02861 (2017)
15. Kingma, D.P., Ba, J.: Adam: a method for stochastic optimization. arXiv preprint arXiv:1412.6980 (2014)
16. Kumar, V., Khattar, D., Gairola, S., Kumar Lal, Y., Varma, V.: Identifying clickbait: a multi-strategy approach using neural networks. In: SIGIR, pp. 1225–1228 (2018)
17. Okura, S., Tagami, Y., Ono, S., Tajima, A.: Embedding-based news recommendation for millions of users. In: KDD, pp. 1933–1942 (2017)
18. Pennington, J., Socher, R., Manning, C.: Glove: global vectors for word representation. In: EMNLP, pp. 1532–1543 (2014)
19. Potthast, M., Köpsel, S., Stein, B., Hagen, M.: Clickbait detection. In: Ferro, N., et al. (eds.) ECIR 2016. LNCS, vol. 9626, pp. 810–817. Springer, Cham (2016). https://doi.org/10.1007/978-3-319-30671-1_72
20. Rendle, S., Krichene, W., Zhang, L., Anderson, J.: Neural collaborative filtering vs. matrix factorization revisited. arXiv preprint arXiv:2005.09683 (2020)
21. Shen, Y., He, X., Gao, J., Deng, L., Mesnil, G.: A latent semantic model with convolutional-pooling structure for information retrieval. In: CIKM, pp. 101–110 (2014)
22. Srivastava, N., Hinton, G.E., Krizhevsky, A., Sutskever, I., Salakhutdinov, R.: Dropout: a simple way to prevent neural networks from overfitting. JMLR **15**(1), 1929–1958 (2014)
23. Thomas, P.: Clickbait identification using neural networks. arXiv preprint arXiv:1710.08721 (2017)
24. Vaswani, A., et al.: Attention is all you need. In: NIPS, pp. 5998–6008 (2017)
25. Yang, Z., Yang, D., Dyer, C., He, X., Smola, A., Hovy, E.: Hierarchical attention networks for document classification. In: NAACL-HLT, pp. 1480–1489 (2016)
26. Zheng, H.T., Chen, J.Y., Yao, X., Sangaiah, A.K., Jiang, Y., Zhao, C.Z.: Clickbait convolutional neural network. Symmetry **10**(5), 138 (2018)
27. Zhou, Y.: Clickbait detection in tweets using self-attentive network. arXiv preprint arXiv:1710.05364 (2017)

Knowledge-Enabled Diagnosis Assistant Based on Obstetric EMRs and Knowledge Graph

Kunli Zhang[1,2], Xu Zhao[1,2(✉)], Lei Zhuang[1], Qi Xie[1], and Hongying Zan[1,2]

[1] School of Information Engineering, Zhengzhou University, Zhengzhou, China
{ieklzhang,ielzhuang,ieqxie,iehyzan}@zzu.edu.cn
[2] Peng Cheng Laboratory, Shenzhen, China
zhaox917@163.com

Abstract. The obstetric **E**lectronic **M**edical **R**ecord (EMR) contains a large amount of medical data and health information. It plays a vital role in improving the quality of the diagnosis assistant service. In this paper, we treat the diagnosis assistant as a multi-label classification task and propose a **K**nowledge-**E**nabled **D**iagnosis **A**ssistant (KEDA) model for the obstetric diagnosis assistant. We utilize the numerical information in EMRs and the external knowledge from Chinese Obstetric Knowledge Graph (COKG) to enhance the text representation of EMRs. Specifically, the bidirectional maximum matching method and similarity-based approach are used to obtain the entities set contained in EMRs and linked to the COKG. The final knowledge representation is obtained by a weight-based disease prediction algorithm, and it is fused with the text representation through a linear weighting method. Experiment results show that our approach can bring about +3.53 F1 score improvements upon the strong BERT baseline in the diagnosis assistant task.

Keywords: Diagnosis assistant · Knowledge graph · Obstetric EMRs · Multi-label classification.

1 Introduction

Health service relations on the health of millions of people, and it is a livelihood issue in our country. Specifically in China, which has a huge population, the

This work has been supported by the National Key Research and Development Project (Grant No. 2017YFB1002101), Major Program of National Social Science Foundation of China (Grant No. 17ZDA138), China Postdoctoral Science Foundation (Grant No. 2019TQ0286), Science and Technique Program of Henan Province (Grant No. 192102210260), Medical Science and Technique Program Co-sponsored by Henan Province and Ministry (Grant No. SB201901021), Key Scientific Research Program of Higher Education of Henan Province (Grant No. 19A520003, 20A520038), the MOE Layout Foundation of Humanities and Social Sciences (Grant No. 20YJA740033), and the Henan Social Science Planning Project (Grant No. 2019BYY016).

M. Sun et al. (Eds.): CCL 2020, LNAI 12522, pp. 444–457, 2020.
https://doi.org/10.1007/978-3-030-63031-7_32

total amount of medical resources is still insufficient. The imbalance between the supply and demand for medical services is still the focus of China's healthcare industry. Although the implementation of China's Universal Two-child Policy in 2016 achieved many benefits, it also leads to an increase in the proportion of older pregnant women and the incidence of various complications [21]. Compared to the overall supply of the medical industry, the lack of obstetric medical resources is prominent.

Since the issue of the Basic Norms of Electronic Medical Records (Trial) [4] by the National Health and Family Planning Medical Affairs Commission in 2010, medical institutions have accumulated many obstetric Electronic Medical Records (EMRs). EMRs are detailed records of medical activities, dominated by the semi-structured or unstructured texts. There is a lot of medical knowledge and health information in EMRs, which is the core medical big data. The first course record in EMRs can be divided into the chief complaint, physical examination, auxiliary examination, admitting diagnosis, diagnostic basis, and treatment plan. In general, there is not a single diagnosis in the admitting diagnosis, it usually includes normal obstetric diagnosis, medical diagnosis, and complications. As a consequence, the diagnosis assistant task based on the Chinese obstetric EMRs can be treated as a multi-label text classification problem, in which the different diagnoses can be regarded as the variable labels. However, the doctor's diagnosis and treatment process are based on comprehensive clinical experience and knowledge in the medical field to make a diagnosis and formulate a corresponding treatment plan. At the same time, they can also explain the corresponding diagnosis basis to the patient in detail. Therefore, rich clinical experience and solid medical knowledge play a vital role in the diagnosis procedure. In order to simulate the diagnosis and treatment process of doctors, we need to introduce external knowledge that is not available in EMRs. The introduction of medical domain knowledge requires formal expression so that it can be easily used in the diagnosis assistant model. To solve this problem, we adopt the Chinese Obstetric Knowledge Graph (COKG)[1] to introduce external medical domain knowledge.

In this paper, we use the BERT (Bidirectional Encoder Representation from Transformers) [6] to generate the text representation of EMRs. The numerical information in EMRs is also important for the diagnosis results, it is being used to enhance the text representation with the multi-head self-attention [19]. For entity acquisition, we compare the bidirectional maximum matching method and the Bi-LSTM-CRF method respectively, and choose the former method to obtain the entity sets from EMRs. Then the entities are linked to the COKG by a similarity-based method. Due to the fact that the negative words in EMRs will have an impact on the semantics, we employ a negative factor to deal with the negative words in EMRs and propose a weight-based disease prediction algorithm to obtain the final knowledge representation. Finally, a linear weighting method is employed to fuse the text representation and

[1] http://47.106.35.172:8088/.

knowledge representation. The experiments on the Obstetric First Course Record Dataset support the effectiveness of our approach.

The main contributions of this paper are summarized as follows:

- In this paper, we propose the KEDA (**K**nowledge-**E**nabled **D**iagnosis **A**ssistant) model to integrate external knowledge from COKG into diagnosis assistant task.
- A weight-based disease prediction algorithm named WBDP is used to limit the influence of negative words in EMRs and generate the final knowledge representation.

2 Related Work

In this paper, we treat the obstetric diagnosis assistant task as a multi-label classification problem. The multi-label classification in traditional machine learning is usually regarded as a binary classification problem or adjust the existing algorithm to adapt to the multi-label classification task [16,18,26,27].

With the development and application of deep learning, CNN and RNN are widely used in multi-label text classification tasks. For example, Kurata G et al. [9] use CNN-based word embedding to obtain the direct relationship of the labels. Chen et al. [2] propose a model that combined CNNs and RNNs to represent the semantic information of the text, and modeling the high-order label association. Baker S and Korhonen A [1] use row mapping to hide the layers that map to the label co-occurrence based on a CNN architecture to improve the model performance. Ma et al. [13] propose a multi-label classification algorithm based on cyclic neural networks for machine translation. Yang et al. [22] propose a Sequence Generation Model (SGM) to solve the multi-label classification problem.

In recent years, the pre-training technology has grown rapidly, ELMo [14], OpenAI GPT [15], and BERT [6] model have achieved significant improvements in multiple natural language processing tasks. They can be applied to various tasks after fine-tuning. However, due to the little knowledge connection between specific and open domain, these models do not perform well on domain-specific tasks. One way to solve this problem is to pre-train the model on a specific domain, but it is time-consuming and computationally expensive for most users. The models in this way are like ERNIE [17], BERT-WWM [5], Span-BERT [8], RoBERTa [11], XLNET [23], and so on. Moreover, if we can integrate knowledge at the fine-tuning process, it may bring better results. Several studies integrate external knowledge into the model. Chen et al. [3] use BiLSTM to model the text and introduce external knowledge through C-ST attention and C-CS attention. Li et al. [10] use BiGRU to extract word features, and use a similar matrix based on convolutional neural network and self-entity and parent-entity attention to introduce knowledge graph information. Yang et al. [20] use knowledge base embedding to enhance the output of BERT for machine reading comprehension.

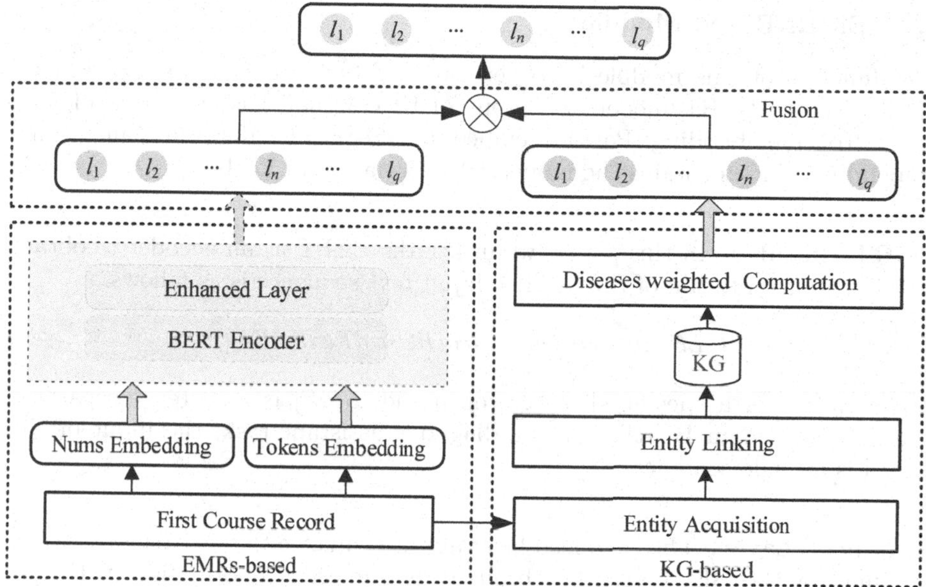

Fig. 1. The architecture of the KEDA model

In terms of the diagnosis assistant based on Chinese obstetric EMRs, Zhang et al. [25] utilize four multi-label classification methods, backpropagation multi-label learning (BP-MLL), random k-labelsets (RAkEL), multi-label k-nearest neighbor (MLKNN), and Classifier Chain (CC) to build the diagnosis assistant models. Ma et al. [12] fuse numerical features by employing the concatenated vector to improve the performance of the diagnosis assistant. Zhang et al. [24] encode EMRs with BERT, and propose an enhanced layer to enhance the text representation for diagnosis assistant.

3 Methodology

3.1 Model Architecture

As shown in Fig. 1, the KEDA model can be divided into three parts: EMRs-based module, KG-based module, and Fusion module. For any given EMR, the EMRs-based module generates the text representation by the BERT encoder firstly, then the numerical information contained in EMR is employed to enhance the text representation. Meanwhile, the KG-based module obtains the entities set and links to COKG through the entity acquisition and entity linking methods. Finally, the final knowledge representation is computed by a weight-based disease prediction algorithm and fused with the text representation through a linear weighting method. The following will introduce the implementation details of this model.

3.2 EMRs-Based Module

The function of this module is to generate the text representation of EMRs. Similar to the BERT model, the input of KEDA model is composed of four parts: Token embedding, Position embedding, Segment embedding, and Nums embedding which contains the numerical information in EMRs.

BERT Encoder. In this paper, we utilize the BERT as an encoder to obtain the text representation of EMRs. The input text sequence is as follows.

$$[CLS]ElectronicMedicalRecordText[SEP]$$

Where $[CLS]$ is a specific classifier token and $[SEP]$ is a sentence separator which is defined in BERT. For the diagnosis assistant task, the input of the model is a single sentence.

Enhanced Layer. The enhanced layer aims to enhance the text representation obtained by the BERT encoder through the numerical information in EMRs. Since the maximum length of the input sequence of BERT is 512, and the average length of EMRs is about 790 characters, we need to reduce the length of the input sequence. The information contained in the EMRs text can be divided into textual information and numerical information. Numerical information usually includes certain examinations or indications characterized by numerical values(For example, it contains the age, body temperature, pulse, respiration, respiration, and so on), which is also important information for diagnosis. So we separately extract the numerical information in EMRs to enhance the textual information, which not only can meet the limit of the input length, but also can better use the numerical information in the EMRs for diagnosis.

Then we adopt a multi-head self-attention proposed in Transformer [19] to integrate the numerical information into text representation of EMRs, as shown in Eq. (1)-(4).

$$Q = K = V = W^S Concat([C]; Num_{1...M}) \tag{1}$$

$$Attention(Q, K, V) = softmax(\frac{QK^T}{\sqrt{d_k}})V \tag{2}$$

$$head_i = Attention(QW_i^Q, KW_i^K, VW_i^V) \tag{3}$$

$$[C^{'}] = Concat(head_i, ..., head_h)W^O \tag{4}$$

Where $[C]$ is the hidden layer state representation of [CLS], $[C^{'}]$ is the text representation after fusing numerical information. $Num_{1...M}$ is the Nums embedding containing M values, which is obtained by standardizing and normalizing the numerical information in EMRs. W^S, W^Q, W^K, W^V, and W^O are trainable parameters, where $Q \in d^{model}$.

3.3 KG-Based Module

Entity Acquisition. Through the analysis of obstetric EMRs, we found that the entities such as symptoms, signs, and diseases in EMRs are high-value information for the intelligent diagnosis, so we mainly identify these entities contained in EMRs.

To achieve better performance, we compared two ways for entity acquisition. One way is a dictionary-based method, the Chinese Symptom Knowledge Base(CSKB)[2], diseases set in ICD-10, and the entity sets of diseases and symptoms in COKG are used as dictionaries. We utilize the bidirectional maximum matching algorithm used in Chinese word segmentation [7] for cntity acquisition, the obtained set includes a total of 9,836 entities. Another way is to use the Bi-LSTM-CRF model for entity acquisition, the texts labeled when constructing COKG is used as the training corpus. The Detailed analysis of experimental comparison results can be found in Sect. 4.

Entity Linking. For the entity sets obtained above, it is necessary to establish a link relationship with the nodes in the knowledge graph. In this paper, the similarity-based approach is used to link the entities in the knowledge graph.

For a given identified entity E_R, we need to find the n entities that are most similar to the knowledge graph COKG, the set of candidate entities is denote as $S = \{E_{K_1}, E_{K_2}, ..., E_{K_i}, ..., E_{K_n}\}$. Then we calculate the similarity between entities r and k, and select the entity with the highest similarity as the entity linked to COKG. The Levenshtein distance, Jaccard coefficient and the longest common substring are used to calculate the similarity respectively, as shown in Eqs. (5)-(7).

$$Sim_{ld} = \frac{lev E_R, E_{K_i}(|E_R|, |E_{K_i}|)}{max(|E_R|, |E_{K_i}|)} \tag{5}$$

$$Sim_{jacc} = jaccard(bigram(|E_R|), bigram(|E_{K_i}|)) \tag{6}$$

$$Sim_{lcs} = \frac{|lcs(E_R, E_{K_i})|}{max(|E_R|, |E_{K_i}|)} \tag{7}$$

These three similarity algorithms measure the similarity of two entities from different angles, and the average value is used as the final score of the similarity of two entities, as shown in Eq. (8).

$$Sim(E_R, E_{Ki}) = (Sim_{ld} + Sim_{jacc} + Sim_{lcs})/3 \tag{8}$$

However, the negative words in EMRs will have an impact on the semantics of components in their jurisdiction. For example, for the descriptions of *There is no discomfort such as vaginal bleeding* (无阴道流血等不适) and *There is involuntary vaginal fluid* (不自主阴道流液) contain the negative words 无 and 不. The first word will change the actual semantics, but the latter word is only a description of *vaginal fluid.*

[2] http://www5.zzu.edu.cn/nlp/info/1015/1865.htm

Therefore, we utilize the negative factor f_{neg} to limit the influence of negative words on semantics. If the negative words that do not change or partially change semantics, the entities described by those words will be linked to COKG, and the negative factor is 1 or 0.5, respectively. For those negative words that will change semantics, their negative factor is -1.

Diseases Weighted Computation. Through entity linking above, we can obtain the symptoms set $S_R = \{s_{R_1}, s_{R_2}, ..., s_{R_i}, ..., s_{R_m}\}$ and the diseases set $D_R = \{(d_{R_1} : f_{R_1}), (d_{R_2} : f_{R_2}), ..., (d_{R_i} : f_{R_i}), ..., (d_{R_q} : f_{R_q})\}$, where f_{R_i} is the frequency of disease entity and $f_{R_1} \leq f_{R_2} \leq ... \leq f_{R_i} \leq ... \leq f_{R_q}$.

Then we propose a weight-based disease prediction algorithm named WBDP. The disease and symptom sets in COKG are denoted as D_K and S_K. Through the matching of tail entities, we can get a set $D_i = \{d_{i_1}, d_{i_2}, ..., d_{i_j}, ..., d_{i_n}\}$ of n candidate disease entities in COKG for symptom s_{R_i}, the disease candidate set corresponding to all symptoms is denoted as D. For each disease d_{ij} in candidate set D, there is a symptom set $S_{d_{ij}} = \{s_{d_{ij}1}, s_{d_{ij}2}, ..., s_{d_{ij}l}, ..., s_{d_{ij}M}\}$ containing m symptoms in COKG associated with it, and $Q_{ij} = S_R \cap S_{d_{ij}}$. The purpose of WBDP is to compute the weight of disease d_{ij}, as shown in Eq. (9).

$$W_{d_{ij}} = \sum_{s_{R_i} \in S_R} \frac{f_{neg} \times p(s_{R_i}, d_{ij})}{\sum_{q_r \in Q_{ij}} p(q_r, d_{ij})} \log_2 \frac{|D|}{|D_i| + 1} \tag{9}$$

where $|D_i|$ and $|D|$ are the number of diseases in set D_i and D, f_{neg} is the negative factor of s_{R_i}, $p(s_{R_i}, d_{ij})$ is the co-occurrence probability of symptom s_{R_i} and disease d_{ij} in COKG.

We adopt two methods to deal with the disease set D_R contained in EMRs. If the disease negative factor f_{neg} is -1, it will be removed from the candidate set. Otherwise, if the candidate set associated with symptoms already contains d_{R_i}, the weight $W'_{d_{R_i}}$ will be computed according to the $W_{d_{R_i}}$ and the frequency f_{R_i}, as shown in Eq. (10).

$$W'_{d_{R_i}} = W_{d_{R_i}} (1 + \frac{f_{R_i}}{\sum_{f_{R_i} \in D_R} f_{R_i}}) \tag{10}$$

If the candidate set associated with symptoms does not contain d_{R_i}, it will be add to the candidate set. Its weight is β times of the average weight, where β is a hyper-parameter and $\beta \geq 1$, the Equation is shown in (11). It is means that the diseases in EMRs have more influence on the diagnosis results than the symptoms.

$$W_{d_{R_i}} = f_{neg} \times \frac{\beta}{|D|} \sum_{d_i \in Dise} W_{d_i} \tag{11}$$

3.4 Fusion Module

The fusion module is aimed to integrate the output of the KG-based module into the output of the EMRs-based module. Inspired by the method proposed by [3],

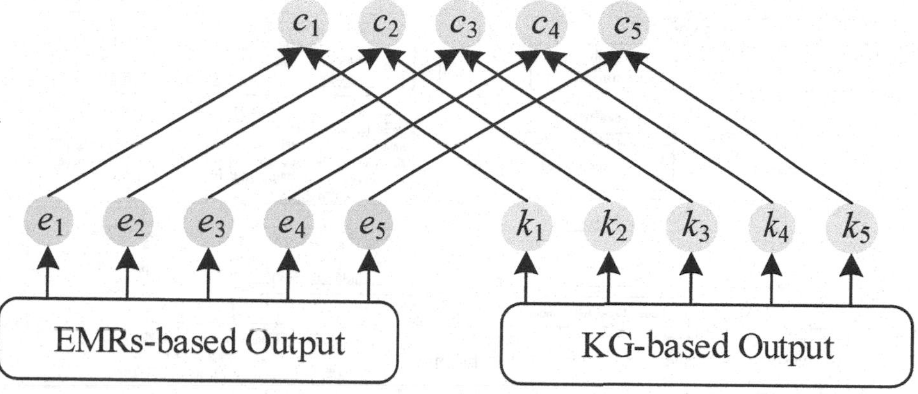

Fig. 2. The fusion module of KEDA model

we employ a linear weighting method to fuse those representations, as shown in Fig. 2.

The output of KG-based module and EMRs-based module is denoted as $K = [k_1, k_2, ..., k_i, ..., k_q]$ and $E = [e_1, e_2, ..., e_i, ..., e_q]$, where k_i is the normalized representation of the weights mentioned above. The fusion process is shown in Eq. (12).

$$c_i = \sigma(\gamma_i e_i + (1 - \gamma_i)k_i) = \frac{1}{1 - \exp(-(\gamma_i c_i + (1 - \gamma_i)k_i))} \tag{12}$$

Where σ is the sigmoid function, γ can be seen as a soft switch to adjust the importance of two representations. There are various ways to set the γ. The simplest one is to treat γ as a hyper-parameter and manually adjust. Alternatively, it can also be learned by a neural network automatically, as shown in Equation (13).

$$\gamma = \sigma(W^T[K; E] + b) \tag{13}$$

Where W and b are trainable parameters.

3.5 Training

To train the KEDA model, the objective function is to minimize the cross-entropy in Eq. (14).

$$\mathcal{L} = -\frac{1}{N}\sum_{i=1}^{N}[y_i \log P_i + (1 - y_i)\log(1 - P_i))] \tag{14}$$

Where $y_i \in \{0, 1\}$, N is the number of labels, and P is the model's prediction.

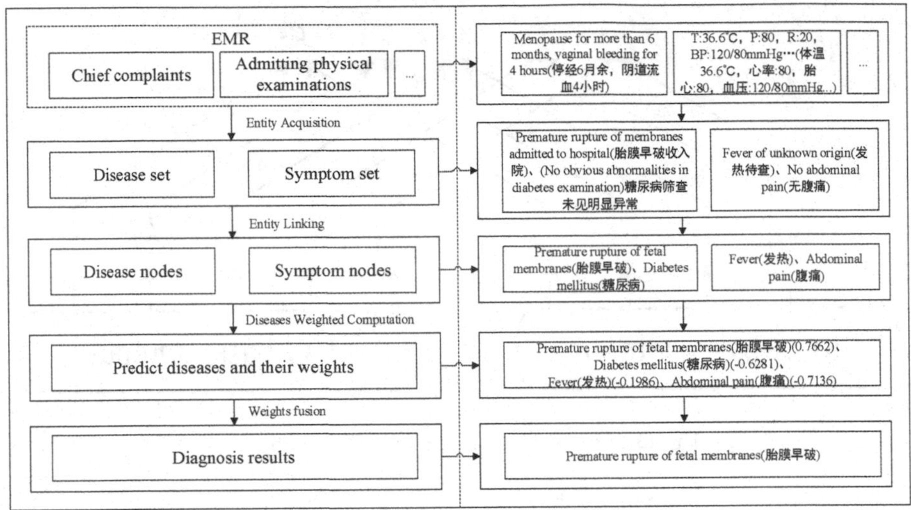

Fig. 3. The procedure of diagnosis assistant

4 Experiments

4.1 The Procedure of Diagnosis Assistant

As shown in Fig. 3, the procedure of diagnosis assistant can be divided into four parts: entity acquisition, entity linking, disease weighted computation, and weights fusion. For any given EMR, we obtain the entity sets through entity acquisition firstly, then the entities in those sets are linked to the COKG by a similarity-based method. As a result, we can get the disease nodes set and symptom nodes set from COKG. The WBDP algorithm is employed to compute the disease weights, and the negative factor f_neg is used to limit the influence of negative words in EMRs for disease or symptom entities. Ultimately, the disease weights are regarded as the final knowledge representation to fuse the text representation so that we can get the diagnosis results.

4.2 Dataset

We conducted experiments on the obstetric first course record dataset and COKG.

Obstetric First Course Record Dataset. The first course records include 24,339 EMRs from multiple hospitals in China. They were pre-processed through the steps of anonymization, data cleaning, structuring, and diagnostic label standardization. 21,905 of them were used for training and 2,434 were used for testing.

COKG. COKG uses the MeSH-like framework as the knowledge ontology to define the entity and relationship description system with obstetric diseases as the core. It contains knowledge from various sources such as the professional thesaurus, obstetrics textbooks, clinical guidelines, network resources, and other multi-source knowledge. COKG includes a total of 15,249 kinds of relations. Among them, 5,790 kinds of relations are semi-automatically extracted, and 9,459 kinds of relations are automatically extracted.

Table 1. The results on obstetric first course record dataset.

Model	F1(%)	Hamming loss	One error	AP(%)
SGM	60.00	0.0200	0.0630	39.00
BERT	79.58	0.0132	0.0961	84.97
BERT+A	80.26	**0.0129**	0.0863	85.42
BERT+A-AP	80.28	**0.0129**	0.0891	85.74
KG-based	53.57	0.0220	0.2417	52.13
KEDA	**83.11**	0.0143	**0.00152**	**88.90**

4.3 Experimental Setup

In this paper, the EMRs are preprocessed by de-identifying, data cleaning, structuring, data filtering, and standardization of diagnostic labels. During the data filtering process, the information that is duplicated and has little effect on the diagnosis is removed. On the one hand, it can meet the limitation of the input length of the BERT model, and on the other hand, it can also retain the useful information. The version of BERT model we used is BERT-base-Chinese, the main parameters are hidden size 768, max position embedding 512, num attention heads 12, num hidden layers 12, maximum input length 512, learning rate 5e-5, batch size 6, training epoch 20. All our experiments are run on an RTX2080ti GPU(12G).

4.4 Results

Experimental results on the obstetric first course record dataset are shown in Table 1. F1 (F1-micro), Hamming Loss, One Error, and AP (Average Precision) were used as evaluation metrics. BERT indicates the results of the baseline Google BERT, SGM is the results of SGM(Sequence Generation Model) [22], BERT+A, and BERT+A-AP are from [24], which experiments are carried out on the same dataset as this paper. The KG-based means only use knowledge graph information, and KEDA is our proposed model.

From Table 1, it can be seen that the improvements in our model over the BERT baseline and other results from [24] are significant and consistent overall evaluation metrics. The AP of KG-based is only 52.13%, which is far lower than the result of KEDA. There may be two reasons for this situation, one of them may be some diagnoses are not obstetric diseases. Another possibility is that COKG is constructed from multi-source texts, which have different levels of detail for different diseases, it may make the number of triples of some diseases insufficient for accurate prediction.

Table 2. The results of entity acquisition.

Method	**F1(%)**	**P(%)**	**R(%)**
Bidirectional maximum matching	89.42	85.20	94.10
Bi-LSTM-CRF	86.53	88.10	85.03

Table 3. The setting of hyper-parameter γ on KEDA.

γ	**F1(%)**	**P(%)**	**R(%)**	**AP(%)**
0.1	62.46	63.25	60.23	64.70
0.3	64.24	65.32	63.68	66.57
0.5	75.30	77.38	74.19	78.95
0.7	77.23	79.86	74.52	80.90
0.9	71.25	73.19	68.26	74.28
Trained	**83.11**	**87.21**	**79.36**	**88.90**

Although the KG-based method does not have an advantage in various indicators, the results of the KEDA are better than BERT and others, indicating that the fusion of knowledge graph can improve the performance of diagnosis assistant. By further analyzing the diagnostic labels in the results, we find that the integration of knowledge graph is more obvious for the improvement of low-frequency labels. For example, the label *Placental abruption* (胎盘早剥) only appeared 5 times in the dataset, due to the scarcity of samples, it is difficult to make accurate predictions using only the BERT-based method. But there are 47 triplets in COKG that describe its symptoms, signs, and related diseases. After introducing the corresponding knowledge graph information, the accuracy of this type of disease has been significantly improved.

4.5 The Results of Entity Acquisition

As mentioned above, in order to choose a better entity acquisition method, we compared the bidirectional maximum matching and Bi-LSTM-CRF on the manually labeled 100 EMRs, the results are shown in Table 2. It is can be seen that the effect of the bidirectional maximum matching method is better than

Bi-LSTM-CRF in testing. Bi-LSTM-CRF is trained on texts such as obstetric teaching materials, national norms, clinical practice, etc.

The differences in training data and test data may have an impact on the effectiveness of the model. The dictionaries of the bidirectional maximum matching method come from CSKB and ICD-10, which are more suitable for the description and content in obstetric EMRs. This may be one of the reasons for its better effect on entity acquisition.

4.6 The Setting of Hyper-Parameter γ

The goal of this part is to verify the effectiveness of the fusion module. Firstly, We manually tune the hyper-parameter γ to explore the relative importance of EMRs-based and KG-based. We adjust γ from 0 to 1 with an interval of 0.2, and the results are shown in Table 3. When γ is equal to 0 or 1, the model will become the KG-based or EMRs-based, its results can be found in Table 1. From these results, the model with $\gamma = 0.7$ performs best. When γ gradually increases, the model performs better, but after 0.7, the performance of the KEDA will decline. This shows that too much introduction of knowledge will also affect the overall performance of the model.

Moreover, the hyper-parameter γ is treated as a trainable parameter to train with the model, the results are shown in the last row of Table 3. Compared with manual adjustment, the way to use γ as a trainable parameter is a better choice.

4.7 Error Analysis

In this section, we analyze the bad cases induced by our KEDA model. Most of bad cases can be divided into two categories.

First, some entities in EMRs are not obstetric disease or symptom, which can not find their corresponding nodes in COKG. For example, those entities like *otitis media* (中耳炎), *glaucoma* (青光眼) and so on, there are not enough descriptions in COKG. Thus, the model can not make the correct diagnosis.

Second, COKG is constructed on multi-source obstetric disease texts, which have different levels of detailed description of different diseases. Among them, the proportion of diseases with less than 10 triplets accounts for more than 60%. If some diseases have fewer triplets in COKG, the model also cannot achieve good performance.

5 Conclusion

In this paper, the obstetric diagnosis assistant task is treated as a multi-label classification problem. We propose a KEDA model for this task, which integrates the numerical information from EMRs and external knowledge from COKG to improve the performance of diagnosis. We utilize the bidirectional maximum matching method to get the entities in EMRs, and the similarity-based approach is used to link the entities in knowledge graph COKG. Then we propose a WBDP algorithm to compute the weights of the entities in the candidate set. Finally, a linear weighting method is employed to fuse the text representation and knowledge representation. The results on the obstetric EMRs support the effectiveness of our approach compared to the BERT model. It turns out that even though the pre-training of BERT involves a large number of corpora, the knowledge graph of the specific domain can still provide useful information.

In the future, we will incorporate more valuable information into deep neural networks to further improve the performance of the diagnosis assistant. We find that some disease entities in EMRs are not included in COKG(For example, the disease entity 'patella fracture' is a diagnosis label in EMRs, but it is not an obstetric disease), to introduce other knowledge graphs that contain more disease entities is an effective feature for diagnosis.

References

1. Baker, S., Korhonen, A.: Initializing neural networks for hierarchical multi-label text classification. In: BioNLP 2017, pp. 307–315. Association for Computational Linguistics, Vancouver, Canada, August 2017. https://doi.org/10.18653/v1/W17-2339, https://www.aclweb.org/anthology/W17-2339
2. Chen, G., Ye, D., Xing, Z., Chen, J., Cambria, E.: Ensemble application of convolutional and recurrent neural networks for multi-label text categorization. In: 2017 International Joint Conference on Neural Networks (IJCNN), pp. 2377–2383. IEEE (2017)
3. Chen, J., Hu, Y., Liu, J., Xiao, Y., Jiang, H.: Deep short text classification with knowledge powered attention. In: Proceedings of the AAAI Conference on Artificial Intelligence, vol. 33, pp 6252–6259 (2019)
4. China's Ministry of Health: Basic specification of electronic medical records (trial). Technical Report 3 (2010)
5. Cui, Y., et al.: Pre-training with whole word masking for chinese bert. arXiv preprint arXiv:1906.08101 (2019)
6. Devlin, J., Chang, M.W., Lee, K., Toutanova, K.: BERT: pre-training of deep bidirectional transformers for language understanding. In: Proceedings of the 2019 Conference of the North American Chapter of the Association for Computational Linguistics: Human Language Technologies, Volume 1 (Long and Short Papers), pp. 4171–4186. Association for Computational Linguistics, Minneapolis, Minnesota, June 2019. https://doi.org/10.18653/v1/N19-1423, https://www.aclweb.org/anthology/N19-1423
7. Gai, R.L., Gao, F., Duan, L.M., Sun, X.H., Li, H.Z.: Bidirectional maximal matching word segmentation algorithm with rules. In: Advanced Materials Research, vol. 926, pp. 3368–3372. Trans Tech Publ. (2014)
8. Joshi, M., Chen, D., Liu, Y., Weld, D.S., Zettlemoyer, L., Levy, O.: Spanbert: improving pre-training by representing and predicting spans. Trans. Assoc. Comput. Linguist. **8**, 64–77 (2020)
9. Kurata, G., Xiang, B., Zhou, B.: Improved neural network-based multi-label classification with better initialization leveraging label co-occurrence. In: Proceedings of the 2016 Conference of the North American Chapter of the Association for Computational Linguistics: Human Language Technologies, pp. 521–526 (2016)

10. Li, M., Clinton, G., Miao, Y., Gao, F.: Short text classification via knowledge powered attention with similarity matrix based CNN. arXiv preprint arXiv:2002.03350 (2020)
11. Liu, Y., et al.: Roberta: a robustly optimized bert pretraining approach. arXiv preprint arXiv:1907.11692 (2019)
12. Ma, H., Zhang, K., Zhao, Y.: Study on obstetric multi-label assisted diagnosis based on feature fusion. J. Chinese Inf. Process. **32**(5), 128–136 (2018)
13. Ma, S., Sun, X., Wang, Y., Lin, J.: Bag-of-words as target for neural machine translation. In: Proceedings of the 56th Annual Meeting of the Association for Computational Linguistics (Volume 2: Short Papers), pp. 332–338. Association for Computational Linguistics, Melbourne, Australia, July 2018. https://doi.org/10.18653/v1/P18-2053, https://www.aclweb.org/anthology/P18-2053
14. Peters, M., et al.: Deep contextualized word representations. In: Proceedings of the 2018 Conference of the North American Chapter of the Association for Computational Linguistics: Human Language Technologies, Volume 1 (Long Papers), pp. 2227–2237. Association for Computational Linguistics, New Orleans, Louisiana, June 2018. https://doi.org/10.18653/v1/N18-1202, https://www.aclweb.org/anthology/N18-1202
15. Radford, A., Narasimhan, K., Salimans, T., Sutskever, I.: Improving language understanding with unsupervised learning. Technical report, OpenAI (2018)
16. Read, J., Pfahringer, B., Holmes, G., Frank, E.: Classifier chains for multi-label classification. Mach. Learn. **85**(3), 333 (2011)
17. Sun, Y., et al.: Ernie: enhanced representation through knowledge integration. arXiv preprint arXiv:1904.09223 (2019)
18. Tsoumakas, G., Katakis, I., Vlahavas, I.: Random k-labelsets for multilabel classification. IEEE Trans. Knowl. Data Eng. **23**(7), 1079–1089 (2010)
19. Vaswani, A., et al.: Attention is all you need. In: Advances in neural information processing systems, pp. 5998–6008 (2017)
20. Yang, A., et al.: Enhancing pre-trained language representations with rich knowledge for machine reading comprehension. In: Proceedings of the 57th Annual Meeting of the Association for Computational Linguistics, pp. 2346–2357 (2019)
21. Yang, H.l., Yang, Z.: Effect of older pregnancy on maternal and fetal outcomes. Chinese J. Obstetric Emergency (Electr. Edn) **5**(3), 129–135 (2016)
22. Yang, P., et al.: SGM: sequence generation model for multi-label classification, pp. 3915–3926 (2018)
23. Yang, Z., Dai, Z., Yang, Y., Carbonell, J., Salakhutdinov, R.R., Le, Q.V.: Xlnet: generalized autoregressive pretraining for language understanding. In: Advances in Neural Information Processing Systems, pp. 5754–5764 (2019)
24. Zhang, K., Liu, C., Duan, X., Zhou, L., Zhao, Y., Zan, H.: Bert with enhanced layer for assistant diagnosis based on chinese obstetric EMRS. In: 2019 International Conference on Asian Language Processing (IALP), pp. 384–389. IEEE (2019)
25. Zhang, K., Ma, H., Zhao, Y., Zan, H., Zhuang, L.: The comparative experimental study of multilabel classification for diagnosis assistant based on Chinese obstetric EMRS. J. Healthcare Eng. **2018** (2018)
26. Zhang, M.L., Zhou, Z.H.: Multilabel neural networks with applications to functional genomics and text categorization. IEEE Trans. Knowl. Data Eng. **18**(10), 1338–1351 (2006)
27. Zhang, M.L., Zhou, Z.H.: Ml-KNN: a lazy learning approach to multi-label learning. Pattern Recogn. **40**(7), 2038–2048 (2007)

Reusable Phrase Extraction Based on Syntactic Parsing

Xuemin Duan[1], Hongying Zan[1(✉)], Xiaojing Bai[2], and Christoph Zahner[3]

[1] School of Information Engineering, Zhengzhou University, Zhengzhou, China
xueminduan@163.com, iehyzan@zzu.edu.cn

[2] Language Centre, Tsinghua University, Beijing, China
bxj@tsinghua.edu.cn

[3] University of Cambridge Language Centre, Cambridge , UK
cz201@cam.ac.uk

Abstract. Academic Phrasebank is an important resource composed of neutral and generic phrases for academic writers. In this paper, we name these neutral and generic phrases reusable phrases, and student writers use them to organize their research articles. Due to the limited size of Academic Phrasebank, it can not meet all the academic writing needs. There are still a large number of reusable phrases in authentic research articles. In order to make up for the deficiency of Academic Phrasebank, we proposed a reusable phrase extraction model based on constituency parsing and dependency parsing to automatically extract reusable phrases from unlabelled research articles. We divided the proposed model into three main components including a reusable words corpus module, a sentence simplification module, and a syntactic parsing module. We created a reusable words corpus of 2129 words to help judge whether a word is neutral and generic, and created two datasets under two scenarios to verify the feasibility of the proposed model.

Keywords: Reusable phrase extraction · Academic phrasebank · Syntactic parsing

1 Introduction

The Academic Phrasebank is a general resource created by the University of Manchester for academic writers. And the items of it are all neutral and generic, which means that you don't have to worry about accidentally stealing someone else's idea when using these items in your academic paper. In this paper, we name these neutral and generic phrases reusable phrases. Reusable phrase means that student writers can use these phrases in their paper without worrying about plagiarism due to they are neutral and generic. The Reusable phrases including the phrases in Academic Phrasebank do not have a unique or original construction, not express a special point of view of another writer.

Now, most of the assisted academic writing research focused on Automated Essay Scoring(AES), but different from the ordinary essay prefer to life and

M. Sun et al. (Eds.): CCL 2020, LNAI 12522, pp. 458–466, 2020.
https://doi.org/10.1007/978-3-030-63031-7_33

social, research article is a scientific record of scientific research result or innovation thinking in theoretical, predictive, and experimental. Research article writing has more rigorous grammar, discourse structure and phraseology. [1] mentioned that the central of designing teaching activities developed by Academic Phrasebank is the purpose of improving the cognitive ability of student writers to potential plagiarism. Learning reusable phrases can effectively help student writers avoid plagiarism, and student writers can use the learned reusable phrases in their own research article writing, so as to improve their academic writing ability. However, Academic Phrasebank does not cover all reusable phrases in authentic research articles. In order to make up for the deficiency of Academic Phrasebank, we propose a new task named reusable phrase extraction. The purpose of this task is to automatically extract reusable phrases from unlabelled research articles, so as to help writers build their own "Academic Phrasebank".

Plenty of research relating to teaching activities about Academic Phrasebank, but little or nothing that concerns extracting reusable phrases automatically. Therefore, in this paper, we introduce a reusable phrase extraction model based on constituency parsing and dependency parsing, which aims to extract similar samples with phrases of Academic Phrasebank from unlabelled research articles. The reusable phrases examples are shown in Table 1.

Table 1. Academic phraseology extraction results.

Sentences	Reusable phrases
This paper have argue that the proposed TDNN could be further improved	This paper have argue that
There have been efforts in developing AES approaches based on DNN	There have been efforts in developing
Further study are required to identify the effectiveness of proposed AES	Further study are required to identify the effectiveness of

In order to analyze the semantics and structure of unlabelled sentences, we first create a reusable words corpus which including all words of the phrases of Academic Phrasebank. Due to the items of Academic Phrasebank all are general and neutral, the word in a given sentence which also belongs to the reusable words corpus can exist in the extraction result.

As there is no relevant study at present, this paper did not select others' baseline for comparison but created two datasets under two scenarios to verify the feasibility of the proposed model. A dataset is completed from the phrases of Academic Phrasebank, which is more standard, while a dataset is annotated from authentic research articles with more complex sentence structure, and the experimental results demonstrate the different effectiveness of the proposed model on a different dataset.

In brief, the main contributions are as follows:

- We propose a new task, named Reusable Phrase Extraction, which contributes to academic writing and provides valuable phrases for student writers to organise their research articles.
- We propose a model by syntactic parsing for Reusable Phrase Extraction, which considers phrase structure, dependency and semantic analysis of the given sentence.
- We collect sentences from authentic research articles and construct a dataset for Reusable Phrase Extraction with human-annotation. In addition, we also collect phrases from Academic Phrasebank and construct a dataset for Reusable phrases Extraction with human-completion.

2 Related Work

Corpus of contemporary American English (COCA) is the latest contemporary corpus of 360 million words developed by [2]. It covers five types of the corpus of novels, oral English, popular magazines, and academic journals in different periods in the United States. Using COCA to study can make up for the lack of students' understanding of vocabulary, and at the same time, it can cultivate favorable conditions for essay writing. However, the COCA is inappropriate to be used as a corpus for judging whether a word belongs to reusable phrases in the process of reusable phrases extraction. So we create a reusable phrases corpus for this paper.

Academic Phrasebank is a general resource developed by Dr. John Morley of the University of Manchester to help student writers writing. [1] has designed some relevant teaching activities developed based on Academic Phrasebank. The research holds that the most important two points of academic writing teaching purpose are to obtain timely writing feedback and improve the cognitive ability of student writer to plagiarism in academic writing. The former means automated research article scoring, and the latter means strengthening students' learning of phrases in Academic Phrasebank and authentic research article. Because the content of Academic Phrasebank is neutral and general, frequent learning of Academic Phrasebank can help students improve their cognitive ability. But the content of Academic Phrasebank is limited. If student writer want to expand their own "Academic Phrasebank", they need to extract reusable phrases from authentic research articles.

The problem of analyzing complex sentences in natural language processing is to make sentences simple to understand, by identifying clause boundaries. Before extracting reusable phrases from a sentence, we choose to simplify the sentence first. [3] provides a survey of predicting clause boundaries while. [4] proposed a rule-based method for clause boundary detection. The latter method is a pipeline that uses phrase structure trees to determine the clauses.

3 Our Approach

In this section, we will introduce our reusable phrases extraction approach. There are three main components in our model, i.e., a reusable words corpus module to help identify whether a word in a sentence belongs to reusable phrases, a sentence simplification module to prevent incomplete reusable phrase from being extracted, and a syntactic parsing module to determine the final results of reusable phrases extraction. We will introduce the details of our reusable phrases extraction approach as follows.

3.1 Reusable Words Corpus

The reusable phrases extraction model is extracting based on the dependency and constituency structure of a sentence, but the final extraction results of two sentences composed of the same dependency and constituency structure are not necessarily the same, because the content of reusable phrases is also related to the semantics of words of a sentence. For example, there are two sentences that only have different subjects, "Further study" and "Bert and transformer". Although they both act as the components of nominal phrases in the sentences, the former can appear in the result, but the latter can not. This is because the content of "Further" and "study" are all neutral and general, but "Bert", "and" and "transformer" have a special word. How to judge whether a word is neutral and general? we need a corpus containing a large number of neutral and general words, reusable phrases, to help us judge.

The reusable words corpus we created contains all words in Academic Phrasebank, which helps us judge whether a word or phrase should appear in the final result. It has a pivotal role in extracting reusable phrases from the unlabelled text. As the phrases in Academic Phrasebank are all reusable phrases, In the process of reusable phrase extraction, the words of a sentence that appear in Academic Phrasebank can all appear in the result of reusable phrases extraction.

We segmented the phrases in Academic Phrasebank and deleted the repeated words to obtain the reusable words corpus. Academic Phrasebank contains 12,451 phrases, and the resulting reusable words corpus we constructed contains 2,129 words.

3.2 Sentence Simplification

English sentences are mainly composed of subject, predicate, object, attribute, adverbial, complement, and other components, in which the predicate component can only be composed of verbs, and the rest of the sentence components can be composed of words or replaced by clauses. English sentences containing clauses are often long and complex, and it is difficult to extract reusable phrases from them. Therefore, for complex sentences, it is necessary to divide them into simple clauses first.

The sentence simplification is to identify more than two English sentences with more than two clauses, mark the boundary of the clauses, and decompose the complex sentences into many simple sentences. In order to improve the accuracy of reusable phrase extraction, we first simplify the complex sentences before extraction and then extracts the reusable phrases from the simple sentences. This kind of syntactic text simplification is non-destructive. It mainly extracts embedded clauses from sentences with complex structures, so as to rewrite them without affecting their original meanings. This process reduces the average sentence length and complexity, making the text simpler. The key point of sentence simplification is to extract the implied clause from the sentence with a complex structure.

In this paper, we identify the relationship between the main sentence and the paratactic or subordinate sentence by constituency parsing, classify the subordinate sentence, determine the optimal clause boundary in the sentence, and extract the clause from the constituency parse tree by using the defined rules.

First, get a constituency parse tree of given complex English sentence, then identify the non-root clausal node of the constituency parse tree (e.g. SBAR, S.) and remove it from the main tree but retain these subtrees, then remove all hanging in the main tree prepositions, subordinate conjunctions and adverbs, the result was simplified sentences. The sentence simplification examples are shown in Table 2.

Table 2. Sentence Simplification Results.

Sentences	Simplified Sentences
The prompt-dependent models can hardly learn generalized rules from rated essays for nontarget prompts, and are not suitable for the prompt independent AES	["The prompt-dependent models can hardly learn generalized rules from rated essays for nontarget prompts.","The prompt-dependent models are not suitable for the prompt independent AES."]
A supervised model is employed to identify the essays in a given set of essays, and it aims to recognize the essays with the extreme quality in the test dataset	["A supervised model is employed to identify the essays in a given set of essays.", "A supervised model aims to recognize the essays with the extreme quality in the test dataset."]
Such relative precision is at least 80% on different prompts so that the overlap of the selected positive and negative essays is fairly small	["Such relative precision is at least 80% on different prompts.", "The overlap of the selected positive and negative essays are fairly small."]

3.3 Syntactic Parsing

Our reusable phrase extraction approach is a rule-based approach using constituency parse tree and dependency tree. By identifying the main verb and determining which nominal phrases of the sentence belongs to reusable phrases by the reusable words corpus, we can easily extract the reusable phrases from the simplified sentence.

The steps for extracting reusable phrases are explained with the help of the following examples: "Further study are required to identify the effectiveness of proposed AES."

Step 1. Obtaining the dependency tree of the given simplified sentence to identify the main verb. The dependency tree is shown in Fig. 1, we can get the main verb is "required".

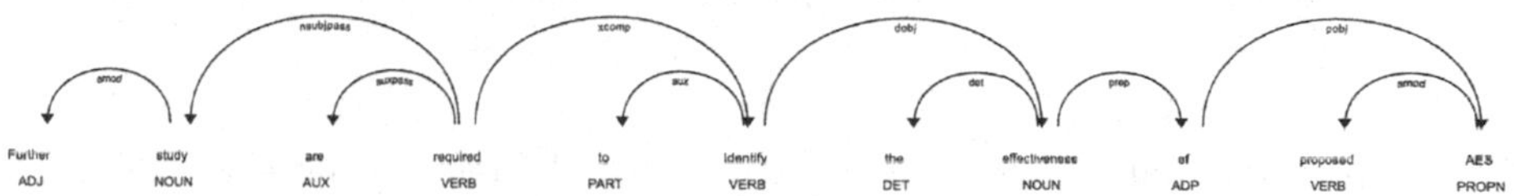

Fig. 1. Dependency parse tree.

Step 2. Obtaining the constituency parse tree to identify all nominal phrases in a sentence and their order. The constituency parse tree is shown in Fig. 2.

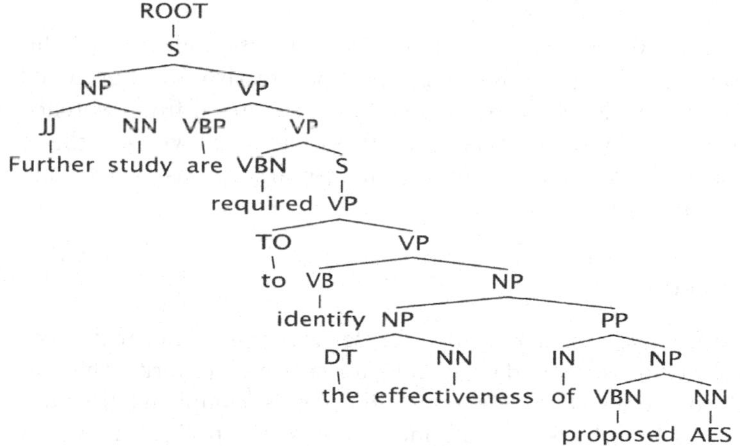

Fig. 2. Constituency parse tree.

Step 3. Taking the main verb as the center and classifying the nominal phrases with left part of verb or right part of verb. Then, using the reusable words corpus to determine whether a nominal phrase is deleted or retained. If the left part of the main verb occupied in the reusable words corpus means that it can be retained. The right part of the main verb is divided into several noun phrases and analyzed from the first one. If the first one belongs to the reusable words corpus, then continue to analyze the next one. If not, delete it and the part on its right, and then finish the analysis. All nominal phrases of this sentence and their determines are shown in Table 3.

Table 3. Nominal phrases judgements.

	Nominal phrases	Judgements
Left part	Further study	Retain
Right part	The effectiveness	Retain
	Proposed AES	Delete

The first nominal phrase, "Further study", can be retained because "further" and "study" all exist in the reusable words corpus. So is the second nominal phrase. The third nominal phrase, "proposed AES", should be deleted since "AES" not exist in the reusable words corpus. According to Table 3, we can get the result of reusable phrase extraction is "Further study are required to identify the effectiveness of ..."

4 Experiments

In this section, we present our experiment datasets and results, which devote to answering the following questions that how effective is the proposed reusable phrase extraction model in extracting reusable phrase from sentences written according to the phrases in Academic Phrasebank and whether this model can obtain a same performance in extracting reusable phrase from real academic papers compared to the former.

4.1 Datasets

Since there are not existing reusable phrase dataset now, we created two datasets under two scenarios, "standard" and "authentic", to verify the feasibility of the proposed model. The "standard" dataset is completed from the phrases of Academic Phrasebank by human. There is no special sentence pattern in sentences completed from Academic Phrasebank, which means that this dataset is more standard. The "authentic" dataset is annotated from authentic research articles by human. There are some special sentence patterns in sentences of authentic research article, which means that this dataset contains many complex sentence patterns that may appear in authentic research paper, such as inverted sentences and accent sentences. It is more "authentic".

We took 1,000 phrases from academic phrasebank and manually completes them into sentences. In addition, we also selected 1,000 complete sentences from authentic research articles and manually annotate their reusable phrases. They are combined together to form the reusable phrase datasets in this paper. The contents are shown in the Table 4.

4.2 Evaluation Metrics

In the process of extracting reusable phrase from sentence, we hope to get more words of our predicted reusable phrases that are the same as those in true reusable phrases. Based on this sense, we calculate Precision, Recall and F score for reusable phrase extraction model.

Table 4. Reusable phrase extraction model datasets.

Datasets	Sentences
Academic phrasebank reusable phrases dataset	1,000
Authentic research articles reusable phrases dataset	1,000

4.3 Results and Analysis

We use the proposed reusable phrase model to experiment with two datasets, the overall experimental results are shown in Table 5.

Table 5. The performance of reusable phrase extraction model on different datasets.

Datasets	Precision	Recall	F1 score
Academic phrasebank reusable phrases dataset	0.96	0.62	0.72
Authentic research articles reusable phrases dataset	0.84	0.99	0.88

From the overall results, we can observe that the performance of the proposed model on Academic Phrasebank Phraseology Dataset is better than on Authentic Research Articles Phraseology Dataset. This is because the reusable phrase extraction model proposed in this paper is designed for the common sentence pattern with the highest frequency in research articles. The Authentic Research Articles Phraseology Dataset has more special sentence patterns, such as inverted sentences and accent sentences.

There is still a lot of room for improvement. If we analyze and modify the proposed academic phraseology extraction model separately for the special sentence patterns that appear less frequently in research articles, the performance of the proposed model on all datasets will be improved.

5 Conclusion

In this paper, we dene a new task in assisted academic writing, Reusable Phrase Extraction, which devotes to providing valuable phrases for student writers to write their research articles. Learning the reusable phrases in the Academic Phrasebank is an important part of improving writing ability. For extracting the similar samples with the phrases of Academic Phrasebank, we proposed a reusable phrase extraction model. The proposed model are divided into three components: reusable words corpus, sentence simplification and syntactic parsing. Experiments on an Academic Phrasebank Reusable Phrases Dataset and an Authentic Research Article Reusable Phrases Dataset validate the effectiveness of our approach.

References

1. Davis, M., Morley, J.: Facilitating learning about academic phraseology: teaching activities for student writers. J. Learn. Dev. High. Educ. (2018)
2. Davis, M.: The corpus of contemporary American English: 450 million words, 1990-present. (2008)
3. Sharma, S.K.: Clause boundary identification for different languages: a survey. Int. J. Comput. Appl. Inf. Technol. **8**(2), 152 (2016)
4. Sacaleanu, B., Marascu, A., Jochim, C.: Rule-based syntactic approach to claim boundary detection in complex sentences. International Business Machines Corp U.S. Patent 9,652,450 (2017)
5. Manning, C.D., Surdeanu, M., Bauer, J., Finkel, J.R., Bethard, S., McClosky, D.: The Stanford CoreNLP natural language processing toolkit. In: Proceedings of 52nd Annual Meeting of the Association for Computational Linguistics: System Demonstrations, pp. 55–60 (2014)
6. Oakey, D.: Phrases in EAP academic writing pedagogy: illuminating Halliday's influence on research and practice. J. Engl. Acad. Purp. **44**, 100829 (2020)

WAE_RN: Integrating Wasserstein Autoencoder and Relational Network for Text Sequence

Xinxin Zhang[1], Xiaoming Liu[1,2], Guan Yang[1], Fangfang Li[3], and Weiguang Liu[1(✉)]

[1] School of Computer Science, Zhongyuan University of Technology, Zhengzhou, Henan, China
weiguang.liu@zut.edu.cn

[2] Henan Key Laboratory on Public Opinion Intelligent Analysis, Zhongyuan University of Technology, Zhengzhou, Henan, China

[3] oOh! Media, North Sydney, Australia

Abstract. One challenge in Natural Language Processing (NLP) area is to learn semantic representation in different contexts. Recent works on pre-trained language model have received great attentions and have been proven as an effective technique. In spite of the success of pre-trained language model in many NLP tasks, the learned text representation only contains the correlation among the words in the sentence itself and ignores the implicit relationship between arbitrary tokens in the sequence. To address this problem, we focus on how to make our model effectively learn word representations that contain the relational information between any tokens of text sequences. In this paper, we propose to integrate the relational network(RN) into a Wasserstein autoencoder(WAE). Specifically, WAE and RN are used to better keep the semantic structurse and capture the relational information, respectively. Extensive experiments demonstrate that our proposed model achieves significant improvements over the traditional Seq2Seq baselines.

Keywords: Language model · Semantic representation · Relational information · Wasserstein autoencoder

1 Introduction

Sequence problems are common in daily life that involves DNA sequencing in bioinformatics, time series prediction in Information science, and so on. NLP tasks, such as word segmentation, named entity recognition(NER), machine translation(MT), etc., are actually text sequence problems. For text sequence tasks, it is required to predict or generate target sequences based on the understanding of input source sequence, so it plays a pivotal role in NLP to deeply understand the generic knowledge representation in different context.

To learn the features of input sequences, probabilisticgraphicalmodels, such as Hidden Markov Models(HMM) and Conditional Random Field (CRF), can

M. Sun et al. (Eds.): CCL 2020, LNAI 12522, pp. 467–479, 2020.
https://doi.org/10.1007/978-3-030-63031-7_34

use manually defined feature functions to transform raw data into features, but the quality of the feature functions directly determines the quality of the data presentation.

Because deep learning can automatically learn the useful and highly abstract features of the data via artificial neural network(ANN), many researchers devoted themselves to using Neural Networks(NNs) to obtain low dimensional distributed representations of input data, especially in language modeling, using AutoEncoder(AE) [1] to retain the text sequence semantic information in different context has shown promising results. These language models are pretrained on large-scale corpus and complex models to obtain the data representation which contains global information and has strong generalization ability, then the latent representation can be adapted to several contexts by fine-tuning them on various tasks. However, these models simply make use of word order information or position information and ignore the implicit relationship between arbitrary tokens in the sequence, resulting in learning inadequately hidden feature representations and obtaining only superficial semantic representation. More recently, studies on attention [2,3] and self-attention [4,5] mechanism demonstrate that it can effectively improve the performance of several NLP tasks by exchanging information between sentences. However, it only calculates the contribution between vectors by means of weighted sum without exploring and taking advantage of the implicit structural relationships among tokens.

In this work, we propose add relational networks(RN) [6] to the Wasserstein AutoEncoder(WAE) [7] on the basis of the Seq2Seq architecture to collect the complex relationship between objects and retain the semantic structure in sentences. Specifically, to keep the relational information and structural knowledge we add RN layer to encoder since RN integrates the relational reasoning structure that can constrain the functional form of neural network and capture the core common attributes of relational reasoning. To better capture the complex relationships and preserve the semantic structure, we use WAE as our encoder because WAE maps input sequences into the wasserstein space that allows various other metric spaces to be embedded in it while preserving their original distance measurements.

The main contributions of our work can be summarized as follows:

1. We put forward an innovative idea to learn more meaningful and structural word representations in text sequneces. We consider relations between objects entail good flexibility and robustness, which are informative and helpful.
2. We propose a WAE_RN model, which integrates WAE and RN to obtain useful and generalized internal latent representations and the implicit relationships in the text sequences.
3. We conducts experimental verification on two text sequence: tasks named entity recognition and machine translation. The experimental results demonstrate our proposed model can achieve better semantic representation.

2 Related Work

2.1 AutoEncoder

Traditional AutoEncoder(AE) maps the high level characteristics of input data distribution in high dimension to the low (latent vector), and the decoder absorbs this low level representation and outputs the high level representation of the same data. Many researchers have been working on how to get better semantic representations of input sequences, methods using AE such as ELMo [8], BERT [9], ALBERT [10], ERNIE [11,12], XLNet [13], etc. have been proven as effective techniques. Each model achieves the optimal effect at that time due to its own advantages, and their corresponding pre-trained word vector can still facilitate many downstream tasks even now. However, the latent representation learned by AE is encoded and decoded just in a deterministic way and with no constraint in the hidden space, resulting in a lack of diversity in encoding results, it was later followed by approaches based on VAE [7,14] and WAE [15].

VAE converts the potential representation obtained by the encoder into a probabilistic random variable and learn a smooth potential space representation, then the decoder reconstructs the input data and outputs the reconstructed original data. The results have shown that VAE performs competitively compared to traditional AutoEncoder, for example, Zhang [16] attempts to use VAE for machine translation, which incorporate a continuous latent variable to model the underlying semantics of sentence pairs. Shah [17] specifies the prior as a Gaussian mixture model and further develop a topic-guided variational autoencoder (TGVAE) model that is able to generate semantically-meaningful latent representation while generating sentences. However, training on VAE often leads to the disappearance of the KL term. In addition, VAE assumes that the latent variables follow a gaussian distribution, so only a gaussian encoder can be used. To solve these problems, VAE is replaced with WAE by researchers.

Wasserstein Autoencoder (WAE) use the Wasserstein distance that measures the distance between two distributions to replace the KL divergence in VAE to prevent the KL term from disappearing and help the encoder capture useful information during training. Besides, the goal of WAE is to minimize the direct distance between the marginal and the prior distribution and does not force the posterior of each sample to match the prior. In this way, different samples can keep a distance from other samples, which makes the results generated are more diverse. For instance, Bahuleyan [18] propose a WAE variant that use an auxiliary loss to encourage the encoder more stochastic, their studies verified the WAE model achieves much better reconstruction performance. Moreover, Wang [19] pointed out that the latent space is so complex that we only use standard Gaussian to assume the prior is not enough, and then they proposed to supplement some geometric properties of input space with Riemannian metric tensor to the latent space to learn more flexible latent distribution.

Furthermore, Wasserstein space is more flexible than Euclidean space, which is helpful for capturing the complex relationships and retaining the semantic structure. Since we focus on capturing the universal semantic representation, we choose WAE as our encoder to generate more meaningful and more flexible latent representation while maintaining the original semantic structure.

2.2 Relational Network

Although WAE has shown its strength in learning meaningful and flexible representations from latent space, it still insufficiently capturing the relational information between input sequences. There is a simple solution proposed to address this issue, that is adding some specific learning modules such as RN to help the model express and learn. RN is a neural network integrated with Relational reasoning structure, which aims to constrain the functional form of the neural network to capture the core common attributes of Relational reasoning. Almost all recent methods focus on using RN to capture the relationships between objects. For example, Zhang [20] introduce RN to learn better representations of the input data and experiments on machine translation demonstrate RN can help retain relationships between words. Chen [21] also use RN to capture the dependencies within a sentence between any two words and verify the effectiveness of their proposed method on two benchmark NER datasets, which all support that the RN can model relations between the input sequences.

Inspired by the success of the RN in learning the relationships between elements, in this paper, we directly incorporate RN into the WAE models, thus to fully learn the semantic representation and keep the relational information and structural knowledge between sequences to the greatest extent.

3 Preliminary

Since the purpose of the proposed method is to better obtain the semantic representation of text sequences, we will focus on the following two issues.

3.1 The Problem of Sequence Prediction

Sequence prediction is the most basic and widely used task, such as word segmentation, part-of-speech(POS) tagging, named entity recognition(NER), dependency analysis, etc. Essentially, it can be viewed as a matter of classifying each element in a linear sequence according to its context representation. That is, after understanding the input sequence and extracting its useful information, the optimal mark is made for each sequence, and then a set of globally optimal marks is selected for a given sequence at one time.

Suppose we have an input sequence $\boldsymbol{x}$ of L elements, and a tag sequence $\boldsymbol{y}$ of the same length, i.e. $\boldsymbol{x} = (x_1, x_2, \ldots, x_L)^T$, $\boldsymbol{y} = (y_1, y_2, \ldots, y_L)^T$, where x_i represents the i-th sequence and y_j represents the j-th tag, it's also requires that the value of y_j is taken from a predefined set of finite tags and i equals j, the final goal is to assign a globally optimal label y_j for each input sequence x_i.

End-to-end learning is directly modeling conditional probabilities $p(y|x)$ and then map the input sequence $x_1, x_2, \ldots, x_L$ to the output sequence $y_1, y_2, \ldots, y_L$, i.e. (1).

$$Y = (y_1, y_2, \ldots y_L) = \underset{y}{\operatorname{argmax}}\, p(y|x, \theta) \tag{1}$$

3.2 The Problem of Sequence Generation

Sequence generation is translating the dataset into a clear narrative of human understanding based on the real understanding of text content, such as machine translation, dialogue generation, abstract generation and so on. We usually decompose the generation probability into the product ofthe generation probability of context-related subsequence, and then use the method of auto-regression to get the text in the form of natural language that human can understand.

Suppose the input sequence is $\boldsymbol{x}$, the goal is to understand the input sequence and generate the corresponding output sequence $\boldsymbol{y}$, i.e. $\boldsymbol{x} = \left(x_1, x_2, \ldots x_{|X|}\right)^T$, $\boldsymbol{y} = \left(y_1, y_2, \ldots y_{|Y|}\right)^T$, where $|X|$and $|Y|$ correspond to the length of input sequence and output sequence respectively. Different from sequence prediction, the purpose of sequence-to-sequence learning is to model the conditional probability $p(y|x)$ with all the sequences before the current sequence as the condition, and then map the input sequence to an output sequence, i.e. (2).

$$Y = \left(y_1, y_2, \ldots y_{|Y|}\right) = \underset{y}{\operatorname{argmax}}\, p(y|x; \theta) = y \arg\max \left(\prod_{i=1}^{|Y|} p\left(y_i|x, y_{<i}; \theta\right) \right) \tag{2}$$

4 Relational Network Based WAE Model

4.1 Architecture of Proposed

In order to obtain universal semantic representations that contain structured knowledge, we propose a Relational Network based Wasserstein AutoEncoder (WAE_RN) model, whichhave the ability to embed the potential structural information contained in sequence into semantic representation. Specifically, a relation network layer is employed to quantify the potential relationships between any two elements in the input sequence, and then these relationships are embedded into the input sequence by WAE to get semantic representation that contains relational information. Finally, the generic representation is sent to different decoders to perform different downstream tasks. Next, we will elaborate our proposed model in detail.

4.2 The Wasserstein AutoEncoder Layer

As shown in the bottom of Fig. 1, the encoder of WAE collects the semantic information of the data, and the RN module learns the relational information

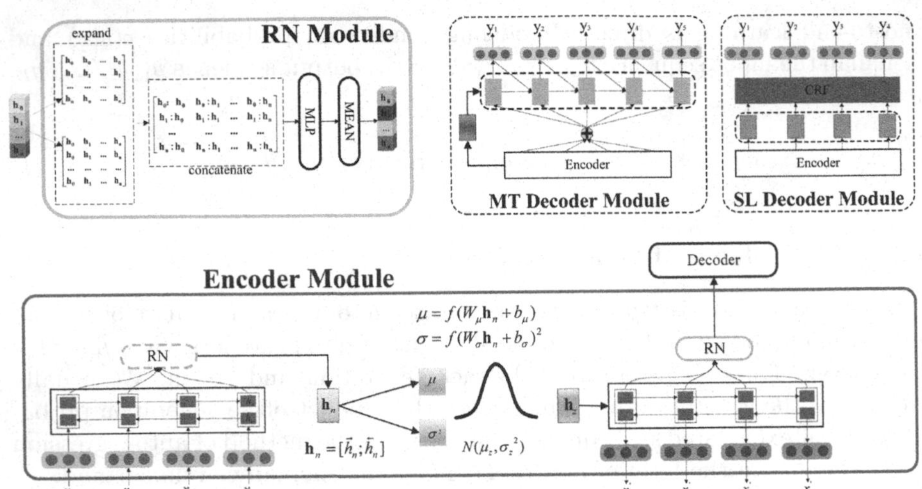

Fig. 1. The architecture of WAE_RN

between the outputs of RNNs, then the context representation is mapped to the Wasserstein space. Compared with embedding data into Euclidean space, which is the most common method, WAE embeds the input data into the Wasserstein space as a probability distribution to can help us capture the complex relationship and retain the semantic structure, so we can obtain the distribution $\mathbf{h}_n = [\boldsymbol{h}_n; \boldsymbol{h}_n]$ that covers both semantic and relational information of input data $x_1, x_2, \ldots, x_L$. Note that the relational network module can be placed either in front of or behind the first encoder, our experiments showed that it is better for the named entity recognition task to put it in the front while for the machine translation task to put it in the back.

After reparameterizing, the reconstructed hidden state $\mathbf{h}_z = N(\mu_z, \sigma_z^2)$ (where $\mu_z = f(W_\mu \mathbf{h}_n + b_\mu), \sigma_z^2 = f(W_\sigma \mathbf{h}_n + b_\sigma^2))$ is sent to the decoder in WAE as its initial state, after that this the latent representation of input data was relearned under the guidance of the hidden state obtained in the previous step, so as to obtain the semantic representation that both follows the source semantic information and retains the structured information. To fully exploit the relational information, we send the representation learned by the second encoder into the relationship network again.

Different from VAE, WAE can use both Gaussian encoder and deterministic encoder. Besides, the goal of Wasserstein distance is to minimize the direct distance between the marginal distribution and the prior, without forcing the posterior of each sample to match the prior, so that different samples can keep a distance from other samples to produce more diverse results.

4.3 The Relational Layer

The architecture of our RN module is shown in the upper left corner of Fig. 1, different from Zhang [20], our RN doesn't use the CNN layer. Besides, to keep the original information of the input sequence to the great extent, we don't use any nonlinear transformations, keeping the dimensions the same. To learn the implicit internal relation between any two elements, we use some transformation between tensors to make objects fully connected and associated with each other, which means, for any vector $C = (\mathbf{c}_1, \mathbf{c}_2, \ldots \mathbf{c}_n)$, after concatenating, its each element $\mathbf{c}_{i,j} = [\mathbf{c}_i; \mathbf{c}_j]$. Then we directly calculate the relationships between any objects: $RN(\mathbf{o}_{i,j}) = f_\phi(W_{MLP}\mathbf{c}_{i,j} + b_{MLP})$. Here, a multi-layer perceptron is used for f_ϕ to find the relationship between all pairwise objects and judge whether and how they are related.

4.4 The Prediction Layer

There is no difference between the decoder used in our model and the traditional decoder. As shown in the upper right corner of Fig. 1, for machine translation tasks, the decoder is the ordinary RNNs with beam search layer, which generates target sequences one by one in an auto-regressive way, while for the named entity recognition task, the decoder is the RNN network with the CRF layer.

4.5 The Objective

For AE, the training objective is the cross-entropy loss or the reconstruction loss, given by $J_{rec}(\theta, \phi, x) = E_{q_\phi(z|x)}[\log p_\theta(x|z)]$. In order to compute the loss of our model, we use MMD (given as $MMD = \left\| \int k(z, ;)dp(z) - \int k(z, \cdot)dq(z) \right\|_{\mathrm{H}_k}$) to approximate Wasserstein distance, where H_k refers to the Hilbert space defined by the kernel k, for high dimensional Gaussian function, k was usually chosen as the inverse quadratic kernel: $k(x, y) = \frac{C}{C + \|x-y\|_2^2}$.

$$L(\theta; \phi; x) = E_{q(x)}\left[J_{rec}(\theta, x) + \alpha J_{task}(\Phi, x)\right] + \beta MMD \tag{3}$$

Thus the loss function (3) of our model consists of three terms: the first is the reconstruction loss, which encourages the encoder to learn to reconstruct data; the second is the Wasserstein distance between the distribution of the encoder $q_\theta(z|x)$ and prior $p(z)$ (usually p is $N(0, 1)$), which measures how much information is lost when q is represented by p; the third is the task loss between the source input $x_1, x_2, \ldots x_{|X|}$ and the generated target sequences $y_1, y_2, \ldots y_{|Y|}$. However, in the experiment we observe that the reconstruct loss has a great influence on the results of our model, resulting in poor performance. To address this problem, we impose a weight α(here α is 2) on the translation loss to balance the influence between the task loss and the reconstruct loss. To achieve better performance, we also give another weight β(here β is 0.0001) on MMD. To the end, our model can be trained in an end-to-end manner by minimizing (3).

5 Experiments

In this section, we aim to investigate our model's performance over NER and MT, where NER belongs to the problem of sequence prediction and MT belongs to the problem of sequence generation. We first present our experimental set up, then compare our method to other baseline systems, finally we give some analyses about our method.

5.1 Datasets

We use two benchmark datasets: OntoNotes5.0 Chinese NER dataset(Onto Notes5.0 Ch-NER) and IWSLT2014 German-English dataset(IWSLT14en-de) for evaluation, the details about these corpora are shown in Table 1.

Table 1. Statistics of OntoNotes5.0 Ch-NER and IWSLT2014en-de

Dataset	Type	Train	Valid	Test
OntoNotes5.0 Ch-NER	Sentences	53.5k	12.8k	4.5k
	Chars	750k	110k	90k
	Entities	62.5k	9.1k	7.5k
IWSLT2014en-de	Sentences	150k	6.9k	6.7k

OntoNotes5.0 Ch-NER. OntoNotes5.0 Ch-NER contains eleven different entity name types(such as PERSON, NORP, GPE, etc.) and seven different value types(DATE, TIME, MONEY, etc.). We use the same OntoNotes data split used for co-reference resolution in the CoNLL-2012 shared task [22] and convert the IOB boundary encoding to BIO tagging scheme (B, I, O). We pre-process by filtering out char-level sentences longer than 150 words and replacing all words that appear less than three times with an <*unk*> token, but for testing data, we use the original dataset.

IWSLT14en-de. IWSLT14en-de contains transcripts of TED talks and translate between German and English in both directions. Following previous works, we use the same data cleanup as Ranzato [23]. We apply the same tokenization and truecasing using standard Moses scripts to both our model and baseline. For training data, sentences longer than 50 tokens were chopped and rared words were replaced by a special <*unk*> token, for testing data, we also use the original version of testing files.

5.2 Experimental Setting

For NER task, we use strong bidirectional Long Short Term Memory with CRF(Bi-LSTM-CRF) baseline, but for MT the baseline is a standard implementation of Bi-LSTM seq2seq model with dot-product attention [2,3] and for decoding we use a beam width of 10 and limit the max sequence length to 100. Detail hyper-parameters can be found in Table 2.

Table 2. Hyper-Parameter Settings

Learning rate	$1e^{-3}$
Learning rate decay	0.5
Batch size	64
Clip norm	5.0
Embedding dim	256
Hidden dim	256
Latent dim	32
Dropout	0.3
Uniform init	0.1
Patience	20

For NER task, we use the entity level accuracy rate, recall rate and F1 value to calculate the score and report standard F1-score for CoNLL NER tasks [22]. For MT task, we adopt BLEU for translation quality evaluation and calculate the BLEU scores on test set using Moses *multi-bleu.perl* script.

5.3 Results and Analysis

In order to enhance the fairness of the comparisons and verify the solidity of our improvement, we train 5 times with random uniform distribution initialization and report average results of our proposed model as well as our re-implemented baselines. Note that we just use simple Seq2Seq architecture as our baseline and don't add any other methods (such as label smoothing, tied embedding, BPE, pre-trained word vector, etc.) to the baseline, because our goal is to demonstrate that our proposed method can yield a more general semantic representation, rather than further boost performance.

Results on Machine Translation. For IWSLT14en-de translation tasks, we use deterministic encoder rather than Gaussian encoder for largely alleviating the training difficulties. We show the test results of different models in Table 3.

The former lines in the table list the performance of previous methods. Shu [24] propose compress word embedding to directly learn the discrete codes via deep compositional code learning, improving the BLEU scores from 29.45%

Table 3. Corpus BLEU scores (%) on IWSLT14en-de translation tasks

	IWSLT14Ge – En(BLEU)	IWSLT14En – Ge(BLEU)
2017RaphaelShu	29.56	-
2018PoSenHuang	30.08	25.36
2019BryanEikema	28.0	23.4
Ours		
RNN_attn(baseline)	27.84	23.74
RNN_attn_RN	28.18(0.3 ↑)	23.95(0.2 ↑)
WAE(d)_attn	28.55(0.7 ↑)	24.24(0.5 ↑)
WAE(d)_attn_RN	**28.87(0.9 ↑)**	**24.46(0.7 ↑)**

to 29.56%. Using SleepWAke Networks (SWAN) that is a segmentation-based sequence modeling method to explicitly model the phrase structure in output sequences, Huang [25] achieves the state-of-the-art results at that time. Eikema [26] use Auto-Encoding Variational NMT model to generate source and target sentences jointly from a shared latent representation, achieving de→en and en→de BLEU scores of 28.0% and 23.4% respectively.

The latter lines show the performance of ours, we can see that our proposed WAE_RN model achieves significant improvement over the baseline system. It demonstrates that our model can capture more useful information and improve the performance of NMT system. In particular, our proposed model outperforms the baseline by 0.9% BLEU points, while only use RN and DAE improves the baseline 0.3% and 0.7% respectively, which effectively illustrate that the combine of RN and WAE can both collect the complex relationship and retain the semantic structure between objects.

Results on Sequence Labeling. For OntoNotes5.0 Chinese NER task, we use Gaussian encoder. As shown in Table 4, the first results is from the CoNLL-2012 Shared Task [27] and the others are ours, we can observe that WAE_RN can significantly outperforms our re-implemented baseline by 0.8, which demonstrates the robustness of our models.

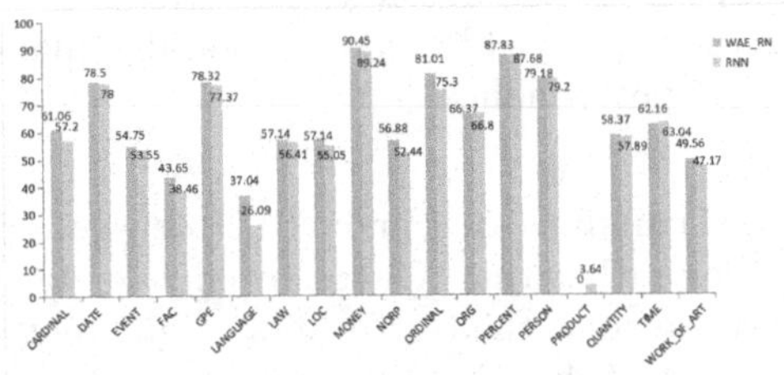

Fig. 2. Performance of our model and baseline on each category.

Table 4. The evaluation results on OntoNotes5.0 Chinese NER task

Method	P(%)	R(%)	F
CoNLL2012	78.20	66.45	71.85
Ours			
BiLSTM_CRF (baseline)	73.08	69.20	71.08
BiLSTM_CRF_SelfAttn	70.93	67.10	68.96
BiLSTM_CRF_LM	72.69	69.59	71.11
BiLSTM_CRF_RN	73.43	69.71	71.52
WAE_CRF	72.79	69.61	71.17
WAE_CRF_RN (best)	73.09	**70.76**	**71.90(0.8 ↑)**

As depicted in Fig. 2, we can see that our method performs well on most categories, such as 'ORDINAL', 'NORP', 'LANGUAGE', etc., and slightly below baseline on the categories of 'PERSON', 'ORG' and 'TIME'. It also should be noted that our model can't find the entity named 'PRODUCT', which is the smallest number of entities in the training dataset. From the results, we can observe that our proposed model does have a positive impact on learning word representation.

Besides, we also conduct experiments using different models to explain the performance promotion of each module, experimental results on NER task confirm the effectiveness of our proposed model, similar as shown in MT tasks.

6 Conclusion

This paper presents a WAE_RN model for text sequence tasks, which aims at learning word representations containing structured knowledge. To be specific, to preserve the semantic structure between objects, we propose use WAE as the model's encoder. To capture the core common attributes of relational reasoning, we introduce RN. Both of which combine well to learn the generic representation that contains relational information. Experimental results on MT and NER tasks demonstrate that the proposed model leads to significant improvements. In the future, we plan to extend the general representation to transfer learning.

Acknowledgements. This work was supported by the Science and Technology Planning Project of Henan Province of China (Grant No. 182102210513 and 182102310945) and the National Natural Science Foundation of China(Grant No. 61672361 and 61772020).

References

1. Rumelhart, D.E., Hinton, G.E., Williams, R.J., et al.: Learning representations by back-propagating errors. Nature **323**(6088), 696–699 (1988)

2. Bahdanau, D., Cho, K., Bengio, Y., et al.: Neural machine translation by jointly learning to align and translate. In: International Conference on Learning Representations (2015)
3. Luong, M., Pham, H., Manning, C.D., et al.: Effective approaches to attention-based neural machine translation. In: Empirical Methods in Natural Language Processing, pp. 1412–1421 (2015). https://doi.org/10.18653/v1/d15-1166
4. Klein, T., Nabi, M.: Attention is (not) all you need for commonsense reasoning. In: Meeting of the Association for Computational Linguistics, pp. 4831–4836 (2019). https://doi.org/10.18653/v1/p19-1477
5. Tan, Z., Wang, M., Xie, J., et al.: Deep semantic role labeling with self-attention. In: National Conference on Artificial Intelligence, pp. 4929–4936 (2018)
6. Santoro, A., Raposo, D., Barrett, D.G., et al.: A simple neural network module for relational reasoning. In: Neural Information Processing Systems, pp. 4967–4976 (2017)
7. Kingma, D.P., Welling, M.: Auto-encoding variational Bayes. In: International Conference on Learning Representations (2014)
8. Peters, M.E., Neumann, M., Iyyer, M., et al.: Deep contextualized word representations. In: North American Chapter of the Association for Computational Linguistics, pp. 2227–2237 (2018). https://doi.org/10.18653/v1/n18-1202
9. Devlin, J., Chang, M., Lee, K., et al.: BERT: pre-training of deep bidirectional transformers for language understanding. In: North American Chapter of the Association for Computational Linguistics, pp. 4171–4186 (2019). https://doi.org/10.18653/v1/n19-1423
10. Lan, Z., Chen, M., Goodman, S., et al.: ALBERT: A Lite BERT for self-supervised learning of language representations. In: International Conference on Learning Representations (2020)
11. Zhang, Z., Han, X., Liu, Z., et al.: ERNIE: enhanced language representation with informative entities. In: Meeting of the Association for Computational Linguistics, pp. 1441–1451 (2019). https://doi.org/10.18653/v1/n19-1423
12. Sun, Y., Wang, S., Li, Y., et al.: ERNIE 2.0: a continual pre-training framework for language understanding. arXiv: Computation and Language (2019)
13. Yang, Z., Dai, Z., Yang, Y., et al.: XLNet: generalized autoregressive pretraining for language understanding. arXiv: Computation and Language (2019)
14. Bowman, S.R., Vilnis, L., Vinyals, O., et al.: Generating sentences from a continuous space. In: Conference on Computational Natural Language Learning, pp. 10–21 (2016). DOIurlhttp://doi.org/10.18653/v1/k16-1002
15. Tolstikhin, I., Bousquet, O., Gelly, S., et al.: Wasserstein auto-encoders. In: International Conference on Learning Representations (2018)
16. Zhang, B., Xiong, D., Su, J., et al.: Variational neural machine translation. In: Empirical Methods in Natural Language Processing, pp. 521–530 (2016)
17. Shah, H., Barber, D.: Generative neural machine translation. In: Neural Information Processing Systems, pp. 1346–1355 (2018)
18. Bahuleyan, H., Mou L., Zhou, H., et al.: Stochastic wasserstein autoencoder for probabilistic sentence generation. In: North American Chapter of the Association for Computational Linguistics, pp. 4068–4076 (2019). https://doi.org/10.18653/v1/n19-1411
19. Wang, P.Z., Wang, W.Y.: Riemannian normalizing flow on variational wasserstein autoencoder for text modeling. In: North American Chapter of the Association for Computational Linguistics, pp. 284–294 (2019). https://doi.org/10.18653/v1/n19-1025

20. Zhang, W., Jiawei, H., Feng, Y., et al.: Refining source representations with relation networks for neural machine translation. In: International Conference on Computational Linguistics, pp. 1292–1303 (2018)
21. Chen, H., Lin, Z., Ding, G., et al.: GRN: gated relation network to enhance convolutional neural network for named entity recognition. In: National Conference on Artificial Intelligence, vol. 33, no. 01, pp. 6236–6243 (2019). https://doi.org/10.1609/aaai.v33i01.33016236
22. Pradhan, S., Moschitti, A., Xe, N., et al.: CoNLL-2012 shared task: modeling multilingual unrestricted coreference in OntoNotes. In: Empirical Methods in Natural Language Processing, pp. 1–40 (2012)
23. Ranzato, M., Chopra, S., Auli, M., et al.: Sequence level training with recurrent neural networks. In: International Conference on Learning Representations (2016)
24. Shu, R., Nakayama, H.: Compressing word embeddings via deep compositional code learning. In: International Conference on Learning Representations (2018)
25. Huang, P., Wang, C., Huang, S., et al.: Towards neural phrase-based machine translation. In: International Conference on Learning Representations (2018)
26. Eikema, B., Aziz, W.: Auto-encoding variational neural machine translation. In: Meeting of the Association for Computational Linguistics, pp. 124–141 (2019). https://doi.org/10.18653/v1/w19-4315
27. Pradhan, S., Moschitti, A., Xue, N., et al.: Towards robust linguistic analysis using OntoNotes. In: Conference on Computational Natural Language Learning, pp. 143–152 (2013)

Author Index

Printed in the United States
By Bookmasters